COEVOLUTION

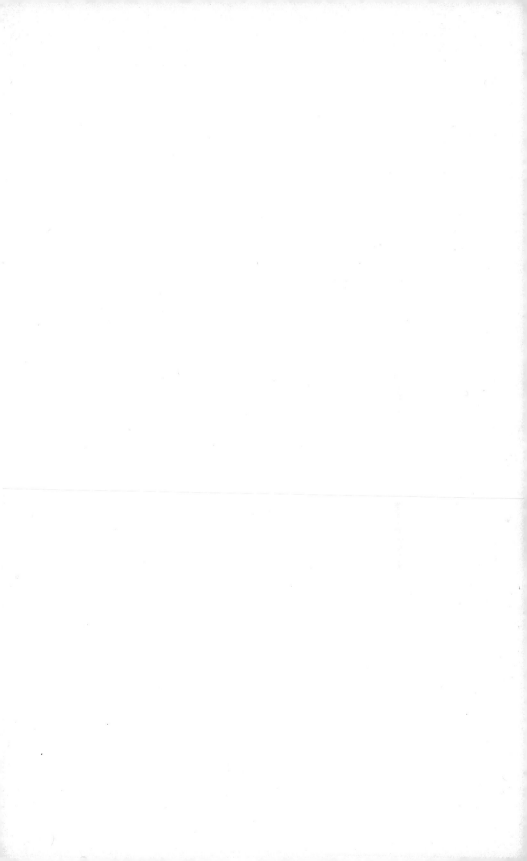

MODELS MIMICS MODELS MIMICS

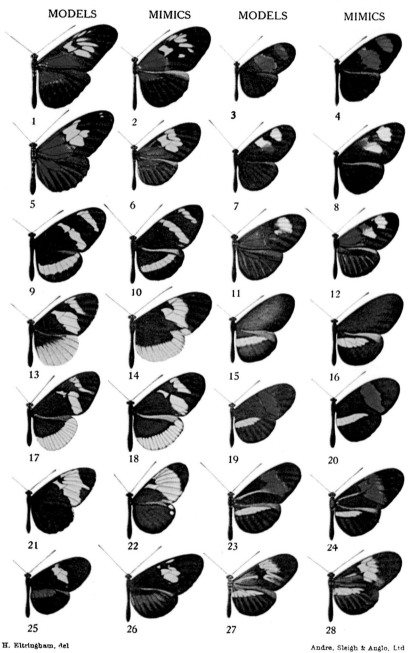

H. Eltringham, del

Andre, Sleigh & Anglo, Ltd

GENUS HELICONIUS. MODELS AND MIMICS.

COEVOLUTION

Edited by Douglas J. Futuyma
STATE UNIVERSITY OF NEW YORK,
STONY BROOK

and

Montgomery Slatkin
UNIVERSITY OF WASHINGTON

With the assistance of Bruce R. Levin
UNIVERSITY OF MASSACHUSETTS
and
Jonathan Roughgarden
STANFORD UNIVERSITY

SINAUER ASSOCIATES INC. • PUBLISHERS
Sunderland, Massachusetts 01375

THE COVER
Harry Eltringham's illustration for his paper on taxonomy and mimicry in *Heliconius* butterflies (Transactions of the Entomological Society of London, 1916). Races of *Heliconius erato* and its close relatives are the "models"; "mimics" are races of *Heliconius melpomene* and its close relatives. The question of coevolution between these groups is considered in Chapter 12. (This illustration is the frontispiece in the hardbound edition.)

Library of Congress Cataloging in Publication Data

Main entry under title:

Coevolution.

Bibliography: p.
Includes index.
1. Evolution.   I. Futuyma, Douglas J., 1942–
II. Slatkin, Montgomery
QH371.C73 1983   575   82-19496
ISBN 0-87893-228-3
ISBN 0-87893-229-1 (pbk.)

COEVOLUTION
For information address
Sinauer Associates Inc.
Sunderland, MA 01375

Printed in U.S.A.

9 8 7 6 5 4 3 2

CONTENTS

Preface   vii

Contributors   ix

1 Introduction   1
DOUGLAS J. FUTUYMA AND MONTGOMERY SLATKIN

2 Genetic Background   14
MONTGOMERY SLATKIN

3 The Theory of Coevolution   33
JONATHAN ROUGHGARDEN

4 Phylogenetic Aspects of Coevolution   65
CHARLES MITTER AND DANIEL R. BROOKS

5 Coevolution in Bacteria and Their Viruses and Plasmids   99
BRUCE R. LEVIN AND RICHARD E. LENSKI

6 Endosymbiosis   128
LEE EHRMAN

7 Plant-Fungus Symbioses   137
JOHN A. BARRETT

8 Evolutionary Relationships between
Parasitic Helminths and Their Hosts   161
JOHN C. HOLMES

9 Parasite-Host Coevolution   186
ROBERT M. MAY AND ROY M. ANDERSON

10 Evolutionary Interactions among
Herbivorous Insects and Plants   207
DOUGLAS J. FUTUYMA

11 Dispersal of Seeds by Vertebrate Guts   232
DANIEL H. JANZEN

12 Coevolution and Mimicry 263
LAWRENCE E. GILBERT

13 Coevolution and Pollination 282
PETER FEINSINGER

14 Intimate Associations and Coevolution in the Sea 311
GEERAT J. VERMEIJ

15 Coevolution and the Fossil Record 328
STEVEN M. STANLEY, BLAIRE VAN VALKENBURGH,
AND ROBERT S. STENECK

16 The Deer Flees, the Wolf Pursues:
Incongruencies in Predator-Prey Coevolution 350
ROBERT T. BAKKER

17 Coevolution between Competitors 383
JONATHAN ROUGHGARDEN

18 Sizes of Coexisting Species 404
DANIEL SIMBERLOFF

19 Convergent Evolution at the Community Level 431
GORDON H. ORIANS AND ROBERT T. PAINE

Epilogue: The Study of Coevolution 459
DOUGLAS J. FUTUYMA AND MONTGOMERY SLATKIN

Acknowledgments 465

Literature Cited 467

Index 541

PREFACE

Every biologist since Darwin and Wallace has recognized that ecological interactions among species have an important influence on their evolution. Until recently there has been little attempt to develop explicit models of the evolution of these interactions. In the past several decades, however, such models have begun to emerge. The evolutionary consequences of interactions among species are now a major area of inquiry, as are the consequences of evolutionary change for the structure of ecological communities. Serving as a complement to the formal theory, evolutionary studies of ecological interactions among populations have come to constitute a major theme in field and laboratory studies as well. Studies of plants and their herbivores, pollinators, and seed dispersers, of hosts and their parasites and pathogens, of predators and their prey, and of mutualists and competitors now commonly take an evolutionary, as well as the more traditional ecological approach.

Coevolution took form as we came to realize that the principles which these different coevolving systems may have in common have hardly been explored; that the literatures of parasitology, of phytopathology, and of insect–plant ecology (for example) are read by few of the same biologists; and that much of the empirical study of ecological interactions has been only slightly influenced by the nascent theory of coevolution, just as the theoreticians have hardly begun to assimilate the vast empirical literature that bears on their theory. In soliciting the essays that make up this volume, we hoped to bring into common focus the diverse lines of study that bear on the evolution of ecological interactions, to describe the features of diverse systems of interacting species that affect their coevolution, and to identify some questions that may help to guide research in this area. We find that there are substantial differences of opinion on how coevolution should be defined, how common it is, and how it proceeds. Moreover, it is clear that a bridge between theoretical and empirical studies is in only the earliest stages of construction.

It will be evident that the development of coevolutionary theory, methods of testing such a theory, and the integration of the theory with empirical studies has barely begun. Thus, although we have attempted to facilitate exchange of ideas among the authors of these essays, they remain highly individualistic in their approach to coevolution. We have not attempted to impose homogeneity of opinion or approach on the authors and, while we attempt a brief overview of their ideas in the Epilogue, it is clear that this volume must be viewed as a first step toward synthesis rather than the synthesis we may hope will someday emerge.

We would like to express our gratitude to Mark Kirkpatrick and Steven Adolph for help with the bibliography of this volume, and to Bruce Levin and Jon Roughgarden for their contributions to its genesis. Our greatest appreciation, of course, is to the authors of these essays for sharing their ideas and knowledge.

DOUGLAS J. FUTUYMA
MONTGOMERY SLATKIN

CONTRIBUTORS

ROY M. ANDERSON, Department of Zoology, Imperial College of Science and Technology, London

ROBERT T. BAKKER, Department of Earth and Planetary Sciences, The Johns Hopkins University, Baltimore

JOHN A. BARRETT, Department of Genetics, University of Liverpool, Liverpool

DANIEL R. BROOKS, Department of Zoology, University of British Columbia, Vancouver

LEE EHRMAN, Division of Natural Sciences, State University of New York, Purchase

PETER FEINSINGER, Department of Zoology, University of Florida, Gainesville

DOUGLAS J. FUTUYMA, Department of Ecology and Evolution, State University of New York, Stony Brook

LAWRENCE E. GILBERT, Department of Zoology, University of Texas, Austin

JOHN C. HOLMES, Department of Zoology, University of Alberta, Edmonton

DANIEL H. JANZEN, Department of Biology, University of Pennsylvania, Philadelphia

RICHARD E. LENSKI, Department of Zoology, University of Massachusetts, Amherst

BRUCE R. LEVIN, Department of Zoology, University of Massachusetts, Amherst

ROBERT M. MAY, Department of Biology, Princeton University, Princeton

CHARLES MITTER, Department of Entomology, University of Maryland, College Park

GORDON H. ORIANS, Institute for Environmental Studies and Department of Zoology, University of Washington, Seattle

ROBERT T. PAINE, Department of Zoology, University of Washington, Seattle

JONATHAN ROUGHGARDEN, Department of Biological Sciences, Stanford University, Stanford

DANIEL SIMBERLOFF, Department of Biological Science, Florida State University, Tallahassee

MONTGOMERY SLATKIN, Department of Zoology, University of Washington, Seattle

STEVEN M. STANLEY, Department of Earth and Planetary Sciences, The Johns Hopkins University, Baltimore

ROBERT S. STENECK, Department of Zoology and Oceanography Program, University of Maine, Darling Center, Walpole

BLAIRE VAN VALKENBURGH, Department of Earth and Planetary Sciences, The Johns Hopkins University, Baltimore

GEERAT J. VERMEIJ, Department of Zoology, University of Maryland, College Park

COEVOLUTION

CHAPTER 1

INTRODUCTION

Douglas J. Futuyma and Montgomery Slatkin

WHAT IS COEVOLUTION?

The word *coevolution* was coined by Ehrlich and Raven (1964) in their discussion of the evolutionary influences that plants and the insects that feed on plants have had on each other. Their use of the term allowed for a variety of interpretations, and it has been used differently by different authors. A restrictive definition is one provided by Janzen (1980a) and is the one adopted by many, but not all, authors in this volume. They use the term to mean that a trait of one species has evolved in response to a trait of another species, which trait itself has evolved in response to the trait in the first. This definition requires specificity—the evolution of each trait is due to the other—and reciprocity—both traits must evolve. A still more restrictive definition would also require simultaneity—both traits must evolve at the same time.

By relaxing one or more of these restrictions, we obtain definitions that other authors have used, either implicitly or explicitly. For example, mimicry among species is often regarded as an example of coevolution. But as Gilbert points out in Chapter 12, Batesian (or for that matter Müllerian) mimicry entails convergence of a mimic toward the phenotype of the model, which may not change in response to the evolution of the mimic. In this case, there is evolution due to the particular trait of the model and the behavior of the predators. Similarly, two competing species may each undergo character displacement, in which case there is clearly coevolution; but if one species diverges while the other remains unchanged, coevolution, according to the most restrictive definition, has not occurred. Yet the distinction seems quite fine and depends on probably unknowable past events.

1

By relaxing the criterion of specificity, the evolution of a particular trait in one or more species in response to a trait or suite of traits in several other species is included. Following Gilbert (Gilbert and Raven, 1975), Janzen (1980), and Fox (1981), we will call this "diffuse coevolution," as distinct from pairwise coevolution, which occurs in interactions between only two species. Many plants have evolved chemical and physical defenses against a diverse suite of insects, and many insects have acquired the ability to detoxify a wide range of plant chemicals. This, like the mutual adaptations of nonspecific pollinators and the flowers they pollinate, would be the result of diffuse coevolution. Yet instances can be discerned in which particular species of plants and herbivores, or plants and pollinators, have formed an intimate pairwise association and have adapted to the specific features of one or a few species. This is pairwise coevolution. Many authors who use the term *coevolution* to discuss generalized responses of groups of species to one another are discussing diffuse coevolution.

Especially when coevolution is diffuse, the criterion of simultaneity may not hold. There are many cases in which the nature of the interaction must have been similar for long periods of time, but coevolution is delayed. For example, the diverse chemical defenses of today's plants may have originated in their Cretaceous ancestors in response to herbivory by Cretaceous insects. The feeding habits of many of today's insects, however, may have evolved much more recently. Although plant evolution was guided by insect herbivory and the insect evolution was guided by the prior evolution of plant defenses, there may have been a succession of adaptive radiations widely separated in time, rather than a continual, closely coupled series of responses of particular lineages to one another. In such cases, the term *coevolution* may be used in a very broad sense, encompassing merely the adaptation of species to features of the biotic environment—features that may remain effectively constant for long periods of time.

But adaptation to an effectively constant feature of the biotic environment does not differ from adaptation to a constant feature of the abiotic environment. Coevolution, too broadly defined, becomes equivalent to evolution. Thus, of the possible definitions of coevolution, we prefer to restrict the definition and say that it has occurred when, in each of two or more ecologically interacting species, there is adaptive response to genetic change in the other(s)—a fairly narrow definition that assumes that the populations at the end of the coevolutionary process are directly descended from those at the beginning. Such a definition would exclude, for example, instances in which plants evolve defenses against beetles and the defenses are later overcome by moths while the beetles remain unchanged or become extinct.

2

However, just as the study of evolution encompasses the analysis of cases in which evolution might fail to occur, the study of coevolution encompasses cases in which reciprocal genetic responses might be expected but nevertheless do not happen. For example, without detailed knowledge of the population density and genetic characteristics of two competing species, we might expect both to undergo character displacement; if only one does, the reasons for the change in one species but not the other are part of the subject matter of coevolutionary studies. Thus, the study of coevolution is the analysis of reciprocal genetic changes that might be expected to occur in two or more ecologically interacting species and the analysis of whether the expected changes are actually realized.

HISTORY OF COEVOLUTIONARY STUDIES

Although the term is relatively new, the idea of coevolution is as old as the study of evolution itself. Darwin's discussion of pollination by insects concludes with "Thus I can understand how a flower and a bee might slowly become, either simultaneously or one after the other, modified and adapted in the most perfect manner to each other" (Darwin, 1859, p. 95). Even before 1859, plant breeders selected crops for resistance to particular parasites, especially molds and rusts (Day, 1974; see Chapter 7 by Barrett).

Some of the early evidence for Darwin's theory was provided by Bates' (1862) description of mimetic complexes. Like many of the observations used to support the theory of natural selection, mimicry had been noticed previously—oddly enough by Charles Lyell (see Lyell, 1881, pp. 417–418)—but its importance was not recognized. It is likely that Batesian mimicry usually entails specific adaptations of the mimics to the models rather than pairwise coevolution, but the possibility of changes in the model to reduce the predation caused by the presence of the mimic cannot be ruled out. Müllerian mimicry (Müller, 1878), on the other hand, may be a result of changes in one or more species in response to one another (see Chapter 12 by Gilbert) and so may be a result of reciprocal coevolution.

Almost as old as the study of coevolution are disputes about whether coevolution occurred or whether the interaction among species was established after the traits of interest evolved for other reasons. A hotly disputed case was the association of various ant species with tropical trees. There was no doubt that such associations existed and little doubt that the ants had adapted to occupy particular plant species (Wheeler, 1910). What was doubted by many botanists and also by Wheeler himself was whether the plants had evolved any traits specifically for the benefit of the ants. Janzen (1966) has since provided strong evidence that traits in each species have evolved specifically to foster their mutualism. However, the general question whether it is possible to show that traits have evolved in response to particular interactions arises frequently in the study of coevolution.

Coevolution has long been recognized as a possibility in the interaction of parasites or pathogens and their hosts. As Mitter and Brooks describe in Chapter 4, systematists have long supposed that parasites and hosts have radiated in parallel. They generally discussed only in passing the question whether or not mutual adaptations of parasites and hosts became refined in the course of their association, but focused on the question whether the similarity of the parasites of different hosts could be used to determine host phylogeny, or vice versa. Mutual adaptations of hosts and pathogens, such as fungi, have been a persistent theme in plant pathology, especially in the last few decades, and numerous cases of genetic adaptations of fungi to new species of hosts have been documented (see Chapter 7 by Barrett). Probably the best-known example of coevolution in action is that of the viral disease myxomatosis introduced to rabbit populations in Australia, England, and France (Fenner and Ratcliffe, 1965). In Australia, the rabbit evolved improved resistance to the virus within a few years after the disease was introduced, and the virus evolved lower virulence, presumably because the more virulent genotypes killed their hosts before the viruses could be carried by mosquitoes to uninfected hosts. Fenner and Ratcliffe (1965, p. 345) also discuss the evolution of plague resistance by rats in India, for which geographic variation in the degree of resistance argued for past coevolution, at least on the part of the rat. In more general discussions of evolution in host–parasite systems, the emphasis has been on the conditions favoring the evolution of parasitic habits by nonparasites (see, for example, Noble and Noble, 1976), although more recently more attention has been paid to reciprocal changes by hosts (Price, 1980).

Brown and Wilson (1956) introduced the term *character displacement* to describe the result of coevolution between competing species. The general principle—that species would evolve to avoid competition by using different resources—was used by Lack (1947), who attributed the diversification of Darwin's finches in the Galápagos Islands to this principle. Using the principle of character displacement as the consequence of competition, Hutchinson (1959), and later MacArthur, Levins, Rosenzweig, Schoener, and other ecologists, developed in the 1960s and early 1970s an extensive theory of the coexistence and resource partitioning of competing species. Such morphological features as body size, head size, or bill size were proposed as measures

4

of resource use and were supposed to conform to minimal ratios among coexisting competitors. More recently, however, the ubiquity of competition and of community-level effects of competition and resource overlap has been questioned (see Chapter 18 by Simberloff) and is the subject of considerable debate at the present.

Since Ehrlich and Raven's (1964) paper, interactions among plants and animals, especially insects, have been the primary foci of the ecological literature on coevolution. The nature of chemical and physical defenses of plants against herbivory, and of insects' adaptations to these features, became a major subject in the 1970s, with contributions by Feeny, Janzen, Dethier, Gilbert, Orians, Cates, Rosenthal, and many others. The major compilation of papers on coevolution between plants and animals, edited by Gilbert and Raven (1975), drew attention to the manifold mutual adaptations of both antagonistic (e.g., plant–herbivore) and mutualistic (e.g., plant–pollinator) relationships. Many of these studies have focused on diffuse coevolution, such as the responses of plants to herbivory in general. Most of the literature concentrates on the adaptations that may have resulted from coevolution rather than on the process itself.

COEVOLUTION AS A SEPARATE FIELD

Defined broadly, coevolution encompasses much of evolution and can hardly lay claim to any distinction as a subject of study. However, progress in understanding particular areas within evolution is often enhanced by drawing attention to them as definable subjects, as the brief history of sociobiology illustrates. Coevolution calls for special attention primarily because it is a major point of contact between evolution and ecology. Evolutionists have long argued that biotic interactions have fostered the evolution of adaptations. Biotic interactions such as competition may prevent a group from radiating or, by providing novel opportunities, may trigger a radiation. Competition among members of such groups as the African cichlids and the Hawaiian honeycreepers may promote adaptive diversity and specialization. Morphological and behavioral innovations may evolve in response to parasites, predators or symbionts. Stebbins (1974) has argued that all of these factors played important roles in the evolution of flowers, fruits, and seeds. Not all adaptations are responses to interactions among species, but many evolutionary events must be understandable only in the context of those interactions.

Evolutionary ecology includes the study of interspecific interactions, and many ecological phenomena are usually thought to be ex-

5

plicable by evolutionary principles. Levins (1968) predicted that the diversity of competing species is controlled by the evolution of characters permitting resource partitioning. MacArthur (1972) concluded that the food webs in species-rich and species-poor biotas are controlled by the evolution of specializations. Ehrlich and Raven (1964) explained the trophic specializations of insects as responses to the chemical features of their host plants. Slobodkin and Sanders (1969) and Southwood (1961) described how evolution will lead to increased numbers and diversity of species in communities. May (1973) argued that communities comprising coadapted or coevolved species tend to be more stable. The idea that coevolution, as well as the impact of immigration and extinction, should lead to predictable ecological relationships among species and to a predictable community structure has led to the question whether independently evolved communities should converge in structure (see Orians and Solbrig, 1978; Cody, 1974; Chapter 19 by Orians and Paine). If community ecology is to be built on an evolutionary theory of species interactions, an understanding of coevolutionary processes is essential.

The study of coevolution forces a different view of genetic evolution than is usually adopted. In population genetics and evolutionary theory, each species is usually considered in isolation, with the environment and associated species relegated to the background, which is assumed to remain unchanged. Coevolutionary theory, as discussed by Roughgarden in Chapter 3, assumes that genetic changes may occur in all interacting species, allowing genetic changes to be driven both by immediate interactions and by the feedback through the rest of the community. The distinctive feature of coevolution is that the selective factor (e.g., a predator) that stimulates evolution in one species (e.g., a prey) is itself responsive to that evolution, and the response should be predictable. In some cases a coevolutionary equilibrium may be established. In other cases there may be no coevolutionary equilibrium, and evolution may continue over longer time scales than are typical for the attainment of gene frequency equilibria as usually treated in population genetic models. For example, as new mutations arise, a prey or host species could evolve new defenses over a long period while its principal predator or parasite evolves new ways to overcome those defenses. This "arms race," as several authors (e.g., Gilbert, 1971; Dawkins and Krebs, 1979) have described it, requires that the rate of origin and the phenotypic effect of new mutations enter the models.

The study of coevolution also requires a different view of the time course of evolution. If one species is considered alone, it would be expected to evolve until it has met whatever challenges it faced and then to stop. For a variety of reasons discussed in Chapter 2, natural selection in a particular direction does not produce continued progress. But

6

if two or more species are evolving in response to one another, then continued progress in each species might occur. A prey or host species could continually produce new defenses while its principal predator or parasite finds new ways to overcome those defenses.

The intensity of ecological interactions can often be measured. If so, then it is possible to estimate the potential strengths of selection in the interacting species. If it is found that species have not coevolved despite pressure to do so, that would call for an explanation. Species might not have been associated for long enough or there might have been insufficient mutations of the appropriate kind. If coevolution turns out to be a rare or sporadic phenomenon, we must ask whether species interactions have been weak or of short duration or whether evolution is strongly constrained by lack of mutations.

Attention to coevolution could raise and help provide answers to many questions about the history of evolution. How often has the adaptive radiation of a group been dependent on the radiation of other groups with which they interact? Does the speciation of hosts and parasites often occur in parallel? Do defense systems of prey become more complex over evolutionary time with the addition of new defenses to the armamentarium, or are old defenses traded for new ones? Do parasites tend toward specialization or toward a benign or even mutualistic relationship with their hosts? In general, how much of the history of evolution must be explained in terms of the evolutionary effects of interspecific interactions?

Finally, the study of some coevolved systems can have direct or indirect practical benefits for agriculture and medicine. If we can understand why insects and pathogens have sometimes been able and sometimes unable to overcome the defenses of plants, we may be able to judge whether the breeding of resistant crops is doomed to failure because of counterevolution by their pests. It would be useful to know whether plant resistance is attained by single defensive features or by complexes of defenses and to know whether pest control can be best achieved by planting monocultures or intercropped polycultures of resistant strains. In medicine, we should like to know to what extent pathogens and parasites are genetically variable for virulence and to what extent populations long exposed to a particular pathogen are more resistant than unexposed populations. What can be learned about the mechanisms of resistance from comparing susceptible and resistant populations or from cross-infection studies of host-specific species? The experiments required to elucidate coevolutionary questions, such as the reasons for host specificity of parasites, could lead to advances on practical fronts.

APPROACHES TO THE STUDY OF COEVOLUTION

In principle, it is rather easy to demonstrate the adaptation of one species to particular features of another. The attraction of crucifer-feeding flea beetles to the allyl isothiocyanate of their host plants, the mimicry of the monarch butterfly by the viceroy, or the venom apparatus of a rattlesnake that subdues its prey are easily seen to be adaptations to particular traits in other species. The demonstration of pairwise coevolution is more difficult, however, because it must be shown that two or more species evolved in response to one another. And diffuse coevolution may consist of events widely separated in evolutionary time and may involve the adaptation of one species to a class of species whose features evolved in ancestors long gone. For example, cyanogenic glycosides of a cherry tree, which tent caterpillars and other insects can detoxify with the enzyme rhodanese (Chapter 10), presumably evolved in an early rosaceous ancestor in response to a wide gamut of insects. It would be difficult to demonstrate or even argue convincingly that cherries evolved cyanogenic glycosides specifically in response to tent caterpillars and that tent caterpillars evolved rhodanese specifically in response to cherries. One could hope to show only that cyanogenic glycosides are rosaceous adaptations to herbivory and that rhodanese is a lepidopteran adaptation to plant toxins or to the cyanogenic glycosides that many plants possess.

It is most likely that evidence can be found to demonstrate pairwise coevolution, the mutual adjustments of two or a few closely associated species to each other. There are several lines of inquiry that could provide evidence of reciprocal coevolution.

Direct observation of genetic changes

An excellent example of the observation of a direct response to interactions is the case of the myxoma virus and rabbits mentioned already (Fenner and Ratcliffe, 1965). Such cases are rare, but a similar one is the appearance of strains of the Hessian fly (*Mayetiola destructor*) that are able to attack a series of sequentially planted resistant strains of wheat (Gallun, 1977). In this case the wheat "evolved" according to a controlled plan, rather than as a natural response to the evolution of the fly. Barrett (Chapter 7) describes similar cases of genetic changes in plants and pathogens.

Laboratory model systems of the coevolution of competitors have been studied by Park and Lloyd (1955) in *Tribolium*, by Moore (1952), Seaton and Antonovics (1967), Futuyma (1970), and others in *Drosophila*, by Pimentel et al. (1965) in muscid flies, and by Chao et al. (1977) in bacteria-bacteriophage interactions (see also Chapter 5 by Levin and Lenski). These experiments provide evidence that rapid

8

coevolution can occur in some cases but not others. Pimentel and his co-workers (e.g., Pimentel and Stone, 1968) have found evidence of genetic changes in both houseflies (*Musca*) and the parasitoid wasp *Nasonia vitripennis* when cultured together, and Hassell and Huffaker (1969) reported increased resistance in the host and increased effectiveness of the parasitoid in a moth–wasp laboratory system. Such studies show, of course, that pairwise coevolution is possible, not that it commonly occurs in nature. In the absence of an actual history of the dynamics of genetic change, the demonstration that each of two interacting species is genetically variable for the characteristics that affect their interaction can at least show the potential for coevolution.

Fossil evidence

The fossil record can provide evidence of coevolution on a vastly longer time scale than is accessible in laboratory and experimental studies. The ideal paleontological evidence would be a continuous deposit of strata in which each of two species shows gradual change in characters that reflect their interaction. For example, Kellogg (1980) provides evidence for character displacement of size in fossil radiolarians. However, the fossil record is seldom complete enough to provide much detailed information, and coevolution is more often inferred than demonstrated. For example, the evolution of hypsodonty in horses is often cited as a coevolutionary response to increased silica in grasses, but we are not aware of evidence that grasses steadily evolved a higher silica content as the horses were evolving hypsodonty. Although the fossil record can show evidence of coevolution only on a geological time scale, it provides the only way to assess the macroevolutionary importance of coevolution.

Taxonomic evidence

Very strong evidence for coevolution of two groups is obtained if the phylogenetic trees of the two groups are congruent or nearly so. We will call this parallel cladogenesis, which does not imply that coevolution is responsible for the diversity of either group, only that the two groups have remained associated with each other. Parallel cladogenesis might be expected for parasites and hosts and has often been claimed in this context. The evidence for parallel cladogenesis (see Chapter 4 by Mitter and Brooks) appears to be clear in some groups and much less so in others.

Taxonomic evidence for the coevolution of adaptive characters

9

need not entail cladogenetic congruence of large taxa. For example, phylogenetic analysis may establish that ancestral and derived states of a defensive character in a host species might be associated with ancestral and derived characters that facilitate exploitation of the host by parasites whose relatives occupy unrelated hosts. In fact, the independence of taxonomic and adaptive patterns may provide strong evidence for the adaptive significance of particular characters.

Functional morphology and ethology

Careful analysis of unusual or unique characters of each of two interacting species can show these features are one species' adaptations to the other species. This evidence is indirect and often not conclusive but may be the only kind available. For example, tannins and other compounds in plants are often cited as defenses against insects, but some authors have raised the possibility that they are primarily defenses against pathogens, and others have questioned whether they evolved for defensive reasons at all—even if they now serve in that role. In other cases, characters appear to be so exclusively designed to mediate ecological interactions that it is hard to imagine any other reason for their evolution. For example, despite Wheeler's (1910) doubts about the evolution of acacias and their associated ants, the proteinaceous bodies on the leaves, on which the ants feed, are restricted entirely to species of acacias that harbor *Pseudomyrmex* ants that live nowhere else. Similarly, the evidence of coevolution is very strong in cases such as aphids that have lost the ability to clean themselves of honeydew, which develops mold if not removed by ants.

Many examples of mutualisms can be understood only as the result of reciprocal coevolution, especially when the interaction entails a very exclusive relationship between the species. Pollinators and the flowers they pollinate show some of the most striking mutual adaptations, but as Feinsinger (Chapter 13) discusses, it is sometimes difficult to argue for the coevolutionary origin of those traits. Janzen (Chapter 11) discusses similar problems that arise for fruits and their dispersal agents.

For hosts and various kinds of parasites, Ehrman (Chapter 6), Barrett (Chapter 7), Holmes (Chapter 8), May and Anderson (Chapter 9), and Futuyma (Chapter 10) find abundant evidence of the adaptation of parasites to specific features of their hosts. The extreme specialization of many parasites to one host or to a few closely related hosts argues for their long-term association. There is less evidence of specific adaptations of hosts to particular parasites. Instead, more generalized defenses against suites of parasites seem to be much more common, as would be expected if hosts were attacked simultaneously by several species of parasites. However, there are numerous exceptions, such as the evolution of resistance by rabbits to myxomatosis, discussed earlier, and the adaptation of *Drosophila* to certain endosymbionts (see Chapter 6 by Ehrman).

Geographical patterns

Another common source of evidence is geographical patterns, exemplified by the study of Brown and Wilson (1956). The strongest evidence comes from the comparison of allopatric populations of each species with sympatric populations. The usual approach is to directly compare the behavior and morphology of species in allopatry and sympatry, but, more recently, Hairston (1980) has shown the power of experimental methods for making the same comparison. Hairston found that where two species of salamanders have very narrow altitudinal overlap, suggesting the operation of strong competition, removal of one species was followed by a pronounced increase in the density of the other. In another mountain range, the altitudinal overlap is much greater, which suggested that the species had diverged in their use of resources. The removal of one species in this zone of overlap had a much smaller effect on the density of the other.

Parallel geographical variation of two broadly sympatric, interacting species may show the adaptive response of one species to another but will seldom be adequate to demonstrate pairwise coevolution (see Chapter 14 by Vermeij). The Müllerian mimic butterflies, *Heliconius erato* and *H. melpomene*, for example, display extraordinary parallel variation (Turner, 1976), but this in itself will not reveal whether one has converged on the other or both to each other (see Chapter 12 by Gilbert). In some cases, a strong plausibility argument can be based on the specificity of the interaction. For example, if the tongue length of a host-specific bee varies in parallel with the corolla length of its host, but the host plant is serviced by many species of pollinators, it is likely that the bee has adapted to the plant but not vice versa (see Chapter 13 by Feinsinger).

If, for a predator–prey or host–parasite interaction, we know only that parasites from each of several localities can exploit only their own host populations, we cannot determine whether coevolution has been reciprocal or unilateral. If parasite populations a and b perform better on their respective host populations A and B than in reciprocal tests, we do not know whether there has been adaptation of the parasites to hosts that are geographically variable for some other reasons or reciprocal evolution of resistance by the host that is insufficient to compensate for the parasite's evolution. It is necessary to perform cross-

11

infections with both "naive" allopatric and "experienced" sympatric populations of both host and parasite. Ideally, one could show, for example, the resistance to malaria of sickle-cell heterozygotes, demonstrate that they are more prevalent in malarial regions, and find that *Plasmodium* from regions of high sickle-cell frequency reproduce more effectively in sickle-cell heterozygotes than do *Plasmodium* from regions of low sickle-cell frequency. There is a paucity of such demonstrations of reciprocal coevolution, although they may not be too difficult to perform.

Theoretical methods

Most of the approaches taken to the study of coevolution, including most of those taken in this book, are empirical. But theory can also play a role by making predictions that can be tested with further empirical studies. Levin and Lenski (Chapter 5), May and Anderson (Chapter 9), and Roughgarden (Chapter 17) have all taken that approach. In all these cases, mathematical models show how the observed patterns can be accounted for by coevolutionary processes. The agreement with a coevolutionary model does not prove the occurrence of coevolution, but it can often add support to the hypothesis that coevolution has occurred and can suggest other experiments or observations that could provide further support.

Although coevolutionary theory, as reviewed by Roughgarden (1979; and Chapter 3) and by Slatkin and Maynard Smith (1980), has played an important role in some coevolutionary studies, there is still a large and obvious gap between the theoretical and empirical aspects of the subject. The reason for this gap is partly that, in many systems, the principal question is whether or not coevolution has occurred, whereas the mathematical models are directed to showing how coevolution will occur under the assumption that it will do so. This gap will be partly closed if field and experimental workers can address the assumptions and conclusions of some of the models and if theoreticians can develop their models with more of an eye for their empirical utility.

Community-level analysis

The approaches described so far concentrate on interactions between pairs of species and the consequences of those interactions. An alternative approach is to look for community-level patterns that might be caused by coevolution. The failure to find such patterns would suggest that coevolution may not be an important process in structuring communities even if some of the component species have coevolved with one another. This conclusion is similar to that reached in some of the discussions of coevolutionary patterns in the fossil record. Although coevolution might have been occurring, it may not be manifest at all levels.

Orians and Paine (Chapter 19) examine the evidence of community level convergence in a variety of marine and terrestrial systems. Simberloff (Chapter 18) describes several recent studies by him and his co-workers on the statistical analysis of community-level patterns. Simberloff's approach has been to generate null models—models generating patterns expected if interactions were not important—and test observed patterns for significant departures from the null models. Although this approach is controversial, the controversy itself will stimulate more detailed studies and more careful interpretation of data.

IN CONCLUSION

The scope of coevolutionary studies, like that of the study of evolution generally, is very broad, ranging from the intricate coadaptation of intimate associations that include even intracellular symbionts to the manifold interactions among species that may play a role in shaping community structure. The chapters in this volume address a wide spectrum of subjects. The authors, who include systematists, paleontologists, ecologists, and geneticists, take different approaches to the study of coevolution and often reach different conclusions. As editors, we have made no attempt to disguise the differences among the various authors or to forge a consensus on the prevalence or importance of coevolution. Our goal was not to produce a synthesis of coevolutionary studies; the field is far too new and diverse for that to be successful. Instead, we have tried to show the breadth and richness of the subject, which may not be apparent from coevolutionary studies of only one group of organisms or of only one type of ecological interaction.

CHAPTER 2

GENETIC BACKGROUND

Montgomery Slatkin

The present state of knowledge about the genetic basis of phenotypic evolution is not complete enough to be summarized by a few simple principles that can be applied in coevolutionary studies. Instead, several factors are thought to govern genetic evolution, and different views of their relative importance are held. In any coevolutionary study, it is usually necessary to adopt one or more of these views, and even when assumptions about genetic evolution are not made explicit, some assumptions are necessarily made.

In this chapter I will discuss the different views of genetic evolution that are most relevant to coevolution. I will generally not advocate one view over another but will emphasize the implications for a coevolutionary study of adopting one or another view. Because coevolution—however it is defined—is the consequence of ecological interactions among species, I will focus on the genetics of phenotypic responses to selection, first by reviewing the theory of selection at different levels, then by examining the different views of the maintenance of genetic variation, and finally by discussing the many factors that can act to constrain evolutionary responses to selection.

LEVELS OF SELECTION

The principal mechanism of Darwin's theory of evolution is natural selection acting on individual differences. As Lewontin (1970) has pointed out, however, any group of objects, from molecules to galaxies, will evolve if they possess heritable differences that are correlated with differences in survival and reproduction. It is not only possible but inevitable for evolution to occur when this condition is met. Evolution can result from selection of individual differences but also from selection of groups of related individuals, populations, species, and entire communities. There are currently strong disagreements

14

about the relative importance of selection at different levels. The population genetic theories of the 1920s and 1930s, which constituted the "modern synthesis" (Huxley, 1942) and which completed the integration of Mendelian genetics into evolutionary theory, focused on individual selection, but many ecologists and ethologists assumed that selection at the level of the population or the species was also important. In the development of sociobiology, selection among kin groups has been invoked to account for many altruistic and social behaviors. The debate over macroevolutionary patterns and their explanation partly depends on the importance of selection among species. Although G. C. Williams (1966) advocated the "principle of parsimony" in evolution—if any adaptation could be explained through individual selection, selection at other levels should not be invoked—a more pluralistic view of levels of selection has emerged. Any coevolutionary study must make some assumption about the level at which selection is operating.

Individual selection

Individual selection is due to differences among individuals in survival and reproduction, which together determine an individual's fitness. An individual's fitness may depend on population size (density dependence) or the distribution of other phenotypes in the population (frequency dependence). In fact, many kinds of ecological interactions probably lead to density-dependent or frequency-dependent fitnesses.

The phenotypic changes due to individual selection depend on both the amount of variation in a population and the intensity of selection. Fisher (1930) formalized this idea and put it in genetic terms with what he described as "the fundamental theorem of natural selection," which states that the rate of evolution of a character at any time is proportional to its additive genetic variance, which is the heritable component of a character's genetic basis. The constant of proportionality is the quantitative measure of the intensity of selection. Intuitively, Fisher's theorem implies that individual selection will tend to make a species better adapted to environmental conditions. Fisher's theorem is exactly true only under rather special conditions and is not true when there are density- or frequency-dependent interactions, linkage among loci, or correlations among characters caused by pleiotropic effects of genes (Roughgarden, 1979; Ewens, 1979). But there are many other cases in which Fisher's theorem is approximately true in the sense that individual selection nearly always leads to an increase in the average fitness of a population (Ewens, 1979). Ecological interactions among members of the same species and of different species would tend to lead to fitnesses that depend partly on total population size. Roughgarden (1971), Charlesworth (1971), and Anderson (1971) have shown that density-dependent natural selection will tend to maximize the total population size of a population. This result is analogous to Fisher's theorem.

One view of the role of individual selection in evolution is that of Darwin and, in genetic terms, of Fisher. This view is that much of evolution consists of the gradual and steady improvement of different characters under the action of natural selection. A character stops evolving when it cannot be improved upon under current ecological conditions. Individual selection is, in this view, both creative and stabilizing.

This view of evolution can be illustrated by an "adaptive topography" (Wright, 1931). As shown in Figure 1, fitnesses of individuals with different measures of two phenotypic characters can be represented by a surface, with the two horizontal dimensions being the ranges of possible values of the characters and the vertical dimension being fitness. The environmental conditions determine the actual shape of the surface, and the surface in turn determines the course of evolution of those two characters. According to Fisher's theorem, evolution will proceed uphill on the surface and stop when a peak is reached.

If there are two or more peaks in the adaptive surface, the peak that a species ultimately reaches is determined by its initial phenotypic composition. In Figure 1, species A, which starts with small values of the two characters, will evolve to peak I, whereas species B, which initially has larger values of the characters, evolves to peak II. Unless environmental conditions change, thereby altering the shape of the adaptive surfaces, Fisher's theorem predicts that the two species will not evolve once they reach their respective peaks, which represent evolutionarily stable combinations of the two characters. It is especially important that species A, at the lower peak, will not evolve to occupy peak II, the higher peak. Even though there is a superior combination of characters, species A cannot attain that combination, according to Fisher's theorem, because it cannot cross the "adaptive valley" that separates the two peaks.

With this view of the role of individual selection, the ecological conditions, including interactions with other species, are the driving forces of evolution. When conditions change (as could occur when other species in the community evolve), the adaptive surface changes, causing the species to evolve to a new peak. Almost by necessity this view is the one usually adopted in coevolutionary studies, because the emphasis in coevolution is on interactions among species.

Although there is no real dispute with Fisher's view of how in-

16

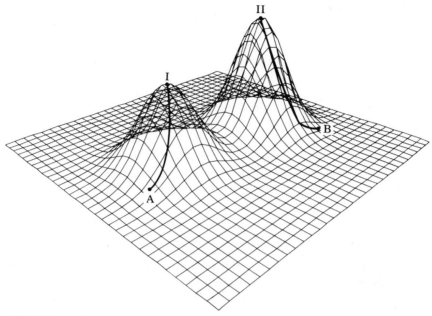

FIGURE 1. Graphical representation of an "adaptive surface." The two horizontal axes represent two phenotypic characters, such as body size and color, and the vertical axis represents the mean fitness of a population that has the particular average values for the two traits. The two peaks, I and II, represent two combinations of these characters that are particularly well coordinated. According to Wright's and Fisher's genetic theories, species will tend to evolve "uphill" on such surfaces. Species A will tend to reach an evolutionary equilibrium at peak I and species B at peak II. Despite the fact that peak II is higher than peak I, indicating that one combination of characters is better than the other, species A cannot evolve to peak II without some other genetic mechanism like genetic drift acting in addition to individual selection.

dividual selection causes genetic changes, alternative views of the role of individual selection arise from different assumptions about how environmental changes affect the adaptive topography. Fisher assumed that evolutionary changes follow environmental changes that alter the adaptive topography. Wright (1931), in contrast, argued that evolutionary changes occurred when a species moved to another adaptive peak through the action of other forces, particularly genetic drift and group selection. In Wright's view, individual selection is more of a conservative force, with evolutionary changes not necessarily being driven by environmental change. Chance events also play an important role.

17

Wright's view of individual selection is not incompatible with Fisher's. They both emphasize the power of individual selection, but Wright sees the need for other mechanisms of genetic changes as well. Wright's view has been adopted by the proponents of the punctuated equilibrium theory of evolution (Eldredge and Gould, 1972; Stanley, 1979). Part of that theory is that individual selection is relatively ineffective in causing phenotypic changes in widespread species. The punctuational theory differs from Wright's because it assumes that phenotypic changes, at least those observable in the fossil record, are associated with species formation rather than phyletic evolution. In the punctuational theory, an entire species does not move from one adaptive peak to another. Instead a species gives rise to a new species which might occupy another adaptive peak. But the punctuationalists agree with Wright that individual selection may play a more conservative and less creative role in evolution than in Darwin's and Fisher's views.

Kin selection and group selection

Kin selection is the term used to describe selection based on differences among collections of related individuals (Maynard Smith, 1964), and *group selection* is the term used to describe selection due to differences among populations of a species (Wynne-Edwards, 1962; Maynard Smith, 1964). Both Fisher (1930) and Haldane (1932) were aware that genetic evolution could occur through effects of an individual on its relatives. Fisher (1930) suggested that the evolution of aposematic coloration would be due to the increased protection of an individual's relatives gained by a predator's learning to avoid a distasteful prey. Parental care, although technically due to kin selection, was regarded as being so obviously adaptive to the parents that it did not require a separate explanation. It was not, however, until the formal development of a theory of kin selection by W. D. Hamilton (1964a,b) and the demonstration that kin selection theory could explain a variety of social phenomena—especially the evolution of eusociality in Hymenoptera—that the potential power of kin selection was recognized.

Group selection was first discussed in genetic terms by Wright (1945), who called it *interdemic selection*. He noted that genetic evolution could occur if there were genetic differences among local populations that caused differences in extinction and migration rates. Lewontin (1962) argued that group selection could be responsible for the observed low frequencies of the *t* allele in *Mus musculus*. Because of meiotic drive, the *t* allele increased in frequency in each population even though the *t*/*t* individuals were sterile. Levin, Petras, and Rasmussen (1969) used a computer simulation to show that local extinction of populations caused by fixation of the *t* allele could account for its low frequency, albeit under rather restrictive assumptions about deme sizes and migration rates.

Many ethologists implicitly assumed the operation of group selection to explain the presence of behaviors that seem to confer no advantage to the individual. For example, Huxley (1966) argued that elaborate display rituals evolved to settle contests among individuals because physical combat would be bad for the entire social group. Wynne-Edwards (1962) argued that many behaviors evolved through group selection as a mechanism to allow individuals to assess their local population densities and limit their reproduction to prevent over-exploitation of resources. G. C. Williams (1966) countered Wynne-Edward's views with the argument that group selection was not needed to explain the presence of most social behaviors. Williams suggested a "principle of parsimony" to be used in evolutionary discussions: if individual selection could account for an adaptation, then there was no reason to invoke group selection. This principle assumes that individual selection is more parsimonious than group selection or selection at other levels.

Numerous theoretical studies have shown that group selection is theoretically possible when populations are extensively subdivided and local extinction is reasonably common (Wade, 1978; Roughgarden, 1979), and Wade (1977) has demonstrated that substantial phenotypic evolution can result from group selection in laboratory populations of *Tribolium*. But what is not known is whether conditions often exist for group selection to produce substantial evolution. Wilson (1980) and Price (1980) contend that many species have population structures conducive to group selection. Parasites in particular have highly subdivided populations and local populations can become extinct by killing their hosts. Group selection is often assumed to be the explanation for the loss of virulence of some disease parasites like the myxomatosis virus in Australian rabbits (Fenner and Ratcliffe, 1965).

Species selection

The term *species selection* was introduced by Stanley (1975), although Fisher (1930), Lewontin (1970), and Eldredge and Gould (1972) all discussed the idea. Species selection can be effective if there are differences among species in extinction and speciation rates. Any character correlated with either or both of these rates would increase in frequency in a group of species considered together. There would tend to be more species with characteristics causing high speciation rates and low extinction rates. For example, it is possible that monophagous insect species have greater tendencies to form new species than do poly-phagous species because of greater opportunities for geographical iso-lation. If species derived from monophagous species are also mono-phagous, then we could expect the species characteristic monophagy to become common purely by species selection. On the other hand, mo-nophagy could tend to increase the extinction rates of species because of the commitment to a single resource, thereby causing monophagous species to become less frequent.

This point is particularly relevant for coevolutionary discussions that rely on comparisons among species. To continue the preceding ex-ample, suppose it were found that there were more monophagous species associated with particular kinds of trees—say, those with large amounts of alkaloids. If only individual selection were considered, the conclusion would be that the presence of alkaloids caused the evolu-tion of monophagy in species associated with those trees. An alter-native explanation is that the alkaloids reduced the speciation rates or increased the extinction rates of the polyphagous species associated with those trees. With that explanation, a polyphagous species feed-ing on alkaloid-rich trees would not necessarily evolve to become monophagous.

Most coevolutionary studies focus on individual selection rather than selection at other levels. If the view of the punctuationalists is correct and species selection has been an important creative force in macroevolution, then, as Stanley (1979) has discussed, many coevolu-tionary studies will have to be reinterpreted. On the other hand, coevolutionary studies, including those described in this book, may provide evidence for the efficacy of individual selection in established species and have some bearing on macroevolutionary theory.

Ecosystem selection

Associations among species in ecosystems could cause each species to have lower extinction rates and higher dispersal rates. The result could be the evolution of associations of species that are mutually beneficial due to the selection of entire ecological communities. This idea has been discussed at various times by ecologists, some of whom have regarded ecological communities as "superorganisms" made up of mutually coadapted species (Allee et al., 1949). Wilson (1980) has more recently generated new interest in the possible importance of selection among ecosystems.

Wilson's (1980) argument is that there are too many features of species in communities that work to the advantage of other species to be explained purely through the action on individual selection in each species. Individual selection (and kin and group selection as well) de-pend on the genetic gain as measured by gene frequencies within the species. A species would not work in the interest of another species, even if that results in the eventual extinction of other species and ultimately itself. For example (Wilson, 1980, p. 5): when earthworms restructure the soil, plant growth is promoted and the long-term availability of food is thereby ensured. However, individual selection would favor traits that make worms more efficient at feeding despite any eventual damage to the plants. This effect of individual selection could be offset by selection among ecosystems: those ecosystems containing earthworms that benefit plants would survive for longer times and would tend to produce colonists for other ecosystems.

The consideration of selection at the level of ecosystems raises two separate questions. One is whether the apparent harmony among species is in conflict with the theory of individual selection. Wilson (1980, Chapter 5) reviews several examples where that may be the case, but in each example there are alternative explanations possible. The second question is whether the observations require selection among ecosystems, as proposed by Wilson (1980) or whether they can be explained by coevolution within the community. As discussed by Roughgarden (Chapter 3, this volume), genetic changes in a species can be affected by the "feedback" through interactions with other species in the community. According to this theory, a species like the earthworm would not be selected to destroy its food supply because the reduced food supply would tend to oppose that selection.

THE PARADOX OF VARIATION

The response to selection at any level depends on the existence of heritable variation among the units being selected. But one of the consequences of most kinds of selection is reduction in the extent of variation. Any view of genetic evolution as primarily due to selection at one level must contain assumptions about how variation at that level is created or maintained.

There is abundant genetic variation among individuals in most populations. Artificial selection of most phenotypic characters in most species readily produces significant changes. The implication for coevolution is that if selection due to ecological interactions is strong enough, then the trait would be expected to evolve quickly. At the genetic level, biochemical studies have shown that numerous loci have more than one allelic state (Lewontin, 1974). The principal question is whether genetic variation among individuals is in itself adaptive or not.

Lewontin (1974) distinguishes two schools of thought about the maintenance of genetic variation. One—the "classical" school—views most heritable variation at any time as being not adaptive. The role of individual selection on most characters is to eliminate deleterious mutations. The genetic model regarded as typical by members of this school is of a locus with mutations to deleterious alleles being balanced by selection against those alleles. When environmental conditions change, alleles that were deleterious might become advantageous, thereby adapting a population to the new conditions. The other school—the "balanced" school—views most genetic variation as being adaptive and maintained in species by individual selection alone. The genetic model viewed as being typical by this school is a locus with two or more overdominant alleles, for which the heterozygotes have higher fitnesses than the homozygotes. When environmental conditions change, a different genetic balance may be reached, but selection would still act to preserve genetic variation.

The extreme views attributed to the two schools are not held by many evolutionary geneticists. Most would adopt a position inter-mediate between these two extremes. But the differences between the views are important, particularly for studies of coevolution. If the view of the balanced school is adopted for a particular character or set of characters, then it is always reasonable to assume that sufficient genetic variation exists to enable each species to respond immediately to new conditions, possibly to conditions created by genetic changes in other species. If the view of the classical school is adopted, then there could be some delay in a species' response to new conditions because the appropriate mutations may not be present. One way to understand the difference between the two schools is to imagine what would hap-pen if the mutation process were somehow stopped. The classical school would predict that genetic evolution would stop quickly and there would be very little potential for response to any new conditions. The balanced school would predict that genetic evolution would con-tinue for some time before stopping. Each species would have a store of genetic variants, and evolution would proceed until that store was exhausted.

There is at the present time no way to determine which view of genetic evolution is more nearly correct. It was thought that the ap-plication of biochemical techniques would resolve the issue when it was shown that there was abundant genetic variation at the molecular level (Lewontin, 1974). But, instead, the issue in population genetics became the question whether the observed genetic variation was selec-tively neutral. For the study of coevolution, the neutral mutation theory of Kimura (1968) and King and Jukes (1969) is not central. But the extent to which phenotypic evolution is limited by a lack of genetic variation at any time is. Coevolution is concerned with evolutionary

22

responses to ecological conditions that can change quickly. If evolutionary changes cannot occur on an ecological time scale, coevolution may not be found even when expected.

Selection at higher levels also depends on variation. Laboratory experiments (Wade and McCauley, 1980) and theoretical studies (Eshel, 1972; Slatkin, 1977, 1981b) showed that genetic drift can act like mutation in creating random differences among local populations. Uyenoyama (1979) showed that random fluctuations in selection intensity in different populations can also lead to local genetic differentiation. But there are no field studies showing that there are characteristics of local populations that are both heritable and also subject to group selection. The extremely subdivided populations of many species, especially parasitic species, provides circumstantial evidence that group selection could be important (Wilson, 1980; Price, 1980), but the critical field studies remain to be done.

For species selection to be a creative force, there must be some mechanism creating differences among species. Stanley (1979), Gould and Eldredge (1977), and Vrba (1980) argue that the fossil record shows that large and unpredictable phenotypic changes are likely to occur when new species are formed. For example, if a new species is formed from a species with a certain average body size, the new species may differ from its parent species in average body size; but it may be larger or smaller. The apparent unpredictability of the changes occurring at speciation do not necessarily imply that speciation is due to random events such as genetic drift (Stanley, 1979; Charlesworth, Lande and Slatkin, 1982). But they may appear random when a large number of speciation events are considered. Stanley (1979) makes the analogy that speciation could act like the mutation process within a population. It could produce variation on which species selection could later act.

CONSTRAINTS ON GENETIC EVOLUTION

Natural selection cannot improve a species indefinitely. Eventually and possibly very quickly, evolution in a particular direction will stop because of one or more constraints. There can be several factors, which are not mutually exclusive, that contribute to limiting the genetic response to selection. All of these factors are important for some species, but there is not wide agreement about their overall importance in evolution because it is impossible to say what is typical of most species or even most species in one taxon.

Pleiotropy

Pleiotropic genes directly affect two or more characters and are commonly found in genetic studies. The *yellow* allele in *Drosophila melanogaster* causes the eye to be yellow and also causes the body to be yellow and modifies the shape of the spermatheca (Dobzhansky, 1970). Dobzhansky (1970) and Wright (1967) have long emphasized the importance of pleiotropy to evolution. Pleiotropy can constrain the evolution of one character because of changes in other characters affected by the same genes. Even if, somehow, environmental conditions favored *Drosophila* with yellow eyes, changes in the spermatheca could cause such a large decrease in reproductive success that the *yellow* allele could not increase in frequency.

At the phenotypic level, pleiotropy is manifested as correlations among characters. Selection on one character results in changes in others. The correlations among characters due to pleiotropy can evolve through modifier genes that alter the pleiotropic effects of other genes or through the increase in frequency of genes that control one of the characters alone. For example, yellow eyes could evolve in *D. melanogaster* if an allele at another locus reduced the effect of the *yellow* allele on the spermatheca or if there were an allele at another locus that produced yellow eyes but did not affect the spermatheca.

Pleiotropy is thought to be an important cause for a phenonenon commonly found in populations subject to strong directional selection. A quantitative character that is strongly selected usually responds quickly for the first few generations. Then a "plateau" is reached after which further progress cannot be achieved. In some cases, no further progress can be made because the selected lines become sterile. The level of the plateau is difficult to predict. Waddington (1960) selected populations of *Drosophila melanogaster* for salinity tolerance, producing populations that could survive on a medium containing 7% salt. Dobzhansky and Spassky (1967), applying the same selection regime to *D. pseudoobscura*, could not obtain populations able to survive concentrations of salt greater than 3%.

Factors other than pleiotropy, particularly linkage and fixation of alleles, contribute to the plateauing in the response to directional selection. But pleiotropy may be the most likely explanation of a second common phenomenon. If directional selection is relaxed, the character selected may return to its original state. Figure 2 shows an example of both the plateauing and the subsequent effect of relaxing selection. The return to the original state can be explained by the effect of individual selection on other characters affected by the pleiotropic genes whose frequencies were altered by the directional selection. When the directional selection is relaxed, the individual selection would tend to return the pleiotropic genes to their original frequencies.

24

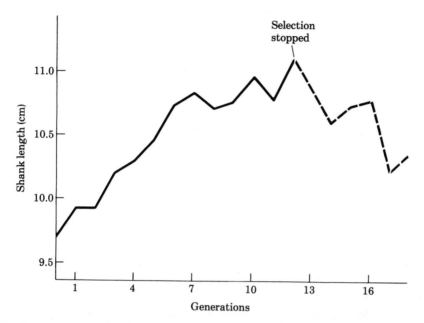

FIGURE 2. The results of directional selection on shank length in chickens for 12 generations with no selection after generation 12. Note that a rough "plateau" is reached by generation 7 and that after selection is relaxed the population average partly returns to the original value. (After Dempster, 1958, Figure 4.10.)

Pleiotropy may be due to a gene's effect on the timing of developmental events. Gould (1977) and Alberch et al. (1979) have emphasized the importance of this kind of pleiotropy both for producing large changes in several characters through genetic changes at a single locus, and for constraining genetic changes at loci that control the timing of development. The potentially large phenotypic effect of such genes may give them an especially important evolutionary role.

Linkage

In general, genes affecting the same character are not closely linked; and, conversely, closely linked genes affect different characters. Linkage between genes affecting different characters can cause selection on one character to cause a correlated change in other characters. For example, selection on color could result in a change in body size. Deleterious changes in the correlated characters could stop the response to selection acting on the first character. Linkage among loci is thought to be another cause of plateauing in selection experiments (Dickerson, 1955).

Although linkage can be shown to be a constraining force in selection experiments, it is not clear how important linkage is in natural populations. Theoretical studies show that unless correlations among genes are actively maintained by selection, they will quickly decay, and the conditions under which permanent correlations are maintained in populations are restrictive (Ewens, 1979). Furthermore, there is little evidence that there are permanent correlations among genes in natural populations, except for genes in inversions (Charlesworth and Charlesworth, 1973; Langley, Smith and Johnson, 1978). Linkage is probably only a temporary constraint on genetic evolution both because correlations caused by linkage will decay on their own and because they can be broken up easily by selection.

Functional constraints

Two or more phenotypic characters may be constrained in their evolution because they are functionally dependent on one another. Adaptations to particular ecological conditions might well be incompatible. For example, ecological conditions experienced by seed-eating mammals may favor individuals able to eat larger and tougher seeds. Natural selection could then favor larger molars with more durable grinding surfaces. But for tooth size to increase significantly, the entire jaw structure must change. Selection on tooth size could be opposed by selection preventing changes in other parts of the jaw.

Some functional constraints are imposed by the laws of physics. Many of the well-known examples of allometry (Huxley, 1932) are a consequence of functional constraints imposed by the relationship between surface area and volume. For example, as an animal becomes larger and increases its length by some factor c, the cross-sectional area of its supporting limbs must increase by a factor of $c^{3/2}$ (Maynard Smith, 1968), unless the strength or number of limbs is increased accordingly. Similar considerations apply to the sizes of other organs such as lungs, heart muscle, and brain (Gould, 1966). In fact, J. B. S. Haldane (1928) suggested that much of functional morphology is the story of the struggle to increase surface area in proportion to volume.

Other functional constraints are due to the use of a character for more than one purpose. For example, the males of many bird species are brightly colored. Bright colors probably do not promote the survival of males, because females and juveniles are not also brightly colored. Yet the bright colors are maintained presumably by female preference for brightly colored males. Whatever advantage there is to duller colors does not change the coloration of the males because of the other function of the bright colors. In a similar way, the talons of birds

26

of prey might be improved as instruments to hold their prey if the talons did not also serve for walking and for grasping perches.

Functional constraints do not entirely prevent evolution of a character for a particular purpose, but they probably slow down the process of adaptation and limit evolutionary changes in any one direction. Allometric relationships can be modified by artificial selection (Atchley and Rutledge, 1980). Several comparative studies have shown that allometric relationships are sensitive to ecological factors (e.g., Baron and Jolicoeur, 1980; Harvey et al., 1980).

Adaptive valleys

The evolution of two or more characters could be constrained by the fact that both would have to be changed significantly before any advantage were gained. In terms of the adaptive surface, this situation is represented by two or more adaptive peaks separated by adaptive valleys (Figure 1). Although one adaptive peak might be higher than another, individual selection alone could not move a population to another peak. For example, if a particular plant species were defended against insect herbivores by two toxic compounds, a species attempting to exploit that plant would have to evolve ways of dealing with both. Even though the insect could develop ways of dealing with both toxins if presented with them one at a time (thereby allowing it to use a new resource), it might never do so because it could never succeed in dealing with both toxins together.

Wright (1931) argued that many major evolutionary transitions were of this type and that genetic drift, group selection, and other stochastic forces were necessary to explain the shift from one adaptive peak to another. This is Wright's "shifting balance" theory of evolution (Wright, 1955). The importance of adaptive valleys in constraining evolution and the importance of genetic mechanisms for crossing adaptive valleys are not agreed upon. Fisher (1930) followed Darwin in saying that the dominant mode of evolution was the continued improvement of characters and combinations of characters. Evolution proceeded up adaptive ridges. Darwin's discussion of the evolution of the vertebrate eye (1859, pp. 187–190) is still a good example of this view. It is often difficult to resolve the difference between this view and Wright's, because of the absence in the fossil record of intermediate forms whose fitnesses can be assessed. The dispute is often over the existence of plausible intermediates; and the failure to envision a sequence of functioning intermediate forms, as described by Darwin for the vertebrate eye, could be due to lack of imagination or to their nonfeasibility.

Changing environmental conditions

A character may fail to evolve because there has not been sufficient time for it to do so. The pattern of environmental variation a species experiences constrains its evolution by allowing adaptation only to those conditions that persist for some time. It is possible that many characters are adapted to past rather than present conditions. Gould and Lewontin (1979) have emphasized that if many characters are in fact not adapted to present conditions, attempts to prove that they are will lead to fallacious and incorrect conclusions. The composition of most communities can change rapidly through large fluctuations in population sizes and through colonizations and extinctions. The potentially short time scale of ecological change may not allow time for the coevolutionary response of most species.

Gene flow

If a species is widely dispersed and experiences different environmental conditions in different parts of its range, then gene flow can oppose the adaptation to local conditions. Sufficiently strong gene flow can completely swamp local adaptations and prevent the genetic divergence of local populations. Theoretical and experimental studies (reviewed by Endler, 1977) show that the swamping of local adaptations is indeed possible, but there is disagreement about how frequently it occurs in natural populations.

Mayr (1963) and more recently Stanley (1979) have argued that gene flow is generally important in preventing local adaptations and thereby causes an entire species to evolve as a single unit. Examples regarded as typical are populations of small mammals that have a dark pelage when living on lava flows in isolated populations but not when the populations on lava flows receive immigrants from populations living on lighter colored rocky substrates (Dice and Blossom, 1937). The view of Mayr and Stanley is that gene flow is generally a strong constraining force to selection in each population.

Ehrlich and Raven (1969) and Endler (1977) argue for the opposite view, namely that gene flow is usually a weak force relative to selection. There are two parts to their argument: first, that measured rates of gene flow are too low to swamp local adaptations; and, second, that isolated populations do not show substantially greater degrees of genetic divergence than do populations receiving immigrants. Ehrlich et al. (1975) describe their study of the butterfly *Euphydryas editha*, which is found both in western California and Colorado but not in the deserts in between. There has been no more phenotypic divergence of populations that have been geographically isolated for tens of thousands of years than there has been of populations known to exchange migrants.

Which view of the role of gene flow is adopted in a coevolutionary study determines the geographical scale that must be examined. If gene flow is relatively weak, then it is appropriate to examine interactions within each community. But if gene flow is relatively strong, then interactions over entire species' ranges would have to be considered.

RATES OF EVOLUTION

Simpson (1953) distinguishes between morphological rates of evolution (as measured by the rate of change in the measure of character) and taxonomic rates (as measured by the rates of turnover of taxa). For coevolution, morphological rates are very important, but they cannot usually be measured directly; therefore, morphological rates are often inferred from estimates of taxonomic rates.

Rates of morphological evolution can be measured precisely in artificial selection experiments. The conclusion from many experiments on many species is that most characters can evolve very quickly and that enough genetic changes occur to produce in a few generations a range of phenotypes that were not found in the initial population. In fact, it is unusual to find a character [such as development time in *Drosophila melanogaster* (Falconer, 1960)] that does not respond to selection for both increased and decreased values. Strong directional selection on a character usually produces correlated changes in several other characters.

Rapid evolution due to strong selection has also been observed in natural populations. The evolution of melanism, insecticide and disease resistance, and various weedy traits of plants have all been caused by human activities. Rapid morphological evolution has been inferred from high taxonomic rates in some groups. At least five species of moths in the genus *Hedylepta* have evolved to specialize on bananas in Hawaii, even though that plant was introduced only 1000 years ago (Zimmerman, 1960). Cichlid fishes have undergone extensive adaptive radiations during the past one or two million years in several African lakes (Stanley, 1979), and they have undergone extensive morphological changes during these radiations.

Rapid evolution, as measured by morphological or taxonomic rates, can occur under natural conditions, but it does not have to occur and certainly has not occurred in all lineages. The so-called "living fossils" such as the horseshoe crab or the alligator are well known but sometimes are regarded as relict species and not typical of species in more successful groups. A different view is expressed by the proponents of the punctuated equilibrium theory of evolution (Eldredge and Gould, 1972; Stanley, 1979). As Stanley (1979) extensively documents, the

29

fossil record of several groups shows that many species do not undergo significant morphological changes once they appear. This observation has led to the theory that the typical pattern of evolution is the morphological stasis of widespread species, with rapid morphological evolution associated with species formation. As discussed by Stanley (1982) and Charlesworth et al. (1982), this pattern of the fossil record is not necessarily incompatible with the view that natural selection is the predominant mechanism causing both the stasis and the rapid changes during species formation, but Gould (1980) and Alberch (1980) have argued for the importance of other mechanisms, including developmental constraints and genetic drift.

If interactions among species are indeed strong, then rapid coevolution could result. The evidence that rapid morphological evolution can take place in nature shows that rapid coevolution is a possibility. But rapid coevolution should not be assumed without considering alternatives. Interactions may not be as strong as supposed and various constraints may prevent significant coevolutionary changes.

OPTIMIZATION AND GAME THEORY

It is rare to study a phenotypic character for which the genetic basis and the ecological role are both well understood. The evident reason for the use of the same few examples of genetic evolution—industrial melanism in *Biston betularia* or banding patterns in *Cepaea nemoralis*—is the scarcity of such examples. Most workers are faced with understanding the evolution of characters whose genetic basis is completely unknown and species for which breeding studies are impossible. This situation is common in the study of coevolution, and it is usual to follow the tradition of Darwin and construct coevolutionary hypotheses without explicit genetic information about the characters of interest. This approach is based implicitly on evidence that there is substantial genetic variation in characters whose genetic bases can be understood.

In the past two decades, several formal theoretical methods have been used to make inferences about phenotypic evolution in the absence of genetic information. Levins (1968) developed a theory of "fitness sets" similar to utility theory in microeconomics (Peterson, 1974) to predict the outcome of evolution in spatially or temporally varying environments. MacArthur and Pianka (1966) developed a theory of feeding behavior based on the principle that animals will choose an optimal diet. That theory has grown into the theory of "optimal foraging strategies" (Schoener, 1971b; Pyke, Pulliam, and Charnov, 1977). Hamilton (1966) implicitly used a game theory approach in defining an "unbeatable strategy" in his analysis of the evolution of senescence; and Maynard Smith and Price (1973) made direct use of

30

game theory to analyze the problem of fighting and displaying in social animals. Maynard Smith and Price (1973) introduced the term *evolutionarily stable strategy* (ESS) to describe the mixture of phenotypes in a population that would resist invasion by any other phenotypes.

These theoretical approaches have had great influence on both evolutionary biology and ecology, and for good reason. They show the common elements of seemingly unrelated systems and provide simple ways to characterize evolutionary trends without requiring detailed knowledge about the traits of interest. Both optimization theory and game theory can be useful for understanding coevolution. For example, optimal foraging theory can lead to predictions about the interactions between a predator and its prey or a parasite and its host. Game theory can be used to understand models of coevolving systems by envisioning each species as a "player" that is trying to maximize its gain through certain kinds of interactions under the assumption that other species are trying to do the same (Slatkin and Maynard Smith, 1979). For example, the coevolution of two competing species can be understood in terms of the strategy of generalization leading to the use of more resources but causing more intense competition, as contrasted with a strategy of specialization leading to the use of fewer resources but also leading to less intense competition (Lawlor and Maynard Smith, 1976).

Optimization theory and game theory also have their limitations, and recently their application to evolutionary problems has been criticized. One criticism is that the characterizations of phenotypic evolution obtained using one of these theories may be incorrect in the sense that there may be no genetic model that is consistent with the characterization. For example, Strobeck (1975) has shown that Levins' theory of fitness sets cannot predict the outcome of temporally varying selection of certain kinds.

Another criticism is that these theories may be correct only if additional assumptions are made. For the application of game theory, Slatkin (1979) and Maynard Smith (1981) have shown that it is necessary to assume that there is sufficient genetic flexibility of the characters of interest before game theory predicts the correct equilibrium states. The application of optimization theory to the evolution of some character requires the assumptions that the character of interest is under genetic control and that there are no constraints of the kind discussed earlier in this chapter that would prevent the optimal state from being reached. The difficulty is not that these assumptions are necessarily invalid but that the theories themselves have no way to test the validity of their underlying assumptions in

31

particular cases. These methods generally lead to some prediction about evolution and coevolution even when they are inapplicable.

An important problem in the application of optimization or game theory to evolution is that the range of possible phenotypes must be specified. The optimal foraging strategy of a predator depends on its speed of movement, its agility, its ability to discriminate prey, and its memory. With no limits on these factors, optimization theory would predict that a predator would move infinitely quickly to any point, possess perfect knowledge of every prey item, and remember all past prey. The actual predictions about the predator depend on what limits are assumed. Levins (1968) called the range of phenotypes the "fitness set" of the species. In game theory, the fitness set is the set of strategies that each player has to choose from. Often, the most difficult part of an evolutionary or coevolutionary model is the specification of the fitness set.

Some criticisms of the use of optimization theory and game theory in evolution are valid. As Gould and Lewontin (1979) point out, ecologists are reluctant to conclude that a particular trait is not optimally adapted to some purpose. Because the prediction of optimization and game theory can depend strongly on the fitness set assumed, it is usually not possible to show that they are inapplicable to a particular trait. Although they are not a substitute for detailed genetic and ecological information, these approaches can provide a useful way to summarize that information.

CONCLUSION

The neo-Darwinian theory of evolution generally provides support for the assumption implicit in most coevolutionary studies that any particular character can evolve quickly if the individual selection affecting that character is strong enough. The neo-Darwinian theory, however, provides additional information that suggests that some caution should be used and some alternatives considered. Genetic changes can result from selection at levels above the individual. Kin and group selection could be effective in a species that is divided into transient local populations. In such populations, individual selection may well not have time to act.

There are also many constraints on the evolution of particular characters. Constraints may be especially important for coevolution because of the specificity of the interactions assumed. The demonstration that species interact in a particular way does not itself imply that coevolution has occurred or will occur. It is possible that studies of coevolution will contribute to our understanding of the importance of selection at different levels and of the roles of genetic constraints in shaping evolutionary changes.

32

CHAPTER 3

THE THEORY OF COEVOLUTION

Jonathan Roughgarden

Coevolutionary theory is one of the newest areas of theoretical ecology. During the last ten years it has provided interesting, and sometimes important, insights into how coevolution works. Yet many questions remain to be explored. This chapter is an invitation to think about coevolution in theoretical terms. It offers a guide to the approaches that have been developed so far. More technical detail may be found in reviews by S. A. Levin (1978) and Slatkin and Maynard Smith (1979) and in books by Christiansen and Fenchel (1977) and Roughgarden (1979).

Coevolutionary theory lies at the interface between population genetics and theoretical ecology. Indeed, models of coevolutionary phenomena tend to highlight differences between the points of view held by scientists of these fields. Geneticists are primarily concerned with the influence of the mechanism of inheritance on the coevolutionary process and with the influence of population interactions on the genetic structure of a population. Ecologists are primarily concerned with the effect of coevolution on the phenotype, especially on traits that determine how populations interact, and with the effect of coevolution on the abundance and distribution of the interacting populations. My orientation is primarily ecological, but I have included references to more genetical approaches.

There are four main parts to this chapter. The first presents some thoughts about the purpose of theory and introduces some basic technical concepts about evolution. The second is an introduction to density-dependent evolution for a single population. The third is a survey of special cases of the coevolution between two species. The fourth is a discussion of how coevolution may influence an entire biological community.

PART 1. PRELIMINARY THOUGHTS

Characteristics and uses of theory

There are two types of mathematical models in ecology. One type presents a simplified and possibly idealized picture of how processes operate. Its purpose is to develop insight. A simplifying model is an analogy rather than an approximation to what is actually going on.

Another type presents a summary of what is known. A summarizing model is useful when numerical predictions are needed, and when large amounts of data need to be rendered in a compact form.

The two types of models need not be mutually exclusive but usually are. Typically, the simplifications in a simplifying model make them too inaccurate for numerical applications, and a summarizing model is too complicated to explore mathematically.

There are no laws about coevolution in the sense of Newton's laws. Theory in ecology does not spring from axioms; instead, it is a small, but growing, collection of models that have proved useful to people.

Most people readily appreciate the value of a summarizing model because there is obvious utility in representing a body of data in a concise and transportable form. Volumes of data on tidal heights at the seashore and on solar illumination during the day are summarized with computer programs that are much easier to use than the original data tables. In contrast, simplifying models often produce a sense of uneasiness; if the assumptions are unrealistic, how can a model be of any value?

But experience shows that simplifying models really are helpful. From models of simple machines like the frictionless pulley and the inclined plane we learn the concept of mechanical advantage. These models are the starting point for the design of real machines. Similarly, in electronics, the formulae for the properties of resistors, inductors, and capacitors are as idealized as the model for a frictionless pulley and yet are used as building blocks in designing electrical circuits. In practice, simplified and idealized models are useful guides to understanding processes even though they use unrealistic and obviously false assumptions like that of no friction.

A conclusion from a model is robust if later research shows that it was originally derived from premises that were unnecessarily restrictive. In fact, almost all theoretical results are first derived in a restricted context that is later generalized to some extent. If the assumptions seem very restrictive, it is possible that the conclusions have limited interest, but it is also possible that the conclusions are more robust than the original assumptions suggest.

34

Absolute fitness

The classic models for evolution are based on an extremely simplified picture of what a population is and how it evolves (see Roughgarden, 1979, pp. 26–30). The life history is assumed to consist of discrete generations, with mating occurring through the random union of gametes. A population with this life history is assumed to live more or less in one place, so that the complexities of migration do not have to be considered. For this model population, let m be the fertility of an adult individual (m is one-half of the gametes that eventually become incorporated into zygotes). Let l be the probability that a zygote survives to be an adult. Then the *absolute fitness* for an individual is the product of the fertility with the survival probability.

$$W = ml \tag{1}$$

The units of W are numbers of individuals. W is zero or positive and is the expected number of offspring that the individual will contribute to the next generation.

Natural selection means that there are *differences* in absolute fitness among the individuals in a population. In population genetics, the *relative fitness*,

$$w = \frac{ml}{W_{max}} \tag{2}$$

is usually used. Here W_{max} is the highest absolute fitness that any individual has. The *relative fitness* of an individual, w, must lie between 0 and 1 and is a unitless number because it is a ratio in which the units of absolute fitness cancel out. Traditional population genetics is based on analyzing relative fitnesses.

For our purposes, the approach of analyzing relative fitnesses is obsolete, because the absolute fitness contains the information needed to predict changes in the population size, and this information is lost when dealing with relative fitnesses. Let $\overline{W}_t$ denote the average absolute fitness in the population at time t. Because $\overline{W}_t$ is the average number of offspring produced per individual, the population size at time $t + 1$ (where time is measured in number of generations) is simply

$$N_{t+1} = \overline{W}_t N_t \tag{3}$$

The average absolute fitness $\overline{W}_t$ depends on the composition of the population—because it is an average over the different kinds of individuals in the population—and also on the population size itself. In a limited environment, the number of offspring left by an individual depends somehow on the resources available to that individual. If the average fitness at time t happens to equal 1, as it will if the population is near a size commensurate with its resources, then the population size at $t + 1$ will equal that at time t.

The fundamental theorem of natural selection

A famous result from traditional population genetics theory is that natural selection brings about the maximization of the average *relative* fitness in the population, that is, the relative fitness as defined in equation (2), averaged over all the individuals in the population. According to this result, the average relative fitness increases after each generation—rapidly if there is a great deal of difference among the individuals in their relative fitness and slowly if most of the individuals have nearly the same relative fitness. Regardless of speed, the average relative fitness continually increases as it approaches, though never attains, a peak value. This finding is called the "Fundamental Theorem of Natural Selection" (see Chapter 2 by Slatkin).

The fundamental theorem is very important, yet it is delicate and easily misunderstood. Evolution involves much more than natural selection; other components to the evolutionary process include mutation, recombination, segregation distortion, genetic drift, mating behavior, age structure, population distribution, correlations between different characters, mechanical constraints on morphology, and so forth. Thus, the fundamental theorem of natural selection is not the fundamental theorem of evolution. Nevertheless, it is approximately correct when selection is much stronger than these other factors.

However, the kind of natural selection to which the fundamental theorem refers is very limited; it pertains to a population in which the *relative fitness* of each individual is *constant through time*. It is not appropriate if the relative fitnesses depend on the population composition (frequency-dependent selection) or size (density-dependent selection). Yet these are precisely the forms of selection that naturally arise in the context of coevolution. There are, instead, replacements for the fundamental theorem that are useful tools in visualizing how frequency-dependent and density-dependent selection operate.

Evolutionarily stable strategy

One approach for frequency-dependent selection is to consider a population all of whose members have a given phenotype and then to ask whether there is a different phenotype that can enter into the population. If there is no other phenotype that can enter the population, then the given phenotype is said to be an *evolutionarily stable strategy*, or ESS (Maynard Smith and Price, 1973). When this approach is used, the set of all possible phenotypes must be precisely

36

defined. The theoretical task is to calculate which phenotype, among all those possible, satisfies the condition that a population of such a phenotype cannot be invaded by individuals with any other phenotype. To do this, one finds the phenotype that has the highest fitness, given that it is the only type in the population. There may be no solution, one solution, or several solutions to this problem, depending on the detailed situation.

The justification for this approach is that fitness relations do predict the initial increase of a gene, even with frequency-dependent selection. The average fitness may decline as the gene continues to spread, but whether the gene enters in the first place depends on its carrier having a higher fitness than the other members of the population at that time (see Roughgarden, 1979, pp. 53–54). This justification also brings out the limitation of the ESS concept. It only refers to the initial increase of a phenotype when it is rare. To determine its subsequent fate, we need more information.

In contrast to the ESS approach that is focused on whether a monomorphic population is stable, Slatkin (1979) has concentrated on a polymorphic population. He has pointed out that a polymorphic population with frequency-dependent selection is, in some models, a state for which the fitness of every phenotype is equal and that there is a tendency toward the equalization of fitnesses.

PART 2.
DENSITY-DEPENDENT EVOLUTION

What is the connection between evolution and abundance? The main result from the theory of density-dependent evolution is that the evolution of traits that bring an increase in fitness causes an increase in abundance. This result seems intuitive, yet it is not generally true for coevolving populations even though it is true for density-dependent evolution in a single population.

Some traits of an organism promote a high growth rate for the population under conditions of low density (high-r traits). Other traits promote a high equilibrium population size for a population under conditions of high density (high-K traits). Density-dependent selection causes the evolution of high-K traits, and density-independent selection, which occurs under low density when a population is expanding, causes the evolution of high-r traits. These results can be derived using an extension of the idea of an evolutionarily stable strategy to density-dependent selection.

When the selection depends on population size, there are two kinds of variables in the system. The first (X) is a variable referring to the trait undergoing evolution. The second (N) is the population size itself, for this inherently changes as the traits whose fitness depend on population size evolve. The absolute fitness for an individual is a function of both variables, that is,

$$W = W(X,N) \tag{4}$$

The idea of an evolutionarily stable strategy with density-dependent selection is that $W(X,N)$ is maximized *with respect to X only,* and the magnitude of W is set equal to 1, as is appropriate if the population size is approximately constant. If a population consists of individuals whose trait maximizes fitness in this sense and if the population size is holding approximately constant through time, then any different type of individual introduced into this population cannot increase when rare. Such a population is at an *evolutionarily stable ecological equilibrium.*

To find the value of a trait that produces an evolutionarily stable ecological equilibrium, write down the fitness function $W(X,N)$ explicitly. Then take the partial derivative of $W(X,N)$ with respect to X and set the result equal to 0. This produces an equation in both X and N. Next, set $W(X,N)$ equal to 1. This is a second equation in X and N. Find the simultaneous solution to both equations. The solution requires that N be greater than 0 and that the second partial derivative of $W(X,N)$ with respect to X be negative. If N is not positive, then the solution is meaningless. If the second derivative is negative, then the solution is a maximum.

For example, suppose the absolute fitness of an individual decreases as a linear function of population size. This assumption leads to the logistic equation, which in discrete time may be written as

$$N_{t+1} = [1 + r - (r/K)N_t]N_t \tag{5}$$

where r is the intrinsic rate of increase and K is the carrying capacity. The expression in brackets is the absolute fitness W. Now suppose, further, that there is a trait whose value, X, influences an individual's r and K, so that we can write $r = r(X)$ and $K = K(X)$. The fitness then becomes

$$W(X,N) = 1 + r(X) - [r(X)/K(X)]N \tag{6}$$

For example, the trait X may refer to body size, and body size may be an indication of the average prey size that an individual uses. If so, it is convenient to take X as the logarithm of body size, and then $K(X)$ must be a function that tends to 0 as X tends to $+\infty$ and to $-\infty$. For simplicity, we may view $K(X)$ as a unimodal function more or less like the familiar Gaussian bell-shaped curve. The peak of $K(X)$ roughly corresponds to the body size that yields the greatest net rate of prey capture. Similarly $r(X)$ depends on body size. What body size should evolve under this density-dependent selection?

The answer is found as follows: First we differentiate $W(X,N)$ with respect to X and set the result equal to 0.

$$\frac{\partial W(X,N)}{\partial X} = \frac{dr(X)}{dX} - N \left\{ \frac{K(X)\,[dr(x)/dX] - r(X)\,[dK(X)/dX]}{K(X)^2} \right\} = 0 \tag{7}$$

Next we set $W(X,N)$ equal to 1,

$$W(X,N) = 1 + r(X) - [r(X)/K(X)]N = 1 \tag{8}$$

Now, equation (8) simplifies to

$$N = K(X) \tag{9}$$

and then equation (7) reduces to

$$dK(X)/dX = 0 \tag{10}$$

Furthermore, taking second derivatives shows that $\partial^2 W/\partial X^2$ is negative if $d^2K(X)/dX^2$ is negative. So, the body size that evolves is the body size that maximizes the carrying capacity. Notice that the relation between X and r is unimportant; it is the relation between X and K that is critical. This point is the essence of what is called "K-selection." Under density-dependent selection what is important is only the equilibrium population size attained and not the rate at which that equilibrium population size is approached.

Evolutionary dynamics

When a population enters a habitat it presumably does not already possess the traits that produce an evolutionarily stable ecological equilibrium there. What happens between the time it enters and the time it achieves an evolutionarily stable configuration? The answer depends on many factors, including the actual genetic mechanism underlying the traits. There is a simple model for the evolutionary dynamics that seems to be useful, although it is based on ignoring the genetic and demographic complexities that are undoubtedly present. There are three main assumptions in the model: (1) evolutionary changes in the traits occur much more slowly than changes in population size; (2) the evolution of the traits can be described by formulae that originate in the study of plant and animal breeding where traits,

39

like body size and shape, are caused to evolve through artificial selection; and (3) the population variance for the traits changes slowly relative to changes in the mean.

In the genetics of plant and animal breeding, the change in the mean value of a trait after one generation of selection is written as

$$\Delta \overline{X} = h^2(\overline{X}_{w,t} - \overline{X}_t) \tag{11}$$

where $\overline{X}_{w,t}$ is the mean among the adults after the selection, $\overline{X}_t$ was the mean before selection, $\Delta \overline{X}$ is the difference between $\overline{X}_{t+1}$ and $\overline{X}_t$, and h^2 is a constant between 0 and 1 that is called the heritability (see Falconer, 1960; Roughgarden, 1979, Ch. 9). Let the selective value of an organism of size X be $W(X)$. If $W(X)$ is expanded to first order about $\overline{X}_t$, we obtain

$$\overline{X}_{w,t} = \overline{X}_t + \frac{\sigma^2}{W(\overline{X}_t)} \frac{dW(\overline{X}_t)}{dX} \tag{12}$$

where σ^2 is the population variance in the trait. σ^2 is taken as a constant because it tends to a limit as long as the mean remains within the bounds of the variation that was originally present (see Roughgarden, 1979, pp. 142–145; Karlin, 1979). Putting these together, we obtain

$$\Delta \overline{X} = \frac{h^2 \sigma^2}{W(\overline{X}_t)} \frac{dW(\overline{X}_t)}{dX} \tag{13}$$

Suppose that the population can approach an equilibrium size, corresponding to a population in which the mean value of the trait equals X. This is found by setting $W(X,N) = 1$ and solving for N as a function of X. Then the model is

$$\Delta X = (\text{const}) \left. \frac{\partial W(X,N)}{\partial X} \right|_{N = \hat{N}(X)} \tag{14}$$

where const is a positive constant. In this model the population size "tracks" the current state of the evolving trait. By this model the trait evolves from any initial condition by climbing the gradient in the fitness function and comes to equilibrium at a value that represents an evolutionarily stable ecological equilibrium according to the previous definition.

This approach assumes that the population dynamics lead to a stable equilibrium abundance for any particular value of the trait. When this assumption is not reasonable, a specialized model for the dynamics must be used in order to characterize the nonequilibrium trajectories.

Other models for the evolutionary dynamics may be based on assuming a particular genetic mechanism for the trait, such as a single locus with two alleles (Anderson, 1971; Roughgarden, 1971; Clarke, 1972). In this particular case, a stable population genetic equilibrium brings about a local maximum to the equilibrium population size, as MacArthur (1962) had originally conjectured. Relevant literature now includes Roughgarden (1976), León and Charlesworth (1976), and Ginzburg (1977). References on r- and K-selection include Cody (1971), Gadgil and Solbrig (1972), and Harper (1977).

PART 3. COEVOLUTION BETWEEN TWO POPULATIONS

Coevolutionarily stable community

Coevolution is the simultaneous evolution of ecologically interacting populations. These interactions include interspecific competition; plant–herbivore and predator–prey interactions; and symbiotic relationships of parasitism, commensalism, and mutualism. For coevolutionary ideas to apply in a particular system, the populations in that system must interact, or have interacted in the past, and must have been together long enough in space and time for the interaction to have had a realistic opportunity to cause evolutionary changes.

We can extend the ideas first encountered in density-dependent selection to the coevolution of interacting populations. Consider two species, with population sizes N_1 and N_2, respectively. Suppose also that there is a trait carried by members of Species 1 whose state is X_1, and also a trait in Species 2 whose state is X_2. Then the fitness of an individual in each species depends on all of these variables. Natural selection within Species 1 responds to the traits of other species but modifies only the traits within Species 1 itself. Hence, suppose we

$$\text{maximize } W_1(X_1,X_2,N_1,N_2) \quad \text{with respect to } X_1 \text{ and set } W_1 = 1$$
$$\tag{15}$$
$$\text{maximize } W_2(X_1,X_2,N_1,N_2) \quad \text{with respect to } X_2 \text{ and set } W_2 = 1$$

A solution to this problem yields a value for the trait in Species 1 and a value for the trait in Species 2 that maximize the fitness within each of those species, respectively, and also yields the population sizes for both interacting species commensurate with their joint use of the resources in the environment.*

A community that is determined in this way is a *coevolutionarily stable community* (CSC). It is stable both ecologically and evolutionarily in the following sense: If every member of Species 1 has a trait whose value is X_1 and if every member of Species 2 has a trait whose value is X_2, and if the population sizes are N_1 and N_2, where

---

*The ecological feasibility of the solution should also be checked and the matrix $\partial W_i/\partial N_j$ should have eigenvalues consistent with ecological local stability.

(X_1, X_2, N_1, N_2) were obtained from the simultaneous solution of the equations in (15), then a mutant with some other value of X_1 cannot increase when rare in Species 1, and also a mutant with some other value of X_2 cannot increase when rare in Species 2.

To obtain a phenotypic model for the coevolutionary dynamics leading to a CSC, we can suppose that the population sizes "track" the current values of the traits in the populations, yielding

$$\Delta X_1 = (\text{const}_1) \left. \frac{\partial W_1(X_1, X_2, N_1, N_2)}{\partial X_1} \right|_{\substack{N_1 = \hat{N}_1(X_1, X_2) \\ N_2 = \hat{N}_2(X_1, X_2)}}$$

$$\Delta X_2 = (\text{const}_1) \left. \frac{\partial W_2(X_1, X_2, N_1, N_2)}{\partial X_2} \right|_{\substack{N_1 = \hat{N}_1(X_1, X_2) \\ N_2 = \hat{N}_2(X_1, X_2)}} \tag{16}$$

This system is used in Roughgarden et al. (1982) to model the dynamics of the coevolution of competitors. The results relate to the coevolution of the *Anolis* lizard populations on islands in the eastern Caribbean as discussed in Chapter 17.

An important limitation of this approach is the assumption of a stable population dynamic equilibrium for every combination of traits in the species; the approach is restricted to the evolution of model parameters where this assumption is true. In multispecies models, there are increasingly frequent findings of parameter sets that lead, asymptotically, to complicated nonequilibrium trajectories (e.g., May and Leonard, 1975; May and Anderson, 1978; and Arneodo et al., 1982).

The coevolution of competitors

How does coevolution influence two competing species? The issue is whether competition causes the evolution of resource partitioning, as indicated, for example, by a difference in the body sizes of the organisms. Biological examples of this idea are considered in more detail in Chapter 17. Here we shall treat this problem theoretically.

Suppose, as before, that the body size of Species 1 is X_1 and of Species 2, X_2. Let $K(X)$ be the carrying capacity of a species as a function of its body size. The new assumption is that the competition coefficient between individuals of different species is assumed to be a function of the difference in their body sizes. Specifically, the competition coefficient for the effect of an individual of Species j against an individual of Species i is

$$\alpha_{ij} = \alpha(X_i - X_j) \tag{17}$$

where $\alpha(X_i - X_j)$ is called the competition function.

To illustrate, we assume the competition function is a Gaussian bell-shaped curve whose peak occurs where $X_i = X_j$. When $X_i = X_j$, we

42

let $\alpha = 1$, indicating that inter- and intraspecific competition are equal. The width of the competition curve is described by the parameter σ_α. The competition function is

$$\alpha(X_i - X_j) = \exp[- (1/2)(X_i - X_j)^2/\sigma_\alpha^2] \tag{18}$$

Similarly, we let the carrying-capacity curve be a Gaussian, with X scaled so that the peak lies at $X = 0$,

$$K(X) = K_m \exp [- (1/2)X^2/\sigma_K^2] \tag{19}$$

For simplicity, we take r for each species as a constant independent of X, but identical results arise if both r_1 and r_2 are functions of X_1 and X_2, respectively.

The fitness functions are

$$W_1(X_1,X_2,N_1,N_2) = 1 + r_1 - [r_1/K(X_1)]N_1 - \alpha(X_1 - X_2)[r_1/K(X_1)]N_2 \tag{20}$$
$$W_2(X_1,X_2,N_1,N_2) = 1 + r_2 - [r_2/K(X_2)]N_2 - \alpha(X_2 - X_1)[r_2/K(X_2)]N_1$$

To find the coevolutionarily stable community based on these fitness functions, we differentiate W_1 with respect to X_1 and set the result equal to 0. Then we set W_1 equal to 1, thereby providing two equations based on W_1. Similarly, we get two more equations from W_2, yielding four equations in four unknowns:

$$N_1 d[1/K(X_1)]/dX_1 + N_2 \partial[\alpha(X_1 - X_2)/K(X_1)]/\partial X_1 = 0$$
$$K(X_1) - N_1 - \alpha(X_1 - X_2)N_2 = 0$$
$$N_2 d[1/K(X_2)]/dX_2 + N_1 \partial[\alpha(X_2 - X_1)/K(X_2)]/\partial X_2 = 0 \tag{21}$$
$$K(X_2) - N_2 - \alpha(X_2 - X_1) N_1 = 0$$

Because of the symmetry in the $\alpha(X_i - X_j)$ and $K(X)$ functions, we know there is a solution of the form $N_1 = N_2 = N$, and $X_1 = - X_2 = X$. The top equation in (21) then becomes

$$N \left\{ d[1/K(X)]/dX + \partial[\alpha(X_1 - X_2)/K(X_1)]/\partial X_1 \Big|_{\substack{X_1 = X \\ X_2 = -X}} \right\} = 0 \tag{22}$$

The last term is written to indicate that the competition function is first differentiated with respect to X_1, and then X_1 and X_2 are set equal to X and $-X$, respectively. If N is positive, then the expression in braces must equal zero. Upon substituting the carrying-capacity and competition functions explicitly, and solving, we obtain

$$X/\sigma_\alpha = + [(1/2)\ln (2\sigma_k^2/\sigma_\alpha^2 - 1)]^{1/2} \tag{23}$$

provided $\sigma_k > \sigma_\alpha$. Meanwhile, the equilibrium population sizes are found to be

$$N = K(X)/[1 + \alpha(2X)] \qquad (24)$$

Thus, the coevolutionarily stable community for these symmetrically competing species consists of one species whose body size is X and another species whose body size is $-X$, and whose population sizes are N where X and N are given by (23) and (24) above. The evolutionary dynamics of the approach to this equilibrium can be modeled with the equations for ΔX_1 and ΔX_2 given previously in (16).

This mathematical example illustrates an approach to determining how coevolution between competitors can cause the formation of resource partitioning. Resource partitioning between species may also be caused by the nonevolutionary mechanism of differential invasion. Hence, the existence of resource partitioning between competitors is not itself compelling evidence of coevolution. Also, the state of resource partitioning predicted in the mathematical example does not need to be attained through evolutionary divergence by the two species. Whether parallel or divergent evolution occurs depends on the initial condition for the coevolution. It is true that the end point can be viewed as a displacement by each species in an opposite direction away from the body size that either would evolve in the absence of the other. However, this view has nothing to say about the actual evolutionary trajectories taken by the coevolving species. These points are discussed further in Chapter 17.

Some of the recent theory on the coevolution of competitors examines the robustness of the result presented in the preceding mathematical example. Key references include Roughgarden (1972, 1976), Bulmer (1974), Fenchel and Christiansen (1977), Slatkin (1980), and Case (1982). A frequent theme concerns the *simultaneous* evolution of average differences between species together with differences among members within each of the species themselves. The discussion concerns traits, like body size, that might be taken as indicators of resource partitioning by individuals between and within species. These papers differ principally in the way that the inheritance of the character is treated. They offer several theoretical arguments that the evolution of resource partitioning between species requires a constraint on the evolution of the variance within each of the species. Furthermore, Roughgarden (1972) and Fenchel and Christiansen (1977) suggest that the evolution of an increase or decrease of the variance of a character within a species is a much slower process than the evolution of a shift in the mean value of the character. Roughgarden (1974), Hespenheide (1975), and Case (1982) argue that the empirical evidence shows the population variance to be evolutionarily conservative relative to the mean of the character. Case (1982) also offers a promising approach

44

toward combining competition theory with the theory of optimal foraging.

If the competition function is asymmetric, then the outcome of coevolution is not necessarily the attainment of niche separation (Roughgarden et al., 1982). Instead, one of the competing species can be driven to extinction by the other species during the course of coevolution. If the community is reinvaded after this extinction, the process can repeat itself. The sequence of coevolutionarily driven extinction followed by invasion, if repeated, produces a long-term turnover in the residents of a community while preserving a total species diversity that is approximately constant. This cyclic process is discussed further below and in Chapter 17.

The extension of the mathematical example to multiple resource axes shows that alternative patterns of resource partitioning can be simultaneously coevolutionarily stable (Pacala and Roughgarden, 1982). This result means that a community can develop a pattern of resource partitioning on either one axis or another, and the result will be coevolutionarily stable in either case. Which pattern actually develops in a particular community depends on historical factors, including any predisposition to a specific axis brought by the original colonists from their ancestral habitat. This result may explain the phenomenon of "niche axis complementarity" discussed by Schoener (1974), whereby the identity of the axis used for resource partitioning varies from place to place. An example of niche complementarity in eastern Caribbean anoles is presented in Roughgarden et al. (1982).

Habitat segregation refers to the occurrence of species in different habitat types. What is being considered are large-scale habitats, not microhabitats. For example, in the Sierra Nevada, Clark's nutcracker occurs at mountain tops and is replaced by the Steller's jay at middle elevations, and by the scrub jay at low elevations. The area of the habitats involved is on the order of thousands to more than one million times the area of the home range of an individual.

If the environment has variation on such a large spatial scale to begin with, we may ask whether the population will tend to occur more in one part of its potential range than another. When genes enter the population that confer higher fitness in one part of the potential range at a cost in fitness in other parts of the range, the population abundance shifts to the favored region of the environment. Thus, even a solitary population in a spatially varying environment can evolve habitat segregation through the accumulation of genes that cause there to be more individuals in one part of the potential range than other parts. If this evolution occurs in a reciprocal manner by two com-

45

peting species, then the resulting pattern is a replacement of competitors along a large-scale gradient like the altitudinal gradient in the Sierra.

To model the evolution of habitat segregation on such a large spatial scale, we need to consider that local migration is possible only between nearby points within the potential range. A standard competition model may be used to provide a picture of the ecology and evolution at local points within the range. Next, the local models for nearby points are coupled by assuming that migration leads to a flow of individuals back and forth. The local population-interaction models have parameters that vary from place to place within the potential range of the populations.

A regional model of this form shows that the spread of a gene, one that increases the abundance of individuals at some points in the range while decreasing it at others, is governed by a threshold (Nagylaki, 1975; Fife and Peletier, 1981; Roughgarden et al., 1982). The threshold is calculated from the geometry of the habitats within the potential range, the degree of difference between the habitats themselves, and the average dispersal distances of individuals. The existence of a threshold means that the evolution of habitat segregation in some regions will not occur at all unless the region satisfies a certain condition.

This result may relate to why some regional systems develop habitat segregation and others do not. Cody and Mooney (1978) have shown that the birds of the chaparral in Chile do not have the degree of habitat segregation that exists among the chaparral birds of California. Also, the *Anolis* lizard populations from some islands in the eastern Caribbean have evolved habitat segregation whereas others have not (Roughgarden et al., 1982).

In summary, the theory for the coevolution of competitors offers results mostly about how competition can cause the evolution of species differences in traits that promote resource partitioning. Other results concern the simultaneous evolution of between- and within-species differences, the invasibility of a coevolved community, species extinction as a result of coevolution with asymmetrical competition, niche axis complementarity, and habitat segregation. There is not nearly as much theory for coevolution involving other ecological interactions. This is not a value judgment on the importance of competition relative to other population interactions; it is a historical accident. I now turn to one of the most important interactions—the predator-prey interaction.

Predator–prey coevolution

The predator–prey interaction is a conflict in which the predators kill prey for food, and yet are dependent upon the continued existence of

46

the prey population for their own survival. Hence, we may ask the following questions. (1) Does evolution of more efficient predators through time eventually lead to predator extinction because of over-exploitation of the prey? (2) Is evolution of increasing predator ability counteracted by a continual evolution of defense, evasion, or toxicity by the prey so that a steady state is maintained? This is sometimes called the "Red Queen" hypothesis: both predator and prey are running hard just to stay even with one another. (3) Does the coevolution of predator and prey result in a coevolutionarily stable community, with an evolutionarily stable expenditure by predators to catch prey and by prey to avoid capture? Each of these questions leads to still more questions in the same vein. Although more work remains to be done, something can be said about these questions now.

The effect of predator–prey coevolution on the stability of the predator–prey interaction itself can be classified for each parameter in a predator–prey population dynamic model. Suppose, for example, that the prey grow logistically and that the predators have a linear functional response (an individual predator's predation rate in relation to the abundance of prey) with a numerical response (population growth rate of the predator in relation to prey abundance) proportional to their functional response (see, for example, Roughgarden, 1979, pp. 440–442). Then the fitnesses for the prey and predator are

$$W_v = 1 + r - (r/K)V - aP \tag{25}$$

$$W_p = 1 - \mu + abV \tag{26}$$

where V and P are prey and predator population sizes, respectively; r is the growth rate of the prey; μ is the density-independent death rate of the predator; a is the slope of the predator's functional response; and b relates the numerical response of a predator to its functional response. There is an equilibrium at which both predators and prey may coexist,

$$\hat{V} = \mu/ab, \quad \hat{P} = (r/a)[1 - \mu/(abK)] \tag{27}$$

This is globally stable for any positive initial condition provided P is positive. The $\hat{P}$ is positive and hence coexistence of predator and prey occurs if $abK > \mu$.

By inspection of W_v we see that natural selection within the prey will tend to increase r and K and to decrease a, everything else being equal. Notice, however, that the evolution of the prey's r and K will have no effect on prey abundance but will raise the predator population size. Evolution that lowers a will increase $\hat{V}$, lower $\hat{P}$ provided

$\hat{V} > K/2$ to begin with, and raise $\hat{P}$ if $\hat{V} < K/2$ to begin with. The highest value of $\hat{P}$ occurs at $a_0 = 2\mu/(bK)$, the value that produces $\hat{V} = K/2$.

Similarly, evolution in the predators will tend to lower μ and increase the product, ab, everything else being equal. The decrease in μ will lower $\hat{V}$ and raise $\hat{P}$, the increase in ab lowers $\hat{V}$ but may lower or raise $\hat{P}$.

With this model we can ask whether the predators will tend to evolve themselves to extinction. Clearly, a positive $\hat{P}$ requires that $abK > \mu$. Since the effect of evolution within the predator is to raise ab and to lower μ, the predator, if initially capable of coexisting with the prey, cannot in this model evolve so as to drive itself to extinction through overexploitation of the prey. (However, this is a deterministic result. If the predator evolves a large enough ab, the prey population size $\hat{V}$ may become very small, so that the prey could become extinct because of random factors. Extinction of the predator would then follow.)

The main destabilization to predator–prey coexistence comes from evolution in the prey. First, as the prey evolve defenses, a becomes lower, so the condition for positive $\hat{P}$, namely $abK > \mu$, may become violated. Hence, the prey will have evolutionarily evaded the predator. Second, as the prey evolve a higher K, the system tends to become more oscillatory. In more complex models (see, for example, Roughgarden, 1979, pp. 443), a stable limit cycle exists for K sufficiently high. In such a situation, the predator or prey population may become extinct, by chance, as population size comes close to 0 during the cycle. This point is the "paradox of enrichment" discussd by Rosenzweig (1971). The paradox is that an experimental augmentation of the prey's productivity may cause the predators to go extinct rather than benefit from the enhancement of the prey's productivity.

The idea that the evolution of increased predator ability is matched by the evolution of increased ability among the prey to defend against and to evade predation is treated in a series of papers beginning with Rosenzweig (1969, 1973) and culminating in Schaffer and Rosenzweig (1978). The argument is very complicated but can, I believe, be fairly restated as follows.

Consider the predation coefficient a in the predator–prey model. We may formally partition the differential of a, representing a small net evolutionary change per generation, da, into two components

$$da = \delta_v a + \delta_p a \tag{28}$$

where $\delta_v a$ is the change to a caused by evolution in the prey and $\delta_p a$ is the change to a caused by evolution in the predators. As before, $\delta_v a < 0$ and $\delta_p a > 0$, everything else being equal. Schaffer and Rosenzweig argue that the overall da is positive if a is sufficiently small and that in some cases da is negative if a has an intermediate positive value. If so, there is at least one value of a such that $-\delta_v a = \delta_p a$, representing a state wherein the continual evolution of increased predator abilities is being matched by counterevolution in the prey. This state results in no net change in the realized predation rate.

The reason for asserting that da is positive when a is low is that in this situation $\hat{V}$ is near K and so there are very few predators available. Hence, the selection pressure within the prey to avoid a rare predator is low, whereas the selection pressure within the predators to increase predation on an abundant prey population is high. Hence, the net effect should be to increase a, if a is sufficiently low.

On the other hand, if a is high, the argument becomes complicated. If a is high enough, there are few prey and also few predators, since for sufficiently high a, both $\hat{V}$ and $\hat{P}$ vary inversely with a. Hence, the net effect on δa is hard to gauge when a is very big. But if a has an intermediate value, one near that producing the highest predator population size, then the selection pressure within the prey to avoid predation will be very high. The value of a producing the highest predator population size is $a = 2\mu/(bK)$. In this case, it may turn out that the strength of this selection to avoid predation exceeds the strength of selection within the predators to increase predation on prey of intermediate abundance. If so, the net effect will be to lower a.

Thus, if a is near 0, da is positive. And if a is near the value producing the highest predator population size, then da may be negative. If so, the system may come to a dynamic coevolutionary steady state where the predation rate parameter a is between 0 and $2\mu/(bK)$ and where each generation the predators improve upon their ability to catch prey but are countered by improvements among the prey in their ability to avoid predation.

This treatment pertains to the accumulation of new or novel mutations. Schaffer and Rosenzweig suppose that in both species there is a finite mutation rate per generation that continually supplies these novel mutations. Since the strength of the selection in each species on these mutations varies with the ratio of $\hat{V}$ to K, it is possible, though not certain, that a steady state results. Obviously, the major unknown is whether mutation rates increasing a in the predators and decreasing a in the prey are high enough for the steady state to be an important coevolutionary phenomenon in nature.

The other approach to predator–prey coevolution deals with rearrangements of the basic phenotypic plan in both predator and prey. It is assumed that there are no novel mutations available to make predators and prey unconditionally better at predation and avoidance.

Instead, for a predator to catch more prey, it must pay a price in terms of other activities. Similarly, the prey must pay a price for the manufacture of defensive toxins and so forth. There is assumed to be in each species genetic variation that permits the rearrangement of the prey's phenotype and the predator's phenotype subject to constraints inherent in the basic body plan of each species. This approach yields a prediction of a coevolutionarily stable community (CSC) for predator and prey in which there is an evolutionarily stable level of effort expended by predators in their search for prey and by prey in their avoidance of predation.

Suppose, for example, that the prey have a trait X_v that can be interpreted as a measure of prey defense against predation and that the predators have a trait X_p that can be interpreted as a measure of the effort expended in catching prey. What is the coevolutionarily stable level of prey defense and predator effort, and what are the abundances of predator and prey at the coevolutionary equilibrium?

As before, the fitness functions in the prey and predators, respectively, are

$$W_v = 1 + r - (r/K)V - aP$$
$$W_p = 1 - \mu + abV \tag{29}$$

Now, as a simple example, assume that r is a decreasing function of X_v in order to indicate that there is a cost to prey defense. The benefit of this defense can be represented by supposing that a is also a decreasing function of X_v. Meanwhile, we can assume that a is an increasing function of X_p, the predator catching effect, and that the death rate μ is also an increasing function of X_p (because of the "cost" of X_p). Hence, a possible model for predator–prey coevolution is provided by the functions

$$W_v = 1 + r(X_v) - r(X_v)V/K - a(X_v,X_p)P \tag{30}$$
$$W_p = 1 - \mu(X_p) + ba(X_v,X_p)V \tag{31}$$

To analyze this model, we first differentiate W_v with respect to X_v and W_p with respect to X_p and obtain

$$\left(1 - \frac{V}{K}\right)\frac{dr(X_v)}{dX_v} - P\frac{\partial a(X_v,X_p)}{\partial X_v} = 0 \tag{32}$$

$$-\frac{d\mu(X_p)}{dX_p} + bV\frac{\partial a(X_v,X_p)}{\partial X_p} = 0 \tag{33}$$

Next, we set $W_v = 1$ and $W_p = 1$ and obtain

$$P(X_v,X_p) = \frac{r(X_v)}{a(X_v,X_p)}[1 - \frac{V(X_v,X_p)}{K}] \tag{34}$$

$$V(X_v,X_p) = \frac{\mu(X_p)}{ba(X_v,X_p)} \tag{35}$$

Now, substituting for P and V into (32) and (33), we obtain a pair of equations for X_v and X_p themselves.

$$\frac{1}{r(X_v)} \frac{dr(X_v)}{dX_v} - \frac{1}{a(X_v,X_p)} \frac{\partial a(X_v,X_p)}{\partial X_v} = 0 \tag{36}$$

$$\frac{-1}{\mu(X_p)} \frac{d\mu(X_p)}{dX_p} + \frac{1}{a(X_v,X_p)} \frac{\partial a(X_v,X_p)}{\partial X_p} = 0 \tag{37}$$

The solution to these equations will give a level of prey defense X_v and predator effect X_p that are coevolutionarily stable, provided the solution represents a maximum, and not a minimum, for each fitness. The CSC abundances are then found from (34) and (35).

To use these equations to obtain a biological result, a submodel for the trade-offs must be introduced. If the predation rate can be written as a product of two factors,

$$a(X_v,X_p) = a_v(X_v)a_p(X_p) \tag{38}$$

the equations become uncoupled, and can be written as

$$\frac{d \ln r(X_v)}{dX_v} - \frac{d \ln a_v(X_v)}{dX_v} = 0 \tag{39}$$

$$-\frac{d \ln d(X_p)}{dX_p} + \frac{d \ln a_p(X_p)}{dX_p} = 0 \tag{40}$$

Each of these can be solved graphically, and the conditions for an acceptable solution will depend on the details of the submodel that is chosen. A possible specific submodel for $a(X_v,X_p)$ is

$$a(X_v,X_p) = a_0(1 + \exp(-\alpha X_v))X_p^{\beta} \tag{41}$$

where $X_v,X_p \geq 0$, and $\alpha, \beta > 0$. Also, we can take

$$r(X_v) = r_0 - \gamma X_v \tag{42}$$

$$\mu(X_p) = \mu_0 + \delta X_p \tag{43}$$

with $\gamma, \delta > 0$. The important point is that by using other choices of submodels, a model for predator–prey coevolution leading to a coevolutionarily stable community can be tailored to specific systems in which empirical studies are being done.

Two other topics in predator–prey coevolution concern the coevolution of the prey with one another caused by the presence of a common predation pressure. In Batesian mimicry, a palatable prey evolves to resemble an unpalatable species. Müllerian mimicry is the coevolution of unpalatable prey species to resemble one another. Theory for the evolution of mimicry has a venerable tradition tracing to Fisher (1930) and earlier (see Chapter 12 by Gilbert). Recent references include Nur (1970), Matessi and Cori (1972), Charlesworth and Charlesworth (1975), Huheey (1976), and Turner (1977, 1981). A related topic concerns coevolution of phenotypic differences among palatable prey, coevolution producing the result that experience by predators in catching one prey type does not enhance their ability to capture other prey species. This topic is known as the evolution of "aspect diversity" among prey. It is treated in Ricklefs and O'Rourke (1975), Endler (1978), and Levin and Segel (1982). The evolution of polymorphism in color within a species caused by predation from visual predators having a search image is known as "apostatic selection" (Cain and Sheppard, 1954; Clarke, 1969). This topic is also discussed in the papers concerned with aspect diversity cited earlier.

Coevolution between symbiotic partners

Symbiosis is a population interaction based on long-term physical contact between individuals of different species. Typically, an individual of one species permanently lives on, or in, a particular individual of another species. The relationship is usually asymmetric in the sense that a member of one species, usually the one that is physically bigger, can be identified as the "host" individual, and the other as the "guest." Traditionally, the symbiosis is considered to be parasitism if the guest exploits and harms the host, commensalism if the guest has little or no effect upon the host, and mutualism if both guest and host benefit from each other.

For examples of symbioses in marine environments see Vermeij (Chapter 14), Hedgpeth (1957), Wright (1973), Bloom (1975), Vance (1978), Losey (1978), and Osman and Haugsness (1981). Terrestrial examples of mutualisms are also accumulating: see Way (1963), Smith (1968), Springett (1968), Bentley (1977), Quinlan and Cherret (1978), Addicott (1979), Brown and Kodrick-Brown (1979), Janzen (Chapter 11), and Feinsinger (Chapter 13). In many of these examples, the mutualist modifies the interaction of the host with another species. The main questions are how mutualism can evolve, and whether a mutualistic association is ecologically stable.

For examples of parasitism, see Levin and Lenski (Chapter 5), Barrett (Chapter 7), Holmes (Chapter 8), and May and Anderson (Chapter 9). Also, Price (1980) has written a monograph on the population biology of parasites. General issues are whether evolution leads to a reduced virulence by the parasite toward its host and whether parasites maintain genetic polymorphism in the host population.

Because the individuals of different species are physically associated with one another for a long time, the survival of at least one member of the association depends upon the survival of the other member. Hence, the fitness of an individual depends on the *fitness* of its associate. This is a very important point. In the preceding examples, the fitness of an individual depended on the *number* of individuals of other species and not on the fitness itself of those individuals. But, with symbiosis, the fitness of an individual in a given species does depend on the fitness as well as upon the number of individuals of the other species.

Let us consider fitness functions for the parties in a symbiotic interaction (Roughgarden, 1975). Let upper case letters denote properties of the host, and lower case letters stand for the "guest." We suppose that an individual can potentially exist in either a solitary or associated state and that the survival and fertility is different for these states. To be in the associated state, a guest must find a host and the host must survive, otherwise the guest is returned to the solitary state. Similarly, to be in the associated state, a host must be found by a guest and the guest must continue to survive.

$$W_g = (pL_a)l_a m_a + (1 - pL_a)l_s m_s \tag{44}$$

$$W_H = (Pl_a)L_a M_a + (1 - Pl_a)L_s M_s \tag{45}$$

where p is the probability that a guest finds a host; L_a and l_a are the survivals of the host and guest, respectively, in an associated state; m_a and m_s are the guest fertilities in associated state and solitary state, respectively; P is the probability that a host is found by a guest; and M_a and M_s are the host fertilities while associated and solitary, respectively. The fertilities and survivals may be density dependent.

It seems natural to view symbiosis in terms of two issues simultaneously. First, to form an association, the guest must expend effort that decreases its fitness as a solitary individual in order to locate and then to enter the host. Also, the host may expend effort that decreases its fitness as a solitary individual either to repel the colonization attempts by potential guests if the guest is harmful to it or to solicit colonization if the guest is beneficial. Second, the guest must assess the degree to exploit the host. To lower the host's fitness sufficiently will cause the host's death and a forfeit of further benefit from the association, whereas not to exploit the host at all would lead to no benefit from the association to begin with. Similarly, the host must assess the extent to which it is advantageous to resist the pressure of a

53

harmful guest once it has colonized, or to facilitate a beneficial guest.

Roughgarden (1975) explored these issues from the point of view of an evolving guest population, assuming the properties of the hosts could be taken as constant. Also, the formation of the association was viewed as occurring before the molding of the association and so these issues were treated separately, not simultaneously. The analysis was organized around the example of the damsel fish-sea anemone association originally studied in the Pacific by Verwey (1930). Roughgarden concluded that mutualism could only evolve in a restricted circumstance because of two considerations.

The potential host individuals must survive sufficiently well for a symbiotic association to form initially; otherwise, a potential guest might have a higher fitness by remaining unassociated with such a host. From this argument it follows that more species of obligate symbionts should have evolved to live on or in long-lived hosts than on or in short-lived hosts, everything else being equal.

The other consideration is that the extent to which a guest should forego exploitation of the host in order to increase the host's survival depends on how much that sacrifice yields in terms of an improvement to the host's survival. If the host already survives sufficiently well, or if the guest can have little impact on the host's survival, there may be little the guest can do to improve upon a host's survival. Thus, the evolution of mutualism by the guest toward the host requires that the host have an intermediate survivorship. It cannot be too low or otherwise the association will not form; it cannot be too high or otherwise there is no benefit to the guest of improving it still further.

If the guest evolves some moderation in its exploitation of the host, it might be called a "gentle parasite," provided the host nonetheless is harmed by the guest. For the guest to be a mutualist, it must actually increase the survivorship of the host above that where it is solitary. The guest is, of course, evolutionarily "unaware" of how well the host survives in the associated state as compared with the solitary state. The evolution of mutualism thus requires a lucky coincidence. The improvement to the host that the guest finds justified from its cost–benefit analysis must happen to be so great that the host finds itself actually aided by the guest. And once this should happen, the host should actively solicit the guest, and reciprocal mutualistic coadaptations can evolve. These considerations were applied to offer an explanation for variation among the relationship between several pairs of damsel fish and sea anemone species.

A problem with the Roughgarden (1975) analysis is that it is not a true coevolutionary analysis. It is a cost–benefit discussion from one party's point of view in a two-party interaction. A coevolutionary formulation would be as follows. Let x_c and x_m be indices of colonization effort by the guests and of the degree of exploitation of the hosts by the guests. Similarly, let X_c and X_m be indices of the host's response to colonization by the guest and of the host's reaction to the presence of a guest that has colonized. Let g and H denote guest and host population sizes. Then the mathematical problem takes the form

$$\text{maximize } W_g\,(x_c,x_m,X_c,X_m,g,H\,) \tag{46}$$

with respect to both x_c and x_m, and set $W_g = 1$; and

$$\text{maximize } W_H\,(x_c,x_m,X_c,X_m,g,H) \tag{47}$$

with respect to both X_c and X_m, and set $W_H = 1$. This is a bivariate maximization in each species together with the requirement of ecological equilibrium. The coevolution of symbiotic populations in this way is unexplored, although the problem would be theoretically tractable if useful submodels can be found for the relationship between the effort indices and the parameters of the fitness function.

Up until now we have only considered individual selection. When the guest is physically much smaller than the host, as is almost a rule with parasites, then the host may contain a group of guests within it. In this context, group selection may become important as well.

Group selection is the genetic evolution of a population (such as a group of parasites within an individual host) through the differential extinction rates and through the differential production of groups that vary in their genetic composition. There are now two well-developed theoretical pictures of how group selection operates. In the first, the whole population is considered to be divided into many small local populations. These go extinct with a probability that depends on the composition of the local population. When a group goes extinct, its place is colonized by a fair sample of some neighboring local population. Also, there is a low level of interchange among all the local populations. Eshel (1972) has shown that it is theoretically feasible for group selection to cause the evolution of a gene whose carriers promote group survival even if such a gene is opposed by individual selection within each of the groups (see Roughgarden, 1979, pp. 283–292, for numerical examples of Eshel's model). Of course, if the gene both promotes group survival and also is favored by individual selection within the groups, evolution will occur even faster than with individual selection alone.

In the second picture of how group selection can work, the population is considered to be divided into small groups during most of the life cycle. All the groups emerge at some point for mating and subsequent redistribution into groups. The groups contribute differentially to the total population, depending on their composition. This approach has been pioneered by D. S. Wilson (1975b, 1977, 1979) who has shown that a gene that confers high group productivity can evolve even if it is neutral to, or opposed by, individual selection within the groups. Wilson (1980, Ch. 2), offers a summary of this work.

The main application of group selection ideas to symbiosis is on the problem of the evolution of reduced virulence by parasites toward their hosts. Day (1974) suggests there is a general evolutionary tendency for viral or bacterial disease to become less virulent, as seems to have happened with the myxoma virus that was introduced to wild rabbits in Australia (Fenner, 1971). The evolution of reduced virulence by the virus may have been accomplished in part through group selection according to the model by D. S. Wilson.

Conditions for the evolution of reduced or increased virulence by a parasite toward its host are derived by Anderson and May (1978). They begin with equations for the population dynamics of the host population, assuming that hosts exist in one of three states: susceptible, infected, and immune. They exhibit conditions whereby genes in the parasites that lower their virulence increase when rare and a comparable condition whereby parasites evolve higher virulence. The treatment is not coevolutionary; the hosts are assumed to be evolutionarily static during the time that the parasites evolve. Also, Levin and Pimentel (1981) have studied the evolution of virulence and focused on the connection between virulence and transmissibility. These studies are reviewed in more detail in Chapter 9 by May and Anderson.

There has also been an explicitly genetic approach to host–parasite coevolution using what are called gene-for-gene models. Flor (1955) found in flax (*Linum usitatissimum*) that there were dominant genes conferring resistance to particular genotypes of the rust (*Melampsora lini*). There is a one-to-one matching whereby a particular allele in flax confers resistance to a corresponding allele in rust. Person (1966) and Day (1974) report that such gene-for-gene systems are widespread among other plant species and their parasites (see also Chapter 7 by Barrett). Models for coevolution in gene-for-gene host–parasite systems have been treated by Mode (1958), Jayakar (1970), Yu (1972), Gillespie (1975), Clarke (1976), and Lewis (1981a). Most of these studies focus on the potential role of parasites in maintaining genetic polymorphism in the host, rather than on the effect of genetic changes on the ecological relationships of the host–parasite system.

Our final theoretical issue concerned with symbiosis is whether or not mutualism should be rare in nature because it is a destabilizing interaction within the community (May, 1973). The mathematical basis for the claim traces to a model of mutualism that is generated by reversing the signs of the competition coefficients in the Lotka-

Volterra competition equations. The fitness functions for this model are

$$W_1 = 1 + r_1 - (r_1/K_1)N_1 + \alpha_{12}(r_1/K_1)N_2$$

$$W_2 = 1 + r_2 - (r_2/K_2)N_2 + \alpha_{21}(r_2/K_2)N_1$$

(48)

The population dynamics are then given by

$$\Delta N_i = (W_i - 1)N_i, \qquad i = 1, 2$$

(49)

If the positive density dependence between the species exceeds the negative density dependence within the species, then coexistence may be impossible. In this case, both populations mutually reinforce one another and, mathematically, tend to infinite abundance in finite time. This issue may be artificial because the fitness function (48) seems to be a poor model for mutualism, because it does not take account of the physical closeness of the mutualists with one another (it is only density-dependent). Hence, it is impossible for mutualism to evolve with these fitness functions under ordinary natural selection. In contrast, natural selection can, under some condition, cause the evolution of mutualism with the fitness functions given by (44) and (45). The population dynamics of the mutualists based on (44) and (45), assuming density dependence in the survivorships and/or fertilities for both associated and solitary states, lead to abundances that are mathematically bounded and that pose no theoretical problem. Several recent papers investigating the stability of a mutualistic interaction include Vandermeer and Boucher (1978), Goh (1979), Travis and Post (1979), Hallam (1980), Heithaus et al. (1980), and Addicott (1981). There are also papers showing that species can have a positive effect on one another through indirect routes (Lawlor, 1979; Vandermeer, 1980).

PART 4.
COEVOLUTION AND COMMUNITY STRUCTURE

There are now some preliminary explorations of how coevolution may influence several interacting species, and possibly even an entire community. The main questions that have been examined are these: How does evolution in one species affect the abundance of that species and other species in the community? Are species groups themselves units of selection? What is the effect of coevolution on the stability of the whole community? Where and how fast does community coevolution take place? And what would be signs that a community was indeed coevolved?

Density-dependent selection acting on a solitary population leads (locally) to the highest equilibrium population size. This result means that for a solitary population the consequence of the spread of traits that confer a higher fitness is that the abundance of the population increases.

This result is not generally true in coevolution, although it is in special cases. When an evolving population interacts with other species, its abundance may or may not increase, depending on the nature and strength of all the ecological interactions of the population with other populations. For example, equations (25)–(27) in the predator–prey model show that evolution in the prey of traits conferring a higher carrying capacity K do not affect the equilibrium prey abundance. This result is a special case that has been treated more generally.

Consider first the situation where there is a trait X_s in each species that affects any, or all, of the parameters for the population dynamics of that species except the interaction coefficients with other species. For example, if species s is a prey population, the X_s might affect r_s and K_s but not the predation coefficient a.

$$W_s = W_s(X_s, N_1, \ldots, N_S), \quad s = 1, \ldots, S \tag{50}$$

At ecological equilibrium, we can implicitly define each equilibrium population size as a function of the value of the trait in a given species. For example, take Species 1: by differentiating $W_s = 1$, for all s, implicitly with respect to X_1, we have

$$\frac{\partial W_1}{\partial X_1} + \frac{\partial W_1}{\partial N_1} \frac{\partial N_1}{\partial X_1} + \cdots + \frac{\partial W_1}{\partial N_S} \frac{\partial N_S}{\partial X_1} = 0$$

$$0 \quad + \frac{\partial W_2}{\partial N_1} \frac{\partial N_1}{\partial X_1} + \cdots + \frac{\partial W_2}{\partial N_S} \frac{\partial N_S}{\partial X_1} = 0 \tag{51}$$

$$\vdots$$

$$0 \quad + \frac{\partial W_S}{\partial N_1} \frac{\partial N_1}{\partial X_1} + \cdots + \frac{\partial W_S}{\partial N_S} \frac{\partial N_S}{\partial X_1} = 0$$

We can then solve for $\partial N_1/\partial X_1$ and all the other $\partial N_s/\partial X_1$ using Cramer's rule. But, by inspection, we can see that if X_1 maximizes the fitness in Species 1, then $\partial W_1/\partial X_1$ is 0 and therefore all the $\partial N_i/\partial X_1$ are 0 also. Hence, the value of the trait in Species 1 that maximizes fitness in Species 1 produces either a local maximum or minimum to the abundance of every species, including Species 1 itself. Now every equation in (51) can be differentiated again to find the second derivatives. The sign of the second derivatives will tell us if the critical points are maxima or minima. In this way we can establish that Species 1 can maximize or actually *minimize* its abundance as a result of the spread of traits that lead to fitness maximization within Species 1. Whether the spread of fitness-increasing traits in Species 1 increases or decreases the abundance of Species 1, depends on the entries in the matrix of derivatives $\partial W_s/\partial N_i$ in the equation above.

There is a condition based on the inverse of the matrix, $\partial W_s/\partial N_i$, which completely determines whether the evolution of fitness-increasing traits within a species maximizes or minimizes the abundance of that species (Levins, 1975; Roughgarden, 1977). This condition can be interpreted in terms of the concept of a "keystone" species.

A keystone species is a species whose removal leads to a still further loss of species from the community. For example, Paine (1966) found that in the absence of starfish, mussels overgrow and eventually exclude barnacles within a local study site. But starfish prefer to feed on mussels, so that in the presence of starfish, barnacles and mussels coexist. The starfish is a keystone species because its removal leaves a community that loses still more species.

There are two distinct methods whereby a species can become a keystone. In the first, the species is necessary for the existence of a positive equilibrium. Paine's example is of this type. Also, there are keystone competitors in this sense (see Roughgarden, 1979, pp. 546–548). What these examples have in common is that the resource arrangement without the keystone does not allow coexistence, but the dynamics are stabilizing in character. I term this type of keystone species a *positivity-enhancing keystone*. In the second method, the species is necessary to stabilize the system. Without the species, the remaining community may have a resource arrangement consistent with coexistence, but the nature and strength of the interactions do not combine to restore perturbations from this equilibrium. I term this type of keystone species a *stability-enhancing keystone*.

A mathematical example of a stability-enhancing keystone is provided by a model of the second species to follow a pioneer species in primary succession. Suppose the pioneer species, by affecting the substrate, makes the habitat better for itself and for the next arrival. If the first species has positive density dependence (it is autocatalytic), then it is unstable by itself. Yet it may stably coexist with a second species if that species brings sufficient negative density dependence to the system. The second species is thus a keystone species, but it is so by virtue of bringing stability to the dynamics. It is easy to construct more complicated examples using more species.

Now the surprising result is this: *if the evolution of fitness-increasing traits minimizes the abundance of a species, then that species is a stability-enhancing keystone.* As adaptive traits accumulate, the popu-

59

lation size of such a species continually declines, until extinction results. A stability-enhancing keystone is, in effect, a time bomb within a community. When it does go extinct, as a result of coevolution, then the remaining community will lose still more species. To the extent that coevolution structures communities, we should expect to see only positively-enhancing keystone species in nature. For a numerical example of a stability-enhancing keystone species evolving a minimum population size, see the pioneer–competitor example in Roughgarden (1979, pp. 470–471).

The preceding discussion concerns the evolution of traits that do not immediately affect the population interactions themselves. To consider traits that do mold the interaction coefficients, the fitness functions become functions of all the traits and all the population sizes,

$$W_s = W_s(X_1, \ldots, X_S, N_1, \ldots, N_S), \qquad s = 1, \ldots, S \qquad (52)$$

If we proceed as before, we immediately find that, in general, $\partial N_i/\partial X_j$ is not 0 in the coevolutionarily stable community. With further analysis one can calculate the sign of $\partial N_i/\partial X_j$, which tells us whether the CSC value of X_j is below or above that which would maximize N_i. Thus, from the standpoint of population abundance, the coevolution of the interaction coefficients always leads to an adaptational mismatch in the sense that the values of the traits that produce a coevolutionarily stable community are virtually never those values that would lead to the highest abundance for any species in the community.

Groups of interacting species as units of selection

The preceding treatment pursues the idea that coevolution is caused by natural selection at the individual level in each of the interacting populations. But if there is a local grouping of members of different species with one another, the structure of the grouping may, in principle, influence the coevolution in ways that parallel the workings of group selection within a single species. D. S. Wilson (1980) has advanced this idea and has begun to develop it theoretically.

A good starting point is the "case of the nasty competitor" introduced by Roughgarden (1976). The trait of pure interspecific nastiness is where a member of one species interferes with a member of another species without directly benefiting from the act. Suppose the trait is in Species 1 and affects Species 2. Using the terminology of the competition equations, the trait X_1 influences $\alpha_{2,1}$ but not $\alpha_{1,2}$. But the fitness of a member of Species 1 depends only on $\alpha_{1,2}$ and not at all on $\alpha_{2,1}$. Hence, the trait is selectively neutral. Yet, if it did happen to spread, then the consequence would be to increase N_1 and to lower N_2. If the members of Species 1 occur in small local groups with members of Species 2, subsequently reassemble into a well-mixed, single-species population, and then redistribute at random into small local populations where both species occur, group selection within Species 1 may cause the evolution of pure interspecific nastiness. The reason is that groups where pure interspecific nastiness is in high frequency contribute a higher production to the population at the mixing phase than groups where interspecific nastiness is rare. Wilson (1980) also focuses on mutualism as possibly evolving in this way.

These ideas are very preliminary. What happens if both species evolve through group selection? That is, we need a true coevolutionary theory of group selection and not an analysis from the point of view of only one party. Do the group structures for both species need to coincide, or can they overlay one another in an uncorrelated mosaic? And so forth. The idea of coevolutionary group selection is promising and needs much more theoretical exploration.

Coevolution and community stability

May (1973) championed the view that complex communities were inherently more unstable than simple ones, a view that was counter to the conventional wisdom at that time. Simple agricultural systems were known often to have large fluctuations in abundance through time and were felt to be more vulnerable to perturbation, especially from pests, than complex natural communities. It seemed that a complex natural community would have checks and balances among the populations that would stabilize it to perturbation. May showed, using models, that it is very touchy to adjust the interactions among the members of a complex community so that the combined effects of the interactions are stabilizing; more often than not, the combined effects of the species interactions are actually destabilizing.

The question then becomes whether the overall consequence of coevolution is to produce communities whose species have just the right characteristics necessary to obtain a complex stable community. Based on the coevolutionary theory so far developed, the answer comes in two parts.

Coevolution itself often, perhaps even typically, destabilizes a community that was stable to begin with. Coevolution frequently leads to species extinction and shapes population interactions in ways that are less stable than the original condition. We have seen three examples where coevolution can lead to extinction. They are (1) in predator–prey coevolution, evolutionary evasion by the prey (i.e., evolution of the predation coefficient a to a value low enough that the condition for predator existence, $abK > \mu$ is violated); (2) in extinction of a competing species during the coevolution of two competitors with asymmetrical competition; and (3) in extinction of stability-enhancing keystone species. Another example of coevolution destabilizing a population interaction is the evolution of a higher K by the prey in a predator–prey situation.

Nonetheless, because a frequent outcome of coevolution is the loss of species, the resulting community tends to be more stable than the original community simply because it is less diverse. Furthermore, one may conjecture that the remaining species in the less diverse community have weaker interactions with one another than an arbitrary community of equal diversity. Coevolution may, so to speak, prune the strongly interacting species from the original community, leaving a community that is both less diverse and has weak interactions among most remaining members. At this time, the idea that coevolution prunes the strongly interacting species from a community is purely conjecture; what is clear is that, in coevolutionary models, the extinction of species is a frequent outcome and that the resulting less diverse community will, for at least this reason alone, generally be more stable than the original noncoevolved community. Thus, it is not by "molding" the interactions, but by causing extinctions, that coevolution may ultimately generate stable communities.

The generalized taxon cycle

E. O. Wilson (1961) advanced the idea that there is a cyclic process in the ant fauna of South Pacific islands such that extinctions occur within an island fauna followed by invasions. The sequence of extinction and invasion leads to a turnover of species identities while preserving the total species number in the fauna at a roughly constant value. It is evident in Wilson's discussion that the extinction results from an interaction between existing members of the ant fauna. According to Wilson, the extinction is not the immediate population-dynamic consequence of the successful invasion of a new form into the fauna; rather, the invasion is made possible by the extinction, together with the withdrawal of resident species from marginal habitat as they shift toward use of preferred forest habitat.

Ricklefs and Cox (1972) have suggested that a type of taxon cycle occurs among the birds of the West Indies. The overall picture is complicated because some bird groups may participate in a cycle and others may not. Pregill and Olson (1981) have expressed doubts about the reality of the taxon cycle in West Indian birds and suggest that long-term changes in the environment are the primary cause of extinctions.

The *Anolis* lizard population on the small islands of the eastern

Caribbean also appear to undergo a cyclic process, as discussed in more detail in Chapter 17 (see also Roughgarden et al., 1982).

The theory of coevolution predicts that there are many situations in which extinction is the consequence of the coevolutionary process, as mentioned previously. If extinctions are followed by invasions, then cycles consisting of alternations between extinction during coevolution followed by invasion might be quite common. I term such a cycle a "generalized taxon cycle."

The idea that frequent extinctions result from coevolution is not inconsistent with a major role being played by climate changes during geologic time as discussed by Pregill and Olson (1981). Climatic changes may effect an extinction that has been "set up" by biotic interactions, including coevolution. There can be a "biological targeting" of the effects of environmental change. For example, in the taxon cycle that seems to be occurring on St. Maarten in the eastern Caribbean, the smaller species *Anolis wattsi* is expected to become extinct in the future. The coup de grace will probably be a 10- to 20-year run of comparatively dry rainy seasons during which the central hills where *A. wattsi* lives will approach the xeric state of the sea-level habitat in which *A. wattsi* is excluded by its competitor, *A. gingivinus*. Thus, the connection between environmental change during geologic time and the extinction of species in the fossil record may be provided by an understanding of community processes, including the role of coevolution in promoting species extinction.

Possible limitation of effects of coevolution to systems of low diversity

The coevolutionary theory of two competitors has been extended, numerically, to more than two competitors (Rummel and Roughgarden, in preparation). From this work it is clear that the time needed for a set of competitors to reach coevolutionary equilibrium increases very rapidly with the number of coevolving species.

The reason that coevolution works slowly in high diversity competition systems is twofold. First, any shift in niche position that decreases competition with one species increases the competition with another species, thereby reducing the net direct selective advantage to such a shift. Second, species influence others through long and indirect pathways, pathways that are longer and whose effects are weaker in high-diversity systems. Weak direct and indirect selection pressures require a long time to produce results.

The relationship between species diversity and the time required for a coevolutionary equilibrium to be attained has not been investigated in other model systems. Nonetheless, it seems reasonable to conjecture that high-diversity communities will generally equilibrate more slowly than comparable low-diversity systems, because the reasons that underlie this relationship in competition communities are not special to the competitive interaction.

Similarly, Case (1982) has pointed out using a two-species coevolutionary competition model that if extinction does not occur the coevolved niche separations are large. They are so large that invasion by other species can readily occur. He suggests that coevolved systems will only be found in locations that are rather inaccessible to invasion, such as distant oceanic islands.

Thus, low diversity and remoteness promote the likelihood that coevolution has been an important contributing cause to the formation of community structure. The possible importance of coevolution in diverse, but very permanent, communities is presently unexplored. However, diverse communities may contain relatively isolated subsystems of low diversity within which coevolution may occur very quickly.

PART 5. CONCLUSION

This chapter testifies to the existence of the beginnings of a comprehensive theory for coevolution among interacting populations. As I see it, the main goal for future work in coevolutionary theory is to develop and to test models that are tailor-made for the particular coevolved systems in which empirical studies can be feasibly done. There is a glaring shortage, even absence, of good models for many kinds of coevolutionary situations, including the plant–herbivore and plant–pollinator interactions. The predator–prey models seem artificial to me and should be replaced with more system-specific models, like those for arthropod predation developed by Hassell (1978). The models based in epidemiology used by Gillespie (1975), May and Anderson (1978), and Levin and Pimentel (1981) also seem well posed. Models for the coevolution of competitors are appropriate for some systems but have unknown generality. We also need models that are natural for the population structure of space-limited marine populations having sessile adults and pelagic larvae. With such tailor-made models, we can increasingly develop coevolutionary predictions that are concrete enough to be rigorously tested under field conditions.

CHAPTER 4

PHYLOGENETIC ASPECTS
OF COEVOLUTION

Charles Mitter and Daniel R. Brooks

INTRODUCTION

Phylogenetics is the discipline that attempts to reconstruct the genealogical relationships among taxa and the sequence of origin of their distinguishing features. In this chapter we will address the way in which phylogenetic methods might be used to study the historical origin of contemporary species interactions and the characteristics that govern them. We will treat two general questions. The first is that of when and how any given association became established. To what extent, we may ask, does a set of currently interacting species represent the descendants of similarly associated ancestors? The second question is that of the consequences of "association by descent": To the extent that interacting lineages have been associated through time, how and to what degree have they influenced each other's evolution? This is the question of "coevolution" as defined in the introduction to this volume.

The first of these issues, that of where associations come from, has received the greater share of attention from systematists. It is the one for which more phylogenetic evidence exists, and it will receive the more detailed treatment here. As relatively sophisticated phylogenetic analyses become more widespread, however, systematics should make an increasing contribution to the study of coevolution.

THE ORIGIN OF ASSOCIATIONS: "ASSOCIATION
BY DESCENT" VERSUS "COLONIZATION"

The notion of association by descent is an old one. A number of early workers in parasitology remarked upon apparent correspondences between the relationships of parasites and their hosts. Kellogg (1896) suggested that avian biting lice gave evidence of relationships among their hosts:

The occurrence of a parasitic species common to European and American birds, which is not an infrequent matter, must have another explanation than any yet suggested. This explanation, I believe is, for many of the instances, that the parasitic species has persisted unchanged from the common ancestor of the two or more now distinct but closely-allied bird species. [p. 51]

Observations of this kind, many of which are summarized by Metcalf (1929), were frequent enough that Eichler (1948) made them the subject of "Fahrenholz's Rule": "the natural classification of some groups of parasites corresponds with that of their hosts." Eichler accounted for this generalization by the theory that "the ancestors of extant parasites must have been parasites of the ancestors of extant hosts, so that the evolution of hosts and parasites has been in correspondence."

How might this assertion be tested by phylogenetic study? We will treat the problem as one of explaining the origin of new host associations during parasite phylogenesis. (It is also possible to approach the subject by tracing the acquisition of parasites during host phylogeny; see Brooks, 1981; Brooks and Mitter, 1983). Accounting for the origin of organisms' features is central to phylogenetics, and it is thus appropriate to analyze ecological associations by logic analogous to that used in phylogenetic reconstruction of character evolution. Before we do this, however, we will review briefly some salient features of phylogenetic methodology.

A précis of phylogenetics

Evolutionary relationships among noninterbreeding entities (e.g., species) can be represented as a phylogenetic tree or cladogram (see, for example, Figure 1). The terminals at the ends of branches (e.g., the *Enterobius* species in the figure) represent taxa on which character data have been gathered. The internal nodes or branch points may be interpreted as most recent common ancestors of all forms lying on paths connecting to and "above" (i.e., closer to a terminal than) them. The line segments thus represent the succession of ancestral and descendant forms within single lineages. Any group of taxa composed of all the descendants of a given ancestor, and only those descendants, is *monophyletic* in the sense of Hennig (1966); the members of a monophyletic group thus share a common ancestor not possessed by any taxon outside that group. For convenience we will often refer to a group by equating it with either its most recent common ancestor or the lineage giving rise to it. Thus, in Figure 1, we may denote the

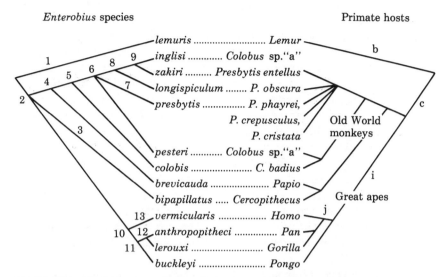

FIGURE 1. Phylogenies of pinworm genus *Enterobius* (Nematoda) and its primate hosts according to Brooks and Glen (1982). Primate phylogeny follows Schwartz et al. (1978), cited in Wiley (1981).

group composed of *Enterobius anthropopitheci*, *E. lerouxi*, and *E. buckleyi* as group 11, or that containing just *Homo*, *Gorilla*, and *Pan* as group j.

A monophyletic group comes by definition into existence when the ancestor of the larger group to which it belongs undergoes speciation. Any two monophyletic groups that together make up a larger one are called sister groups and are, by definition, of equal age. In Figure 1, groups 1 and 2 are sister groups, and both are older than any of the groups included within group 2.

A widely adopted approach to the construction of phylogenetic trees follows the "phylogenetic systematics" of Hennig (1966). We will presuppose a set of homologous characters, general descriptors (e.g., number of cervical vertebrae), whose character states or particular conditions (e.g., seven, the number of cervical vertebrae found in mammals) have been scored for each taxon under study. Under Hennigian analysis, one seeks the most parsimonious genealogical "explanation" for the distribution of character states over taxa, that is, the phylogenetic tree requiring postulation of the fewest character state changes. The smallest conceivable number of changes for a specified character is clearly just one less than the number of states, but this minimum will only be realized if all occurrences of each state can be accounted for by a single origin. If a particular state (e.g., the possession of seven cervical vertebrae) is found in all and only the members of some putative monophyletic group (e.g., the animals currently classified as "mammals"), we need invoke just a single origin of that state, in an ancestor unique to that group. In contrast, if a state (e.g., homeothermy) occurs only in widely separated parts of a genealogy (e.g., in birds and mammals, among the amniotes), we may be required to postulate more than one origin for it; such states are said to show low *consistency* (Kluge and Farris, 1969) on the specified phylogeny. Several proposed methods assign ancestral states to each monophyletic group so as to minimize the total number of character changes. Fitch's (1971) method assumes that any state may be transformed directly into any other state. Farris' (1970) method uses a predefined "transformation series," that is, a rule that postulates an evolutionary ordering of the character states. Thus, the transformation series a↔b↔c stipulates that the character cannot be transformed directly from state a to state c, without passing through b. Both of these methods produce an estimated historical ordering of character states during evolution (Mickevich, 1978, 1981).

Mickevich (1981) has argued that a cladogram corroborates a particular transformation series if states adjacent in that series occur in taxa that are adjacent on the cladogram. (The best-fitting transformation series under this "nearest-neighbor" criterion also gives maximal consistency under the Farris procedure.) Theories about the nature of character transformation can thus be evaluated by the fit to cladograms of the transformation series deduced from them (see detailed treatment in Mickevich, 1981).

The parsimony criterion by itself does not place restrictions on which state is most ancestral (see Farris, 1970, 1979), and thus additional criteria (e.g., concordance with stratigraphic position) must be invoked to fix the "roots" of genealogies and character histories once their branching forms have been established (see, for example, Lundberg, 1972). Finally, we should point out that some workers have rejected the application of philosophical parsimony to phylogenetic inference and proposed other approaches (see, for example, references in Mitter, 1980).

Association by descent

The preceding discussion implies that a theory about the evolution of parasites' host associations must make a prediction about the order of origin of particular habits. It is appropriate to point out the close formal resemblance between the issue of "association by descent" versus "colonization" in historical ecology, and that of "vicariance" versus "dispersal" explanations for disjunct distributions in biogeography.

In biogeography, the geographical "disjunction" between related forms may have arisen simultaneously with (and as a result of) the separation between the areas they occupy, or later, as a result of colonization. Similarly, related parasites may occupy different hosts because they speciated in concert with the hosts or because the parasite of one became transferred to ("colonized") the other. Our discussion will therefore draw heavily on the recent literature of phylogenetic biogeography, particularly Mickevich (1981; also Nelson and Platnick, 1981). However, although there is agreement on broad principles, there is as yet no consensus on a general method for specifying the systematic consequences of "disjunction" theories.

Under the hypothesis of association by descent, origin of new host associations of parasites occurs solely through divergence of ancestral hosts into daughter species. The ancestral host of any monophyletic group of parasites must therefore have been the host ancestor giving rise to all hosts in which those parasites occur. In that case, the transformation series for the host association "character" of the parasites should look like the host phylogeny, with the addition of a state corresponding to the ancestor of each host group (see Nelson, 1974). For example, in Figure 6A, the transformation series $(1 \rightarrow 2 \rightarrow 3)$ expresses the history of host association in the lineage leading to the parasite of *Caiman* (Cai), state (1) being the common ancestor [Croc, All, Cai] of the three crocodilian genera, state (2) the common ancestor [All, Cai] of *Alligator* and *Caiman*, and state (3) *Caiman* [Cai]. Thus, the minimum number of host association changes that must have occurred is one less than the number of states. However, when ancestral host-association states are assigned to the nodes (common ancestors) in a parasite cladogram under the hypothesis of mutual descent of hosts and parasites, the number of host association changes that must be postulated will often be greater than the minimum that would be required if the host and parasite cladograms had been perfectly congruent. The ratio of the latter quantity to the former, called the consistency index (Kluge and Farris, 1969), will be higher the greater the correspondence between the ordering of hosts on the parasite phylogeny and their relative age according to the host phylogeny.

We will illustrate the approach just outlined using the genus *Enterobius*, a set of primate-infesting pinworms (Nematoda). Figure 1 depicts a phylogeny for these parasites, based on analysis of their morphology (Brooks and Glen, 1982), along with a widely accepted view of the genealogy of their hosts. Let us examine the distribution of several hosts over the parasite phylogeny for compatibility with an "associa-

69

tion by descent" explanation. From the host cladogram we would predict, for example, that any event isolating a pinworm lineage on *Lemur* should occur before disjunctions separating parasites on individual genera of higher primates, because the latter did not exist until after the origin of *Lemur*. This expectation is in accord with the occurrence of the oldest *Enterobius* species in *Lemur*.

Within group 10, association by descent predicts that the lineage isolated on *Pongo* should be older than the one on *Homo* and that isolation on the latter should precede restriction to either *Gorilla* or *Pan*. To reconcile this expectation with the observed "reversal" of position between *Pongo* and *Homo*, we must postulate some history like that in Figure 2, which represents inferred ancestral hosts for each parasite group as the collection of all that host's descendants (Nelson, 1974). The independent derivation of parasite groups 12 and 13 from forms living in the same ancestral host could be explained by postulating, for example, that an ancestor of group 10 underwent a divergence event not affecting host lineage i, producing two sister parasites (10′ and 10″ in Figure 2) infesting i. Both diverged later along with group i, but representatives of group 10′ are missing from *Pongo* (as well as from *Pan* and *Gorilla*) due to extinction, failure to establish, inadequate sampling, etc., and group 10″ is absent for similar reasons from *Homo* (see discussion in Nelson and Platnick, 1981). Thus, if the phylogenies of the parasites and their hosts had been perfectly congruent, there would have been six changes in host association of the parasites, corresponding to the six line segments (internodes) in the phylogeny of the four hominoid genera from their common ancestor (Figure 1). Under the assumption of mutual descent, the *actual* phylogenies of the hosts and their parasites require seven changes in host association, described in Figure 2 by the solid internodes emerging from the two nodes marked [G, Pa, H, Po]. Thus, the consistency index is 6/7.

Although specific points of conflict can be found in the *Enterobius* example, a high level of overall agreement with the predictions under association by descent seems evident. It is important, however, to ask whether these observations might be accounted for equally well by a colonization model under which the host group diversifies completely before its members are sequentially invaded, with subsequent divergence, by the parasites. How might we choose among alternative possible histories for such colonization? In this case the states of the "host association" character are presumed to give rise directly to one another, so we could simply optimize this character according to the Fitch criterion. For group 10 of Figure 2 there are several equally good solutions, each requiring the minimum of three disjunction events. For example, the lineage ancestral to group 10 could have lived on *Pan* and then colonized in succession *Homo, Pongo*, and *Gorilla*, or it could

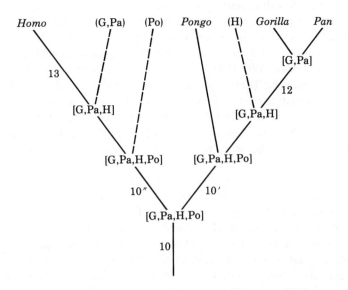

FIGURE 2. Hypothetical history for hominoid pinworms (group 10) of Figure 1, compatible with association by descent. Dashed lines represent lineages failing to establish in, or extinct from, the hosts in parentheses. Host ancestor inhabited by each parasite ancestor is represented by brackets containing extant hosts descended from that ancestral host. H, *Homo*; G, *Gorilla*; Pa, *Pan*; Po, *Pongo*.

equally well have started on *Pongo*, colonized *Homo*, and then given rise to a colonist of *Pan*, which subsequently colonized *Gorilla*.

Because some such colonization history can always be found that will maximize fit to the observations, how can we ever reject such an explanation? Association by descent accounts for disjunctions by a single principle, host phylogeny, whose consequences for the distribution of host associations can be specified and then tested. By contrast, a colonization hypothesis chosen simply to maximize fit to the parasite phylogeny is purely ad hoc. Colonization hypotheses are not inherently ad hoc, but to be maximally testable they too must make independent predictions about disjunction sequences, based on some property of the hosts such as geographical proximity or ecological similarity (see Mickevich, 1981, for extended discussion).

For most of the examples to be discussed, it is difficult at present to propose concrete alternatives to association by descent. The information required to construct such models, which would have to come

71

from studies of life histories and ecological interactions, is harder to obtain than the morphological data on which a phylogeny can be based. Thus, we will generally be limited here to evaluating association by descent hypotheses in isolation, comparing "expected" disjunction sequences (derived from host phylogenies) to observed ones ("phylogeny-like" sequences fitting the data best; see Figure 6; also Mickevich, 1981). We shall assess only qualitatively whether the agreement between these seems too close for coincidence; there has been no resolution of the issue of how (or even whether) sampling distributions for either transformation series or cladograms should be constructed, or what a reasonable null hypothesis in biogeography/coevolution might be (see, for example, Farris, 1981; Simberloff et al., 1981).

In fitting association by descent to our example so far, we have accounted for apparent deviations by assuming "extra" host divergences not supported by observation. It seems plausible that both mutual descent and colonization may have contributed to any observed set of associations, even when the overall pattern suggests strong agreement with one or the other mode of explanation. For example, it seems possible that the occurrence of *Enterobius inglisi* in a species of *Colobus* (Figure 1) might represent a host shift from *Presbytis*, because under mutual descent this association requires several "extra" disjunctions. There would be little point in abandoning a broadly corroborated mode of explanation just to achieve better fit in particular instances unless independent support could be found for the alternative theory. In zoos, however, pinworms are able to invade hosts not used in the wild (Brooks and Glen, 1982); the *Colobus* species is sympatric with *Presbytis*; and cooccurrence of two species (i.e., *inglisi* and *pesteri*) in the same host is otherwise unknown in *Enterobius*. Taken together these facts suggest that the recorded association of *E. inglisi* and *Colobus* is a recent one that may not even represent the typical host of this parasite. These partners are, however, descended from lineages that probably were associated for a significant fraction of the joint history of the groups to which they belong, i.e., up to the time of divergence between *Presbytis* and *Colobus*. Thus, mutual descent and colonization are not mutually exclusive; rather, we must inquire as to the relative contribution of each to the history of a parasite–host assemblage.

The example above raises other currently unresolved issues. First, how do we regard members of the host group that harbor no parasites? Thus, gibbons (*Hylobates* spp.) are generally regarded as the sister group of the rest of the great apes, but *Enterobius* has not been recorded from them. It can be argued that because the state "*Hylobates*" has not been observed for the pinworm host-association "character," a transformation theory for that character need no more concern itself

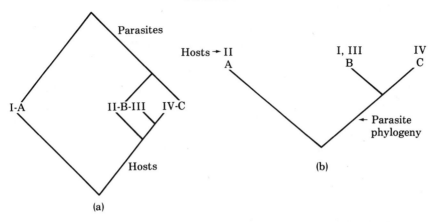

FIGURE 3. Hypothetical examples with multiple hosts for some parasites. (a) Pattern fully compatible with association by descent. Parasite ancestor (A,B) fails to differentiate when host ancestor (II, III, IV) splits to give II and (III, IV). Subsequent isolation splitting IV from III is accompanied by parasite speciation, isolating C on IV, B on II and III. (b) Same phylogenies as in a, but distribution of parasite over hosts not readily explained by continuous mutual descent.

with this than with any other nonobservation. Or it can be argued that the origination and subsequent divergence of a pinworm lineage restricted to the ancestor of great apes including *Hylobates* is required under association by descent and that any host transfer theory, say, that would place *Hylobates* at the end of instead of the middle of the transformation sequence, should gain credence as a result.

Second, the same parasite is frequently found in more than one host, as is true for *E. presbytis*. The general problem of how such observations are to be accommodated in disjunction theories is unsolved. Occupation of multiple phylogenetically adjacent hosts (not necessarily sister groups; see Hennig, 1966), resulting from host divergence without concomitant parasite speciation (see Figure 3a), seems fully compatible with mutual descent. Occupation of cladistically distant hosts by the same parasite, in contrast (Figure 3b), seems more difficult to explain by this process.

EVIDENCE REGARDING ASSOCIATION BY DESCENT

In this section we will review some of the systematic evidence that bears on the issue of how ecological associations typically originate. We will use a variety of examples to address two major questions: (1)

73

At what level (if any) in the genealogical hierarchies of associates is each of the two modes of origin most evident? (2) How do these patterns vary among different types of association?

For very few groups on which there is substantial ecological information have explicit phylogenetic trees been proposed, and for fewer cases still are there phylogenetic trees for both associated lineages. Moreover, it is frequently not possible to determine from the information supplied just how a published genealogy was derived. We have tried to choose cases that appear to have a reasonable grounding in morphological evidence, excluding, for example, those in which parasite phylogenies have been based partly on extrapolation from their hosts' relationships; in the examples below, we shall generally take the authors' taxonomic conclusions as given and shall merely examine their consequences for the questions raised above.

Animals living in, on, or with other animals

Some recent phylogenetic studies of animal host–parasite associations support the notion that these generally have long histories. In the *Enterobius* pinworms, as we have seen, there is good agreement at the highest taxonomic levels (i.e., involving the earliest divergences) with the expectation from mutual descent, but several points of departure within these larger groups. These include the "reversal" of position between parasites in *Homo* and *Pongo*, the possible "colonization" of *Colobus* sp. "a" by *E. inglisi*, and the placement of *E. brevicauda* with group 5, which is inconsistent with the requirement that any disjunction between *Cercopithecus* and *Papio* not precede the separation of these from the other old-world monkeys (Figure 1). But these are minor rearrangements, involving taxa relatively nearby on the tree; there is a strong overall association between the position of a host in the host phylogeny and its position in the disjunction rule that would fit best the parasite phylogeny. Given the number of taxa involved, the agreement seems too good to be accidental, and this case may be the best-documented instance of long-term mutual descent.

A phylogenetic analysis of digenean trematodes of the subfamily Acanthostominae (Brooks, 1980b) presents more complexity. A comparison with the vertebrate phylogeny shown in Figure 5 suggests that the Acanthostominae might have developed contemporaneously with their hosts, because the first split in both basal lineages (2 and 4) separates a teleost-inhabiting line from one in crocodilians. This hypothesis requires us to suppose, however, that the two trematode lineages have both diverged with lower vertebrates and yet left no descendants in any of the hosts that lie phylogenetically between crocodilians and perciform fishes (Figure 5). Moreover, the occurrence of group 12 in fishes would necessitate postulation of extra disjunctions to account for the restriction of earlier-diverging lineages 7 and 9

74

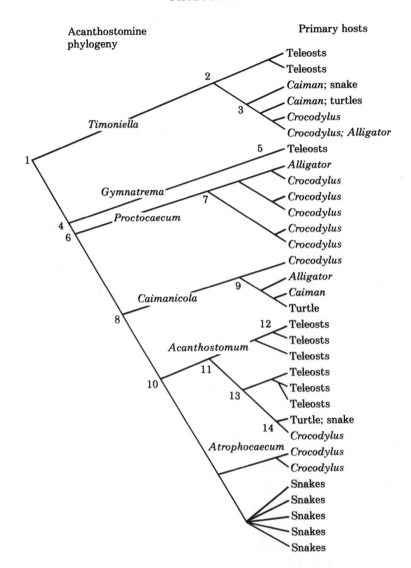

Acanthostomine
phylogeny

Primary hosts

FIGURE 4. Phylogeny and host associations of digenean trematodes of the subfamily Acanthostominae (Cryptogonimidae), simplified from Brooks (1980b).

to crocodilians. Because all of the hosts of the acanthostomines are either fish or piscivores (Brooks, 1980b) or both, making parasite transfers among them plausible, it seems reasonable to suppose that their occupation at least of the distantly related major groups of hosts

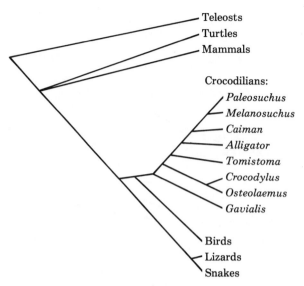

FIGURE 5. A partial phylogeny of vertebrates, including hosts of Acanthostominae (Figure 4), following Romer (1966), Sill (1967), and Neill (1971).

involved host transfer. If we divide the hosts into "teleosts" versus "reptiles" and assume that the nearest relatives of the acanthostomines, like the great majority of species in the family to which they belong, occur in fish, then the most parsimonious scheme of host transfer would show independent invasions of crocodilians by groups 3 and 6. Within group 11, we could either regard groups 12 and 15 as showing independent reversions to a fish-inhabiting habit or regard group 14 as an independent invasion of reptiles descended from a fish-parasitizing ancestor (11). With this part of ecological history provisionally established, we may investigate the possibility that *within* the major invasions, new associations have arisen predominantly by descent. Within group 3, the associations with turtles and snakes, like similar ones in *Caimanicola* and *Acanthostomum*, almost surely represent host transfers, given the wide phylogenetic separation of these hosts from crocodilians. What about the possibility that group 3 has otherwise undergone mutual descent with its predominant and presumably original host group, crocodilians? If we ignore the several host genera not harboring group 3, we can depict the disjunction sequence predicted from the generally accepted host phylogeny as in Figure 6A. Because host ancestors cannot reappear after splitting and because the oldest parasite species occurs in the youngest host, all three parasite ancestors must have lived in the oldest host ancestor and must have undergone subsequent mutual descent with the host group

76

independently. Because a different, but unacceptable, host phylogeny (e.g., Figure 6B) would give much better fit, we may be justifiably skeptical about the notion of association by descent with crocodilians for group 3. By contrast, for group 6, which we shall interpret as representing a single invasion of crocodilians (with various derivatives colonizing other hosts), the best "association-by-descent" transformation series is the one corresponding to host relationships, requiring 11 disjunctions as opposed to 13 under the next-best possibility. Thus, although a much more formal and rigorous treatment is desirable, the data on acanthostomines would appear to rule out a history of continuous association in general with their hosts but offer limited support for mutual descent of particular subgroups of hosts and parasites.

Other cases involving internal parasites of vertebrates (see, for example, Brooks, 1977, 1978b, 1979a; Brooks et al., 1981) fall into the broad categories suggested by these examples. Long-term mutual descent seems to be a reality, as evidenced by groups like *Enterobius*, in which the order of "host disjunctions" corresponds well to host phylogeny. But there is strong evidence for host transfer in many cases, namely those, like some acanthostomines, requiring large numbers of ad hoc assertions under association by descent and in which ecological factors such as cooccurrence in the same habitat appear to be better explanations of the origin of host associations. Finally, there are some cases in which the data are compatible with association by descent, but to a degree ambiguous enough that more detailed analyses, with rigorous investigation of plausible alternatives, will be required before more definite statements can be made about them.

One approach to the resolution of "ambiguous" cases, urged by a number of earlier workers (e.g., Szidat, 1956) and elaborated by Brooks

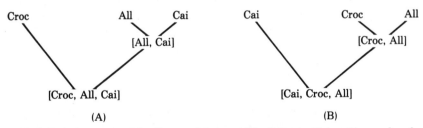

FIGURE 6. Two possible "mutual-descent-like" host disjunction rules for group 3 of Acanthostominae (Figure 4). (A) Cladogram that corresponds to actual relationships among hosts; (B) cladogram that does not, but fits better to parasite phylogeny. Croc, *Crocodylus*; All, *Alligator*; Cai, *Caiman*. For interpretation of brackets, see Figure 2.

(1981), is to broaden the inquiry to the entire complex of different parasite groups associated with the same hosts. The sum of these independent cases might suggest that the parasite fauna as a whole has developed along with the hosts and might suggest associated descent of individual groups that in themselves provide but weak evidence for it. Brooks (1981) concluded from such an analysis that the trematode fauna of crocodilians, including the groups discussed above, shows a general pattern of mutual descent.

We have concentrated on recent, explicitly phylogenetic studies because we consider these to provide the most definite evidence. However, many earlier workers, investigating a wide diversity of groups, have addressed these issues. Indeed, an elaborate series of "rules" for host–parasite evolution has developed. Some of the most important of these rules are as follows:

1. Fahrenholz's Rule (see Eichler, 1941a,b; Stammer, 1957; Cameron, 1964; Dogiel, 1964; Hennig, 1966; Ashlock, 1974): *Parasite phylogeny mirrors host phylogeny.*
2. Szidat's Rule (Szidat, 1956, 1960a,b): *The more primitive the host, the more primitive the parasites it harbors.*
3. Manter's Rules (see Manter, 1955, 1966; Inglis, 1971): (a) *Parasites evolve more slowly than their hosts.* (b) *The longer the association with a host group, the more pronounced the specificity exhibited by the parasite group.* (c) *If the same or two closely related species of host exhibit a disjunct distribution and possess similar faunas, the areas in which the hosts occur must have been contiguous at a past time.*
4. Eichler's Rule (see Eichler, 1941a,b, 1948; Inglis, 1971): *The more genera of parasites a host harbors, the larger* [i.e., the more speciose] *the systematic group to which the host belongs.*

Detailed reviews by Eichler (1948), Stammer (1957), and others drew the conclusion that conformity to Fahrenholz's Rule was widespread. However, most of the evidence cited for this generalization consisted only of the restriction of particular taxa of symbionts to particular taxa of hosts, without consideration of phylogenetic relationships within or among groups. To be sure, such correlations are a necessary consequence of mutual descent, and repeated failure to find them, especially when coupled with evidence for the importance of ecological factors in the evolution of host selection, would suggest a predominant role for host transfer. Thus, extensive studies on the gregarine sporozoans inhabiting various invertebrates (Stammer, 1957) showed that, whereas most of the species are restricted to more or less related host species, most gregarine genera are scattered across several host families, orders, or even classes. Supporting evidence for the prevalence of host transfers in these symbionts is provided by the fact that encounters among different host species frequently result in "straying" of gregarines into the "wrong" hosts. A similar pattern obtains for many mites parasitic or commensal on insects, mammals, and birds (Stammer, 1957; but see Regenfuss, 1967). In the case of syringo-philid feather mites, Kethley and Johnston (1975) propose that the variable most strongly predicting host associations among related mites is not bird taxonomy but quill size.

In contrast to the examples just cited, there are many instances in which pairwise associations of host and symbiont taxa are strongly evident. A widely cited case is that of the wood-digesting flagellates inhabiting the guts of lower termites, reviewed by Kirby (1937; see also Honigberg, 1970). Some groups of flagellates, for example, *Trichonympha*, are very widespread among the termites, with a distinct species group occurring even in the primitive wood roach *Cryptocercus*. Others, however, are restricted to particular termite groups. For example, the well-defined subfamily Pyrsonymphinae is found only in *Reticulotermes*. Termite flagellates are passed between host individuals by proctodeal feeding and seem incapable of forming resistant cysts. Coupled with the usual extreme social isolation of individual termite colonies, this makes the prospect of transfer among termite species in nature seem remote, although such transfers can be carried out artificially. Thus, many authors have concluded that present-day associations represent continuous mutual descent since the origin of blattoid insects, with subsequent loss of protozoa from higher roaches and termites.

Pairwise associations of particular host and symbiont groups by themselves cannot, however, be taken as strong evidence for association by descent, without more detailed examination of cladistic sequences either within or among those pairs (see Hennig, 1966). Such correlations might reflect only a tendency for host transfers to occur among near neighbors on the host phylogeny (Figure 7). Phylogenetic analyses of the many cases similar to the one just discussed (e.g., Rühm, 1956) would seem a fruitful area for future coevolutionary studies.

Szidat (1956) considered Szidat's rule a necessary long-term corollary to Fahrenholz's rule and described several cases of parallels between host and parasite groups in relative "primitiveness" (i.e., branching order as inferred from a presumed transformation series of one or more characters). For example, 11 subfamilies of paramphistonid trematodes, ordered according to the position of the testes, corresponded to the presumed sequence of origin of their fish, am-

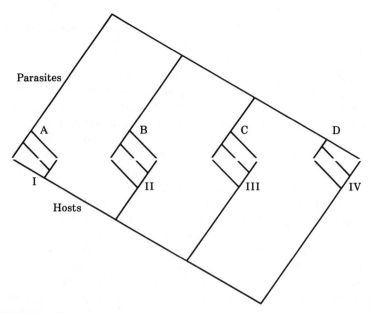

FIGURE 7. An artificial example in which pairs of higher parasite and host taxa are strongly associated. However, phylogenies among and within these taxa rule out association by descent.

phibian, sauropsid, and mammalian hosts (Szidat, 1939, cited in Eichler, 1948). A more recent example of such correspondence is that described by Radovsky (1967), involving 15 genera of macronysiid and laelapid mites and the bat families on which they live. It will be of great interest to see whether the conclusion of associated descent, which is strongly suggested by such cases, is supported by phylogenetic analysis. Some groups postulated in the literature to show mutual descent have nevertheless been described, somewhat paradoxically, as failing to obey Szidat's rule. For example, Kirby (1937) stated that although it seemed clear that many symbiotic flagellates had evolved together with their termite hosts, there was no overall tendency for the primitiveness of a flagellate fauna to reflect that of its host. This need only mean that a number of lineages of flagellates, some more primitive than others, had originated before the diversification of the extant host groups. Presumably one would detect the operation of Szidat's rule *within* any such lineage.

Coevolution and historical biogeography

Systematists have traditionally been more interested in the utility of parasites for resolving problems such as host phylogeny and biogeographical history than historical ecology per se, and have raised

80

the possibility that biogeographical and host parasite studies might be used in conjunction with one another (see Manter's Rules) to "reciprocally illuminate" issues of both kinds (see Hennig, 1966). A typical case from the early literature involves the "southern frogs" (family Leptodactylidae). These were regarded by some as necessarily polyphyletic because their disjunct distribution across the southern hemisphere was in conflict with the widespread theory that animals found presently in the southern hemisphere migrated there from the north under duress of competition (see Nelson and Platnick, 1981). Metcalf (1929, pp. 3–4) argued as follows in response:

In the recta of Australian and American southern frogs occurs a characteristic ciliate protozoan, *Zelleriella*, one of the Opalinidae, and some of these Australian and South American ciliates are almost if not quite specifically identical. This genus of ciliates is absent from the old world (except Australia) and in the New World is southern. . . . The parasites . . . indicate seemingly beyond question that the Australian and American southern frogs are related and also that they arose in the Southern Hemisphere and passed by some southern route from one to the other of their southern habitats. It might be possible, however unlikely, that the southern frogs of Australia evolved from very ancient ancestors in a way parallel to that of the South American frogs, though almost always in cases of parallel evolution there are found some genetic criteria to distinguish such resemblance from that due to genetic relationship. But no one can for a moment believe that, along with the parallel evolutions of the American and Australian hosts, there was also a parallel evolution of their opalinids.

Metcalf used the congruence of two independently derived hypotheses of monophyly as evidence favoring both and argued further that the "linkage" of the same disjunct land areas by two independent sister groups of animals suggested a similarly close historical relationship between these areas. Since the publication of his papers, the evidence of plate tectonics has shown that these areas were, in fact, in contact.

From the point of view of the present chapter, the potential utility of broadening phylogenetic coevolutionary studies to include biogeography is that we might thereby increase our ability to distinguish between hypotheses of colonization and of association by descent. Consider a case involving only a few "host disjunctions," each fitting equally well both host phylogeny and some plausible theory of host transfer. For example, the African ostrich and the South American rhea share parasite taxa (species or genera) of several kinds, including lice, roundworms, and tapeworms (see Hennig, 1966, p. 179; Eichler, 1948). One might suppose that, were the hosts not geographically isolated, parasites should transfer with relative ease between them.

However, it is highly unlikely that the parasites of one or the other could have crossed the south Atlantic in the absence of a host. If a parasite phylogeny is compatible with both a plausible host transfer sequence and a host phylogeny, additional concordance of the latter with a geologic sequence may resolve the issue in its favor. In the case of the parasites of the ostrich and the rhea, mutual descent seems likely, because these birds are thought to have diverged when Africa and South America became separated by continental drift.

In closing this section we will cite one last case that serves to illustrate the great variety of animal associations challenging the ecological phylogeneticist. Among the numerous organisms found associated with termite nests are 100-odd genera of staphylinid beetles, representing approximately 11 independent invasions of this habitat (see review by Kistner, 1969). The ecology of the interaction is poorly known, but there is indirect evidence of marked adaptations of the beetles to life in the nest. For example, many forms exhibit a remarkable convergent enlargement of the abdomen, known as physogastry, whose utility is unknown. A number of other species are marked by a rounded, flattened "limuloid" body shape. Various authors have observed mutual grooming or licking between staphylinids and termites, and many of the beetles have become strongly dependent on the microclimate and food sources provided by the nest. The relationship is best regarded as commensal, rather than mutualistic, because the termitophiles are relatively rare and many nests lack them entirely.

Most termitophilous staphylinids are strongly host specific, and their distributions generally parallel those of their hosts, even when the latter are spread across several continents. These facts suggest that individual beetle lineages, once having taken up this way of life, might have subsequently undergone associated descent with their hosts. Seevers (1957) examined this issue in detail and concluded that most evolution in the termitophiles had indeed occurred in association with that of their hosts. Two of the phylogenies depicted by Seevers are presented in Figure 8. If we consider just the hosts occupied by the subtribe Corotocina (Figure 8A), there is some support for association by descent, in that of the three possible arrangements of the host genera, the one best fitting the beetle phylogeny corresponds to termite relationships as presently understood. However, there remains the unresolved problem of the large number of phylogenetically intervening but unoccupied host groups. Although some of these absences may simply reflect limited sampling, it would seem important to search for geographical or ecological factors predicting the absence of termitophiles from particular host lineages. The Termitogastrina, on the other hand (Figure 8B), show no strong evidence of mutual descent. The termitophilous staphylinids present just

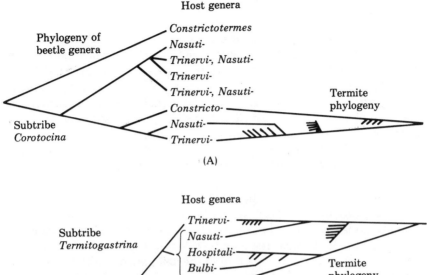

FIGURE 8. Phylogenies of two groups of staphylinid beetles obligately associated with termite nests, following Seevers (1957), together with host termite phylogenies according to Krishna (1970). Small, unlabeled branches on termite cladograms represent genera not known to harbor members of beetle group. All termite names end in -termes.

enough suggestion of a coevolving fauna to make further investigation highly desirable. We must note, finally, that for each case discussed in this section, associated descent may not be evident overall, yet could occur *within* the lowest-level taxa being considered, that is, on a shorter time scale than that involved in the establishment of host association differences among these taxa. Over very short time spans, *most* associations are undoubtedly "inherited." The question to be

83

asked is how long such evolutionary continuity of association typically persists.

Phytophagous insects

Much of the current surge of interest in coevolution can be traced to a paper about butterflies and their host plants (Ehrlich and Raven, 1964). It therefore seems appropriate to review some of the evidence regarding the degree to which phytophagous insect groups and their hosts have evolved in association. The model implicit in Ehrlich and Raven's discussion and developed explicitly by Benson et al. (1975) involves not strict, that is, continuous, mutual descent of hosts and herbivores, but alternating periods of diversification of each. A hypothetical phylogenetic pattern of the sort that might be expected under their model is shown in Figure 9. The two long branches in the host phylogeny represent lineages that have developed some novel form of immunity to attack by herbivores; this development allowed them to undergo a period of relatively rapid diversification. Each long branch of the herbivore tree may be taken to represent a line that has evolved some method of overcoming the defenses of a hitherto-immune plant group, a method that enabled it in turn to radiate across those already-diverse hosts. Each pairwise association is a result most immediately of colonization, but host phylogeny will still be a significant, albeit imperfect, predictor of the evolution of host associations. Even though there is no correspondence of phylogenies within "radiations," earlier-originating herbivore groups will tend on the average to be found on earlier-originating plant groups, if only because these were the sole hosts present when those herbivores were differentiating. The higher-level pattern of mutual descent will be most apparent when the colonizers of each recently diversified host taxon are derived from forms previously occupying the nearest relatives, and hence some not-too-distant ancestor, of that host group (Figure 9). If no such stricture applies, the pattern may be all but completely obscured, meaning that the points of historical contact between any two associated lineages (i.e., the periods during which respective ancestors giving rise to them were associated) have been few. Then the opportunity for mutual historical influence between particular lineages would have been limited, even though the evolution of hosts and herbivores might have been broadly contemporaneous. Each lineage may in this case be regarded as adapting to a background composed of numerous actual or potential associates, a process that has been termed "diffuse" coevolution.

How might one distinguish between the "reciprocal radiation" model of Ehrlich and Raven and a strict hypothesis of either colonization or mutual descent? Neither the methodology nor the evidence required are yet available, but a brief inquiry into this issue may il-

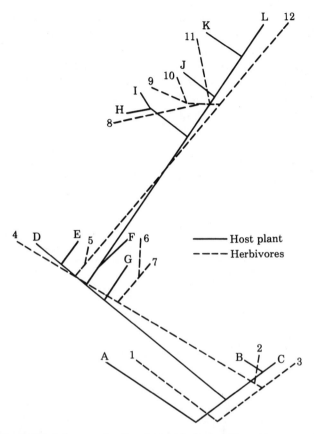

FIGURE 9. Artificial example showing a phylogenetic pattern possible under "reciprocal adaptive radiation" model of coevolution. For explanation, see text.

lustrate some of the difficulties involved in deciding among similar historical theories. The "reciprocal radiation" model leads to several expectations. First, there should be a nonrandom fit overall to association by descent. Second, the deviations from this fit should themselves be predictable, occurring within but not among groups of hosts that differ in antiherbivore innovations that can be recognized on independent grounds. Thus, the best single predictor of the evolution of host preference should be the ordering of "defensive novelties" specified by the host phylogeny. The expectation so established, however, might also be compatible with a simple host-shift model (the "sequential evolution" of Jermy, 1976b), under which each host is colonized from the host with the most similar "defense" characteristics. The more nearly the evolution of defenses is strictly divergent, the less distinguishable these alternatives will be.

The two theories may be separable, however, if unrelated hosts frequently have converged in their defenses, as have rutaceous and umbelliferous plants that produce the same essential oils. The strict host-shift theory predicts a close match to similarities in defenses, because all hosts are available simultaneously for colonization. Under the "reciprocal radiation" model, by contrast, host transfers should be more closely tied to phylogeny, because the convergent host may be unavailable (e.g., not yet evolved) when a particular transfer is occurring.

Although Ehrlich and Raven (1964) held that "plants and phytophagous insects have evolved in part in response to one another . . .

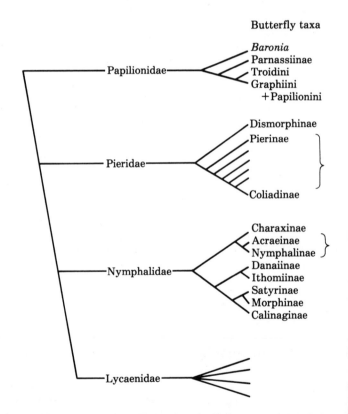

FIGURE 10. Summary of recent ideas on butterfly phylogeny. Arrangement follows Ehrlich (1958), except as modified by Kristensen (1975) and as follows: phylogeny of Pieridae after Klots (1931); arrangement of Papilionidae after Munroe and Ehrlich (1960). Hosts listed are major ones according to Ehrlich and Raven (1964). Slash (/) separating hosts indicates close relationship. Ar-

in a stepwise manner," they drew relatively few conclusions about the temporal order of "radiation" of particular groups. Their major concern was to establish that many higher taxa of butterflies are largely restricted to sets of hosts sharing a property (namely, secondary chemistry) known to strongly influence ecological interactions. Many of the hosts so grouped are phylogenetically distant, suggesting that host transfer has been a major mode of establishment of new butterfly habits. The possibility that there has nevertheless been broad-scale coevolution between butterflies and their food plants is not strongly supported by the sketchy outlines of butterfly and angiosperm phylogenesis currently available. To be sure, one might argue from Figures 10

Major hosts

Fabaceae
Aristolochiaceae
Aristolochiaceae
"Woody Ranales"   Rutaceae, Umbelliferae,
                              many others
                  Rosaceae
Fabaceae

Cruciferae/Capparaceae

Fabaceae and others

"Woody Ranales" and many others
Violales/Passiflorales; Urticales/
        Euphorbiales; and many others
Asclepiadaceae/Apocynaceae
Asclepiadaceae/Apocynaceae → Solanaceae
Grasses and other monocots
Variety of both dicots and monocots
Urticales

(Very large and poorly known group; great
diversity of feeding habits, large-scale
patterns rarely evident; many associated
with ants)

rows indicate probable secondary feeding habits. Ithomiinae feed mostly on Solanaceae, but earliest ones probably ate Apocynaceae (see Edgar et al., 1974; Gilbert and Ehrlich, 1970). "Woody Ranales" include families in both Magnoliales and Laurales (see Figure 11).

87

and 11 that, on average, the earliest-originating lines within major butterfly taxa feed on relatively ancient hosts. Thus, for example, the aristolochiaceous and "woody ranalian" hosts fed on by the main line of papilionids are older (or have more primitive secondary compounds) than either the rutaceous, umbelliferous, or rosaceous plants (among others) representing more derived habits within this family, or the dilleniid plants that are widely fed on by more recent groups such as Pieridae and Nymphalidae. However, major exceptions to this generalization are apparent. For example, the earliest diverging lineages of both Papilionidae and Pieridae feed on legumes (Fabaceae), one of the most recent and highly evolved of dicot families. The fact that many more ancient lepidopteran groups (see below and Figure 12) are associated with advanced, as opposed to primitive, dicots also suggests that the ancestral butterflies fed on advanced hosts. Much additional work on both papilionoid and angiosperm relationships, particularly among the Nymphalidae and their dilleniid hosts, will be required to fully settle the issue, but the evidence to date argues against a significant role for mutual descent in establishing the host preferences of the larger groups of butterflies (see also Vane-Wright, 1978).

Even if the Papilionoidea as a whole invaded their host taxa largely after the major taxa of plants had evolved, "reciprocal radiation" might nevertheless have operated over much longer or shorter time scales than those of the divergence of butterfly families. Perhaps, for example, the diversification of angiosperms, with subsequent colonization by butterflies, represents just one episode in a large-scale coevolutionary association of land plants and the order Lepidoptera as a whole. In the much-condensed view of lepidopteran relationships depicted in Figure 12, the first two host disjunctions are compatible with such a notion, although one would need to account for the absence of any ancient moth lineages from both ferns and monocots (on both of which several advanced groups of Lepidoptera feed), as well as the considerable disparity in the estimated ages of the two groups. Even if the very earliest lepidopteran and land plant ancestors evolved in concert, however, there is little indication that the main line of evolution giving rise to the major groups of extant Lepidoptera was associated with successive plant radiations. Some of the oldest angiosperm-inhabiting moth groups (e.g., the Eriocraniidae) are associated primarily with relatively young host groups, whereas primitive hosts characterize only a few, relatively recent, higher lepidopteran taxa (e.g., Papilionidae). A similar pattern was described earlier for the major groups of termite-inhabiting flagellates, however, and there remains the possibility that diversification of early lepidopterans on early angiosperms gave rise to a multitude of lineages (e.g., butterflies), each of which subsequently diversified in tandem with angiosperm phylogenesis

Plant taxon

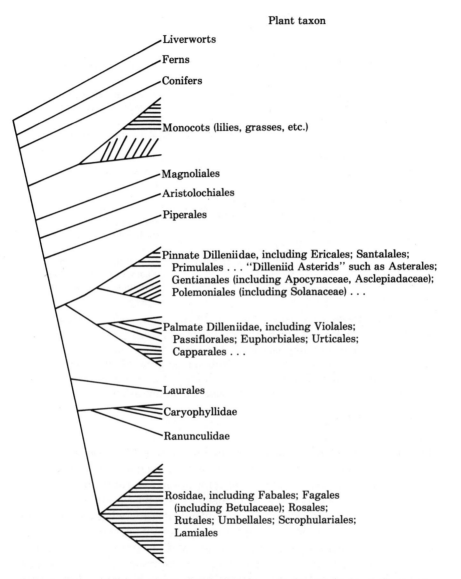

FIGURE 11. An informal phylogeny of some lepidopteran host plants. Arrangement of angiosperms (monocots and above) follows Hickey and Wolfe (1975). Numbers of branches within larger taxa indicate approximate rank of number of orders included. Recent evidence will necessitate some changes in this scheme (L. J. Hickey, personal communication), and other workers would give somewhat different arrangements.

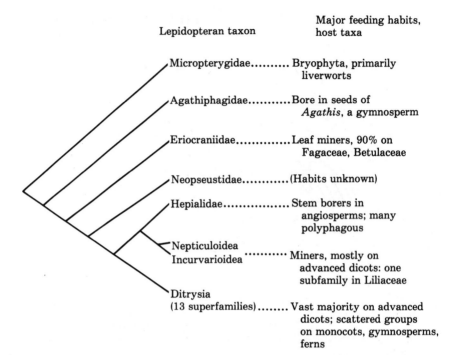

| | Major feeding habits, |
| Lepidopteran taxon | host taxa |

Micropterygidae.......... Bryophyta, primarily
liverworts

Agathiphagidae........... Bore in seeds of
Agathis, a gymnosperm

Eriocraniidae.............. Leaf miners, 90% on
Fagaceae, Betulaceae

Neopseustidae............ (Habits unknown)

Hepialidae................. Stem borers in
angiosperms; many
polyphagous

Nepticuloidea
Incurvarioidea .... Miners, mostly on
advanced dicots: one
subfamily in Liliaceae

Ditrysia
(13 superfamilies)........ Vast majority on advanced
dicots; scattered groups
on monocots, gymnosperms,
ferns

FIGURE 12. Early evolutionary history of Lepidoptera, following Kristensen and Nielsen (1980). Food plant information from Powell (1980). Ditrysia contain vast majority of living species. "Advanced dicots" means (largely) Dilleniidae and Rosidae (see Figure 11). (Note added in proof: More recent information suggests that Micropterygidae will feed on a variety of plant material found in the litter layer; D. Davis, personal communication.)

(Powell, 1980). Much greater resolution of relationships within the groups depicted here will be required to test such a theory. Available fossil dates suggest roughly contemporaneous radiation of Lepidoptera and Angiospermae. Suppose, as seems likely, that future investigation fails to demonstrate mutual descent with angiosperms within each of the major lepidopteran lineages. This could mean that, whereas the oldest ancestors of the hosts and phytophages were themselves associated, host transfers have in the meantime been so pervasive that present-day associations have no significant joint history (see Powell, 1980). It could also mean that extinction has obliterated the pattern. (See caption to Figure 12 for information added in proof.)

As we have seen, many relatively low-level lepidopteran taxa (e.g. tribes and genera) are narrowly restricted to particular angiosperm groups. Even if these associations themselves were established primarily by host transfers, mutual descent may have been important

90

in their subsequent elaboration. In very few instances, however, is there enough systematic information to examine this question.

Perhaps the best-studied case is that of the heliconiine nymphalid butterflies, a largely Neotropical group of approximately 70 species in 11 genera, all of which feed on Passifloraceae. This association was examined from a systematic and coevolutionary point of view by Benson et al. (1975), on whose discussion ours will rely heavily. A phylogeny and associated character information for the heliconians were presented by Brown (1981; see Figure 13); the somewhat different arrangements depicted by Brown (1972) and Emsley (1965) lead to the same conclusions to be drawn here. The hosts listed in Figure 13 are, exception as noted, those designated by Benson et al. (1975), as the primary ones for each butterfly taxon; many of the species have been also recorded from other passionflower groups. Benson et al. arranged the genera of new-world passiflors according to presumed degree of "evolutionary advancement" with respect to a number of floral and vegetative characters (following Killip, 1938). Although recognizing, as did these authors, the danger of equating degree of advancement with cladistic sequence, we can use this assumption to obtain a rough estimate of the prevalence of mutual descent in this assemblage. The evidence at hand appears to support the hypothesis of coevolution only weakly: the first and second divergences, isolating groups on the relatively primitive *Adenia* and *Passiflora* section *Astrophea*, agree fairly well with this explanation. However, on the rest of the heliconian phylogeny, primitive and advanced hosts seem to follow each other in no discernable order, and many quite different transformation series for the host association "character" would fit the observations equally well. Although a more rigorous analysis is clearly desirable, one cannot rule out the possibility that most (perhaps all) of the evolution of heliconians occurred after the diversification of the Passifloraceae, including that of *Passiflora* into its various sections.

Benson et al. (1975), whose largely implicit analysis seems to have treated the heliconian phylogeny not as a whole but as a series of independent radiations and regarded the "primitive" forms of each as indicating the ancestral feeding habits despite their sometimes nonbasal cladistic positions, reached the opposite conclusion. The "reciprocal radiation" theory would gain credence if just the apparent exceptions to it were especially likely to represent host transfers. Thus, Benson et al. (1975) argued that the *sara-sappho* Heliconius were able to "reradiate" onto the primitive *Passiflora* section *Astrophea* because they used a different part of the plant (meristems versus mature leaves) from earlier heliconians using the same hosts. Work in progress on

91

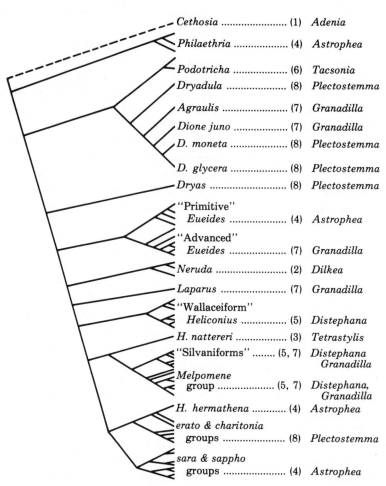

Heliconiines		Major host taxa

FIGURE 13. Phylogeny of heliconiine butterflies, simplified from Brown (1981). Names in quotes after Benson et al. (1975). Hosts listed are "preferred plant taxa that are assumed to have been responsible for past radiation" (Benson et al., 1975) of each butterfly group, except that host for *Podotricha* is from Brown (1981) and for *Cethosia* from Corbet and Pendlebury (1978). Numbers in parentheses indicate rank in "degree of advancement," and hence approximate age, of host groups, according to Benson et al. (1975). (*Adenia*, not treated by Benson et al., placed by their criteria.) Ranks 4–8 are sections of *Passiflora*; all others separate genera.

both the systematics and ecology of this system (see, for example, review in Brown, 1981) may eventually resolve these problems. The heliconians and their relatives might be regarded as a test case for mutual descent: the dominant hosts for many of these butterflies are

92

the families Passifloraceae, Violaceae, Turneraceae, and Flacourtiaceae, an assemblage coherent on phylogenetic grounds but for which Ehrlich and Raven (1964) could identify no secondary chemical similarity, though they predicted that one would be found.

Other phytophagous groups can be found for which mutual descent also seems plausible. Thus, Chemsak (1963) postulated parallel evolutionary development of the longhorn beetle genus *Tetraopes* with its milkweed food plants (*Asclepias*), although he did not present a phylogenetic analysis. An especially intriguing candidate is the association of the hundreds of species of figs and their highly specific wasp pollinators, family Agaonidae (see Wiebes, 1979; Chapter 13 by Feinsinger). Ramirez (1975) presented cladograms showing complete agreement with mutual descent for a group of agaonid genera and their host figs. Until a number of such cases have been examined phylogenetically, it is not possible to state whether mutual descent, including "reciprocal radiation," plays a significant role in the evolution of herbivorous insects and their hosts at this taxonomic level. It will also be of great interest to determine whether mutual descent operates at the lowest level accessible to phylogenetics, that is, the differentiation between races or species, as postulated for cynipid oak gall wasps by Cornell and Washburn (1979). The species in the two genera that have been studied in most detail, *Cynips* and *Neuroterus* (Kinsey, 1923, 1930, cited in Cornell and Washburn, 1979), exhibit large numbers of "varieties," the most closely related of which generally occur on most closely related oaks, which tend to be allopatric. This would appear to be a test case for species-level associated descent, and further investigation is greatly to be desired.

THE STUDY OF COLONIZATION

We chose for the preceding examination several cases in which mutual descent seemed most likely. In many other groups of phytophagous insects, however, the taxonomic separation between hosts of even sister phytophage species allows one to reasonably discount mutual descent out of hand (see Chapter 10 by Futuyma for examples). Indeed, the evidence available suggests that host transfer is by far the predominant mode of establishment of new host associations among insect phytophages, as it is for many other kinds of association. For this reason, it is appropriate to consider in more detail what generalizations might be made about colonizations.

Of the many questions that can be asked about host transfer, the most germane is that of which host characteristics explain the evolutionary sequences of host associations inferred from symbiont cladograms (see Mickevich, 1981). Among the many factors that might affect the likelihood of transfer between pairs (or sets) of hosts, we will distinguish between those that affect the likelihood of encounter of a given host by residents of another and those that determine whether such encounters will result in the evolution of a new association.

The probability of encounter, including such factors as physical proximity of different hosts, has been proposed as a dominant influence in the evolution of host associations for a variety of symbionts. Many phytophagous insects, for example, are closely bound to particular types of habitat, often more so than their hosts. In such cases the diets of oligophagous or closely related monophagous herbivores frequently include plant species whose major commonality appears to be their habitat association. Many examples could be cited; Hering (1954) discussed the case of the moth *Euspilapteryx phasianipennella*, which feeds, in moist habitats, on the related families Chenopodiaceae and Polygonaceae, but also on *Lysimachia* (Primulaceae) and *Lythrum* (Lythraceae), which are hosts that are taxonomically distant from the former pair but are characteristic of wet places. Similarly, the shift of an originally rodent-infesting group of fleas onto burrowing owls (Rothschild and Clay, 1952, cited in Ross, 1962) is readily explained by the hopping of parasites from prey to predator.

Intrinsic differences in hosts' ability to attract and support the growth and reproduction of particular symbionts are also likely to determine host transfer sequences, especially in very vagile groups for which different hosts are readily accessible. Several examples were cited earlier, including the apparent importance of quill size in the colonization of bird species by feather mites (Kethley and Johnston, 1975) and of plant secondary chemistry in the evolution of phytophagous insect diets.

Although a number of workers have assessed in some fashion the importance of one or more effects in the evolution of a particular set of associations, there has been almost no attempt at fitting models combining factors of different types to host transfer sequences inferred from symbiont cladograms. In both methodological and empirical respects, the phylogenetic study of host shift sequences is in its infancy. Thus, Ehrlich and Raven (1964) argued the importance of plant chemistry in butterfly evolution on the basis of broad associations of butterfly taxa with sets of chemically similar plants; in almost no cases do they describe the kind of multistep sequence needed to determine the degree to which invasion of new hosts is truly "stepwise" with respect to host chemistry. (A possible example of such "stepwise" colonization is the evolution of the umbellifer-feeding habit in some papilionids from an ancestor originally on the chemically dissimilar "woody Ranales"; the "gap" may have been bridged by a transfer first to Rutaceae, some of which share alkaloids with "woody Ranales" but also attractant compounds with the umbellifers; Ehrlich and Raven, 1964.) The phylogenetic study of colonization "rules" presents many of the same problems encountered in our discussion of mutual descent. In particular, there are likely to be so many causes operating simultaneously that sorting out their contributions within any one group may be very difficult. A "faunistic" approach, as in our earlier example of the parasite community on crocodilians, might be used to advantage here as well.

We have been concerned above with the question of constraints on the evolutionary "accessibility" of hosts from the point of view of symbionts, in much the way that systematists ask whether particular states of a morphological character must lie on a historical path between certain others. It is possible for such constraints to exist and yet not obviously correspond to any generalization about host properties. Thus, Ehrlich and Raven (1964) identified a number of butterfly host shifts that did not make obvious "sense" on chemical grounds. These "exceptions" may be explainable by some other as-yet-unstudied general factor, but they might also be due to highly idiosyncratic preadaptation (see Mitter and Futuyma, 1983). It is also possible that, at least within a restricted set, all hosts may be more or less equally colonizable from one another, host choice evolution being guided only by localized and temporally varying ecological factors. Thus, Gilbert and Smiley (1978) argued that the potential food plant range (as judged by larval fitness) of many *Heliconius* species is much greater than that expressed in nature and suggested that geographical variation in host preference in these butterflies may be due to "ecological monophagy," that is, localized selection pressure on female oviposition choice imposed by competition, parasitism, etc. (see also Fox and Morrow, 1981). When host transfer sequences appear to follow no generalization, experimental study (on preadaptation or on the ecological circumstances surrounding recent host transfers) may help to decide between the alternatives of "no intrinsic constraints" and "idiosyncratic constraints" on colonization histories.

Many other questions can be asked about evolutionary host transfer. For example, there has been much interest in the evolution of varying degrees of feeding specialization among phytophagous insects. An oft-repeated suggestion in the earlier literature (see, for example, Brues, 1920) was that, since many "primitive" herbivores, namely, grasshoppers and their relatives, tend to have relatively broad diets, there may have been an overall trend for the evolution of monophagy from polyphagous beginnings, perhaps because of the increasing diversity of plant defenses. It is now recognized, however, that the phytophagous habit has been acquired independently by a number of major insect groups, for example, the Lepidoptera. Within this group, the observed trend is, if anything, opposite to Brues' suggestion (Figure 13). The oldest lineages that feed on seed plants (e.g., Agathiphagidae, Nepticuloidae, Eriocraniidae) are mostly composed of mono- or oligophages, and polyphagy appears to be a derived condition that has developed independently in a number of primitively specialized-feeding lineages. Closer examination might reveal instances of the reverse trend, and very little is known about smaller-scale patterns in the evolution of this trait.

Distinct from the problem of what *sequence* the evolution of association will follow, there are the issues of how often and under what circumstances host transfers should occur. Although phylogenetic evidence clearly bears on the latter as well, limitations of space prevent its consideration here (but see, for example, Bush, 1975; Gilbert, 1979; Mitter and Futuyma, 1983; Ross, 1972). The two questions are not wholly separate; for example, one would like to know how and why the rules governing host shift sequences vary among lineages (see, for example, Gilbert, 1979).

EVOLUTION IN ASSOCIATED LINEAGES

We have thus far been concerned mainly with how and when ecological associations originate. A phylogenetic approach can also aid in the study of the evolutionary consequences of such association. Our discussion of this topic must be brief and largely hypothetical because of the rarity of studies providing the necessary combination of systematic and ecological information.

Separating "common causes" from coevolution

The fundamental postulate of coevolution is that interacting species influence each other's evolution. This hypothesis can be tested by systematics by finding some form of temporal relatedness of evolutionary events (character changes or divergences) between associated lineages. However, the evolution of a set of such lineages could be governed entirely by the independent responses of each to a shared history of geographical isolating events, producing matching temporal patterns of speciation and perhaps even of rates of character change, with no requirement for "host tracking" or any other form of coevolution *sensu stricto* (see Brooks, 1979b). Thus it is necessary to distinguish "correlated evolution" due to coevolution (the "coaccommodation'" of Brooks, 1979b) from that due to noninteractive causes.

Phylogenetic studies might contribute in several ways to the evidence for adaptation caused by coevolution (see Chapter 1 by

96

Futuyma and Slatkin). Experiments designed to demonstrate a fitness differential capable of accounting for the evolution of a characteristic compare that feature with some alternative type, often constructed artificially (see, for example, Bentley, 1976). The use of phylogenetic evidence to select a form closely resembling the ancestral morph from which the putative adaptation evolved may allow a better description of the actual historical change. For example, Benson (1972) performed an elegant test of Müllerian mimicry between the butterflies *Heliconius erato* and *H. melpomene* in Costa Rica, by comparing the fitness of control *erato* to that of specimens modified to look like the race endemic to a region of Colombia (see also Chapter 12 by Gilbert). Suppose, however, that we wished to determine experimentally whether the selection for Müllerian mimicry could have been discriminating enough to account for the evolution of the Costa Rican morph from its immediate ancestor. Then the most appropriate "experimental" form should be one constructed to resemble as closely as possible the phylogenetically inferred ancestral type [see Turner (1981) for phylogenies of some of the morphs of these two species].

Although it may often be difficult to resolve a particular question of adaptation considered in isolation, it may nevertheless be possible to show that the case is part of a clear-cut larger pattern, involving analogous features evolved independently under similar circumstances. Such a finding would be strong evidence that the particular analog was an adaptation of some sort. In other words, convergence, which phylogenetic methods are designed to detect, can be a strong source of evidence on adaptation. An illustration is provided by the phenomenon of physogastry in termitophilous insects, which occurs independently as a derived character in a number of unrelated lineages (see Kistner, 1969) and is not present in earlier-originating termitophilous forms. From these observations we can rule out the possibility that physogastry is a "preadaptation" that allowed its possessors to invade termite nests or that it reflects a purely environmental influence of termite nests on termitophile development. (For an argument of similar kind, see Chew and Robbins, 1982.) Therefore we can have considerable confidence that physogastry evolved as an adaptation to termitophily, even though its actual function is poorly understood.

There does not seem to be any criterion for judging hypotheses of "reciprocal coevolution" apart from those used to test the putative individual adaptations they entail. Moreover, tests of adaptation may be made difficult in coevolving systems by the extinction of intermediate stages. However, in a sufficiently elaborately coevolved system in which intermediates are extant, it may be possible to apply the criteria of temporal ordering and repeatability of change to make inferences about the course of adaptation. To take a hypothetical example, suppose we wish to show that in a host–parasite system (1) some hosts have features that evolved because of parasite pressure and (2) some parasites have traits that evolved because they permitted circumvention of evolved defenses. Phylogenetic evidence for the first of these would comprise the appearance of some type of feature (e.g., a new secondary compound in the case of plants) preferentially in lineages primitively exhibiting a high level of parasite attack. Evidence for the second would comprise preferential occurrence of novel states of some host-related character in parasites feeding on hosts bearing evolved defenses, as opposed to their relatives exhibiting more primitive habits. We are not aware of any case presenting all of these features, but some apparent instances of "stepwise" evolution of secondary compounds in plants attended by successive restriction of herbivore faunas [e.g., the appearance of cucurbitacins in some Cruciferae, normally characterized only by mustard oils (Chew and Rodman, 1979)] merit further study. A number of authors have speculated on the long-term course that such reciprocal adaptation should take (see, for example, Futuyma, 1979, and Chapter 10; Wilson, 1980).

Associates evolving together may affect each others' rates of diversification, in addition to evoking reciprocal adaptations. For example, one implication of the model envisioned by Ehrlich and Raven (1964) is that the temporary escape from parasite attack conferred by a novel defense should allow the host lineage bearing it to become more diverse than its sister group (see Figure 1). These authors speculated that the enormous diversity of angiosperms, as compared to other seed plants, might have resulted from the acquisition of the secondary compounds now found in their primitive representatives. A similar trend should hold for parasite groups colonizing new host taxa. Such seemingly straightforward phylogenetic predictions have yet to be quantitatively evaluated for any parasite or host group.

CONCLUDING REMARKS

In this essay we have attempted to identify some major questions to be addressed by the phylogenetic study of ecological associations, to outline some of the logic useful in answering them, and to sample the range of ecological–phylogenetic patterns found in nature. It is clear that both the evidence and the methods required for such inquiries are in a primitive stage of development. It is thus too early to tell either how much regularity ecological history exhibits or how much resolution its phylogenetic study will permit. It is our conviction, however, that the combination of ecological and systematic approaches pioneered by Ehrlich and Raven (1964) may ultimately lead to a sophisticated understanding of community evolution.

CHAPTER 5

COEVOLUTION
IN BACTERIA
AND THEIR VIRUSES
AND PLASMIDS

Bruce R. Levin and Richard E. Lenski

INTRODUCTION

Populations of bacteria can be parasitized by a variety of independently replicating genetic molecules. Based on functional rather than phylogenetic considerations, these replicons are classified as *plasmids* or *viruses*. Plasmids are extrachromosomal molecules of DNA that are present in one or more copies per cell. They replicate at the same average rate as the host chromosome and, in the course of cell division, are transmitted to descendants of infected cells with high frequency. In addition to this capacity for vertical transmission, some plasmids have specific adaptations for horizontal (i.e., infectious) transmission. By processes that require cell–cell contact, copies of these *conjugative* plasmids may be transmitted from *donor* cells to *recipient* cells. For reviews of the basic biology of plasmids, see Meynell (1973), Falkow (1975), and Broda (1979).

Bacterial viruses (*bacteriophage*, commonly contracted to *phage*) differ from plasmids in that they can exist outside the cell, encapsulated in a protein coat that both augments their extracellular term of survival and enables them to attack sensitive cells. Phage infection commences with the adsorption of the virus to specific receptor sites on the bacterium and the passage of its genetic material (DNA or RNA) into that cell. For *virulent* bacteriophage, replication is necessarily by a *lytic* cycle, which terminates with the death of the

99

host and the release of large numbers of phage particles. *Temperate* bacteriophage may also go through this lytic replication cycle, but there is a certain probability that following adsorption the viral DNA will be maintained as a *prophage* that is stably inherited by the descendants of the originally infected cell. This prophage may be integrated into the host chromosome or, in the case of some temperate bacteriophage, exist as a plasmid-like extrachromosomal element. At rates that depend on environmental conditions, individual bacteria carrying prophage—*lysogens*—will be *induced* and go through a lytic cycle that terminates with the death of the host cell and the release of phage particles. For reviews of the basic biology of bacterial viruses, see Adams (1959), Stent (1963), and Stent and Calendar (1979).

Bacterial viruses and plasmids are parasites in the sense that they have no host-free mode of reproduction and do not unconditionally increase the likelihood of their hosts surviving and reproducing (a definition of parasite somewhat broader than that in Chapter 9 by May and Anderson). Indeed, in the case of virulent bacteriophage, the cost of infection for individual cells is quite dear, and even the carriage of seemingly innocuous plasmids may impose a burden on bacteria. On the other hand, for many plasmids and some phage, the association between these autonomous replicons and their hosts is more that of mutualists. Resistance to antibiotics and heavy metals; the capacity to produce restriction enzymes, toxins, bacteriocins, and antibiotics; the ability to ferment certain carbon sources; and the production of structures for the invasion of specific habitats are characteristics that are often determined by plasmid-borne, rather than chromosomal, genes. (For reviews of the various kinds of plasmid-determined bacterial phenotypes, see Novick, 1974; Chakrabarty, 1976; Broda, 1979; and Davey and Reanney, 1980.) Although the abundance and diversity of phage-determined bacterial phenotypes is less than that of plasmid-determined phenotypes, there are some cases where antibiotic resistance and toxin production are due to genes borne on the prophage of temperate viruses.

Conjugative plasmids and temperate and sometimes virulent bacteriophage also play a role in bacterial adaptation and evolution by serving as vehicles for the exchange of genetic material. In the course of their infectious transmission, these autonomous replicons may pick up chromosomal genes or non-self-transmissible plasmids from one bacterium and may transmit them to another. Because the host ranges of bacterial plasmids and viruses often exceed "species" bounds and because there are mechanisms for recombination in the absence of close genetic homology (insertion sequences and transposable genetic elements; see Calos and Miller, 1980), the range of gene exchange mediated by plasmids and phage can encompass very phylogenetically diverse groups of bacteria.

100

It seems clear that the association between bacteria and their viruses and plasmids is a very ancient one. Phage and/or plasmids have been found in virtually every species of bacterium that has been examined for their presence. More than 90% of genetically distinct clones of *Escherichia coli* carry at least one plasmid, and the majority of these carry more than one (see, for example, Caugant et al., 1981). It has been estimated that nearly 100% of naturally occurring members of the genus *Pseudomonas* are lysogenic for some temperate virus (Holloway, 1979). For *E. coli* and closely related Enterobacteriacea, more than 200 plasmids and more than 80 species of phage have already been described (Novick, 1974; Reanney, 1976).

Plasmids and phage (as well as their bacterial hosts) can accumulate genetic variability through mutation, recombination, and the acquisition or loss of transposable elements. Because plasmids and phage cannot reproduce outside their bacterial hosts and because these replicons influence their hosts' survival and reproduction, coevolution must be very significant to the overall evolution of bacteria and their plasmids and phage. In this chapter, we will consider the nature and consequences of this coevolution. Using simple models tailored to the specifics of the interactions (Slatkin and Maynard Smith, 1979; Chapter 3 by Roughgarden) between these replicons and their hosts, we will predict the direction of selection on the parameters governing the systems. We will compare these a priori considerations with empirical results obtained from experimental and natural populations of bacteria and their plasmids and phage.

COEVOLUTION IN VIRULENT PHAGE AND THEIR HOSTS: A PRIORI CONSIDERATIONS

A model

A schematic representation of the association between populations of virulent phage and host bacteria is presented in Figure 1. The mathematical model, also presented in the figure, is a modified version of one employed by Levin, Stewart, and Chao (1977) and is analogous to that developed by Campbell (1961). The model assumes that bacteria and phage are thoroughly mixed in a liquid habitat of constant volume; bacterial resources enter and populations are washed out at a constant rate ϱ.

In the absence of phage infection, the bacteria multiply (via binary fission) at a rate ψ, the intrinsic rate of increase of the bacterium under specified environmental conditions. Environmental conditions critical to bacterial growth include temperature and resource concentrations

101

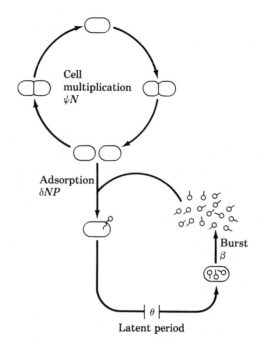

FIGURE 1. Schematic representation of the dynamics of a virulent phage and its bacterial host. N, Density of uninfected host cells; P, density of free phage; ψ, rate of cell multiplication; δ, phage adsorption rate parameter; θ, time between adsorption and burst of infected cells; β, number of phage produced per infected cell; ϱ (not shown in figure), rate of flow through the habitat and concomitant dilution of cell and phage populations. This model can be expressed as time-delay differential equations:

$$dN/dt = \psi N - \delta NP - \varrho N$$
$$dP/dt = \beta e^{-\varrho\theta}\delta N'P' - \delta NP - \varrho P$$

where N' and P' refer to the densities of cells and phage θ time units before the present.

(see Monod, 1949). Phage adsorb to sites on bacterial membranes at a rate that is proportional to cell density N and to the adsorption rate parameter δ. Infected bacteria are removed from the growing cell population because, as a result of infection, they cease multiplication and are fated to die. After a latent period of duration θ (during which the phage multiply in the host cell), a fated cell lyses and bursts, releasing β phage particles. Infected and uninfected cells and free phage are washed out from their populations at a rate ϱ that is independent of their densities.

Heritable variation in the parameters governing the growth and phage infection properties of the bacterial and phage populations can

102

result in differential rates of survival and reproduction, that is, natural selection. Therefore, by considering the effects of changes in the various parameters of the model, we can make inferences concerning the direction and intensity of selection in these populations.

Selection in the bacterial population

From the perspective of the host bacteria, selection should act to increase the rate of cell multiplication ψ, regardless of the density of phage. Because adsorption leads to cell death, selection acting on the cell population should reduce the rate of adsorption δ; the intensity of this selection will, however, be dependent on phage density. Thus, at low phage density, a mutation rendering a cell resistant to the phage may be disadvantageous if that mutation engenders a significant reduction in the cell's intrinsic rate of increase. The same mutation will likely be advantageous, however, when phage are abundant.

Because neither the burst size β nor the lag time θ directly enters the equation for bacterial growth in the mass-action model presented above, changes in these parameters should not be the direct result of selection acting on the bacterial population.

Selection in the phage population

The growth rate of the phage population is directly related to the burst size β and to the rate of adsorption to uninfected cells δ, and hence any increase in either of these parameters should be favored by selection acting on the phage.

Selection acting on the phage should reduce the latent period θ, although an examination of the time-delay differential equation for phage population growth does not immediately reveal this. An increase in the latent period effectively reduces the burst size, as a result of the washout of infected cells. Because "progeny" phage are subject to the same rate of washout ϱ, whether they are in infected cells or free, the advantage of reducing the latent period is *not* related to ϱ. Instead, the advantage of shortening the latent period lies in the earlier opportunity it provides for progeny phage to infect new host cells and further multiply.

Antagonistic selection and persistence

Unilateral selection in the bacterial population could result in the elimination of the phage from the habitat. If more resistant cells (those with lower δ) replace the more sensitive and if the phage are unable to increase when rare in a population of these more resistant cells, the phage would be eliminated. In cases where resistance is complete (i.e., $\delta = 0$), the resistant cells could completely displace the sensitive cells *only* if the former have an equal or greater rate of increase ψ. Thus, *if resistance engenders some cost in the competitive performance of the bacteria and if the sensitive cells are able to maintain a stable association with the phage, the evolution of resistance will not lead to the elimination of the phage from the habitat* (Campbell, 1961; Levin et al., 1977).

Unilateral selection in the bacterial population can actually augment the density of the phage population and possibly stabilize its association with the phage. This can be seen most readily by inspection of the equations for the equilibrium density of bacteria and phage. As long as the bacteria and phage can maintain their populations in the habitat, there will be an equilibrium with

$$\hat{P} = \frac{\psi - \varrho}{\partial}$$

$$\hat{N} = \frac{\varrho}{\partial(\beta e^{-\varrho\theta} - 1)}$$

(Levin et al., 1977). Thus, partially resistant bacteria are likely to have a selective advantage in cultures with phage and sensitive bacteria, and their evolution would result in an increase in the equilibrium density of the phage. The net effect of this would be a community that is further from the inelastic boundaries of $N = 0$, $P = 0$.

Selection in the phage population is necessarily antagonistic to the bacteria and can lead to the demise of the phage population. This, too, can be seen by an examination of the above equilibrium equations. Selection in the phage population would favor increases in the adsorption parameter δ and burst size β and reductions in the latent period θ. The effect of these changes in the infection parameters is a reduction in the equilibrium density of the bacteria. An increase in the adsorption rate parameter δ would also reduce the equilibrium density of the phage. The interested reader may wish to contrast the expectations derived from this model with those based on the predator–prey model presented in Chapter 3 by Roughgarden.

COEVOLUTION IN VIRULENT PHAGE AND THEIR HOSTS: EMPIRICAL CONSIDERATIONS AND EXTENSIONS

Resistance and persistence

From a priori considerations, we anticipate that selection in the bacterial and phage populations is antagonistic. The bacteria would be selected for resistance (reductions in δ), whereas the phage would be selected for higher levels of virulence (increases in δ and β and reductions in θ). It is clear that the potential for this type of antagonistic coevolution exists. One can readily isolate bacterial mutants that are fully resistant to phage to which other members of their clone are sensitive, and it is frequently possible to isolate *host range* phage that can attack these resistant bacteria as well as the sensitive cells. Indeed, bacterial resistance to phage was the phenotype used in the original demonstration of the randomness of mutation (Luria and Delbruck, 1943), and host range phage mutants played a very significant role in the early studies of bacteriophage genetics (Luria, 1945; Hershey, 1946).

This antagonistic selection in bacteria and their virulent phage has been observed in the various studies that have been done with experimental populations (Paynter and Bungay, 1969; Horne, 1970; Levin et al., 1977; Chao et al., 1977). These studies used a variety of different strains of *E. coli* and species of T phage, but in most cases phage-resistant mutants evolved and became the dominant clones. The evolution of these resistant mutants changes the continuous culture populations from a phage-limited to a resource-limited state. In the study by Chao et al. (1977) with *E. coli* B and the phage T7, there were at least three bacterial clones: the original sensitive, one mutant resistant to the original phage, and another mutant resistant to the original phage and to a host range mutant of that phage. At least two phage clones were present in these cultures: the original clone and a host range mutant of that clone. In spite of the antagonistic coevolutionary changes in these experimental populations, they persisted for extended periods of time (more than 80 weeks in the Horne, 1970, study).

It is of particular interest to ask why these "predator–prey" systems are stable (*sensu* persistence). If the bacteria are selected for a resistant mutant and if that mutation is not countered by a host range phage, why would the phage population not be eliminated? In the absence of mutants that are resistant to the phage and with continuous selection for higher levels of virulence, why would the phage not eliminate all of the bacteria? There are a variety of mechanisms that can account for the observed stability of the phage/host system. As demonstrated in a theoretical study by Levin et al. (1977), there are parameter values that specify stable states of co-existence for sensitive bacteria and virulent phage in the absence of genetic changes in their populations. Once one allows for evolutionary changes in these populations, there are at least three ways the association can continue to persist: (1) continuous selection for resistant hosts and counterselection for host range phage; (2) lower competitive performance of the resistant bacteria (relative to sensitive ones), and lower competitive performance of host range phage (relative to wild type) on sensitive hosts; and (3) the evolution of partially resistant bacteria with rates of adsorption that are still high enough to support the population of virulent phage. We believe that the second and third mechanisms are the primary ones accounting for both the short-term and long-term persistence of the associations between virulent phage and bacteria.

As demonstrated in theoretical studies by Campbell (1961) and Levin et al. (1977), sensitive bacteria and virulent phage can maintain a stable association as long as the phage-resistant cells are at a competitive disadvantage to sensitive cells in the absence of phage. This was observed in the experimental portion of the study by Levin et al. (1977). The introduction of T2-resistant clones of *E. coli* K-12 into chemostat populations of T2-limited *E. coli* B resulted in the ascent of the K-12 clone, with the persistence of the phage and of the sensitive *E. coli* B population. In phage-free competition, the T2-resistant K-12 clone was at a selective disadvantage relative to the T2-sensitive *E. coli* B. A similar result was obtained by Chao et al. (1977), but in that case T7-resistant clones of *E. coli* B evolved in that culture. Both the first-order resistant and second-order resistant *E. coli* in these experiments were, in the absence of phage, at a competitive disadvantage to the sensitive cell population from which they were derived. Chao et al. also found that the host range phage had a selective disadvantage when competing with the wild-type T7 for sensitive hosts.

Constraints on antagonistic selection

From these theoretical and empirical considerations, we postulate that to a great extent the evolutionary stability of phage/host associations can be accounted for by costs associated with resistance and host range shifts. Based on physiological considerations, one would expect that phage resistance imposes a cost on the bacteria (see discussions of constraints in Chapter 2 by Slatkin and in Chapter 3 by Roughgarden). The receptor sites to which the phage adsorb are membrane organelles that are likely to have other functions. Changes in their structure associated with resistance could impair these functions. In addition to the study by Chao et al. (1977), other evidence supports the view that mutations to phage resistance impose a cost. Demerec and Fano (1945) performed 50 pairwise competition experiments with a phage-sensitive clone of *E. coli* B and various mutants that were resistant to one or more T phage. In 34 of these phage-free experiments, the ratio of sensitives to resistants at the end of the experiment exceeded that at the beginning ($p < 0.05$). An analogous physiological argument could be put forth for the anticipated competitive disadvantage of host range phage relative to wild type, but perhaps more appealing is an a posteriori argument. If host range phage were as fit as or fitter than wild-type phage when competing for sensitive clones, there would be no wild-type phage. Unfortunately, save for the limited study of Chao et al. (1977), we are unaware of any experimental analyses of the relative competitive abilities of wild-type and host range phage. If there is generality to the observations of lower competitive ability for resistant bacteria and for host range phage and if physiological constraints prevent these disadvantages from being readily overcome, then a continuous progression of resistant and host range mutations is not only unnecessary for stability, but unlikely.

In addition to mutations that render bacteria absolutely resistant to phage, there are also those that result in quantitative reductions in the rate of phage adsorption. Although little consideration has been given to these "partially resistant" mutants, both theoretical considerations and informal results suggest they may play an important role in the evolution of stable host/virulent phage associations. One class of partially resistant mutants of *E. coli* excretes a mucilaginous substance (giving their colonies an aesthetically unappealing character). When cultures of *E. coli* K-12 (but not *E. coli* B) are challenged by a variety of different virulent and temperate phage, these mucoid colony types appear in high frequency among the surviving cells.

Cultures of these mucoid cells are able to maintain high density populations in the presence of phage and at the same time support a high density of the phage (B. R. Levin and P. Gidez, unpublished observations). We attribute this to a reduction in the rate parameter of phage adsorption associated with the mucoid phenotype. If these partially resistant mutations have only a small effect on cell growth rate, then their presence could preclude the evolution of fully resistant types. In that way, they might also restrict the evolution of host range mutants.

COEVOLUTION IN TEMPERATE PHAGE AND THEIR HOSTS: A PRIORI CONSIDERATIONS

A model

A schematic representation of the association between populations of temperate phage and host bacteria is shown in Figure 2. Uninfected cells are multiplying at a rate ψ_N, exclusive of losses to the phage and washout. As with virulent phage, temperate phage adsorb to host cells at a rate that is the product of cell density N and to the adsorption parameter δ. In contrast to virulent phage, however, not all infected cells are subject to cessation of growth and lysis. Instead, some frac-

107

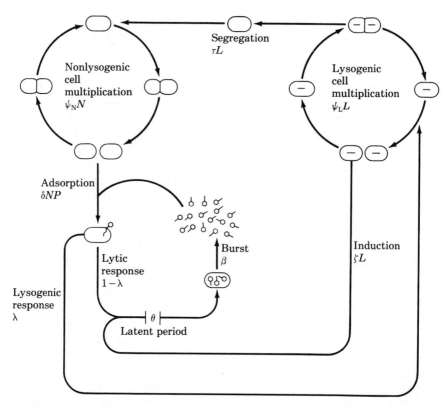

FIGURE 2. Schematic representation of the dynamics of a temperate phage and its bacterial host. N, L, Densities of nonlysogenic and lysogenic cells, respectively; ψ_N, ψ_L, rates of cell multiplication; λ, probability of lysogenic response given adsorption; ζ, rate of induction; τ, rate of segregation; P, δ, θ, β, ϱ, see Figure 1. This model can be expressed as time-delay differential equations:

$$dN/dt = \psi_N N + \tau L - \delta NP - \varrho N$$
$$dL/dt = \psi_L L + \delta\lambda NP - \tau L - \zeta L - \varrho L$$
$$dP/dt = \beta e^{-\varrho\theta}[\delta N'P'(1 - \lambda) + \zeta L'] - \delta NP - \varrho P$$

where N', L', and P' refer to population densities θ time units before the present. (After Lwoff, 1953.)

tion λ of the infected cells incorporate the phage genome as a *prophage*. These *lysogenic* cells of density L continue to multiply at a rate ψ_L, which may differ from the growth rate of the nonlysogenic cells. The remainder $(1 - \lambda)$ of the infected cells exhibit the lytic response, just as if the adsorbed phage were virulent.

The lysogenic cells are immune to subsequent infection by phage particles of the same type. (At high levels of superinfection, this immunity may break down. This effect is not present in our model but has been considered in a model by Noack, 1968.) However, at a rate ζ,

108

lysogenic cells are induced to exhibit the lytic response and thus enter the fated cell population. In addition, some lysogenic cells lose the phage genome through segregation at a rate τ, thereby entering the nonlysogenic cell population. As in the previous model, cells and phage are washed out of the habitat at a constant rate ϱ.

Let us again consider the consequences of changes in the various parameters of this model on the growth rates of cell and phage populations. In this case, however, we cannot view selection as acting independently on the sensitive cell, lysogenic cell, and free phage populations. Both the bacterial and phage genomes exist in two states, the former as sensitive cells and lysogens and the latter as prophage and free phage. Thus, in examining selection in this system, it is necessary to consider the phage and bacterial genomes at large rather than separately treat the different states in which they exist.

Selection on the bacterial genome

As in the previous model, any increase in the rate of cell multiplication is favored, for both lysogenic and nonlysogenic cells. Selection acting on the bacterial genome is expected to increase the likelihood of lysogeny (given adsorption), because any infected cell that does not become a lysogen is fated to death. Similarly, selection acting on the bacterial genome should minimize the rate of induction ζ, because induced cells are also fated to lysis.

Selection acting on the bacterial genome is somewhat more complex with respect to the parameters δ and τ. The direction of selection on the segregation rate will depend on the relative growth rates of lysogenic and nonlysogenic cells and on the relative death rates due to lysis of the two cell populations. If $\psi_L - \zeta > \psi_N - \delta P(1 - \lambda)$, then the phage genome is an advantage to its host and selection acting on the bacterial genome should minimize τ. Conversely, if the expected net growth of the nonlysogenic population is greater, then τ should be increased by selection acting on the bacterial genome. As with virulent phage, selection on the host genome should tend to reduce the rate of phage adsorption δ, *unless* the net rate of lysogenic cell multiplication is sufficiently greater than that of nonlysogens to offset the risks of lysis.

Selection on the phage genome

As with virulent phage, selection in temperate phage will be intense for increased rates of adsorption δ, because phage cannot reproduce outside their hosts. Similarly, the burst size β will be maximized and the latent period θ minimized.

The nature of selection acting on the phage genome for the likelihood of the lysogenic response λ is clearly critical to our understanding of the adaptive value of a temperate mode of existence (a $\lambda = 0$ renders a phage virulent). Therefore, it is important to examine the relative contributions of adsorbed phage that exhibit the lysogenic and the lytic responses. After time θ, the lytic response yields β free phage, whereas the lysogenic response nets $e^{\psi_L \theta}$ prophage (cells growing at rate ψ_L for time θ, assuming ζ and τ are near zero). Given realistic values for β (e.g., 100), θ (e.g., 0.5 hours), and ψ_L (e.g., 0.7/hour), the *short-term* dynamics are such that the lytic contribution will be far greater than the lysogenic contribution. However, if we compare the fates of the *progeny* free phage and the *progeny* prophage, we obtain a different conclusion. If nonlysogenic host cells are very rare, then free phage produced via the lytic response will have very low rates of subsequent reproduction, because adsorption to a new host is infrequent. In contrast, the progeny prophage resulting from lysogenic cell multiplication do not need to find a new host and can continue to multiply at the modest cellular rate indefinitely. If the product δN is sufficiently small to offset the short-term advantage of the lytic response, then selection acting on the phage genome should increase the probability of lysogeny (see also Campbell, 1961). *Thus, from the perspective of the phage, temperance appears to be an adaptation to low host-cell densities.*

Selection acting on the phage genome with respect to the rate of lysogen induction ζ will be of opposite direction and similar intensity to selection on λ, because induction is essentially a reversal of lysogeny. Under all conditions, selection in the phage population would be to minimize the rate of prophage loss by vegetative segregation, that is, to minimize τ.

If conditions are such that selection has favored phage temperance, an opportunity exists for the development of a mutualistic relationship with its host. As long as the cost in lower β or δ or higher θ is small, it would be to the advantage of the phage to carry genes that enhance the growth rate of lysogenic cells.

COEVOLUTION IN TEMPERATE PHAGE AND THEIR HOSTS: EMPIRICAL CONSIDERATIONS AND EXTENSIONS

Resistance and immunity

As with virulent phage, selection in temperate phage and their hosts *could* be antagonistic, that is, for more resistant hosts and more virulent phage. In accord with the model, selection would favor hosts that are resistant to the phage; but as long as the rate of mutation to resistant types is less than the probability of lysogeny, immune lysogens are likely to precede resistant clones. When one challenges sensitive *E. coli* with the temperate phage Lambda, the vast majority of surviving cells are Lambda lysogens rather than Lambda-resistant mutants. Indeed, Lambda-resistant clones are most readily isolated by challenging the sensitive bacteria with virulent mutants of Lambda. Thus, although resistance can evolve, the primary evolutionary response to infection with temperate phage should be the rise of the lysogenic population in which selection on the phage and bacterial genomes could be complementary.

At this time, we are aware of only one experimental study that has been directed at the evolutionary response of populations of sensitive bacteria to infection by temperate phage (J. Arraj and B. R. Levin, unpublished). The results of that study, with *E. coli* K-12 and Lambda in chemostats, support the hypothesis that the primary response of the bacteria is the rise of a lysogenic population. However, these results also suggest that the situation is somewhat more complex. Following the rise of Lambda lysogens, clones that are both lysogenic *and resistant* to Lambda appear and achieve substantial frequencies. It may seem redundant to be resistant as well as immune; however, immunity becomes ineffective at high levels of superinfection, whereas resistance apparently does not. The evolutionary importance of resistant lysogens has not yet been explored.

Selection for prophage-determined host phenotypes

In accord with the view that the direction of evolution in temperate phage is toward a mutualistic association with their host, there are a variety of prophage-determined characters that augment the growth rate of their hosts. The most obvious of these is, of course, immunity to subsequent infection by phage of that type, that is, *superinfection immunity*. There are also restriction enzymes coded for by prophage (Arber and Linn, 1969) and prophage-borne resistance to antibiotics (Williams Smith, 1972). Although the mechanism is not immediately apparent, we would anticipate that the diphtheria toxin coded for by the Beta phage of *Corynebacterium diphtheriae* (Uchida et al., 1971) enhances the fitness of the bacterium.

There is some evidence suggesting that prophage genes that augment host fitness are quite general. In a series of pairwise competition studies with *E. coli* in chemostats, Edlin and his colleagues have shown that under some culture conditions lysogens have a competitive advantage over nonlysogenic, resistant cells. They have obtained this result for a variety of different phage: Lambda, P1, P2, and Mu (Edlin et al., 1977; Lin et al., 1977). At this time, it is not clear how these prophage enhance the competitive performance of their hosts.

111

To us, the most intriguing problem concerning coevolution of bacterial viruses and their hosts is the conditions under which selection favors a temperate rather than a virulent mode of replication. We see three distinct hypotheses: (1) lysogeny enhances the stability of the phage–host association; (2) lysogeny enhances the fitness of the bacteria; and (3) lysogeny is an adaptation to low host densities. Although these hypotheses are not mutually exclusive, we consider the last mechanism to be the most important. We expand on this below.

It seems likely that the associations between temperate phage and their hosts are more stable than those of virulent phage and their hosts, that is, temperate phage are unlikely to drive their hosts to extinction. A number of authors have suggested differential extinction as a mechanism to account for the evolution and maintenance of temperance (e.g., Dove, 1971; Echols, 1972). This mechanism requires that selection operate at the level of the group (interdemic selection), the more stable phage–host associations having a group advantage. However, if virulent phage have an advantage *within* these groups, then the conditions for interdemic selection to favor temperate phage are likely to be very restrictive (Levin and Kilmer, 1974). If temperate phage have an advantage over virulent phage within these groups, then interdemic selection is not necessary.

If the prophage code for characters that enhance the fitness of their hosts, then selection could favor higher probabilities of lysogeny and lower rates of induction, that is, evolution in the direction of greater temperance. However, it seems unlikely that such "niceness" could be the *primary* selective pressure leading to lysogeny. For phage to express genes that enhance the fitness of their hosts, the phage genome would have to be maintained by host cells, that is, form some sort of prophage. Thus, from the perspective of the phage, they would already have to *be* somewhat temperate to *become* temperate. From the perspective of the host, selection would, of course, favor any mechanism that reduces the likelihood of a lytic infection by phage, even if that mechanism results in the maintenance of the phage genome. However, it is unlikely that the evolution of the temperate phage was through unilateral selection in the host population; although there is variation in the probability of lysogeny and rate of induction among bacteria, the proteins involved in insertion and excision (lysogeny and induction) are coded for by phage genes and *not* host genes.

The hypothesis that the temperate mode of phage existence evolved as an adaptation to low densities of sensitive cells seems, to us, the most parsimonious of the three. When sensitive hosts are rare, free phage produced via lytic infections would have low rates of subsequent reproduction, because adsorption to a new host is infrequent. On the other hand, prophage replication does not require a quest for new hosts. Although we know of no evidence to either support or refute this hypothesis, it should be amenable to direct experimental tests.

COEVOLUTION IN PLASMIDS AND THEIR HOSTS: A PRIORI CONSIDERATIONS

A model

In Figure 3 we present a schematic representation of the association between populations of conjugative (i.e., self-transmissible) plasmids and their host bacteria. The model presented in the figure is identical to that employed by Stewart and Levin (1977). The plasmid-free (P-F) and plasmid-bearing (P-B) cells grow at rates ψ and ψ_+, respectively.

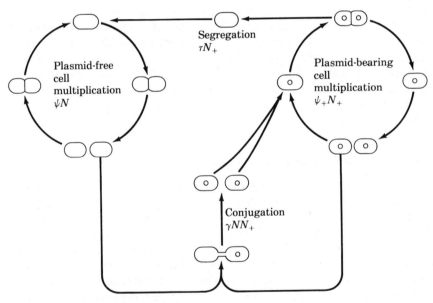

FIGURE 3. Schematic representation of the dynamics of a conjugative plasmid and its bacterial host. N, N_+, Densities of plasmid-free and plasmid-bearing cells, respectively; ψ, ψ_+, rates of cell multiplication; γ, conjugational transfer rate parameter; ϱ, τ, see Figures 1 and 2. This model can be expressed as differential equations:

$$dN/dt = \psi N + \tau N_+ - \gamma NN_+ - \varrho N$$
$$dN_+/dt = \psi_+ N_+ + \gamma NN_+ - \tau N_+ - \varrho N_+$$

113

P-B and P-F cells encounter one another at random with a frequency that is proportional to the product of their densities. The proportion of encounters resulting in the transfer of a copy of the plasmid from a P-B cell to a P-F cell is governed by the conjugational transfer rate parameter γ. P-B cells can lose the plasmid, and thereby enter the P-F population, via vegetative segregation, at a rate τ.

Selection on the host genome

For both P-B and P-F cells, selection should favor an increase in the exponential growth rates ψ and ψ_+. The intensity of selection on the two growth rates will be independent of densities and relative frequencies of the component populations. Selection acting on the host genome will favor an increase in the segregation rate parameter τ if and only if the growth rate of the P-F cells exceeds that of the P-B cells.

With respect to the conjugative rate parameter γ, selection acting on the host genome is somewhat more complex. Any mutation arising in the P-B cell genome that increases *donor ability* would probably be selected against, because there is likely to be some cost associated with the mechanics of conjugative transfer (e.g., plasmid replication and the synthesis of structures called conjugative pili that are required for conjugation). This prediction is independent of the relative magnitudes of the growth rates of the P-B and P-F cells. However, a mutation that increases the *receptivity* of P-F cells will be favored if the plasmid augments the growth rate of P-B cells. The intensity of this selection for the recipient's contribution to γ will depend on the density of P-B cells.

Selection on the conjugative plasmid genome

Plasmid-borne genes that increase the growth rate or decrease the death rate of cells carrying that plasmid (those determining a higher ψ_+) would be favored by selection. In this model, the intensity of the selection for plasmid-borne genes that enhance host fitness would be independent of the frequency of P-B cells and the density of the population. Under all conditions, selection would favor plasmid-borne genes that augment the stability of the plasmid in the host, that is, that reduce the segregation rate τ. The intensity of this selection would be independent of the density of the population or the frequency of P-B cells.

Selection would favor plasmid-borne genes that increase the rate of infectious transfer of the plasmid. The intensity of this selection for higher γ would be directly proportional to the population density and the frequency of P-F cells. At very low densities of P-F cells, infectious transfer would make a negligible contribution to the rate of increase of the plasmid. Thus, if there were a significant cost associated with conju-

114

gative ability, selection in the plasmid could actually favor reductions in the rate of infectious transmission when recipients are rare.

Selection on the nonconjugative plasmid genome

From one perspective, nonconjugative (non-self-transmissible) plasmids can be considered as a limiting case of the conjugative factors (where the rate constant of transfer $\gamma = 0$). However, this simple interpretation is not really sufficient. Nonconjugative plasmids, like segments of the host chromosome, can be infectiously transmitted by being "picked up" by either conjugative plasmids or phage, a process known as *mobilization*. In some cases mobilization can be quite effective (e.g., Levin and Rice, 1980).

In modeling the population biology of nonconjugative plasmids, one must consider the population dynamics of the mobilizing replicon(s), as well as that of the nonconjugative plasmid. This is an exercise we shall refrain from in this forum (but see Levin and Stewart, 1980). The primary issue of concern here is that even for non-self-transmissible plasmids, there could be selection for changes in the rates of infectious transfer by mobilization. The nature and direction of this selection would be similar to that considered for the rate constant of conjugative plasmid transfer.

Plasmids coding for allelopathic substances

In the model presented in Figure 3, the plasmids may increase the fitness of cells by augmenting the growth rate of their immediate hosts. There is, however, an important class of plasmid-determined characters for which the plasmid is to the *disadvantage* of the individual host carrying it but which enhance the fitness of the P-B population at large. The most extensively studied plasmid-determined characters of this type are the *bacteriocins* (Reeves, 1972). These are proteins that kill sensitive bacteria of the same or closely related species. In addition to coding for the production of these *allelopathic* molecules, bacteriocinogenic plasmids also confer immunity to these agents. The individual cells carrying these plasmids are at a disadvantage because bacteriocin synthesis and release is lethal to the host cell. At any given time, however, only a small minority of the bacteriocinogenic population is induced to synthesize and release bacteriocin. If there is a cost associated with the carriage of plasmids coding for other antibiotics (such as those produced by the streptomycetes; Hopwood and Merrick, 1977), then these too fall into this class of plasmid-determined phenotypes.

115

Although the model presented in Figure 3 is not an accurate analog of the population biology of allelopathic plasmids, the major components of the population dynamics of these types of replicons have been mimicked by simple models (Chao, 1979). The primary conclusion drawn from Chao's model of nonconjugative bacteriocinogenic plasmids is that selection for these elements would be frequency-dependent. Cells carrying them could only increase when they are relatively common. The reasons for this are rather straightforward. Due to the lethal synthesis and other costs associated with these factors, bacteriocinogenic cells would be at a disadvantage when competing with sensitive P-F cells, *unless* the concentration of that allelopathic agent is high enough to kill sufficient numbers of P-F (i.e., sensitive) cells to make up for that competitive disadvantage. The latter will occur only when the P-B cells are at relatively high densities.

COEVOLUTION IN PLASMIDS AND THEIR HOSTS: EMPIRICAL CONSIDERATIONS AND EXTENSIONS

Selection and persistence: plasmids are not just "selfish DNA"

Plasmids, unlike phage, have no free state; they are parts of cells, like chromosomes. Thus, one might conclude that their evolution would be toward an increasingly mutualistic relationship with their hosts. There are, however, possible exceptions to this mutualistic form of coevolution. If the plasmids impose a cost on their hosts (i.e., reduce their rates of growth), selection in the plasmid and host populations would be antagonistic. The plasmids would be selected for higher rates of infectious transmission and the hosts for resistance to infection by plasmids and higher segregation rates. Thus, in considering the nature of coevolution in plasmids and their hosts, it is first necessary to ask (1) whether plasmids are likely to impose a fitness cost on their hosts and (2) whether plasmids can be maintained by infectious transfer alone.

In the absence of selection favoring plasmid-borne genes, cells carrying plasmids have a competitive disadvantage relative to identical cells without those extrachromosomal genetic elements. This is, in fact, what one would anticipate on physiological grounds. The additional DNA and protein synthesis associated with the carriage of plasmids must impose some cost on a bacterium. The magnitude of this cost for conjugative plasmids is surprisingly high. In competition experiments between P-B and P-F *E. coli* in conditions where there was no selection for plasmid-borne genes, Levin (1980) reported a 10% lower growth rate for cells carrying the conjugative plasmid R1. Whether plasmids generally impose this high a cost on their hosts remains to be seen.

Using a mathematical model like the one presented here for conjugative plasmids, Stewart and Levin (1977) demonstrated the existence of a broad set of conditions under which infectious transfer can overcome segregational loss and substantial levels of selection against P-B cells and thereby lead to the maintenance of the plasmid in high frequency. Although these conditions for the maintenance of plasmids deleterious to their hosts can be met under laboratory conditions (with high density cultures and mutant plasmids with very high transfer rates), Levin et al. (1979) and Levin (1980) suggest that these conditions are unlikely to obtain in natural populations of bacteria. This idea can be stated as a hypothesis: *for plasmids to be maintained in natural populations of enteric bacteria, they must carry genes that (under at least some conditions) enhance the fitness of their immediate hosts or that of cells carrying that plasmid in the population at large.* "There are no neutrals there" (Reece, 1932), that is, plasmids are not just "selfish DNA" (Doolittle and Sapienza, 1980).

If the above hypothesis is valid and general, then for all plasmids and their hosts, coevolution would necessarily be mutualistic. The existence of a diverse array of plasmid-determined phenotypes is clear evidence in support of this interpretation. Most characters coded for by plasmid-borne genes can enhance the fitness of the bacteria carrying that element. Just how significant plasmid-determined characters are for bacterial adaptation is very dramatically (and from a clinical perspective, frighteningly) demonstrated by the rise of antibiotic resistance. Most clinically important antibiotic resistance in bacteria is determined by plasmid-borne genes; and since the start of the antibiotic era in the late 1940s, the frequency of bacteria carrying antibiotic resistance (R) plasmids has increased enormously. At present, the majority of clinical isolates for some pathogenic bacteria like *Salmonella typhimurium* (gastric enteritis) and various species of *Shigella* (bacterial dysentery) carry at least one R-plasmid (see reviews by Anderson, 1968; Mitsuhashi, 1971; Falkow, 1975).

In addition to the increase in the frequency of bacteria carrying R-plasmids, the number of antibiotic resistance genes carried by single R-plasmids also increased during this period. When these plasmids were first discovered in the late 1950s (Watanabe, 1963), the majority isolated from enteric bacteria conferred resistance to one or two antibiotics. The average number of resistances determined by single R-plasmids increased rapidly during the early 1960s (Anderson, 1968). Currently, it is not uncommon to isolate from enteric bacteria R-plasmids that have four or five resistance genes.

The rapid rate by which plasmids acquire additional antibiotic resistance genes can be attributed to the fact that these genes are often present as parts of transposable genetic elements (i.e., transposons; Falkow, 1975; Broda, 1979; Campbell, 1981). Genes on transposons do not require sequence homology for recombination and insertion into chromosomes and plasmids. In the course of their travels among different hosts, conjugative plasmids pick up transposons with different resistance genes. As long as bacteria are confronted with a variety of different antibiotics, plasmids that confer resistance to more antibiotics would have a selective advantage.

In accord with this hypothesis of mutualistic coevolution, we would anticipate that if plasmids augment the fitness of their hosts, there would be selection on the host to increase the receptivity to plasmids and reduce the rate of segregational loss. At this juncture, we know of no evidence to either support or refute this interpretation. However, because one can select hosts that are refractory to conjugative plasmids (Reiner, 1974), it should be possible to test this hypothesis.

The allelopathic plasmids

In our a priori considerations, we suggested that selection acting on plasmids that code for allelopathic substances would be frequency dependent. The results of studies that have been done with colicinogenic plasmids in experimental populations support this interpretation (colicins are bacteriocins that affect *E. coli* and closely related species). Using a variety of different Col plasmids, Zamenhof and Zamenhof (1971), Adams et al. (1979), Chao (1979), and Chao and Levin (1981) did competition experiments with colicinogenic and sensitive *E. coli* in chemostats. In all of these experiments, selection favored the colicinogenic cells only when they had initial frequencies in excess of 1%.

Because the sensitive bacteria are being killed by the colicin, it seems reasonable to assume that in these experiments, there would be selection for colicin-resistant cells. This was, in fact, observed in the colicin E1 study of Adams et al. (1979). Chao and Levin (1981) also reported the existence of mutants that were resistant to the colicin E3 they were studying, but in their cultures, these resistant cells did not achieve substantial frequencies. They attributed this to a marked reduction in competitive performance associated with the Col E3 resistance mutation.

The importance of physically structured habitats

Frequency-dependent selection for allelopathic plasmids raises a question about the evolution and maintenance of these types of plasmids. If they cannot increase in frequency *until* they are relatively common, how does one account for their evolution and persistence? One possible explanation is "hitchhiking," that is, bacteriocin-determining genes

118

are on plasmids that code for characters that enhance the competitive performance of the individual cells carrying them (e.g., antibiotic resistance). An alternate hypothesis, which we favor, was suggested by Reeves (1972) and expanded upon and tested by Chao (1979) and Chao and Levin (1981). According to this hypothesis, bacteriocins (and, by extension, other allelopathic substances like antibiotics) are adaptations for interference competition (Gill, 1974) in *physically structured* habitats. Although cells carrying the Col E3 plasmid could not increase when rare and when competing with P-F sensitive cells in liquid (mass) culture, they could increase when competing in a soft agar matrix.

In a physically structured habitat, the bacteria grow as colonies rather than as individual cells (as they do in mass culture). The allelopathic agents synthesized by induced cells of bacteriocinogenic colonies diffuse out into the environment and kill sensitive cells in the vicinity of that colony. The net effect is reduced competition for limiting resources for colonies of cells carrying plasmids coding for allelopathic substances, and the production of more P-B cells. As a result of this *resource sequestering*, cells carrying allelopathic plasmids have an advantage over sensitive cells at all *frequencies* (although only at high *densities*, where competition is important). Presumably, this same type of mechanism would also favor colonies that are lysogenic for temperate phage.

Constraints on infectious transmission

Because it is copies of plasmids rather than the plasmids themselves that are transmitted by conjugation, it is reasonable to assume that selection would always favor plasmids with higher rates of infectious transmission. But empirical considerations suggest that this is not the case. There are a vast number of apparently viable plasmids that are non-self-transmissible, and among these there are some that are extraordinarily difficult to mobilize with conjugative plasmids. Moreover, most conjugative plasmids isolated from natural populations are repressed for conjugative pili synthesis (Meynell, 1973). At any given time, only a minority of the P-B cell population produces conjugative pili and is capable of transmitting the plasmid. Shortly after the receipt of a plasmid, a cell produces conjugative pili and is capable of transmitting that element. As time proceeds, a plasmid-coded repressor protein accumulates in that cell and its descendants and competence for transfer of that plasmid declines. For the wild-type (repressed) plasmid R1 in steady-state P-B populations, the rate constant

119

of plasmid transfer γ is three orders of magnitude lower than that for its permanently derepressed mutant, R1-drd-19 (see Levin et al., 1979).

We see two (not mutually exclusive) hypotheses for the evolution of repressible conjugative pili synthesis and the generally lower-than-possible fertility of wild-type plasmids. Anderson (1968) suggested that repression of conjugative pili synthesis is a mechanism to avoid infection by donor-specific bacteriophage that adsorb to conjugative pili. The second hypothesis, which we prefer on grounds of parsimony and generality, asserts that repressible conjugative pili synthesis and the existence of non-self-transmissible plasmids is a consequence of the limited opportunities for transfer in natural populations and the high cost of self-transmissibility. As we pointed out in our a priori considerations, the intensity of selection for infectious transmission is dependent on the density of potential recipients. If the latter is low, as a result of a low overall population density or a low relative frequency of possible recipients, and if there is a cost associated with infectious transmission, selection would favor lower rates of infectious transmission.

The mechanisms for conjugative pili synthesis require a substantial number of genes: approximately 15 megadaltons of DNA or approximately ¼ of the genome of a plasmid like R1 (Willetts, 1972). As a result of this additional DNA and the synthesis of conjugative pili, the burden imposed by conjugative plasmids would even be greater than that for nonconjugative elements. The results of the limited number of experiments we have done on this problem are consistent with this view. In competition experiments with the nonconjugative plasmid pCR1, P-B cells had a disadvantage of less than 5% (relative to P-F cells) as compared to 10% for the conjugative plasmid R1 (Levin et al., 1979). Moreover, cells carrying the permanently derepressed mutant, R1-drd-19, had a disadvantage of between 15 and 20% relative to P-F cells of their type (Levin, 1980). Thus, unless the density of recipients is substantial, one would not anticipate selection to favor higher rates of infectious transmission.

DEFENSE MECHANISMS AND MEASURES TO COUNTER THEM: A POSTERIORI EVIDENCE FOR COEVOLUTION

Compelling evidence for the long-term coevolution of plasmids, phage, and their hosts is the existence of specific systems to prevent and/or limit infections by parasitic DNAs and mechanisms to counter these barriers to their replication. Host defense mechanisms operate in three basic ways: (1) preventing novel DNAs from entering cells, or *exclusion*; (2) destroying novel DNAs that do enter, or *restriction*; and (3) preventing foreign DNA from replicating at a rate sufficient for persistence, or *incompatibility* (Bennett and Richmond, 1978). In the following section, we briefly consider the nature of these defense mecha-

120

nisms and how they serve as evidence of coevolution among these replicons and their hosts or, in some cases, among different replicons.

Exclusion

The cell envelope is probably the most important barrier to infections by plasmids and phage. To some extent this may be coincidental, perhaps like our "resistance" to Dutch elm disease. However, in other cases it is clear that an exclusion mechanism evolved for the specific purpose of limiting invasions by parasitic replicons. The existence of mutations conferring resistance to bacteriophage surely stands as evidence for the potential for the evolution of exclusion. Analogously, the existence of host range mutations serves as evidence for the potential of phage to evolve mechanisms to overcome cell envelope defense. Some exclusion mechanisms are determined by replicons themselves. Many conjugative plasmids code for mechanisms that prevent entry by closely related plasmids (Novick, 1969). It seems reasonable to assume that these evolved for plasmids to prevent competition for replication within individual bacteria.

Restriction

As is the case with the microparasites of multicellular organisms, once past their host envelope the parasitic DNAs of bacteria have to contend with "immune" systems of their hosts. Some of these intracellular mechanisms to prevent replication of invading DNAs are rather limited in the range of DNAs upon which they can operate, for example, the repressors responsible for superinfection immunity for temperate phage. In other cases, the range of foreign DNAs acted upon is quite broad. The *restriction modification* systems are perhaps the prime example of "immune" systems of the latter type (Arber and Linn, 1969; Stent and Calendar, 1978).

Novel DNAs entering a cell are cut at specific sequences of bases by *restriction endonucleases*. As is the case for the generalized "immune" systems of higher organisms, there is a need to distinguish "self" from "nonself." This is accomplished by modifying bases within the cleaving sequences recognized by the restriction endonuclease, by adding methyl groups to the cytosines or adenines in these cleaving regions. The modification methylases that catalyze the latter reactions are coded for by genes that are closely linked to those for the restriction endonucleases.

The effectiveness of restriction varies considerably among the vari-

121

ous phage and plasmid DNAs penetrating the cell envelope. In some cases, it is clear that restriction systems are very effective, reducing the likelihood of cells succumbing to a mortal infection by a phage by four or five orders of magnitude. However, as one might anticipate, the modification system necessary for recognizing self limits the effectiveness of restriction. Phage or plasmid DNA that evade restriction are modified and recognized as self DNA. These modified replicons are fully effective against bacteria with that restriction modification system. Some phage, such as T-even coliphage, have 5-hydroxymethylcytosine instead of cytosine as part of their normal genome and are therefore relatively immune to most restriction enzymes. Could it be that the atypical bases of these phage evolved as a mechanism to overcome host restriction?

Those investigators using specific endonucleases for DNA manipulations (e.g., gene splicing) have shown that there are many different restriction enzymes with different cleavage sites. These enzymes are coded for by host, plasmid, and temperate phage genes and are present in a very phylogenetically and ecologically diverse array of bacterial groups. Included among these is *Thermoplasma acidophilium* (McConnell et al., 1978), an Archaebacterium that lives in coal piles and is most readily cultured between pH 1 and 2 at 59°C (Searcy et al., 1981). If restriction modification systems are, as they appear to be, defense against novel DNAs, the *Thermoplasma* situation clearly indicates just how universal the problem of coping with parasitic DNA is to bacteria.

Incompatibility

The successful passage through the gauntlet of exclusion and restriction defenses does not ensure maintenance of a parasitic DNA in a bacterial lineage. For stable inheritance, it is necessary for that DNA to replicate at a rate at least as great as that of the host chromosome and, upon cell division, to be transmitted to both daughter cells. Temperate phage and plasmids have evolved a variety of mechanisms to ensure this vertical transmission. The incorporation into the host chromosome by prophage and the production of multiple copies by plasmids clearly augment the likelihood of vertical transmission. Many autonomous single-copy replicons are also very stable (with vegetative segregation rates of 10^{-5} per generation or less) and some rather extraordinary mechanisms have evolved to ensure their stability (see Austin et al., 1981).

Operating against the stable inheritance of replicons are a variety of incompatibility systems. In some cases, the invading DNA remains intact and expresses its genes but fails to replicate at all (Stocker, 1956; Hayes, 1968). In other cases, replication does occur, but the seg-

122

regation rate is high and the replicon, usually a plasmid, can only be maintained in a lineage by continuous selection for the genes carried by that element. In some cases these incompatibilities may be coincidental; there is no specific mechanism to preclude replication, but the replicon simply has not evolved a mechanism to allow for its stable inheritance. This may well be the case where the plasmid is transferred to a species that is phylogenetically very distant from the donor. There are, however, incompatibility systems that have clearly evolved as mechanisms to preclude the stable inheritance of invading replicons. The best studied of these are coded by plasmid rather than host genes and operate most effectively against plasmids of the same or closely related types (Falkow, 1975). Why do these and other replicon-determined defense mechanisms fit the cliché of competition being most intense among closely related species?

COEVOLUTION AND BACTERIAL SEXUALITY:
MUCH ADO ABOUT VERY LITTLE

Save for the incorporation of free DNA (a process known as *transformation*), plasmids and phage are the sole vectors for the exchange of genetic material between bacteria. Thus, it might seem that a good deal of the coevolution of these organisms would be directed toward the role of plasmids and phage as vehicles of recombination. We suggest that this has not been the case. Although gene exchange mediated by these vectors is unquestionably important to bacterial adaptation and evolution (Reanney, 1976; Bennett and Richmond, 1978; Davey and Reanny, 1980), it is unlikely that natural selection has acted *directly* to increase the effectiveness of these replicons as vectors for recombination.

In natural populations of bacteria, recombination appears to be an extremely rare event from the perspective of an individual bacterium. To be sure, high rates of recombination can be obtained with laboratory strains such as *E. coli* K-12 and permanently derepressed F plasmids that incorporate into the host chromosome (*Hfr*: see Hayes, 1968). However, most naturally occurring plasmids are repressed for conjugative pili synthesis (Meynell, 1973) and do not readily incorporate into the chromosomes of their hosts (Holloway, 1979). Based on estimates of the rate constants of plasmid transfer and phage adsorption, on the likelihood that these vectors will pick up and transfer host genes, and on the densities of natural populations, Levin (1981) suggests that for *E. coli* in their natural habitat the per capita rate of gene exchange by plasmid- and phage-mediated recombination is as low as or lower than by mutation (10^{-6} per cell per generation or less). The results of electrophoretic studies of structural gene diversity in natural populations of *E. coli* (Selander and Levin, 1980; Caugant et al., 1981) are consistent with this interpretation. Gene complexes are maintained for extended periods of time without being broken down by recombination, and natural populations are far from linkage equilibrium.

The fact that in laboratory culture one can select for plasmid and phage that are effective vehicles for host gene recombination and that vectors of this type do not exist in natural populations is, of course, a posteriori evidence for the absence of selection for high rates of recombination in natural populations. This, however, begs the question of *why* high rates of phage- and plasmid-mediated recombination have not been selected for. We believe that part of this answer lies in the fact that being a vehicle for the transfer of host genes is likely to be a disadvantage for a plasmid or phage. This is clearly the case for some of the plasmids and phage used for recombination analyses in genetic studies. In the case of the F plasmid incorporated into the chromosomes of *Hfr*s, the F replicon is not transmitted in entirety until the whole host chromosome is transmitted (Hayes, 1968). Thus, that plasmid would usually gain neither from the advantages of being in the recombinant nor from the mobility of infectious transfer. For many general transducing phage, the individual virus responsible for recombination contains few or possibly none of its own genes. In the course of replication in the donor cell, it "accidentally" picks up a headful of host genes. Whether all high frequency recombination plasmids and phage used in laboratories are less fit than the native replicons from which they were derived remains to be seen.

A more general explanation for the low frequency of plasmid- and phage-mediated recombination arises from the limitations of natural selection for altering the frequency of occurrence of intrinsically rare events. This can be seen if we take the rather extreme view that the receipt of random genes necessarily augments the fitness of the recombinant (a view we would not want to defend), for example, increasing its growth rate by 10%. If the basal probability of recombination is 10^{-6} and if a mutation doubles the probability of a cell becoming a recombinant (or doubles the probability of a vector transmitting host genes), then for cells of the mutant type the expected growth rate is $\psi(1+2\times 10^{-7})$ as compared to $\psi(1+1\times 10^{-7})$ for the nonmutant type; this is a very low selective differential and one that is likely to be overridden by stochastic factors or periodic selection (Atwood et al., 1951; Levin, 1981).

In discussing the evolution of sex in bacteria, we have intentionally neglected group or interdemic selection. In fact, we do not believe *any*

124

selection is necessary to account for plasmid- and phage-mediated recombination in bacteria. It is most parsimonious to assume that these low rates of recombination are the results of errors in replication and infectious transfer for the vector plasmids and phage. Although these errors are not products of natural selection, they play an important role in the adaptation and evolution of bacteria, perhaps approaching the significance of the errors responsible for mutation.

AN OVERVIEW

We have attempted to portray the various types of plasmids and phage as functionally similar genetic elements, that is, as parasitic replicons. Our intent was to deemphasize the differences in their modes of vertical and horizontal (infectious) transmission and offer a more unifying view of this phenomenon. However, because we succumbed to the convenience of separate treatment, we fear the reader may have failed to appreciate a more comprehensive interpretation. For this reason, some emphasis, explanation, and expansion seems warranted.

By suggesting that the various kinds of plasmids and phage are a single type of genetic element, we *do not mean* to imply that they have a common ancestry. This is clearly not the case for bacteriophage and unlikely to be so for plasmids. Bacteriophage have a variety of different genetic molecules: some single-stranded DNA, some double-stranded DNA, and some RNA; and these molecules replicate in a number of fundamentally distinct ways (Stent and Calendar, 1978). Although all known bacterial plasmids appear to be covalently closed circles of double-stranded DNA, this may be the result of convergent evolution due to the constraints of vertical transmission. There are at least two mechanisms of control for the replication of plasmid DNA (Falkow, 1975), and it may well be that the assay methods used may fail to detect plasmids of other types of genetic molecules. Because of these differences, we conclude that extant bacterial plasmids and phage are polyphyletic. Furthermore, it is likely that many plasmids and phage do not have unique ancestries but are chimeras composed of components from a variety of plasmid, phage, and host lineages.

It seems reasonable to suppose that all lineages of bacterial plasmids and phage were originally derived from the DNA or RNA of bacteria or possibly that of higher organisms and that incipient plasmids and phage are continually being generated from these sources. It also seems likely that some plasmids and phage evolved from each other. It is clear that the lines separating the different modes of replication and horizontal transmission are neither sharp nor insurmountable. Some prophage, such as P1, replicate as plasmids (Ikeda and Tomizawa, 1968). With modest genetic changes, temperate phage can be made virulent (Lwoff, 1953; Ptashne, 1971), conjugative plasmids can be made nonconjugative (Willetts, 1972), and temperate phage can be made into plasmids (Signer, 1969).

Viewing the various types of plasmids and phage as a single type of genetic element is useful for evolutionary and ecological considerations. In their nascent phase, as they emerge from cellular DNA or RNA, all of these elements would be autonomous replicons without specific mechanisms to assure either their maintenance in the descendants of their original host cell or their infectious transmission. They are likely to be confronted with some physiological mechanism of selection by the host to rid itself of the burden of foreign DNA or RNA. To survive in this hostile climate, the incipient plasmid or phage would have to evolve some mechanism for "over-replication" (Campbell, 1981). We see two nonexclusive ways for this to occur: (1) "niceness," that is, acquiring genes that enhance host fitness; and (2) infectious transmission. As long as there is a finite rate of segregation, becoming innocuous (selectively neutral) and replicating with high fidelity would not be sufficient for maintenance.

The route that is taken for over-replication will depend on the genetic and physiological constraints on the replicon and its host and on the environment of the bacterial population. With respect to the latter, we believe that *population density of the host is the primary factor in determining the form of over-replication.* In high density populations of bacteria, replicons with effective mechanisms of horizontal transmission would be favored. The extreme of this would be virulent phage and the resulting antagonistic coevolution. As the density of the host population declined, so would the intensity of selection for infectious transmission. Niceness would become increasingly important for the persistence of the autonomous replicon, and mutualistic coevolution would result. In bacterial populations of intermediate density or in populations with high amplitude oscillations in density, temperate phage and conjugative plasmids would flourish. In populations with sustained low densities, the costs associated with the mechanisms for infectious transmission could not be overridden, and niceness would serve as the only means available for the over-replication of parasitic genetic molecules; under these conditions, nonconjugative plasmids would be favored.

We have made a number of general and specific statements about the nature and direction of coevolution in bacteria and their viruses and plasmids. Although somewhat legitimatized by mathematical modeling and selected facts, most of these statements about how things came to be are no more than microbial "just so stories." As is the case with other evolutionary phenomena, there is no way to for-

126

mally demonstrate that the suggested pathways are indeed the actual ways things came to be. However, in the case of bacteria and their plasmids and phage, these evolutionary hypotheses can be readily tested with experimental and natural populations. We hope researchers will find some worthy of testing.

CHAPTER 6

ENDOSYMBIOSIS

Lee Ehrman

A fascinating variety of adaptations have developed by coevolution, from the nitrogen-fixing bacteria (*Rhizobium*) that live in specialized nodules on the roots of legumes to the nudibranch—a sea slug that starts life as a larva entrapped in jellyfish tentacles, is then engulfed by its jellyfish, but ends its cycle with the slug full grown and the medusa a tiny vestigial parasite residing on the ventral surface of the snail's mouth (Thomas, 1979).

These designs for living are produced by symbiosis—a condition in which organisms of different species live together in a state of mutual influence. There are several types of symbiotic relationships: parasitism, in which one species lives at the expense of the other; commensalism, where one species benefits from the association while the other is neither harmed nor benefited; and mutualism, where both species benefit from their relationship.

Broadly defined though they are, these affinities have many levels of integration. The continuum may range from nearly autonomous partners to components that are so merged that the identity and function of one of the organisms is difficult to ascertain. These patterns of association and the levels of integration of the partners provide valuable clues to the history of the coevolution of the symbionts and how their relationships have shaped their metabolic and behavioral interactions.

The most intimate of these coevolved systems is endosymbiosis, in which one of the organisms incorporates the other—an intricate cohabitation of two species in which a symbiont exists in the cells of a host for at least a discrete portion, if not all, of its partner's life cycle.

Margulis (1970) postulates that endosymbiosis explains the origin of the eukaryotic cell, which houses its genetic material in a nucleus. The serial endosymbiosis theory stipulates that the incorporation of free-living prokaryotes, represented now by the mitochondria and

128

chloroplasts, by a host cell is the paradigm for the emergence of the eukaryotic cell and led to one of the most important phylogenetic distinctions—that between prokaryotes and eukaryotes. Thus, endosymbiosis is considered one of the great primeval events in our evolutionary history. Margulis (1976) notes that "the biosphere is conspicuous for the frequency and diversity of associations between organisms that share only remote ancestry. It can be argued in some cases that the origin of certain higher taxa was made possible by merging alliances able to perform functions that individual partners could not."

From an abundance of literature on endosymbiotic liaisons, I shall describe two cases that provide us with some insights into the evolutionary dependence of host and endosymbiont and the implications of these symbiotic systems as bravura evolutionary events.

BACTERIAL SYMBIONTS OF *PARAMECIUM AURELIA*

Among the ciliated Protozoa, the *Paramecium aurelia* complex (of 14 species) houses intracellular bacterial symbionts belonging to the genus *Caedobacter*. Some of these endosymbionts are characterized by the fact that they produce toxins. The paramecia that bear them are called killers. Sonneborn (1938) discovered the killer trait in *P. aurelia* and provided evidence that it was cytoplasmically inherited. Five years later Lindegren and Altenburg separately suggested the microorganismal nature of the killer trait. All were correct.

The inhabitants of *P. aurelia*, known as kappa, are now classified as *Caedobacter taeniospiralis*. These bacteria in turn often carry one or more viruses or bacteriophages. Thus, we have endosymbiosis with at least three participants. As the hosts and their endosymbionts have evolved, the relationships have become highly specific, so that the hosts are intolerant of the endosymbionts borne by other hosts. Quackenbush (1977) has proposed that *C. taeniospiralis* (kappa) comprises more than one species.

When two strains of paramecia of complementary mating types are mixed together and the result is death of one strain, the surviving strain can be assumed to harbor an agent (or symbiont) lethal to the moribund strain. The results of mixing two strains of paramecia vary according to the type of kappa, the type of phage, the sensitivity of the paramecia, and environmental conditions (Preer, 1975). Thus, even within a single species of paramecium, different genotypes carry different kinds of kappas, and a paramecium is sensitive to the strains of kappa carried by other genotypes.

In most species of the *P. aurelia* complex, the bacteria cannot infect via the medium but are transmitted strictly by heredity (Preer, 1975). If a paramecium does not harbor kappa, it is termed "sensitive" and is vulnerable to kappa's toxin. Each genotype of *P. aurelia* is resistant to the kappa it may house, but other effects of kappa on its host are obscure.

Kappas employ glycolysis and respire (Kung, 1970, 1971). They contain citric acid cycle enzymes and possess cytochrome. Kappas have neither mitochondria nor nuclear membranes. The shapes and sizes of their cell walls resemble those of other bacteria, and like many other bacteria, they harbor lysogenic phages. Nearly half of the kappas are distinguished by the presence of refractile bodies (R bodies), which are proteinaceous, ribbon-like, tightly wound (but capable of unwinding) structures thought to be products of defective phages. It is believed that the phages housed by kappas specify the toxin (Preer et al., 1974).

In 3 of the 14 known species in the *Paramecium aurelia* complex, it has been shown that the paramecia–hosts excrete a poison of kappa particles into the culture media. For a paramecium to be toxic, it must inherit its cytoplasm from a killer strain with at least one nuclear dominant gene K, which provides tolerance to the kappa. The recessive allele k precludes the maintenance of kappa in the cytoplasm. Thus, tolerance can be bred out of resistant paramecia, because when Kk paramecia reproduce sexually, the issue may be KK (resistant) or kk (sensitive). However, if a KK paramecium does not actually maintain kappa in its cytoplasm, it will find the kappas in its food to be toxic. The kappas that had been liberated into the medium enter through the gullet, attack the food vacuoles, and induce blebs (blisters) on the cuticle. The presence of the ingested kappa with its toxin-producing phage proves lethal not only for kk paramecia, but for the kappaless K genotype as well. Thus, kappa cannot become established in a new host that does not already harbor kappa (J. Preer, personal communication).

Killer stocks of paramecia can be made sensitive through a number of laboratory techniques that remove their kappas: exposure to high temperatures, X rays, nitrogen mustard, chloromycetin, or inadequate media. Also, the paramecia can be induced to multiply faster than the kappas can; this procedure results in new paramecia that have not inherited any kappa. At least some of these "cures" must sporadically occur in nature, but the number of sensitive paramecia does not increase because of their resultant vulnerability.

As different species of bacterial symbionts with kappa-like activity have been discovered, other Greek letters have been used to identify them. *Mu* (mate-killers) is a species that cannot be transmitted via the medium but needs cell-to-cell contact to kill. *Lambda* and *sigma* are rapid lysis killers, usually showing their lethal effects on sensitive

130

paramecia in less than 30 minutes. *Delta* has been found in all paramecia known to carry endosymbionts. It was originally reported to be
a killer but does not show that property now in laboratory-maintained
stocks.

Preer (1977) considers these bacterial endosymbionts to be very ancient. New infections by free-living bacteria have never been observed
in the laboratory and transfer at conjugation has been shown to be
unlikely. Because symbiont-bearing cells contribute more progeny to
the next generations than cells that do not bear symbionts, it appears
that kappas confer advantages upon their host.

The prime one is via the inhibition of competitors for resources (see
Gill, 1974). Preer (1975) notes that "the presence of the symbiont
renders its host resistant to the toxin which it produces. The endosymbionts obviously profit from the association with their hosts, acquiring
from them all necessary nutrients, a place of abode and a buffer between them and the external environment . . . . The endosymbionts are
dependent on specific nuclear genes of *Paramecium* for their maintenance. These genes appear to be highly specific in respect to the particular endosymbiont whose maintenance they control."

The paramecium–kappa relationship is thought to be an ancient
one established before geological movements isolated today's continents (Preer, 1977). Their widespread distribution in North America,
Scotland, Europe, India, Japan, Australia, and Africa would support
this thesis as paramecia are killed by seawater. They do not form cysts
and need reasonably fresh moisture for survival, so transport to their
present locations was probably effected by continental drift. The
emergence of a single primordial endosymbiont would therefore be sufficient to populate an ancient land mass during geological times when
the continental land masses were adjacent or connected by land
bridges; then through geographical isolation the strains or races developed that we find today.

DROSOPHILA PAULISTORUM

The endosymbionts of *Paramecium aurelia* are an interesting example
of a common biological phenomenon—a relationship between two en-
tities, one of which is transmitted by inheritance. Such transmission
occurs with intracellular parasites, symbionts, and cases of infectious
heredity. The genus *Drosophila* provides another case of the coevolution of host and an endosymbiont, in this case a microorganism that is
known as CWD (cell wall deficient) until it is more formally named.

All six of the known semispecies of *Drosophila paulistorum*—Cen-

131

troamerican, Amazonian, Orinocan, Transitional, Andean–Brazilian, and Interior—have been found to harbor CWDs that seem essential to the host's welfare but cause male sterility when crossed to a different semispecies. As their names imply, these six semispecies of *D. paulistorum* are evolutionarily separated by their geographical distribution, an extrinsic isolating mechanism. Three intrinsic isolating mechanisms are also operative: sexual (behavioral or ethological) isolation, hybrid sterility, and hybrid inviability. Of these isolating mechanisms, the behavioral one is the most effective. This is seen in the few cases where the geographical distributions of two or three semispecies overlap. Where this occurs, they do not interbreed.

The genes responsible for sexual isolation are numerous: they are scattered throughout the three pairs of chromosomes and are apparently additive in their action (Ehrman, 1961, 1965; reviewed by Ehrman and Parsons, 1981). When hybrids are forced into being in the laboratory, hybrid sterility and inviability are observed. The F_1 hybrid progeny consist of fertile females and sterile males. The F_1 females will again produce fertile females and sterile male progeny when backcrossed to males of parental strains. Clearly, the isolating mechanisms prevent gamete wastage in nature.

For his final oeuvre, Dobzhansky recorded a series of phenomena that he interpreted as evidence for an incipient species, originating in the laboratory. Some years before, he had captured a gravid female of the Interior semispecies and had established an experimental population. With deference, it is perhaps best to let Dobzhansky describe what happened (Dobzhansky et al., 1976, p. 211):

A strain descended from a single female captured in the Llanos of Colombia in 1958 produced fertile hybrids of both sexes when crossed to strains of the Orinocan semispecies of *Drosophila paulistorum*. Since 1963, however, this strain, now referred to as New Llanos, gives sterile hybrid males with Orinocan strains. New Llanos gives fertile hybrids with strains of the Interior semispecies. Since this latter effect was discovered only in 1964, there is no way to ascertain whether the original Llanos would have been interfertile with Interior in 1958.

New Llanos shows little or no ethological isolation from either Orinocan or Interior semispecies, although these latter exhibit a fairly strong isolation from each other. From 1966 to 1974, artificial selection was carried on to erect an ethological isolation barrier between New Llanos and an Orinocan strain. For this purpose, two nonallelic recessive mutants were used. Homogamic matings were yielding phenotypically recognizable homozygotes, and heterogamic matings yielded wild-type heterozygotes. These latter were destroyed, while the homozygotes served as parents of subsequent generations. The selection was partially successful: strains showing a pronounced preference for homogamic matings, but not a complete ethological isolation were obtained. The experimentally induced ethological isolation is about as strong as the weakest isolation observed between semispecies of this group that occur in nature.

One possible explanation for the change in compatibility is that somehow the omnipresent CWD underwent changes induced by laboratory culturing of the host *Drosophila*.

The sterility of hybrid *D. paulistorum* is caused by a cell wall-deficient microorganism, CWD, in the testes of the hybrid males. CWD is an infectious agent that rapidly proliferates in the testes with a concomitant breakdown of spermatogenesis. This microorganism is destroyed by low pH, lipid solvents, UV, and exposure to 56°C for 30 minutes. It appears to be sensitive to tetracycline and insensitive to penicillin. CWD is also pathogenic in an unnatural host, the larva of the Mediterranean meal moth, *Ephestia kuehniella*, where it possesses killing power. It can be serially passed through *Ephestia* and retain its specificity for its original *D. paulistorum* semispecies when later reintroduced into *D. paulistorum* (Gottlieb et al., 1981).

No untreated ("uncured") *D. paulistorum*, male or female, has ever been observed to be free of CWD, and the flies do not survive long after their CWD is removed by heat shock and antibiotics. Each semispecies of *D. paulistorum* has its own highly specific endosymbiont without which it cannot thrive. CWD appears to pay its rent by providing some essential factor or service—perhaps the synthesis of vitamins. The benefits of this endosymbiont to its specific host are obvious, but the exact nature of the benefits remains in the realm of conjecture.

In the wild state, CWD is strictly inherited through the egg's cytoplasm. Laboratory-induced infections of foreign CWD produce the same fertility patterns as hybridization. Extracts of testicular CWD from one semispecies injected into a female of another semispecies will result in fertile daughters and sterile sons, showing that the hybrid genotype is not necessary for this phenomenon.

Adult females do not seem to suffer from foreign CWD, but its effects can be seen in larval inviability. Electron microscopy shows gross morphological differences between hybrid and nonhybrid embryos. When foreign CWD is carried in the egg's cytoplasm, the embryo develops abnormally and cytoplasmic blebs form (Ehrman and Daniels, 1975). Many larvae are thus lost before sexual differentiation.

The taxonomy of these CWDs has so far not been satisfactorily resolved. Ultrastructural studies reveal central fibrous networks surrounded by peripheral granulation, as described for *in vitro* cultures of *Mycoplasma hominis* (Kernaghan, 1971; Ehrman and Kernaghan, 1971). Razin (1973), in a review of mycoplasmal physiology, points out that although the internal structure and overall morphology of the *D. paulistorum* endosymbionts show "striking resemblances" to classic mycoplasmas, the presence of a duplex membrane suggests that they may be rickettsia or chlamydia.

Another infectious agent is a microorganism that causes the sex ratio (SR) condition in *Drosophila willistoni*. As with some other species of *Drosophila*, females of the *willistoni* group collected from natural populations will occasionally produce progeny that are entirely or almost entirely female. This condition continues in the strain indefinitely if "maintainer" males from other strains are used as fathers. The SR agent is, like CWD, maternally transmitted through the egg cytoplasm. Williamson and Poulson (1979) comprehensively reviewed what is currently known of SR in the *willistoni* group.

The SR trait was first studied in a *D. willistoni* strain from Jamaica and a *D. paulistorum* strain from Colombia (Malogolowkin and Poulson, 1957; Malogolowkin, 1958). It has since been found in *D. nebulosa* and *D. equinoxialis*; but curiously, it has not been found in *D. tropicalis*, a sibling species. Although documentation of the geographical range of the trait in each species is very poor, it is clear that the SR trait is quite widespread in the Caribbean and South America. The frequency of females with SR in a population usually ranges from 0.5 to approximately 10% (Williamson and Poulson, 1979; Marques and de Magalhaes, 1973).

One of the first indications that SR in the *willistoni* group was caused by a microorganism came from injection experiments (Malogolowkin and Poulson, 1957; Malogolowkin et al., 1959, 1960). Supernatant fractions from homogenates of SR females, when injected into non-SR strains, could produce the SR trait in recipient strains. Once established, the trait was passed on through the egg cytoplasm. The highest concentration of infective material came from adult hemolymph (Sakaguchi and Poulson, 1960, 1961). Interspecific transfer of the SR trait is possible outside the *willistoni* group, for example, to *D. melanogaster* and *D. pseudoobscura* (Williamson, 1965; Williamson and Poulson, 1979). In the natural host, death of male embryos occurs early in embryogenesis in the egg (Counce and Poulson, 1962). When transferred to other hosts, for example, *D. melanogaster*, death of males may occur later in the larval or pupal stage (Counce and Poulson, 1966).

Staining reactions and examination by light microscopy implied that the causative agent of SR was a spirochete resembling the genus *Treponema*. Electron microscopic studies, however, indicated that the organism lacked certain features of spirochetes (Williamson and Whitcomb, 1975; Williamson et al., 1977). Rather, it is morphologically identical and serologically similar to spiroplasmas (Williamson and Poulson, 1979; and for details concerning the spiroplasmas, related to mycoplasmas and CWD, see Barile and Razin, 1978).

A variety of studies indicate that there is more than one type or species of spiroplasma involved. For example, mixed infections of organisms from two species of *Drosophila* produce interference as indicated by an interruption of the SR condition in the host (Sakaguchi et al., 1965). Mixtures of hemolymphs *in vitro* can cause clumping of spiroplasmas if the hemolymphs are from two different species of *Drosophila* (Williamson and Poulson, 1979). Thus, unlike the interaction between CWD and its host, in which the endosymbiont is benign and even beneficial to the specific semispecies that carries it, different species of *Drosophila* harbor different spiroplasmas but are not fully adapted to them, because the endosymbiont continues to cause mortality of male flies.

Finally, viruses harbored by spiroplasmas vary in different species. Oishi and Poulson (1970) detected a lysogenic virus associated with *D. nebulosa* spiroplasma. When this virus is injected into a fly carrying spiroplasma derived from *D. willistoni*, the virus lyses the spiroplasma and the fly strain is "cured" (Oishi, 1971). At least six different viruses that vary in their lytic properties have been characterized (Williamson and Poulson, 1979). It appears that not only do different strains and species of flies harbor different spiroplasmas, but each type of spiroplasma has its own unique virus, to which it is tolerant. More details concerning the viruses and their morphology may be found in Williamson et al. (1977).

CONCLUSION

Endosymbiosis provides a bonanza for the scientist interested in charting coevolution. The multiple-tiered systems are especially intriguing. Research can be approached from several points of view; genetic, taxonomic, coevolutionary, ecological. Future analyses will undoubtedly elucidate the remarkable biochemistry that allows the endosymbiont to be helpful to its coevolved host and harmful to the host's near relatives. The endosymbioses that are described here—paramecium/bacteria/phage or fly/CWD/virus—exemplify *par excellence* the points made by Grun and by Margulis.

For as Grun (1976) has stated in his review of cytoplasmically inherited endosymbionts, "The evolutionary coadaptation between the cytoplasmic factors of a population of plants or animals and their nuclear genes is a specific product of the evolutionary selection pressures to which both are subject. The result is that the interaction supplies coherence to the organisms of the population." It is interesting to speculate that the dominant modes of the coevolution of endosymbionts are association, cooperation, integration, and coalescence and not "nature red in tooth and claw." To return to the thesis (Margulis, 1970) with which this chapter began—that endosymbiotic

135

processes were responsible for the origin of the eukaryotic cell—one wonders if the coevolution of these particular symbionts will progress to a further phase in the life cycle of their hosts. Like the free-living organisms that became the chloroplasts and the mitochondria of cells with which they integrated, will (have?) these endosymbionts (kappa, CWD) become organelles of the creatures they inhabit? Only time, selection, and coevolution, in concert, will tell.

CHAPTER 7

PLANT–FUNGUS
SYMBIOSES

John A. Barrett

In its original sense, symbiosis meant that two species were found in close association for a considerable part of their life cycles [De Bary: quoted by Lewis (1973a) and Cooke (1977a)]. This definition is very broad and covers a range of associations, from those in which one member is totally dependent on the other for its existence (parasitism) to those in which both symbionts cannot exist in the absence of the other (mutualism). Plants and fungi form a wide range of symbiotic associations, and this chapter will examine the ways in which these associations have evolved. Evidence of evolutionary change can be gleaned from two main sources. The first is the taxonomic relationships between different groups of organisms, from which inferences can be made about the adaptations that characterize each group. The second source is the observation of microevolutionary change at the population level.

In the first part of this chapter, I will examine the coevolutionary processes in associations between plants and parasitic fungi. Most of this evidence is derived from studies of microevolutionary changes in fungal parasites of crop plants. In the second part, I will examine the evidence for coevolution in "mutualist" associations such as lichens and mycorrhizae in which genetic evidence is lacking and inferences have to be made about the form of the reciprocal adaptation from physiological, morphological, and taxonomic studies.

137

PARASITIC SYMBIOSES

Is there any evidence for heritable variation in plants with respect to their association with parasitic fungi?

Ever since humans have made records of their activities, the destruction of crops by disease has been a recurrent theme in these records. For example, the first recorded epidemic of ergot caused by the fungus *Claviceps purpurea* was made in AD 857 in the Rhine Valley; the role of barberry (*Berberis vulgaris*) as the alternate host of wheat rust (*Puccinia graminis*) was sufficiently well established by the seventeenth century that laws requiring the eradication of barberry were enacted in France (Horsfall and Cowling, 1978). At the beginning of and during the nineteenth century, systematic examination of cereal crop species had enabled the early plant breeders, such as Le Couteur, Cooper and Shirref, to make selections and establish improved varieties. In 1815 Knight recommended that only those selections that showed reduced symptoms of disease (resistance) should be used to establish new varieties. Among many of the examples quoted by Darwin to support his ideas on variation within populations and natural selection can be found observations of the differing susceptibility of cultivated plants to disease, e.g., "Cuthill's Black Prince (a strawberry variety) evinces a singular tendency to mildew." So from the early stages of methodical plant breeding, there are observations that lines and varieties may differ in their susceptibility to disease, and these observations imply that this variation is in part inherited (Barrett, 1981).

However, not until the "rediscovery" of Mendel's work at the beginning of the twentieth century was any progress made in the elucidation of the inheritance of disease resistance. Considering the economic and social importance of disease on crops, it is perhaps not surprising that some of the earliest papers describing Mendelian inheritance should be on disease resistance. In 1905 and 1907, Biffen published two papers that described the Mendelian basis of resistance of barley to mildew (*Erysiphe graminis*) and wheat to yellow rust (*Puccinia striiformis*). Moreover, he was able to demonstrate that "immunity is independent of any discernable morphological character, and it is practicable to breed varieties morphologically similar to one another, but immune or susceptible to attacks of certain parasitic fungi" (Biffen, 1907). This demonstration that disease resistance is inherited led to much work on the mode of inheritance of resistance to economically important diseases in most of the major crops of the world. By 1974 Day estimated that over 1000 papers had been published describing the inheritance of disease resistance. By and large, most of the papers describe resistance as being controlled by one or few genes (oligogenic resistance), but this may be an artifact, as most of the experiments have been carried out on cultivated species, and characters controlled

138

by one or few genes are easier to handle during breeding programs. However, many cases of quantitative inheritance of disease resistance (polygenic resistance) have been described. In some crops, resistance to a parasitic fungus has been demonstrated to be controlled both by single genes and polygenically (e.g., resistance to *Puccinia sorghi* in corn; Day, 1974).

Prior to intensive cultivation, each farming area had a range of "varieties" characteristic of that area. These local "varieties" (*land-races*) consisted of mixtures of different genotypes that had been maintained under mass selection by generations of growers. The early stages of methodical plant breeding involved the extraction of "improved" true-breeding lines from land-races and it is from this source that some of the resistance factors used in modern varieties were originally isolated, e.g., gene *Mlg* controlling powdery mildew (*Erysiphe graminis* f. sp. *hordei*) resistance in barley (see Barrett, 1981). However, with increasing pressure on plant breeders for even better varieties, the wild progenitors and near relatives of crop species have become increasingly exploited as sources of useful characters for incorporation into new varieties, not least as sources of disease resistance. Among the crosses in which Biffen first described the inheritance of powdery mildew resistance in barley, a resistant line of the wild barley *Hordeum spontaneum nigrum* was used. In the search for resistance to late blight of the cultivated potato caused by *Phytophthera infestans*, it was found that lines of the wild potato *Solanum demissum* showed resistance and a number of resistance genes were successfully transferred from *S. demissum* to the cultivated potato *S. tuberosum*. [See Day (1974) and Van der Plank (1963) for discussion.]

Is there evidence of heritable variation in fungal parasites with respect to their plant hosts?

During 1917 and 1918 Stakman and his colleagues published what have since become classic papers of plant pathology (Stakman et al., 1917, 1918a,b). They noted that isolates of wheat stem rust (caused by *Puccinia graminis tritici*) that were taken from different wheat varieties and used to infect a range of varieties tended to produce most disease on the varieties from which they were originally isolated. This phenomenon they termed "physiologic specialization"; although this term is not strictly biologically correct, it is still useful for describing the phenomenon. The fact that such isolates retained their specificity for the host variety was at least an indication of a genetic component of this specialization. The use of sets of varieties (differential sets) to

139

dissect the population structure of parasite populations has become established as a basic tool of plant pathologists, and much effort has been devoted to cataloging pathogen populations using this method in most economically important diseases. Proof of the genetic basis of the pathogenicity (i.e., the ability to infect host plants) was not long in coming for Waterhouse (1929) and Newton and Brown (1930) demonstrated segregation of single loci controlling pathogenicity in stem rust [see also Johnson (1953) for review]. The elucidation of the genetic control of pathogenicity is inextricably bound to the process of resistance breeding in the host crop species. Consequently, the genetic basis of pathogenicity has arisen as a by-product of the study of the inheritance of resistance.

Is there evidence of complementary heritable variation in parasitic plant–fungus symbioses?

On the basis of observations that different genotypes of crop plants varied in their ability to resist disease-causing fungi, that this variation had a genetic basis, and that, conversely, different isolates of disease-causing fungi varied in their ability to incite disease on different "lines" of host plants, Flor began a series of experiments to investigate the genetic relationships between plants and their fungal parasites.

Flor used different varieties of flax (*Linum usitatissimum*), which could be classified as either resistant or susceptible to a defined "race" of flax rust (*Melampsora lini*), and different "races" of flax rust, which were either "virulent" (i.e., able to attack) or "avirulent" (i.e., unable to attack) on a certain variety. By crossing the different varieties of flax and testing the F_2 generation with the races to which the parental lines were susceptible, he was able to demonstrate that resistance in flax was inherited as a dominant character controlled by a single locus. Similarly, by crossing rust races that differed in their ability to infect different varieties and testing the F_2 generation on the same varieties, he showed that "virulence" in the flax rust was inherited as a recessive character controlled by a single locus (Flor, 1942, 1955, 1956). From these series of experiments Flor came to the conclusion that "for each gene conditioning rust reaction there is a specific gene conditioning pathogenicity in the parasite" (Flor, 1956). The idea of complementary genetic systems in both host and parasite first demonstrated by Flor has become known as the "gene-for-gene hypothesis." Since the publication of Flor's papers, other systems have been demonstrated to exhibit gene-for-gene relationships (see Day, 1974).

Although the "gene-for-gene hypothesis" neatly summarizes the phenomenon described by Flor, not all systems so described conform exactly to Flor's usage. Flor's original definition implied that for each locus controlling resistance in the host there is a complementary locus in the parasite controlling pathogenicity. However, current usage is to describe any system in which both resistance–susceptibility and virulence–avirulence segregate in a simple Mendelian fashion as "gene-for-gene" systems, even if the allele for "susceptibility" to a given fungus "race" is in fact an allele for resistance to another "race" of the fungus. For example, further research on the flax–flax rust system has revealed that multiple alleles at a number of loci controlling resistance exist but that the corresponding virulence factors to each of the resistance alleles are inherited at different loci, that is, in these cases resistance may be monogenic (one locus) but the complementary pathogenicity is controlled by multiple loci (Lawrence et al., 1981a,b). Notwithstanding the fact that the genetic interaction between host and parasite may not be on a strictly locus-for-locus basis, the fact remains that in many cultivated species and their fungal diseases, reciprocal adaptation appears to be controlled by a small number of genes.

The simplicity of the gene-for-gene hypothesis has, perhaps, blinded many workers to the possibility that the interaction may be more complex. A resistant variety is not resistant just because of the effects of a single locus substitution; expression of the character "resistance" is also dependent on the genetic background in which the resistance is placed. Within a group of varieties, each carrying the same major resistance factors, differences can be observed between isolates taken from each of the varieties in their ability to infect other varieties carrying the same resistance gene(s) (Wolfe and Schwarzbach, 1978; and personal communication). So although the basic "gene-for-gene" relationship remains, it is overlaid by variation due to differences in genetic background of the host plants to which the parasite can also become adapted.

Because all of the experiments on complementary genetic systems in hosts and parasites have been carried out on cultivated species, there always remains the possibility that the observation of gene-for-gene systems and the emphasis on their importance is an artifact of agriculture and plant breeding (Day, 1974). Faced with a severely diseased crop, a plant breeder will be looking for large effects in his search for resistance characters to incorporate into his breeding programs. Moreover, the screening techniques used in breeding will often only give unequivocal classification when large resistance effects are present. When varieties incorporating apparently total resistance to a parasite are exposed to a relatively heterogeneous parasite population on a large scale, it is likely that only a substantial change in the

141

parasite will enable it to attack the new variety. It appears that in the majority of cases in fungi parasitic on crop plants the phenotypic change is outside the normal range of quantitative variation present in the parasite population, and so the change is more easily accomplished by changes at one (or few) major gene loci. This is not to say that quantitative variation has no role in the response. Once a genotype that can survive on a new host variety is present, genes that modify the effects of the major gene(s) and improve its (their) performance on that variety will be favored by natural selection. For example, the barley variety *Keg*, which carries major gene resistance to powdery mildew, was grown on an experimental plot (after it had been introduced commercially, but before it was grown on a large scale) and exposed to natural infection by powdery mildew (*Erysiphe graminis* f. sp. *hordei*). Isolates taken from the youngest leaves at the end of the growing season showed a reduced variance in pathogenicity when compared with isolates taken from older leaves that had become infected earlier in the season, when tested under standard laboratory conditions (M. S. Wolfe, personal communication).

Two further factors contribute to the demonstration of major gene resistance in most plants. First, characters controlled by simple Mendelian factors are easier to handle in breeding programs. Second, experiments on the inheritance of resistance are often carried out under controlled environmental conditions, and hence any environmental contribution to phenotypic expression will be reduced; Ellingboe (1978) has argued that by carefully adjusting environmental conditions, all host–parasite interactions can be shown to segregate in a Mendelian fashion.

By the same token, generalizations about the dominance relations of resistance–susceptibility and virulence–avirulence must be treated with caution. Following the arguments of Fisher (1928a,b, 1930), if selection is not too strong, an advantageous gene spreading through a population will acquire modifiers that enhance its expression and so such genes are more likely to be dominant than recessive. In screening breeding material for resistance, very large populations cannot be tested; therefore, only resistance genotypes present at reasonably high frequencies are likely to be detected. Because such resistance is likely to have been advantageous in the recent past of the screened population, the resistance genes detected are more likely to be more dominant than recessive. Furthermore, the breeding and screening techniques involved in the production of new varieties will themselves tend to select genetic backgrounds that enhance expression of the resistance genes. Consequently, the final commercial variety is more likely to carry resistance genes that are more dominant than recessive. When a new variety is introduced into cultivation, it is resistant because the corresponding pathogenicity is very rare, possibly because it is disadvan-

142

tageous, and in a diploid parasite, more likely to be recessive (Fisher, 1928a,b, 1930). On exposure to the new resistant variety, all genotypes except the rare recessive homozygote will die, and consequently, on the new resistant variety, the virulence gene becomes effectively fixed in one generation; there is, therefore, no possibility of heterozygotes being formed and dominance evolving (for discussion, see Sved and Mayo, 1970).

No doubt, on the basis that a simple explanation is intellectually more appealing, some host–parasite interactions have been described as exhibiting gene-for-gene relationships on the basis of single locus changes in the host plant and a phenotypic correspondence in the parasite (or *vice versa*) without any genetic tests being carried out (Rodrigues et al., 1975; Hiura, 1979). Indeed, it may well prove impossible to elucidate the genetic basis of the ability of parasites to attack hosts where the sexual stage of the fungus is not known and no parasexual cycle has been demonstrated. Even in species such as *Verticillium* spp., where sexual stages are unknown and parasexuality is known, little work has been carried out on the genetics of pathogenicity, although physiologic specialization has been demonstrated (Day, 1974; Puhalla, 1979). Thus, although there is evidence that complementary genetic systems can exist in plant host–fungal parasite systems, the fact that most of this evidence is derived from the interaction between cultivated crop plants and their parasites means that the simple genetic basis of these interactions may be an artifact.

Is there evidence of genetic change in host populations when exposed to fungal parasites?

The potato was introduced into Europe in the sixteenth century, but *Phytophthera infestans*, the causal fungus of potato late blight, was not introduced with it. Potato cultivation prospered, eventually becoming the staple food crop of the peasantry of Europe. Sometime during the 1840s *P. infestans* was introduced into Europe and gave rise to the Great Potato Blight of 1845, which led to the deaths of many thousands, perhaps millions, of people who depended on the potato. After these great epidemics, the severity of the disease declined, but it was still prevalent in western Europe. In 1833 a consignment of potatoes had been taken to Basutoland (Lesotho) and cultivated there in the absence of *P. infestans*. These potatoes were shown to be very susceptible to potato late blight when compared experimentally with early twentieth century varieties (Van der Plank, 1963). It would appear that, if the potatoes taken to Basutoland were representative of early nineteenth century European potato stocks, then these stocks were very susceptible to potato late blight and that this susceptibility may have accounted for the devastating effects of the disease when it arrived in Europe. However, the early twentieth century varieties with which they were compared were derived from postblight stocks and, because little methodical breeding in potatoes was practiced during the immediate postblight period, it seems likely that the resistance found in the later varieties was derived from resistant genotypes that had survived the Great Potato Blight.

Another American staple crop, corn, achieved worldwide distribution after the discovery and exploration of the Americas by Europeans in the fifteenth and sixteenth centuries. Corn was successfully introduced into West Africa where it was grown extensively by both commercial growers and subsistence farmers. In 1949 the rust fungus *Puccinia polysora* became established in West Africa, and the corn in that area succumbed to the disease. When lines of corn from areas where *P. polysora* was endemic were grown in West Africa alongside local lines, the local lines were more susceptible (Van der Plank, 1963; Robinson, 1976). This demonstrated that the average resistance of the crop cultivated in the absence of disease was very low; whether this was a chance effect because the lines originally introduced into West Africa carried little or no resistance or whether resistance had declined because of the absence of selection by *P. polysora* cannot be determined. Within five years, the epidemic had begun to decline as farmers planted the survivors from the previous season's epidemic and resistance slowly accumulated in the crop to a level sufficient to produce a reasonable yield.

Thus, although there is no direct evidence that fungal parasites can generate an evolutionary response in their host plant populations, there is circumstantial evidence that permits this inference to be made.

Is there evidence that evolutionary changes can occur in parasite populations in response to genetic variation in host populations?

In 1916 the North American wheat belt was devastated by an epidemic of stem rust (*Puccinia graminis* f. sp. *tritici*). As part of the reaction to this disaster, a breeding program for resistant varieties was initiated and led to the development of the first modern variety incorporating major gene resistance. However, in 1935, when a large part of the wheat-growing area of the Midwest was planted with the variety *Ceres* incorporating this resistance gene, a second epidemic hit the area. When the rust population was examined using differential sets, it was found that the population consisted predominantly of one "race" (phenotype), race 56, which was able to attack *Ceres* (Van der Plank, 1963; Barrett, 1981).

144

As agriculture has developed during the twentieth century, the following sequence of events has become more common. Drawing on the fact that genetic variation for resistance exists, plant breeders produce new varieties that are resistant to the prevailing parasite population. Because the resistant varieties produce higher yields, in part because of disease resistance, the new varieties are grown on a large scale. This increased scale of cultivation selects out of the parasite population those genotypes that can overcome the resistance; these genotypes increase in frequency, as they are the only forms of the parasite that can infect the "resistant" varieties, and "resistance breakdown" occurs. Because the benefits of resistance have now been lost, the varieties fall from favor with the growers and are usually replaced with new "resistant" varieties, and the process repeats itself. This process of producing varieties with increased resistance followed by adaptation of the parasite population and the fall from favor of the variety is known as the "boom and bust" cycle and has become more and more common wherever intensive crop husbandry is practiced, especially in cereals (for more examples, see Barrett, 1981).

The social and economic consequences of "resistance breakdown" can be very dire indeed. Perhaps one of the best documented cases of recent years is that of the southern corn leaf blight epidemic in the southern states of the United States in 1970 (Anon. 1972). Southern corn leaf blight, caused by *Helminthosporium maydis*, had long been known as an important disease of corn, and over the years breeders had gradually improved the resistance of corn varieties, by selection of polygenically controlled resistance, until the severity of the disease was of manageable proportions. Yet in 1970 an epidemic caused substantial damage; statewide yield losses of 50% were not uncommon. The cause of the epidemic was a new race of *H. maydis*, which was able to attack most of the cultivars in use at the time. At first sight this was surprising because most commercial cultivars of corn are either single- or double-cross hybrids and genotypically fairly heterogeneous. However, to improve the efficiency of the crossing programs that give the commercial hybrids, cytoplasmic factors that confer male sterility are used. The one common factor in all of the varieties that had been attacked by *H. maydis* during this epidemic was that they all possessed the same cytoplasmic male-sterility factor, Texas male-sterile cytoplasm (*Tms*). It has been estimated that at the time of the 1970 epidemic approximately 10^{15} cytoplasmically identical plants were growing in the United States (Browning, 1972). It was possession of this cytoplasmic factor that made the corn plants susceptible to a toxin produced by a new race of *H. maydis*, race *T*. After the epidemic had receded and the postmortem on its cause was carried out, it was found that in a preserved collection of *H. maydis* samples, from well before the epidemic, genotypes were present that could attack cultivars carrying *Tms* cytoplasm. The existence of these races had been noted at the time of introduction of the *Tms*-carrying varieties, but these races were predominantly of a mating type different from that that eventually caused the great epidemic of 1970 and did not appear to have the ability to attack the corn plants very severely. However, the selection imposed by the extensive use of these varieties shifted the population of *H. maydis* toward increased ability to attack these varieties.

Pearl millet is a staple crop of subsistence farmers on poor land in India. In order to increase production, hybrid varieties were introduced and total yields more than doubled over a period of 20 years. However, in 1971 the pearl millet crop was devastated by an epidemic of downy mildew (*Sclerospora graminicola*). On investigation it was found that this epidemic was a repeat of the southern corn leaf blight–*Tms* story. In most of the varieties that succumbed to the disease, just one source of cytoplasmic male sterility (Tift 23A) had been used (Safeeulla, 1977). So, changes in the genetic composition of crop populations can induce evolutionary responses in parasite populations that can have disastrous consequences.

VARIATION AND EVOLUTION IN NATURAL ECOSYSTEMS

The constant breakdown of resistant varieties in agriculture has forced plant breeders to continuously screen for resistance genes in wild populations of the progenitors and near relatives of cultivated species. So, to a large extent the study of wild populations has been to identify and conserve resistance genes and other useful characters as sources for plant breeding. However, a few studies have been devoted to a close examination of the variation present in natural populations; little or no work has been carried out to explore the genetic interactions in plant host–fungal parasite systems unrelated to crop species or to follow the dynamics of such systems under genuinely "wild" conditions. The most extensive studies of this type have been carried out on the wild barley *Hordeum spontaneum* and its fungal parasite *Erysiphe graminis* and on the wild oat species *Avena sterilis* and *Avena barbata* and their fungal parasite *Puccinia coronata* (crown rust) in Israel (Dinoor, 1974, 1977; Eshed and Dinoor, 1981; Fischbeck et al., 1976; Segal et al., 1980; Wahl, 1970; Wahl et al., 1978). By testing plants grown from seed collected in the wild with known genotypes of the fungal parasites, it is possible to detect the presence of major resistance genes. By testing isolates of the fungus on a range of different varieties carrying different resistance genes, it is possible to

146

determine whether or not the isolate is carrying the corresponding gene for pathogenicity. These studies have revealed that substantial variation exists for resistance in the host population and for pathogenicity in the parasite population; these traits are controlled both by major genes and by the genetic background. Within host populations, substantial levels of susceptibility exist when defined races of the parasite are used. Wahl and his co-workers have also shown that in areas where the conditions are not suitable for epidemic development until late in the season, the proportion of the barley populations exhibiting resistance is lower than in areas where conditions are more suitable for development of powdery mildew. This latter observation is also supported by studies in the wild relatives of the cultivated potato in Mexico (Niederhauser, 1961). In the arid highlands of northern Mexico, where inclement conditions prevail for the development of *Phytophthera infestans*, the native *Solanum* species carry little or no resistance to this fungus. In the more cool and humid conditions of the central highlands, conditions are more compatible with the development of blight, and all potato species can be demonstrated to be carrying resistance genes of varying effectiveness.

How specialized are fungal parasites to their hosts?

Parasites are just one of the factors that constitute the environment of their hosts. As host species have evolved in response to other environmental pressures and formed new species, it would not be unreasonable to suppose that the parasites too will have tracked these changes. On this assumption, coevolution of parasite and host would lead to similar phylogenetic relationships between parasite species and between host species. In addition, the sharing of a parasite between different host species could then be taken as evidence of close taxonomic affinity between the host species (e.g., see Meeuse, 1973). This approach has been taken as axiomatic in some taxonomic and phylogenetic studies (Savile, 1968). On the other hand, this view ignores both the general biology and ecology of the host and parasite species. The diversity of potential host plants in an ecosystem will offer a range of opportunities to parasites, and consequently the host range of a parasite may then be more dependent on the diversity of host plants available to it than on the phylogenetic relationships among these plants. An appreciation of the "host range" of parasites is an important consideration in our understanding of the coevolution and ecology of host–parasite relationships.

The taxonomy of the fungi relies to a great extent on morphological

147

characters. However, different populations of apparently morphologically identical fungi may not infect the same host species. Where populations can be distinguished in this manner, they are identified by their morphological species name plus a further name that identifies the *forma speciales*, or host specialization group, to which they belong. For example, *Fusarium oxysporum* has forms that can attack banana, f. sp. *cubense*; flax, f. sp. *lini*; tomato, *lycopersici*; and peas, f. sp. *pisi* (Day, 1974). While this naming convention is taxonomically convenient, it may impose a certain rigidity on the appreciation of the diversity in a fungal group or overlook some biologically important facts. The use of a nomenclature that uses formae speciales as taxonomic units does carry with it a certain circularity; a fungus species is known to have a form, f. sp. *a*, which can attack host species *A*; therefore any isolate of this fungal species taken from host species *A* is forma speciales *a* and any host plant morphologically similar to host species *A* and attacked by f. sp. *a* is host species *A*. Indeed, parasitic fungi have been used to define taxonomic groupings of host plants (see, for example, Meeuse, 1973; Eshed and Dinoor, 1981). But a number of observations indicate that the simplicity of this logical process may be erroneous. When parasitic fungi that have morphologies similar enough to cause them to be classified as a single species are taken from different crop host species but exhibit specificity for the host species, the classification into formae speciales is usually unambiguous. For example, *Erysiphe graminis hordei* does not attack wheat, rye, or even other barley species when taken from cultivated varieties in the United Kingdom. However, when Wahl et al. (1978) tested samples of *E. graminis* isolated from cultivated barley *Hordeum vulgare* in Israel, they found that these samples could attack 37 species of grasses from 18 genera and 4 tribes. When the same tests were carried out with the formae speciales *E. graminis tritici* and *E. graminis avenae*, the number of species attacked were, respectively, 47 species from 16 genera and 31 species from 20 genera within the same four tribes. Similarly, Eshed and Dinoor (1981) took a series of formae speciales of crown rust and tested them on a range of different host species. They found that "there is no relation between the classification of grass genera into tribes and the host range of any form of *Puccinia coronata*." In no case was the ability to initiate an infection restricted to just the host species from which the fungus had originally been isolated. Gerechter-Amitai (1973) also demonstrated that under field and greenhouse conditions the host ranges of *P. coronata avenae* and *P. coronata tritici* were very wide (see also Browning, 1980). The evidence thus far indicates that some parasitic fungi may have a wide host range when tested under experimental conditions. However, this demonstration does not mean that a wide host range will characterize the parasite under natural conditions, as there may be ecological factors that restrict infection between the different host species. Indeed, the data of Gerechter-Amitai indicate that the host ranges of *P. coronata avenae*, *P. coronata tritici*, and *P. striiformis* are much reduced when the tests are carried out under field conditions rather than in the greenhouse.

The story is further complicated by the phenomena of "induced susceptibility" and "induced resistance" (Chin, 1979; Ouchi et al., 1979; Wheeler, 1975). In some fungi, for example, *Erysiphe graminis*, inoculation of a host plant by a fungal isolate compatible with it, followed by inoculation with an isolate incompatible with the host, can often give rise to infection by both fungal isolates—"induced susceptibility." This can occur both within a forma speciales, between different formae speciales (e.g., *E. graminis hordei* and *E. graminis tritici*), and even between related fungi (e.g., between *E. graminis hordei* and *Sphaerotheca fuliginea*) (Ouchi et al., 1979). Reversal of the inoculation procedure, that is, incompatible isolate first followed by compatible isolate, can prevent infection by the compatible line thereby providing "induced resistance."

Not all fungi show the close adaptation to their hosts that gives rise to formae speciales or races. *Verticillium dahliae* can attack a wide range of plant species; however, although there are differences in pathogenicity, there is no specialization to different host species. *Septoria nodorum*, primarily a parasite of wheat, shows no specialization to different varieties of wheat. More recently, however, *S. nodorum* has become increasingly important as a parasite of barley and while there does seem to have been some adaptation to barley, successful cross infections between wheat and barley are still possible (Puhalla, 1973; Gair et al., 1979). Whether the lack of specialization to host species in these parasitic species represents an inability of the fungus to evolve such adaptation, an absence of selection for close adaptation, or strong selection against close adaptation because of other ecological factors has not been determined. On the other hand, the introduction of some crops into new areas has in some cases led to the evolution of parasitic fungi previously unknown on the crop. For example, the introduction and development of banana cultivation in the Caribbean and Central America selected a new form (f. sp. *cubense*) of the soil-borne wilt fungus *Fusarium oxysporum* capable of attacking banana plants (Simmonds, 1959).

One of the earliest genetic tests on parasitic fungi was carried out by Goldschmidt (1928, quoted by Day 1974), who carried out crosses between different races of *Ustilago violacea* with different host ranges within the family Caryophyllaceae and demonstrated that host range

149

seemed to be under simple Mendelian control. Since then a number of studies have been carried out to determine the genetic basis of host specificity and formae speciales (e.g., Hiura, 1979). Crosses between *E. graminis tritici* and *E. graminis agropyri* suggest that at least four loci are involved in the adaptation to wheat and wheat grass. In some parasitic fungi, notably ascomycetes such as *Puccinia graminis* and *Puccinia coronata*, the asexual stage occurs on a range of hosts that form the basis of the classification into formae speciales. The sexual stage takes place on a second host (the alternate host), which may be taxonomically distant from the host of the asexual stage. The alternate host is shared by all of the formae speciales, and so it is possible for different formae speciales to interbreed and produce novel genotypes on the alternate host. Similarly, although formae speciales may be easily identified on crop plants, they may still share an ability to infect wild species and consequently be able to hybridize. That this can occur has been demonstrated by Burdon et al. (1981), who have shown that a new genotype of *Puccinia graminis* in Australia probably arose by somatic hybridization between the formae speciales *P. graminis tritici* and *P. graminis secalis* on a wild grass. Not all attempts to cross formae speciales of *E. graminis* sexually have been successful, but *E. graminis* is heterothallic and so this may be due to mating type incompatibility rather than adaptive reproductive isolation (Hiura, 1979).

So there is some evidence that the ability to attack different host species does vary within each fungal species, although some fungi do not show this adaptation, and that the extent of the specialization has undergone evolutionary change according to the selection imposed by the host plant population structure; in heterogeneous wild populations containing many plant species, parasitic fungi seem to have a wider host range than under the more homogeneous agricultural systems.

MUTUALISM

Two species can be said to exhibit mutualism when both have their survival, growth, or reproductive ability (i.e., fitness) enhanced by the presence of the other. In plant–fungus symbioses, a range of such interactions can be found, from those in which neither plant nor fungus can survive in the absence of the other to those in which both plant and fungus are capable of an independent existence but both have increased probability of survival and reproduction when they grow in close association.

LICHENS

Although lichens are superficially a distinct taxonomic entity, a close examination of their structure and morphology reveals that they consist

150

of two organisms, an alga (the phycobiont) and a fungus (the mycobiont), in close association. Over 70,000 species of fungi have been described and of these about 15,000–20,000 are involved in lichen symbioses (D. C. Smith, 1975; Ahmadjian, 1970). Twenty-six different genera of alga have been found in lichens, but this figure is somewhat misleading as approximately half of all lichens described appear to involve just one genus of alga, *Trebouxia* (Ahmadjian, 1970). The diverse taxonomic composition of the lichens almost certainly points to a polyphyletic origin (Ainsworth, 1971). Lichens can be found in virtually all terrestrial habitats from the polar regions to the tropics. The fact that lichens can be found in habitats where algae and fungi would find difficulty in surviving alone indicates that there is perhaps something special about the association and that the two organisms have coevolved to permit survival in these conditions.

Is there any evidence that the algal and fungal components of lichens vary in their mutual dependence?

Lichens by definition are dual organisms and so one is faced with the immediate problem of recognizing the components in a free-living stage, because the physiological interaction of the two organisms may alter the morphology of each of the components in the dual organism (Ahmadjian, 1966, 1980; Ahmadjian and Heikkila, 1970; Hawksworth, 1973).

Fungi are classified primarily by their morphology and in particular by their reproductive structures. In those lichens in which sexual reproductive structures are produced, the taxonomic affinities of the mycobiont can be determined. On this basis, it appears that the vast majority of mycobionts are close relatives of the free-living ascomycetes (Ainsworth, 1971). Although the sexual structures appear to be homologous to the same structures in free-living fungi, there is only indirect evidence that the spores produced are sexual products (Ahmadjian, 1967, 1980; Hale, 1967). But spores are produced, and these may be dispersed and form mycelial mats that may be capable of an independent existence at least for a short time (e.g., *Strigula complanata*), during which contact with algal cells could give rise to a new association. On the other hand, a mycelium developing from an ascospore could be capable of an autonomous saprotrophic existence (i.e., as a decomposer) and be indistinguishable from free-living fungi. For example, when grown on agar, the mycobiont *Buellia stillingiana* is morphologically indistinguishable from the free-living saprotrophic fungus *Sporidesmium folliculatum* (Hale, 1967). It has already been pointed out that the majority of lichens involve the alga *Trebouxia*; of all the algae involved in lichen symbioses, it is the only genus that is not found in the free-living state. However, small microcolonies have been described in close proximity to lichen thalli; Ahmadjian (1980) argues that these colonies may be escapes of zoospores from the thallus and are probably fairly short-lived and hence do not constitute a true free-living stage. On taxonomic grounds Ahmadjian (1970, 1980) has further argued that *Trebouxia* and *Pseudotrebouxia* are closely related to the free-living alga *Pleurastrum* but have become specialized to a symbiotic existence.

Is there any evidence of specific interactions between mycobionts and phycobionts?

There are two ways in which the specificity of the interaction between the alga and fungus can be tested: (1) by isolating and identifying the mycobiont and phycobiont from different lichens, and (2) by taking lichen components and attempting to resynthesize lichens using different combinations of fungus and alga. Both methods are extremely laborious, and the environmental conditions have to be controlled very rigorously to even resynthesize a lichen from its dissociated components. However, evidence is accumulating not only that taxonomically distinct lichens share the same mycobiont but differ in the phycobiont but also that some species can share the same phycobiont but differ in the fungal component (Ahmadjian, 1980; Hildreth and Ahmadjian, 1981). However, only two lichen species have been described that form associations with two different genera of algae (Ahmadjian, 1980). By attempting to synthesize lichen using mycobionts and phycobionts isolated from different lichens, Ahmadjian and his co-workers (Ahmadjian et al., 1978, 1981; Ahmadjian, 1980) have been able to demonstrate that the formation of the initial contact between fungus and alga may be quite indiscriminate. Appresoria were formed by the mycobionts on a wide range of different phycobionts, but these appresoria did not terminate the fungal hyphae. Indeed, the fungi were demonstrated to form similar structures with glass beads of the same size as algal cells. In most cases the algal cells were destroyed by the fungus, but in some cases structures characteristic of the early stages of lichen thallus development were produced (see also D. C. Smith, 1976). The destruction of algal cells in incompatible combinations indicates that the mycobionts have some parasitic ability and offers support for the possibility that lichen symbioses evolved from a parasitic association. James and Henssen (1976) have described a cline in morphology between two morphologically distinct lichens, with different light and humidity requirements on a single boulder in New Zealand. On close examination it was found that the cline did not consist of a cline of individuals with slightly different morphology but rather of a continuous mat of mycelium and a cline in the proportions of green and

152

blue-green algae within this mycelium. Thus, it appears that environmental factors may be important in determining which combinations of fungus and alga will develop into lichens.

Is there any evidence that would suggest that coevolution had occurred between the members of lichen symbioses?

If a mutualist association requires total interdependence of the two members, then it could be argued that independent reproduction would be disadvantageous, because there could be no easy way for the propagules to reestablish the association. Although the mycobionts in some lichens do form structures similar to sexual structures in free-living fungi, evolution should tend either to reduce sexual reproduction in the mycobiont or produce mechanisms that ensure that both fungal and algal propagules are dispersed simultaneously. Examples of both strategies can be found in lichens (Ahmadjian and Heikkila, 1970; James and Henssen, 1976).

Although it has become conventional to regard lichens as being composed of two organisms each acting to the other's benefit (e.g., Cooke, 1977b), there is little evidence to support this interpretation. Algal cells tend to leak metabolites into their surrounding medium. The presence of a mycobiont can further stimulate the release of metabolites (Smith et al., 1969; D. C. Smith, 1975), which are then taken up by the mycobiont. Conversely, it has proved exceedingly difficult to demonstrate that any mycobiont products are taken up or utilized by the phycobiont. D. C. Smith (1975) has argued that "if two-way flow of nutrients is to be regarded as an essential definition of mutualism, lichens will have to be excluded until there is appropriate evidence for nutrient flow from fungus to alga." Instead he argued that by a slight semantic rearrangement and by describing lichen symbioses as "mutual exploitation," the "altruistic" connotations of mutualism can be avoided (see also Alexander, 1971; Meyer, 1966). Indeed, it may well be that the "gain" of the phycobiont is in its "ecological amplitude," the enclosing fungal hyphae protecting the algal cells from damaging environmental factors, such as dessication and excessive insolation (Scott, 1973; Lewis, 1974; Ahmadjian, 1970).

More recently, Law and Lewis (1982) have argued that the apparent lack of specificity in lichen symbionts may itself be evidence of coevolution. They have classified symbionts in mutualistic associations into two groups; the "inhabitant" (in the case of lichens, the phycobiont) and the "exhabitant" (the mycobiont), which encloses it. If the two symbionts are truly dependent on each other, then there will

153

be constraints on the rate of evolution at the interface between the two symbionts. Because the inhabitant is enclosed within the exhabitant, it is effectively buffered from many environmental influences, whereas the exhabitant is exposed to these factors. Consequently, one would expect to see a greater range of morphological and physiological variation between exhabitants than between inhabitants. The rate of phylogenetic change should then be faster in the exhabitant than in the inhabitant. This hypothesis could account for the apparent lack of specificity between many phycobiont and mycobiont taxa. It would also follow that the superficial similarity between many mycobionts and free-living fungi may be the result of parallel evolution rather than taxonomic affinity; this conclusion would lend support to Ahmadjian's suggestion that the *Trebouxia*- and *Pseudotrebouxia*-containing lichens are of monophyletic origin (Ahmadjian, 1970, 1980). Although the hypothesis proposed by Law and Lewis is speculative, it does receive support not only from the lichens but also from a range of other mutualist symbioses. In summary, however, there is little evidence that would unequivocally demonstrate that the algae and fungi involved in lichen symbioses have coevolved a mutualist association.

MYCORRHIZAE

The root systems of many plants have fungal hyphae associated with them. These associations are not primarily parasitic, because if the "host" plants are grown under sterile conditions so that there are no fungi associated with the roots, the plants do not thrive as well as when the fungi are present (see Zak, 1964; Marx, 1972; Cooke, 1977b). Where such a nonparasitic relationship can be demonstrated, the fungus–root associations are known as *mycorrhizae*. Mycorrhizal symbioses appear to have evolved independently in a number of different groups of plants and fungi. Although they may be conveniently classified into two broad categories, based on gross structure and morphology, the groupings probably include different types of interaction. *Ectomycorrhizae* form a sheath of hyphae around the roots of their "host" plants and root hair production is suppressed. Some penetration by hyphae between the epidermal cells of the root may occur. A broad spectrum of interactions between ectomycorrhizae and plants can be discerned. The least specialized are capable of a free-living saprotrophic existence but may form mycorrhizae with suitable hosts. A second group can exist as free-living forms but are normally found in mycorrhizal associations; they tend to have wide host ranges. The vast majority of ectomycorrhizal fungi have been found only in mycorrhizal associations, but within this group of fungi species with both narrow and wide host ranges can be found.

The ectomycorrhizal fungi do not usually produce large quantities

154

of extracellular degradative enzymes, and this implies that they derive most of their carbon requirements from their "host" plants (Cooke, 1977a); however, those species that are capable of an independent saprotrophic existence may not be capable of competing with more specialized saprotrophs and become ecologically constrained to existence as mycorrhizae (Lewis, 1973a). Apart from differences in host range there is little information available on the mechanisms of interaction or of variation in the ability to form mycorrhizal associations. So far no compound has been isolated from roots that could act as a specific recognition signal for a mycorrhizal fungus. Although some root extracts will often elicit a growth response from fungi, no compound has yet been isolated from roots that is specific to the mycorrhizal fungi of a particular host species (Cooke, 1977a).

In *endomycorrhizae*, there is no sheath of fungal hyphae and root hair production is not impaired. These mycorrhizae can be subdivided into two further groups, the vesicular–arbuscular (V-A) and ericaceous–orchidaceous mycorrhizae. V-A mycorrhizae are by far the most common form of mycorrhizal symbiosis and can be found on the majority of plants both in agricultural and natural ecosystems. In V-A mycorrhizae, the hyphae penetrate between the root cells and form clusters of fine branches (arbuscules) and oil–rich vesicles. The arbuscules appear to have a definite life span and then disappear, possibly by digestion by the "host" plant. The fungi involved in these symbioses have not been cultured on artificial media and so they appear to be obligate biotrophs (i.e., parasitic on living host tissue). As the name implies, ericaceous–orchidaceous mycorrhizae are found on ericaceous plants (heathers and heaths) and orchids. In these associations, the fungal hyphae penetrate the host plant walls, where they are digested. Thus, the host plant appears to be exploiting the fungus necrotrophically (i.e., killing the fungus cells and digesting the dead remains) (Cooke, 1977a,b). However, it must be remembered that the division of endomycorrhizae into these two groups is merely one of convenience and makes no assumptions about the form or evolution of the interaction.

In the relationship between the fungi and the orchids, and some other plant species, the ecological roles of the symbionts are reversed in that the fungus acts as a carbon source for a heterotrophic plant. The fungi in orchidaceous mycorrhizae show some ability to break down lignin and cellulose, so they are partially, at least, saprotrophic (Meyer, 1966). In some symbioses with achlorophyllous plants such as *Monotropa*, successful growth of the plant relies on the establishment of a further mycorrhizal association with a second, green, plant (Cooke,

155

1977a,b). Most orchid seedlings lack chlorophyll but derive their carbon by digestion of fungal hyphae that have penetrated their cells. As development proceeds, those orchids that have chlorophyll in the more mature stages become less dependent on the fungus as a carbon source.

Many species of fungi excrete compounds that are toxic to bacteria and other fungi (antibiosis). Although these compounds appear to be defensive, a necrotrophic role cannot be ruled out. It may be that a plant that forms a mycorrhizal association with a fungus that excretes these compounds may be protected from infection by other pathogenic organisms (Marx, 1972; Gadgil and Gadgil, 1971; Zak, 1964; Lewis, 1973b). All in all, virtually nothing is known about the processes that have brought about the evolution of mycorrhizal symbioses. As in the lichens, the physiological and nutritional interactions that have been assumed to underlie the association have proved very difficult to demonstrate.

Is there any evidence that mutualism is the result of coevolution?

Although there is little direct genetic evidence of coadaptation between plants and fungi involved in symbioses that appear to be mutualistic and although there is little evidence of the physiological basis of their association, the facts that the symbioses appear to be ecologically successful and that within the taxa involved in these symbioses the symbionts appear to show a range of ability to form symbioses imply that coevolution could have brought about these associations. The evidence that does exist seems to indicate that the basis of mutualism is reciprocal exploitation rather than the anthropomorphic concept of the symbionts "acting for each other's benefit." On this interpretation, it seems likely that mutualism evolved out of initially parasitic associations.

THE EVOLUTION OF PLANT–FUNGUS SYMBIOSES

Any discussion of the origin and evolution of plant-fungus associations is fraught with difficulties because any account relies heavily on the origin, evolution, and taxonomy of the fungi, and these fields are, perhaps, some of the most contentious areas of biology. As extremes of the opinions offered, I shall summarize two interpretations of the evidence.

1. The ancestors of terrestrial fungi were parasitic on marine algae. As the algae colonized the margins of the seas and terrestrial forms evolved, the parasitic fungi coevolved with them, tracking the radiation of terrestrial plants. Or, alternatively, the ancestors of the fungi colonized the land as chlorophyllous algae, growing as

epiphytes on larger plants; biotrophic parasitism then evolved from these epiphytes and chlorophyll was lost. In both of these hypotheses, parasitism is seen as the primitive condition and necrotrophy and saprotrophy evolving from them (e.g., Cain, 1972; Raper, 1968; Savile, 1968).

2. The ancestors of the terrestrial fungi evolved initially as saprotrophs and parasitism evolved from these free-living forms, tracking the evolution and radiation of the terrestrial plants (e.g., Lewis, 1973a).

The difficulties of working with the fungi and the dearth of fossil evidence make it possible that no definitive account of the evolution of the fungi will ever be obtained; "the best that can be hoped for is a consensus of plausibility which must be tempered by an appreciation of probability" (Raper, 1968). Moreover, given our ignorance of the actual mechanisms involved in plant–fungus symbioses, even an elucidation of evolution within a small group of symbioses seems to lie a long way in the future.

MICROEVOLUTION IN PLANT-FUNGUS SYMBIOSES

General ecological models of host–parasite systems, e.g., the Lotka-Volterra equations, predict that both stable and unstable equilibria can exist in host and parasite population numbers. Where unstable equilibria are produced, stable limit cycles are often produced in which both host and parasite population numbers fluctuate (see Chapter 3 by Roughgarden). However, ecological models assume that there is no genetic variation in either host or parasite population with respect to the interaction between the symbionts.

Evolution in host and parasite populations can be modeled using simple selection models if it is assumed that the other symbiont is an environmental factor and that there is no genetic feedback between the two populations. But for coevolution to occur, genetic feedback must be included in the models. The demonstration of gene-for-gene relationships has on the one hand made genetic modeling of coevolution easier but on the other hand grossly oversimplified the processes involved in host–parasite symbioses in natural ecosystems (see previous sections). The essence of most genetic models of host–parasite systems can be found in an essay by Haldane (1954): "In a natural population of hosts and pathogens composed of many genotypes there will be selection for pathogens adapted to common biochemical types, and against those only adapted to rare ones. The selection among host plants will be in the opposite direction. Thus host and pathogen will constantly alter their prevailing genotype, in so far as it affects the host–parasite relation. . . . This process will favour a diversity in the host and probably in the pathogen also." The models that have so far been developed tend to be based on gene-for-gene systems with alleles in the host conferring resistance or susceptibility to each parasite genotype and those in the parasite conferring increased pathogenicity or reduced pathogenicity to each host genotype (see Chapter 9 by May and Anderson).

A host infected by a parasite may have its fitness reduced by either a reduction in survival or fertility (or fecundity). On the other hand, the fitness of an obligate parasite depends not only on its ability to survive and reproduce but also on the survival of its host.

As in the ecological models, the results of the genetic models indicate that stable equilibria, unstable equilibria, and limit cycles in gene, genotype, or phenotype frequencies can exist depending on the genetic and breeding systems built into the models (e.g., Clarke, 1976; Leonard, 1969, 1977; Leonard and Czochor, 1980; Mode, 1958, 1961; Yu, 1972; Person et al., 1976; Groth and Person, 1977). The results indicate that stable or dynamically stable polymorphism can be maintained by reciprocal frequency-dependent selection as predicted by Haldane. Clarke (1979) has further argued that because successful parasitic infection will depend to a greater or lesser extent on the ability of a parasite to "tune-in" to the metabolism of the host, high levels of biochemical polymorphism may be a consequence of the interaction between hosts and their parasites and thus offer an explanation for the biochemical polymorphism exposed by the technique of electrophoresis over the last two decades (see also Ewens and Feldman, 1976; Ayala, 1976; Koehn and Eanes, 1978).

One further piece of evidence points toward the existence of a dynamic genetic interaction between hosts and parasites. Often when a fungal parasite has its asexual stage on an annual plant, it has a sexual stage either synchronized with maturation of the annual host, as in *Erysiphe graminis*, or on the alternate host, which is often a perennial plant (as in *Puccinia graminis*). Thus, changes in the genetic composition of the host population wrought by the parasitic infection will be counterbalanced to some extent by recombination in the parasite to produce a wider range of variation in the succeeding season and permit the parasite to "track" changes in genetic composition of the host population more easily. The cycling of genes in a population is not the only possible outcome of this type of interaction. If the resistance genes can spread quickly enough so that they approach fixation before any substantial response from the parasite, a second resistance gene may start to spread through the population in response to the erosion of resistance by the parasite. In this case there will be a runaway or cascade process in which the host population accumulates resistance genes as the parasite tracks the former changes in the host population in a manner somewhat analogous to the proposed "arms races" of behavioral

158

ecologists. There is probably a limit to how far this process can proceed, but without building further ad hoc assumptions into the models, for example, a physiological cost for each additional "resistance" gene, there seems to be no stable end point for such a process.

Just as ecological models are simplifications and ignore genetic variation, so genetic models tend to ignore the population dynamics of the host and parasite populations. The dynamics of populations can have substantial effects on the evolutionary outcome of host–parasite interactions. For example, Barrett (1978, 1980) has demonstrated that in an agricultural crop, the ability of a fungal parasite to evolve genotypes capable of attacking a range of different resistance genotypes in the host population depends critically on the rate of migration of fungal spores between different plants in the host population. It is unlikely that the genetic models of host–parasite interactions will be equally applicable to all plant–fungus parasitic symbioses because of differences in their ecology.

Also, because most genetic models of microevolutionary change are based on assumptions derived from the study of agroecosystems, it seems likely that their conclusions may only be applicable to natural ecosystems in a crude qualitative way. The rapid shifts in the genetic composition of host and parasite populations observed in cultivated plants are a consequence of cultivation and will not be typical of natural ecosystems. Even in simple models of coevolution that produce limit cycles with biologically reasonable parameters, the cycling frequency is so slow that it is unlikely that it would be detected experimentally.

One of the problems in examining any parasitic symbiosis is that what tends to be observed is the severity of infection or infestation of the host. This is most clearly seen in the investigation of pathological conditions of humans and of crop animals and plants. As has already been pointed out, the fitness of a parasite is not only a function of its ability to grow and reproduce but also a function of host survival. This effect will limit the level of pathogenicity possible in a parasite because there will be a trade-off between pathogenicity and fitness. Although highly pathogenic genotypes of a parasite may be detectable and isolated, they will fail to spread because "the main problem that a parasitic species has to solve if it is to survive is to manage the transfer of its offspring from one individual host to another" (Burnet and White, 1972). This difference between what one can observe (and measure) and what natural selection is acting on may give the appearance of an apparent evolution of attenuation of pathogenicity. Whether individual selection or group selection arguments are needed to account for the evolution of attenuation of pathogenicity in genetic models will

159

depend on the biology of both the host and the parasite. At the moment there is little direct evidence of the evolution of attenuation in plant–fungus symbioses, as most microevolutionary changes have been described in agroecosystems where the population dynamics and experimental methods would tend to work against its evolution or detection. However, the classic example of the evolution of attenuation, in the myxoma virus–European rabbit (*Oryctolagus cuniculus*) association (Fenner, 1965; Mead-Briggs, 1977), occurred when a novel parasite was introduced, so it is possible that some of the recent examples of devastating disease caused by a novel parasite being introduced into a new habitat, for example, the Jarra dieback epidemic caused by *Phytophthera cinnamomi* in Australia, may offer an opportunity for its observation in a plant–fungus system in the future (Cooke, 1977b). If mutualistic symbioses have evolved from initially parasitic associations, then the evolution of attenuation would offer a reasonable model for the establishment of mutualism.

Because any ecosystem that we can observe today is a survivor of a range of systems that could have occurred, any host–parasite or mutualistic association that we can observe is but a sample of those that could have occurred during evolutionary time. It is only those that have evolved or been *de novo* dynamically balanced that have persisted. Those systems in which the host plants have not been able to evolve in response to the challenge of parasites will have led to the extinction of the host and possibly the parasite also, unless attenuation has evolved. Those hosts that have evolved complete immunity to a parasite will not leave any evidence of the original association. Our understanding of the process of coevolution in plant–fungus symbioses is based on very fragmentary evidence. On the one hand we have superficially stable or quasi-stable symbioses about which we have taxonomic and ecological evidence but little or no evidence of the process of symbiotic interaction. On the other hand we have direct evidence of microcoevolutionary change from grossly disturbed agroecosystems; this evidence may indicate the processes that underlie coevolution but may, to some extent, be an artifact of human activities. The inference that coevolution in plant–fungus symbioses has occurred is a plausible and probably the most parsimonious explanation of their existence, but we must remain eternally vigilant against the dangers of teleological arguments and adaptive storytelling in describing the symbioses that do exist.

CHAPTER 8

EVOLUTIONARY RELATIONSHIPS BETWEEN PARASITIC HELMINTHS AND THEIR HOSTS

John C. Holmes

INTRODUCTION

Price (1977, 1980) has concluded that extensive reciprocal coevolution with the host is the rule for parasites. Most of his evidence comes from studies of the relationships between plants and their pathogens (see Day, 1974) or from plant–insect (see Gilbert and Raven, 1975) or insect-insect (see Price, 1975) systems. Waage (1979) came to similar conclusions in a review of vertebrate–insect systems. In this chapter, I will assess the importance of coevolution to animal–helminth systems, with occasional references to other animal–parasite systems when they demonstrate features less well known, or only suspected, for helminth systems.

Before proceeding, a word about the relative evolutionary capacities of hosts and helminths is in order. The conventional wisdom, as expressed for example by Noble and Noble (1976, p. 502), is that helminths evolve more slowly than their hosts. However, two modes of evolution may be distinguished (M. J. D. White, 1978): the origin of new genetic lineages through splitting of an existing lineage (cladogenesis), and the change within an existing lineage without splitting (anagenesis). The traditional view, based largely on the relative taxonomic differentiation of hosts and helminths, focuses on clado-

161

genesis. There is abundant evidence that many helminths diversify less rapidly than their hosts. For example, marine fishes on the Atlantic and Pacific sides of Panama have almost all diversified into separate species, whereas many of their trematode parasites have not (Manter, 1955). Frogs of the genus *Rana* in North America have diversified far more than their lung trematodes, *Haematoloechus* (M. J. Kennedy, 1980). Similarly, the radiation of species of glypthelminth trematodes and proteocephalid tapeworms parallels that of families, not species, of the host anurans or salamanders (Brooks, 1977, 1978a, respectively).

A very different view is espoused by Price (1977, 1980), who points out that the short generation times, large populations (relative to those of the hosts), and isolated subpopulations of parasites combine to produce very high evolutionary potentials, and that parasites may track evolutionary changes in their hosts or adapt to new conditions very quickly. This view may apply to cladogenesis in some groups of helminths, such as the hymenolepidid tapeworms of waterfowl or trichostrongylid nematodes of ungulates, but seems most applicable to anagenetic evolution. One example is the development of resistance to thiabendazole in *Haemonchus contortus*, the stomach nematode of sheep (LeJambre et al., 1979); after three generations of selection, treatment with therapeutic doses did not significantly reduce worm numbers. Another is the evolution and spread of resistance to chloroquin (and other drugs) in malaria (see Beale, 1980; Ferraroni et al., 1981). These examples indicate the capacity to evolve quickly in response to anagenetic pressures.

This compound view of parasite evolution was recognized earlier by Brooks (1979a) in his differentiation between cospeciation (parallel cladogenesis of hosts and parasites) and coaccommodation (anagenetic adaptation). It must be kept in mind when evaluating host–parasite coevolution. Long evolutionary contact, as indicated by congruent cladistic relationships, is not necessary for effective coevolution. Considerable evidence of anagenetic coevolution may be seen even in such a recent host–parasite system as rabbits and the myxoma virus, both recently introduced into Australia (Fenner and Ratcliffe, 1965).

PARASITE ADAPTATIONS

Most parasitic helminths are obligate parasites—they cannot exist without their hosts. In the simplest life cycles, sexually reproducing adults in the definitive host alternate with free-living egg or larval stages. In the more complex life cycles, additional larval stages (which may reproduce asexually or parthenogenetically) in one or more intermediate hosts may be interposed. Because, in all cases, one or more other organisms (the hosts) constitute an integral part of the hel-

162

minth's environment, evolutionary adaptations to those organisms must be expected. Such adaptations are frequently intricate and highly specific. They may involve the morphology, physiology, or behavior of the parasite.

The surface topography of the mucosa of the spiral intestine of elasmobranchs of the genus *Raja* differs markedly between species (H. H. Williams, 1960, 1966) and from anterior to posterior within the same individual (H. H. Williams, 1961, 1966; Carvajal and Dailey, 1975). Tapeworms of the genus *Echeneibothrium* are each almost entirely specific to individual species of *Raja*. The tapeworms are distinguished largely on the basis of the morphology of the scolex (H. H. Williams, 1966), the details of which are closely correlated with the morphology of the mucosa of a particular region of the intestine of their specific host (H. H. Williams, 1960, 1966; Carvajal and Dailey, 1975). Such close morphological adaptations, correlated with narrow host specificity, appear to be good evidence for a long history of coevolution between host and parasite.

Many parasites respond to host hormone levels, usually to regulate the timing of their life cycles. Although examples are known from helminths (e.g., regulation of reproduction in polystomatid monogeneans; Bychowsky, 1957; Stunkard, 1959), the most clearly defined example is the regulation of maturation, copulation, and oviposition of the rabbit flea, *Spilopsylla cuniculi*, in response to hormone levels in the host, *Oryctolagus cuniculus* (summarized by Rothschild and Ford, 1973). Fleas moving randomly between rabbits concentrate on pregnant does in response to moderate levels of estrogens and corticosteroids. The ovaries of female fleas mature in response to the high levels of these hormones in does during the last 10 days of pregnancy. At parturition, most of the fleas transfer to the newborn nestlings, stimulated, at least in part, by the major changes in hormone levels in the doe. Fleas on the nestlings, or in the nest, but not those remaining on the doe, mate and produce eggs, stimulated, at least in part, by airborne kairomones emitted by the nestlings or in their urine. (The system is actually more complex; see Rothschild and Ford, 1973; Rothschild, 1975.) This flea, and others showing similar responses to changes in host hormone levels (Rothschild and Ford, 1972, 1973; Rothschild, 1975) seem "hormone bound" to their hosts. The hormones to which the fleas respond are not host specific, but the kairomones, and the ecological context in which the controls are provided, may be.

Helminths can be bound to their hosts by specific behavioral responses as well, as exemplified by the responses of the monogenean,

Entobdella soleae, to exudates from the skin of the common sole, *Solea solea* (reviewed by Kearn, 1971, 1976). The host is nocturnally active and spends the day immobile and partly buried in sand. Most eggs of *E. soleae* hatch during the first 4 hours of daylight, the beginning of the host's inactive period. Hatching, at any time of the day, is stimulated by urea or ammonia in the mucus of *S. solea* or other fishes. The newly hatched larvae are specifically attracted to diffusible compounds in sole mucus, and they selectively attach to exposed skin of the sole, avoiding the skin of other flatfishes. If transferred to other flatfishes, they cannot survive. Other parasites that actively select their hosts show similar suites of adaptive behavior patterns. MacInnis (1976) outlines the strategies involved.

That the tactics involved in those strategies are under precise selective pressure is indicated by the work of Shiff (1974). The phototactic responses of miracidia of *Schistosoma haematobium* were dependent on temperature in such a way as to bring the miracidia to the habitat occupied by the host snails at different times of the year. The miracidia showed responses to temperature that were identical to those of the snails.

Most filarid nematodes are transmitted from one definitive host to another by a blood-feeding vector, such as a biting fly, which ingests microfilariae with its meal of host blood. The microfilariae develop in the fly's tissues and are transmitted to a new host at a later blood meal. The microfilariae show a variety of behavioral traits specific to the vector. Microfilariae that live in the blood frequently show marked diel cycles of abundance in peripheral blood. These cycles, as a result of active migration patterns of the microfilariae (see review by Hawking, 1975), are directly correlated with the biting habits of the vectors. For example, over most of its range, the vectors of *Wuchereria bancrofti* are night-biting mosquitos; in these regions, numbers of microfilariae in peripheral blood peak during the night and are reduced to virtually indetectable numbers during the day. In some areas, particularly in the South Pacific, the vectors are day-biting mosquitos; in these regions, the microfilariae show an obvious, but less exclusive, peak during the day. Hawking (1975) reviews many other examples of both diel and seasonal cycles, both tied to vector activity.

Microfilariae that inhabit tissue fluids rather than blood often show preferential concentration in regions that are the preferred feeding sites of their vectors (Eichler, 1971; Mellor, 1974). Each of several species found in the same host species (or individual) is concentrated in a different region, where they are usually fed on by different vectors (Mellor, 1974; Schulz-Key and Wenk, 1981). The microfilariae may have to undertake a long migration to reach their preferred site, as in *Onchocerca tarsicola* in red deer (*Cervus elaphus*), which are produced by adults around the joints of the lower legs and migrate to the skin of the ears (Schulz-Key and Wenk, 1981). Microfilariae also are attracted to the feeding vectors, an attraction that may be specific to the strain level in both parasite and vector (DeLeon and Duke, 1966; Duke et al., 1967).

In an earlier review (Holmes, 1976), I pointed out that the strategies adopted by these filarids to reach their vectors and those adopted by other parasites to reach their intermediate hosts are similar to those adopted by plants that depend on animals for seed dispersal. However, the filarids depend on the host to provide nutritional rewards and attractants for the vector, so that opportunities for reciprocal coevolution of parasite and vector seem limited. Other helminths may provide the nutritional reward and attractants, either directly, as in the glycogen- and fat-rich proglottids of tapeworms (Bartel, 1965) or the dried body of a female nematode (Lee, 1957), or indirectly, as in the host-produced slime balls stimulated by dicrocoelid trematodes (Krull and Mapes, 1952). All are avidly collected by foraging ants and fed to their larvae. In such systems, opportunities for reciprocal coevolution seem greater than for the vector-borne helminths.

Strategies for infecting definitive hosts are adapted to general features of predator–prey interactions (see reviews by Holmes and Bethel, 1972; Holmes, 1976). Adaptations include the pulsation of the colorfully banded sporocysts of the trematode *Leucochloridium* in the tentacles of their snail hosts, which make the snails very conspicuous and "lure" ingestion by the avian definitive hosts (Ulmer, 1971). The behavior of infected intermediate hosts may also be altered so as to make them more susceptible to predation by the definitive host (Bethel and Holmes, 1973, 1977). Most of these adaptations appear to be relatively nonspecific, increasing chances of predation by a variety of predators. Some, however, are more specific, in that they increase the overlap between the infected intermediate host and the feeding niche of the definitive host (see Holmes and Bethel, 1972, for examples).

The adaptations outlined above, and myriad others to be found in any book on parasites, are all specific adaptations to characteristics of their hosts. Such adaptations are obvious in helminths and are obviously dependent on the host's constituting a major part of the environment of the parasites, but only rarely appear to exemplify reciprocal coevolution.

HOST ADAPTATIONS

Where helminths show such specific adaptive traits, one might expect hosts to show specific counteradaptations. An example is the use of water by the Hadza tribe of Tanzania: they rarely wash or play in schistosome-infected waters and instead collect drinking water from holes dug in dry river beds. As a consequence, they are practically free from the schistosomiasis common in neighboring tribes (Bennett et al., 1970). Altmann and Altmann (1970, p. 159) suggest that "avoidance of immersion, the digging of drinking holes . . . and even the minimal contact between lips and water surfaces while drinking" in baboons are similar adaptations to minimize schistosome infections.

Other behavioral adaptations appear to be more general. Waage (1979) discusses several such adaptations to ectoparasitic insects, including grooming, emigration from areas of intense attack, desertion of heavily infested nesting areas, and aggregation to minimize attacks. Special areas for defecation, especially for burrow-inhabiting rodents; the avoidance of fecal deposits by grazing ungulates; wandering movement patterns of social animals, the behavioral equivalent of pasture rotation (Freeland, 1976, 1980); and even the size of social groups (Freeland, 1979a) may be similar adaptations to minimize helminth infections.

Two major classes of adaptation to parasites, genetic adaptations and immune responses, deserve special attention. Although recent studies on the genetics of the immune system (see Nobel lectures by Benacerraf, 1981; Dausset, 1981) indicate the interconnection of the two, they will be treated separately.

Genetic factors

Intraspecific variation in susceptibility to helminths has been recognized for a long time, both in vertebrate and in invertebrate hosts. In many cases, this variation has been shown to be due to physiological differences due to age, sex, or diet of the host animal, but in most cases, it is genetically based (see review by Wakelin, 1978a).

A very important concept has emerged from the long history of breeding crop plants for resistance to plant pathogens—the gene-for-gene concept of reciprocal coevolution of plants and their pathogens. The concept is discussed more fully by Barrett in Chapter 9. In systems exhibiting the gene-for-gene relationship, hosts with specific resistance genes can be invaded only by parasites with counteracting virulence genes; host genes are predictably codominant, linked, or allelic and tend to be manifested as polymorphisms; and parasite genes are predictably recessive and nonallelic (Person, 1967; Ellingboe, 1978).

Barrett has pointed out that the simple system in crop plants may be an artifact of strong artificial selection; more complex systems are to be expected under more natural conditions. However, the kinds of selection pressures envisaged should make two features widespread in coevolved host–parasite systems: polymorphic resistance factors in

166

host populations and infection that is dependent on compatible phenotypes of host and parasite.

The best example of a host genetic polymorphism that conveys resistance to a parasite is the malaria–hemoglobin S (sickle-cell hemoglobin) system in humans. A single amino acid difference (a negatively charged glutamic acid in normal hemoglobin is replaced by an uncharged valine) allows deoxygenated hemoglobin S to polymerize into tight helices, which distort (sickle) the red blood cell. In this case, a gene, normally lethal in the homozygous state, can be advantageous in the heterozygote by reducing in severity the malaria caused by *Plasmodium falciparum*. Asexually reproducing stages of the parasite produce a specific receptor (on knobs on the infected red blood cell) that specifically bind to host endothelial cells (Udeinya et al., 1981), trapping the infected blood cells in capillary beds, where oxygen levels are relatively low. Metabolism by the parasite lowers the pH of the infected cells; in heterozygotes, under these conditions approximately 40% of the cells sickle, a process that kills the enclosed parasites (see brief summaries by Maugh, 1981; Allison, 1982).

Several other polymorphisms are known to affect resistance to various species of malaria (reviewed by Motulsky, 1975; Allison, 1982) and to *Schistosoma mansoni* (Constant-Desportes et al., 1976; Trangle et al., 1979) or other helminths (reviewed in Wakelin, 1978a). These and other examples led Clarke (1976) to suggest that selective pressures exerted by various kinds of parasites may be responsible for the genetic variation in proteins so common in a wide variety of organisms. Such polymorphisms not only would protect some host individuals but also would affect parasite population dynamics by effectively lowering the susceptible host population size and by providing a "sink" for loss of any parasites that attempt to infect protected individuals.

Only hemoglobin S appears to be maintained by heterozygote superiority; for other polymorphisms, frequency-dependent selection seems likely (see discussion by Hamilton, 1982). The advantages of recombination for providing temporal variation in such systems has been suggested as a major factor in the evolution of sexual reproduction (Bremermann, 1980; Hamilton, 1980, 1982).

The necessity for compatible phenotypes in host and parasite has emerged from a variety of studies. Studies reviewed by Wakelin (1978a) have established the necessary variation in host populations. Variation in the helminths have been shown by electrophoretic studies (e.g., Fletcher et al., 1981), by hybridization studies (e.g., LeJambre, 1981), and by studies on the immunogenicity of different batches of cercariae of *Schistosoma mansoni* (Smith and Clegg, 1979). In addition, passag-

167

ing a helminth through a new host has been shown to alter the infectivity of that helminth for the new host (Laurence and Pester, 1967) or the old one (Saoud, 1965), presumably by selecting for helminth genotypes more compatible with those of the new host. Bayssade-Dufour (1979) has demonstrated that the arrangement of sensory setae on the cercariae differs between strains of *S. mansoni* of human and rodent origin. Laboratory passage of human strains through mice alters the arrangement toward the condition in rodent strains and increases the infectivity to mice. In a study of the immune response of different mouse strains to different isolates of *S. mansoni*, Dean et al. (1981) have demonstrated that the response depends on the genetics of both host and parasite. The best evidence, however, comes from studies of schistosome–snail interactions (reviewed by Wright, 1971; Basch, 1975; Jordan et al., 1980). Three patterns are particularly relevant. (1) Geographical isolates of *S. mansoni*, highly infective to snails from the same locality, are generally less infective to snails from other localities. (2) Laboratory stocks bred for resistance may be infected occasionally, especially when exposed to large numbers of miracidia. (3) Stocks bred for susceptibility may encapsulate an occasional miracidium (a process associated with resistance).

The coevolutionary consequences of genetic variation and the necessity for compatible phenotypes are twofold. First, coevolutionary relationships should be closest at the level of the deme, not the species. Local helminth populations are adapting to the genetic patterns in local host populations as well as to local environmental or ecological conditions. An extension of this concept to patterns of host specificity will be discussed later. Second, where susceptibility in host populations (and presumably, in responses of the helminth populations) is polygenic (the majority of cases, at least in vertebrate hosts; see review by Wakelin, 1978a), differential compatibility will provide one mechanism producing the usual clumped distribution of individual helminths within a host population. Because damage done by helminths is usually a function of their number (Anderson, 1978), the system provides a powerful selective pressure for coevolution.

Immune responses

Both vertebrate and invertebrate animals mount defensive responses against invading organisms, and the mechanisms employed (cellular or humoral) show some similarities. There is a major difference, however, in that invertebrate systems do not appear to be inducible: the host either reacts or it doesn't, regardless of prior exposure (see review by Lackie, 1980). Invertebrate–parasite systems, therefore, appear to be regulated largely by genetic factors. On the other hand, the induction of response is the hallmark of vertebrate immune systems. It is the inducible response that is treated in this section.

168

The immune response to helminths is very complex and still imperfectly understood (see recent reviews by Wakelin, 1978b; Mitchell, 1979a,b). The system is based on two types of lymphocytes (B and T cells) that are specialized to recognize "foreign" (= "non-self") antigens. These cells interact with each other, with other specialized cells, or with specialized serum proteins, such as those of the complement or clotting systems. A simplified diagram, modified from Wakelin (1978a) and including additional mechanisms reviewed by Mitchell (1979a), is shown in Figure 1. B cells produce antibodies, which are released into the blood serum and act alone or in concert with other humoral or cellular elements to attack the helminth, neutralize its enzymes, or affect its environment. T cells are more complicated. They can help or inhibit the response of the B cells, attack helminths directly, or interact with humoral elements (including antibodies) or cellular elements to attack the helminth, encapsulate it, or affect its environment.

Bloom (1979) and Mitchell (1979b) have emphasized that successful parasites must adapt so as to evade the immune system partially but not completely. (This is not necessarily so. I will return to this point later.) Both authors indicate three basic ways a parasite may do so: reduction of antigenicity, modification of macrophage function, and modulation of the immune response. Table 1 enumerates a variety of mechanisms, giving an example for those known or suspected for helminths (see Bloom, 1979; Damian, 1979; Mitchell, 1979b; for additional examples and references).

Each of these mechanisms may be considered to represent reciprocal coevolution because each is a response (or series of responses) by the parasite to a specific trait in the host that has evolved, at least in part, as a response to parasites. Each can be very complex in detail (see Mitchell, 1979b), several may be involved in any one host–parasite system, and various mechanisms may interact. The net result of any one specific mechanism may be to protect the host, to protect the parasite, or to produce immunopathology, or it may even be irrelevant (Mitchell, 1979a). The opportunities for coevolution are obviously extensive, but the extent to which the opportunities are realized has been largely unexplored. However, Mitchell (1979b) has emphasized that the more severe disease produced by "natural" mouse parasites in nude mice (congenitally hypothymic, hence deficient in T cells), plus the amelioration of those diseases by injection of T cells, suggest that T cell-dependent mechanisms have been important in the development of balanced host–parasite relationships. (Other parasites, found in mice only in the laboratory, do not necessarily show these patterns.)

169

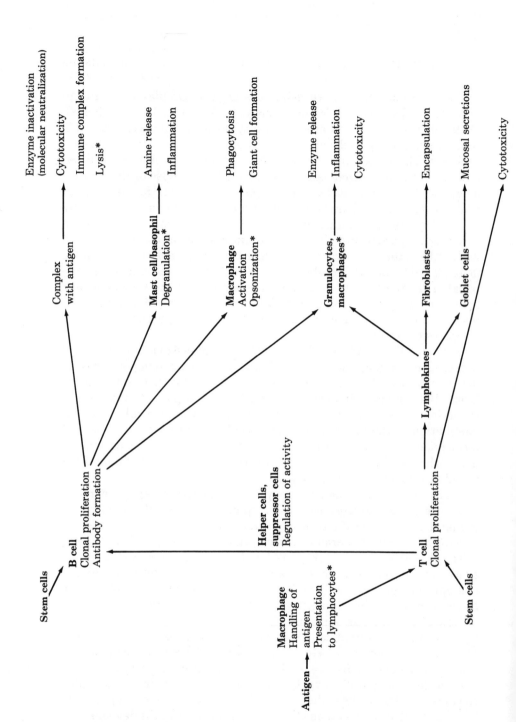

FIGURE 1. Simplified diagram of known or postulated antihelminth immune responses (modified from Wakelin, 1978a). Asterisks (*) indicate systems dependent on complement.

170

TABLE 1. Mechanisms of evasion of host responses by parasites (modified from Mitchell, 1979b), with examples of those known or postulated for helminths.

Mechanisms	Parasite[a]	Host	†
REDUCED ANTIGENICITY			
Anatomical inaccessibility	*Trichinella spiralis**	Mammals	(1)
Antigenic variation	*Fasciola hepatica*	Sheep	(2)
Modulation of antigens	*Nippostrongylus brasiliensis*	Rat	(3)
Molecular mimicry	*Schistosoma mansoni*	Mouse	(4)
Blocking antibodies	*Taenia taeniaeformis**	Mouse	(5)
Non-Ig masking of antigen	*Schistosoma mansoni*	Mouse	(6)
Loss of MHC antigens on parasitized cell			
MODIFICATION OF INTRAMACROPHAGE ENVIRONMENT			
MODULATION OF HOST IMMUNE RESPONSE			
Immunosuppression			
Lymphoid tissue disruption	*Mesocestoides corti**	Mouse	(7)
Clonal deletion	*Echinococcus granulosus**	Mouse	(8)
Mitogens, antigenic competition, and increased antibody turnover	*Nippostrongylus brasiliensis*	Rat	(9)
	Heligmosomoides polygyrus	Mouse	(10)
Nonspecific suppressor cells	*Trichinella spiralis*	Mouse	(11)
Specific active suppression	*Schistosoma mansoni*	Human	(12)
Cytotoxic parasite molecules	*Echinococcus granulosus**	Cattle	(13)
Effector cell blockade	*Schistosoma mansoni*	Rat	(14)
Degradation of antibodies	*Schistosoma mansoni*	Rat	(15)
Anticomplementary effects	*Taenia taeniaeformis**	Rat	(16)
Antiinflammatory effects	*Trichinella spiralis*	Rat	(17)

[a]Asterisks indicate larval stages.
†References: 1. Mitchell (1979b); 2. Hanna (1980); 3. Ogilvie (1974); 4. Damian (1979); 5. Richard (1974); 6. Smith and Kusel (1974); 7. Mitchell and Handman (1978); 8. Ali-Khan (1978); 9. Jarrett and Bazin (1974); 10. Brown et al. (1976); 11. Barriga (1980); 12. Ottesen and Poindexter (1980); 13. Annen et al. (1981); 14. Mazingue et al. (1980); 15. Auriault et al. (1980); 16. Hammerberg et al. (1980); 17. Castro et al. (1980).

The central role of helper and inhibitor T cells in immune regulation should be noted.

One mechanism for which coevolution has been explicitly assumed is molecular mimicry, a term applied to those systems in which antigenic determinants produced by parasites have been selected to resemble host antigenic determinants, reducing the host's ability to respond to them. Damian (1979) provides an extensive discussion of the history of the concept, its relationship with other mechanisms, such as antigen masking or molecular camouflage, and its extension to selection of other parasite-produced compounds that mimic host products such as hormones. Two features reviewed by Damian (references there) are particularly relevant in the present context. Both show that helminths may produce antigens that mimic specific host proteins (some of which regulate the immune system), thus potentially modifying the host's abilities to respond to other antigens. First, schistosomes maintained in a strict baboon–snail cycle (with no exposure to mice for many generations) were shown to possess an antigen similar to mouse α_2-macroglobulin. Both the mouse's "model" protein and the parasite's mimic affect T-cell regulation of B-cell production, that is, are immunoregulatory. Second, in another system, mouse–*Nematospiroides dubius*, the nematode was shown to share antigens associated with the major histocompatibility complex (MHC) of a susceptible strain of mice. (Wassom et al., 1979, showed that MHC genes also affect susceptibility to another nematode, *Trichinella spiralis*.) The MHC is that portion of the genome that regulates cellular interactions, including self-recognition and cell-mediated immune responses (see readable reviews by Benacerraf, 1981; Dausset, 1981). Damian goes on to review speculation that mimicry of MHC antigens by parasites has been responsible for selection for diversity in MHC genes, hence for diversity in immune capabilities. He concludes that the hypothesis cannot be ruled out.

It is still unclear whether the immune system originated in response to invaders or to aberrant host cells (i.e., cancer). However, it is clear that invaders, including helminths, have responded by evolving means of evading the immune system. It is highly probable that the complexity of the immune system, including its effectors, has arisen to counteract those evasive mechanisms. These interactive features and those of the genetic systems covered in the preceding section provide the strongest available evidence for reciprocal coevolution of vertebrates and their helminth parasites.

HOST SPECIFICITY

Helminths vary greatly in the extent to which they are host specific. A recent checklist of the parasites of the fishes of Canada (Margolis and Arthur, 1979) provides data to illustrate the variation (Table 2).

Specialists might be expected to be found in a small number of species in a single family of hosts, generalists in a wider variety of hosts. From the data, it is clear that most of the Monogenea are specialists, most Acanthocephala are generalists, and the other groups include substantial numbers of each.

TABLE 2. Host specificity of helminths of Canadian fishes (data from Margolis and Arthur, 1979). Values are number of parasite species with that host distribution (* = not counting species reported only once).

Number of families infected	NUMBER OF HOST SPECIES INFECTED					
	1	2	3-4	5-8	9+	Percentage*
Monogenea						
1	109(35)*	30	14	5	1	92
2		4	1	1	1	8
Percentage*	38	37	16	7	2	
Digenea						
1	72(29)	15	6	3	4	40
2		15	12	8	1	25
3-4			6	13	1	14
5+				6	22	20
Percentage*	20	21	17	22	20	
Cestoda						
1	41(9)	11	9	3		46
2		12	3	1	1	24
3-4			2	3	6	16
5+				4	6	14
Percentage*	13	33	20	16	19	
Acanthocephala						
1	6(0)	5				21
2		1	2			13
3-4			1	6		29
5+				1	8	38
Percentage*	0	25	13	29	33	
Nematoda						
1	23(11)	4	1	1	1	39
2		2	2	2		13
3-4			6	2	1	20
5+				4	9	28
Percentage*	24	13	20	20	24	

In addition, some groups, such as the Monogenea (Llewellyn, 1957; Rhode, 1978b) or the tapeworms of elasmobranchs (Euzet, 1957; H. H. Williams, 1966), show narrow host ranges wherever they are found. Others, such as the trematodes of marine fishes (Rohde, 1978b), show latitudinal patterns, with the degree of host specificity increasing toward the tropics.

The preceding analyses were based entirely on the number of host species from which the parasite has been recorded, an imperfect representation of the actual specificity of the parasite (Holmes, 1979; Rohde, 1980). A more adequate measure of specificity should take into account the frequency of infection in different species of host, the numbers of parasites found in each (as in Rohde, 1980), the reproductive rates of the parasites in each, and the relative abundance of each of the host species (Holmes et al., 1977). An analysis of this kind is essential to determine the relative selective forces different host species exert on a given parasite. A host species that is rarely infected, that is infected with few individuals, or in which the parasite rarely matures would not be important in selection for traits of that parasite. Coevolution with one species of host could be extensive even in the face of an extensive host range, if the majority of the reproducing parasites are found in that host species.

Hairston's (1962) analysis of *Schistosoma japonicum* in the Philippines demonstrates such a situation. There were four definitive hosts present: humans, dogs, pigs, and native rats. According to Hairston's calculations, the large population of rats present, and the high rate of exposure to cercariae, made them the only required definitive host, despite a low prevalence of infection with adult worms and a low egg production per female worm (both due to the short life span of the rats). The rats, therefore, exerted the greatest evolutionary pressure on the worms. Bayssade-Dufour (1979) was able to distinguish between human and rodent strains of *S. mansoni* by the sensory structure of cercariae (see preceding section); his data on strains of *S. mansoni* in Guadeloupe suggest a similar situation there.

Earlier in this chapter, I pointed out that local populations of parasites adapt to local genetic patterns in their hosts. Local populations may also adapt to very different host complexes. For example, Leong (in Holmes et al., 1977) showed that the acanthocephalan *Metechinorhynchus salmonis* in Cold Lake, Alberta, was found, often in large numbers, in every species of fish in the lake. However, the helminth matured essentially only in salmonid fishes, which harbored an estimated 99% of the gravid females. Two species, *Coregonus clupeaformis* (52% of the gravid worms) and *C. artedii* (37%), were important hosts (Table 3). In Lake Michigan, however, Amin and Burrows (1977) found gravid worms in 13 of 14 species of fishes examined, including six species of nonsalmonids (Table 3). Assuming that the

174

TABLE 3. Host specificity of *Metechinorhynchus salmonis* in two lake systems.

Host	COLD LAKE[a]			LAKE MICHIGAN[b]		
	n^c	Abund[d]	% of ♀[e]	n^c	Abund[d]	% of ♀[e]
Clupeidae						
Alosa pseudoharengus				149	0.2	0.1
Salmonidae						
Coregonus artedii	757	2.8	37.0			
C. clupeaformis	836	169.2	51.8	8	22.0	1.8
C. hoyi				79	12.5	16.4
Oncorhynchus kisutch	291	189.5	1.8	15	140.3	30.2
O. tshawytscha				6	53.3	6.2
Salmo gairdneri				8	38.1	4.3
S. trutta				16	16.4	3.8
Salvelinus namaycush	35	421.0	8.4	7	81.9	10.9
Osmeridae						
Osmerus mordax				446	3.1	21.7
Esocidae						
Esox lucius	62	25.6	0.2			
Cyprinidae						
Notropis hudsonius				15	0.0	0.0
Catostomidae						
Catostomus catostomus	12	25.7	0.0			
C. commersoni	36	0.3	0.0	24	0.9	0.1
Gadidae						
Lota lota	29	174.0	0.9	1	18.0	0.3
Gasterosteidae						
Pungitius pungitius	1038	0.3	0.0			
Percidae						
Perca flavescens				11	0.8	0.1
Stizostedion vitreum	12	2.0	0.0			
Cottidae						
Cottus cognatus				37	6.8	4.1

[a]Data from Leong and Holmes (1981).
[b]Data from Amin and Burrows (1977).
[c]Number of fish examined.
[d]Number of worms per fish examined.
[e]Percentage of gravid females in that species of host; for method of calculation, see Holmes et al. (1977).

relative numbers of fishes examined roughly approximate their relative abundance in the lake, calculations suggest a more even distribution of gravid females, with *Oncorhynchus kisutch* having the most (30%) and a smelt, *Osmerus mordax*, the second largest proportion (22%). Clearly, *M. salmonis* is acting as a relative specialist in Cold Lake, but as a generalist in Lake Michigan. These different patterns of host specificity in the two systems are likely to produce vastly different coevolutionary pressures.

How do such differences in host specificity arise? Continued association with a given host species is likely to select for those parasite genotypes that are compatible with that host. Also, there may be selection against genotypes compatible with other hosts, if there is any complementarity of genotypes (as suggested by the work of Bayssade-Dufour, 1979) or if there is a cost in maintaining adaptations to other hosts. Increases in the relative population of that host would be likely to increase these selective pressures. Conversely, any factor, such as overfishing or the successful introduction of a new fish (e.g., *O. kisutch*), that reduces the relative population size of a major host would be likely to reduce them. Alteration of host community structure, therefore, would be expected to affect the coevolutionary selection patterns of any parasite having substantial reproducing populations in more than one host species.

As indicated earlier, some of the adaptations of helminths involve mechanisms for increasing the efficiency of infecting their hosts. Such adaptations may also result in exposure of the helminth to new potential hosts. Selection for any genotypes capable of reproducing in the new host will result in a "host capture" (Chabaud, 1959) and an expansion of the helminth's host range. For example, gammarids infected with the cystacanths of *Polymorphus paradoxus* are attracted to light and, when disturbed, tend to cling to objects at the water surface, a behavior that makes them more susceptible to predation by mallards (Bethel and Holmes, 1977). The same behavior also increases their exposure to muskrats and beaver, allowing *P. paradoxus* to be the only species in the genus to be found regularly in mammals.

Examples like the last two emphasize that coevolution may lead to decreased as well as increased host specificity.

MODELS OF HOST–PARASITE COEVOLUTION

There are three contrasting models of the way hosts and parasites may coevolve. In one (the mutual aggression model), the two are regarded as taking part in an evolutionary arms race, with each species continually evolving in an aggressive manner toward the other. Selection in the parasite is always for greater exploitation of the host and, in the host, always for more efficient exclusion of the parasite. The gene-for-

176

gene hypothesis assumes this type of coevolution, as does Van Valen's (1973) Red Queen hypothesis. In the second (the prudent parasite model), selection in the parasite is for characteristics that limit the damage done to its host; this effect allows a longer life span for the host and for the parasite. As a consequence, the selection pressure the parasite exerts on the host and, therefore, on the host's response (assuming some cost), should be reduced. This is the traditional view, envisaging a deescalating aggressive system that evolves from a relatively pathogenic to a relatively benign one. The third model (incipient mutualism), suggested by Lincicome (1971), Davies et al. (1980), and Wilson (1980), regards coevolution as actively cooperative, with both host and parasite evolving mutualistic attributes, so as to promote the continued presence of the other.

Table 4 summarizes the major selective pressures on parasites and hosts for each model. The primary differences between the models appear to be due to the actions of the parasites, especially the extent to which they exploit the hosts; in each case, the host responds appropriately to the selective pressures provided by the parasite. The major dichotomy between the mutual aggression model, in which the parasite emphasizes exploitation of the host, and the other two, in which the parasites emphasize coexistence with the hosts, implies a trade-off between degree of exploitation and length of life. Heavy ex-

TABLE 4. Selection pressures under differing models of host–parasite coevolution. Levels in parentheses indicate strength of the selection pressure.

Characteristic subjected to selective pressure	MODEL		
	Mutual aggression	Prudent parasite	Incipient mutualism
Parasite characteristics			
Infectivity	Maximize	Moderate	—
Exploitation of host	Maximize	Moderate	Minimize
Benefits to host	—	—	Maximize
Host characteristics			
Susceptibility	Minimize (high)	Minimize (low)	Maximize
Defense against effects of parasite	Maximize (high)	Maximize (low)	Minimize

ploitation of the host may enable the parasite to produce more off-spring per unit time, but also increases the parasite's pathogenicity, reducing the host's (and the parasite's) life span (for examples, see Ractliffe et al., 1969; Chapter 9 by May and Anderson). Because the net reproductive rate of the parasite is a function of both its reproduction per unit time and its length of life, under what circumstances should a parasite be selected to emphasize the former?

One factor, suggested by most models of epidemic disease agents, is the presence of a large population of susceptible hosts, favoring transmission and an expanding population. Under such circumstances, early and rapid production of propagules is more valuable than production of propagules over a longer period of time (Lewontin, 1965; Mertz, 1971). This factor is associated with parasites producing epidemic (or epizootic) outbreaks and is more likely to be important for microparasites (those characterized by small size, short generation times, and direct reproduction within their host, i.e., viruses, bacteria, and protozoans; Anderson and May, 1979) than for helminths. However, some monogeneans that share some of these characteristics with microparasites and that can also cause epizootics (e.g., *Dactylogyrus vastator*; Paperna, 1964) may be influenced by this factor.

Another set of circumstances is obviously that in which the death of the host does not mean death of the parasite but is necessary for transmission of the parasite to the next host. The eggs of a few helminths (e.g., the liver nematode *Capillaria hepatica*; Freeman and Wright, 1960) are released in this way, but the main examples are those larval stages that are transferred to definitive hosts by ingestion in an intermediate host. Such larval stages frequently "regulate" the intermediate host so as to predispose it to predation (see section on parasite adaptations).

A major factor is likely to be a host with a short life span and rapid population turnover. A parasite of such a host would have to be an "r" strategist, emphasizing reproduction instead of longevity. Immediate and extensive exploitation would be more advantageous than lesser exploitation over a somewhat longer period of time. The castration of invertebrate (and fish) hosts (Baudoin, 1975), the extensive regulation of host functions by parasitoids (Vinson and Iwantsch, 1980; Stamp, 1981), and the modification of intermediate host behavior mentioned earlier are all examples of such exploitation.

A few host–helminth systems show a pattern of single-factor, dominant genetic resistance, of the type predicted for the mutually aggressive gene-for-gene system. One example is the resistance of the mosquito *Aedes aegypti* to various larval filarid nematodes. Susceptibility to groups of filarids is genetic, based on allelic or linked recessive traits (Macdonald and Ramachandran, 1965; Zielke, 1973; Terwedow and Craig, 1977). Paige and Craig (1975) have demonstrated an interesting pattern of susceptibility of East African strains of *A. aegypti* to *Brugia pahangi* (and presumably to the human filarid, *Wuchereria bancrofti*, which is in the same susceptibility group). "Domestic" strains, taken from inside houses or from domestic water supplies, were refractory; "sylvan" strains, taken from areas remote from human habitations, were at least partially susceptible, and "peri-domestic" strains, taken near but not immediately around habitations, were intermediate. Paige and Craig interpreted this pattern as indicating a strong selection against susceptibility to human filarids, which develop in the flight muscles and impair ability to fly. Van Valen (1973) has suggested that coevolution of predator-prey systems via the arms race route is inherently unstable, and its breakdown may cause extinction. If the premise that a simple genetic resistance is indicative of a mutual aggression type of coevolution is correct, then Paige and Craig's observations suggest that this type of host–parasite coevolution can also result in extinction, at least locally.

Nelson (1979) has indicated that coevolution of plants and fungi involve not only the responses of each partner to the other, but also the responses of one or both to other conditions. He regards the latter type of response to be important in stabilizing host–parasite systems. The importance of similar responses to stabilizing an exploitative host-helminth system is suggested by the work of Tokeson and Holmes (1982). *Gammarus lacustris* parasitized by the cystacanths of the acanthocephalan *Polymorphus marilis* are castrated and have higher over-winter mortality than uninfected gammarids or those harboring small early larval stages. Young gammarids are infected in late summer; larval worms are arrested by low or decreasing temperatures in early fall; in the spring, when water temperatures reach 10°C, the worms resume their development and castrate their hosts. However, the gammarids, which are univoltine annuals, mature during the winter, then breed and oviposit at spring breakup, at water temperatures below 10°C. As a result of the different responses to temperature, the gammarids bring off one brood before being sterilized. Because up to 60% of the adult gammarid population may be infected during midsummer, such a response is very important in maintaining both host and helminth populations.

The traditional view of host–parasite coevolution extends such indirect adaptations to the direct interaction between the two. Given frequent infections, parasites would be expected to moderate their exploitation of the host, allowing their hosts (and themselves) to live longer. The moderated exploitation of the host may be selected for by two methods. The better established method is a direct selection

179

against more pathogenic lines of the parasite, which would kill more hosts (and themselves) earlier in the infection, reducing the number of their offspring. This method emphasizes the time available for reproduction (or transmission). The Australian rabbit–myxoma virus example cited in Chapter 1 is a good example. (It also demonstrates that such selection may lead, not to a nonpathogenic state, but to an intermediate level of pathogenicity; see Levin and Pimentel, 1981; Chapter 9 by May and Anderson.) Another example is the inadvertent selection in the laboratory for reduced infectivity of larval stages of the tapeworm *Hymenolepis citelli* to laboratory cultures of the beetle *Tribolium confusum* (Schom et al., 1981). In the latter system, mortality of the beetle was due to multiple infections; most occurred before the cysticercoids became infective to the final host. Over some 16 generations in the laboratory, the helminths actually became more pathogenic, but less infective, to the beetles. The net result was lower mortality in the beetles. Because the progeny of the successful helminths, but not those of infected beetles, were used to establish successive cultures, all evolution must have been in the helminths. Levin and Pimentel (1981) used the rabbit–myxoma system as an example of interdemic (group) selection; the beetle–tapeworm system appears to be a second example. Where parasites occur singly (or as a single clone; see Bremermann, 1980) in their hosts, individual selection could accomplish the same result.

A second method has been suggested in the literature. Sprent (1962) suggested that parasites could adapt by reducing the harm done to the host, so that selection on the host would not increase, and might even decrease, the host's response to the parasite. [In subsequent papers, for example, Sprent (1969), he switched his attention away from this "moderation" response toward mechanisms whereby parasites might avoid the immune system. The latter is more properly considered an aggressive response of the parasite. The "moderation" response does not seem to have attracted much further consideration; but see Randolph (1979).] This method assumes that overcoming host responses costs the parasite something, and it emphasizes the lower costs (to the parasite) associated with deescalation of aggressive interactions, a result that enables greater reproductive efficiency in the parasites. If the host responded directly, and individually, to the stimuli provided by the individual parasite, this "adaptation tolerance" could be selected for directly. An alternative method is suggested by the trait group selection model of Wilson (1980): infrapopulations (parasites in a single host individual) that contain relatively nonpathogenic parasites would stimulate less host resistance, hence may be able to produce more offspring per unit time.

I know of no clear-cut example of the second method for any host–helminth system. Two suggestive examples are the relatively

180

nonpathogenic nature of the filarid nematode *Elaeophora schneideri* and the meningeal worm *Parelaphostrongylus tenuis* in their normal hosts and their severe pathogenesis in abnormal hosts. Developing *E. schneideri* occur primarily in leptomeningeal arteries; in mule deer (the normal host) they migrate out to the common carotid arteries before maturing, but in elk and sheep they remain in the leptomeningeal arteries and block the smaller cranial arteries, producing extensive, frequently fatal, damage (Hibler and Adcock, 1971). Similarly, a difference in pathway and speed of migration within the spinal cord is associated with the neurological disease produced by *P. tenuis* in moose, caribou, or other ungulates, but not in their normal host, white-tailed deer (Anderson, 1972). These differences in migration do not appear to be due to host defenses.

One way for the parasite to reduce the net harm done to a host is by providing some off-setting benefit—the "incipient mutualism" model of coevolution. Lincicome (1971), in a provocative, but poorly known, paper entitled "The goodness of parasitism," reviewed three host–parasite systems in which parasites produce micronutrients important to the hosts under conditions of dietary nutritional imbalance. In at least one of the systems (rat–*Trypanosoma lewisi*) the host was shown to produce a product stimulating growth of the parasite.

Similarly, Smith (1968) has demonstrated that under certain conditions the presence of a cowbird brood parasite can be advantageous, because of the tendency of the nestling cowbirds to feed on ectoparasitic fly larvae, the chief mortality factor for nestlings of the icterid host species. N. G. Smith (1979) also showed that there were two types of host colonies: one that sought out and nested in association with colonies of wasps or stingless bees (which provide some protection against the flies), the other independent of the hymenopterans. The former did not tolerate cowbirds or their eggs; the latter did.

Barbehenn (1969) has suggested that parasites may also be of benefit to their hosts as "weapons of competition." He hypothesized that the outcome of competition between two species could be determined by a parasite well adapted to (coevolved with?) one species but pathogenic to the other. In such a case, selection would be for the host–parasite combination as a unit, presumably also leading to selection for adaptations in the host that encouraged the continued presence of the parasite. Barbehenn gave, as an example, the dramatic dieoffs of moose in areas of overlap with white-tailed deer, dieoffs attributable to infections with the meningeal worm, *P. tenuis*, maintained by the deer. Further discussion of this example, and others, can be found in Holmes (1979).

181

Finally, an existing complex of well-adapted parasites may benefit the host by protecting it from mortality due to invasion by less well adapted parasites or to high populations of a single species of parasite. Such protection is well known for bacterial assemblages of vertebrates (see Silva et al., 1981; Weinack et al., 1981, for examples), as is the severe pathogenicity that can result from overgrowth of one of the normal bacterial species (see Hungate, 1975). The nutritional interactions of *Ascaris lumbricoides* and *Plasmodium falciparum* (see following section) show that helminths can take part in such interactions. That a normal helminth fauna can protect against invasion by less adapted species is suggested by observations on mute swans. In a zoo, where the swans lacked their normal tapeworm fauna, they acquired large numbers of *Dicranotaenia coronula* (a well-known pathogen; see Soulsby, 1965) and *Diorchis stefanskii* (Kotecki, 1970). Neither was present as more than a rare, immature worm in wild swans, which have a well-developed fauna of specialist tapeworms (Czaplinski, 1975).

This brief and incomplete review indicates that there is evidence that each of the proposed models of host–parasite coevolution may apply to systems involving helminths. May and Anderson (Chapter 9) emphasize that coevolution between hosts and parasites may take many pathways and may lead to very different patterns of pathogenicity. They point to the form of the complex relationship between pathogenicity (virulence in their terms) and transmissibility as a major determining factor. Although the models they use apply to micro-parasites, not helminths, the rationale behind the models, and their conclusions, do apply. Coevolution between helminths and their hosts should be expected to be complex enough to stimulate interest for some time to come.

COEVOLUTION OF CO-OCCURRING HELMINTHS

Vertebrate species, and usually individual animals as well, are normally parasitized by more than one species of helminth. Some of the helminth species (heirloom parasites, using the terminology of Sprent, 1969) have coevolved with, or adapted to, their host for some time, and could have coevolved with one another as well. Others (the souvenir parasites of Sprent) are obtained by exchange from ecological associates (see Leong and Holmes, 1981, for an analysis of parasite exchange between fish species) or are generalists, adapted to a variety of host species. These others would have had less opportunity to coevolve with the host or with each other.

Coevolution between parasitic helminths has been examined largely in the context of site segregation. Sogandares-Bernal (1959; see also Martin, 1969) suggested that differences in site selection by co-occurring trematodes were coevolved, driven by selection to prevent hybridization. Holmes (1973) also concluded that such differences were coevolved but suggested they were driven by selection to avoid competition. Rohde (1979) denied that coevolution played any role in site selection by gill parasites, arguing instead for site limitation as a result of independent selection for greater contact for reproduction. [But see Llewellyn (1956) for another mechanism for monogenean site selection independent of coevolution and Lebedev (1977) for an interpretation that involves coevolution.] Price (1980) reviewed evidence on parasite communities and concluded that coevolution among competing species is rare; most communities were considered to be assemblages of independently adapted specialists.

Price (1980, p. 143) indicated that most of the studies demonstrating interactions between helminths involve acanthocephalans (and cestodes), which absorb soluble nutrients and "may be particularly susceptible to exploitative competition." It is this group of helminths that is the object of study in our laboratory. The detailed results will be published elsewhere, but Bush (1980), Butterworth (1982) and T. M. Stock (unpublished) have demonstrated that (1) heirloom helminths occupy predictable locations in the intestine of their host; souvenir helminths may or may not; (2) the ranges occupied by heirloom helminths are positively correlated with their abundances; ranges of other helminths may, or may not, be so correlated; (3) the abundances of heirloom helminths in individual hosts are generally positively intercorrelated; others are generally not; (4) despite the greater ranges occupied in birds with more abundant helminths, there is no increase in overlap between the ranges occupied by heirloom helminths; instead, the individual species adjust their locations. Others may or may not show increased overlap. These features appear to be those of a coevolved complex of heirloom species.

Another way in which parasites may interact was suggested by Murray et al. (1978). They found that a nutritional deficiency consequent to heavy infections with the intestinal nematode *Ascaris lumbricoides* suppressed malaria caused by *Plasmodium falciparum.* They outlined several ways in which the nematode may deprive the protozoan of a nutrient essential for its growth and suggested that such a mechanism "may be more than fortuitous and may represent an ecological balance for optimum co-survival of the host and the two pathogenic parasites."

Two features identified earlier in this chapter as ways in which parasites coevolve with their hosts—molecular mimicry and modulation of the immune system of the host—also suggest additional ways in which parasites may interact. The extensive antigenic cross-reactions between parasites with the same host may indicate convergent mimicry of host antigens (Damian, 1979). More likely is the

183

possibility that the modulation of the host immune system by one species of parasite may affect other parasites.

The immunosuppression produced by *Nematospiroides dubius* has been shown to prevent (or reduce) immune expulsion of *Nippostrongylus brasiliensis* (Jenkins, 1975; Della Bruna and Xenia, 1976), *Trichuris muris* (Jenkins and Behnke, 1977), or *Trichinella spiralis* (Behnke et al., 1978). An analogue of immunosuppression—interference with a genetically based resistance mechanism—has been shown to be essential in some insect–parasitoid systems. Virus particles, carried by some adult parasitoids and injected into the host with the egg, block the normal encapsulation defense of the host (Vinson, 1977; Edson et al., 1981). The block can protect other parasitoids as well (Vinson, 1977). Gotz et al. (1981) have described an interesting interaction involving a somewhat different use of such interference. A nematode (*Neoaplectana carpocapsae*) pathogenic to insects maintains a symbiotic bacterium (*Xenorhabdus nematophilus*) in its intestine. After invading an insect, the nematode produces an inhibitor that selectively destroys components active against the bacterium. The bacteria are then released into the insect hemocoel, where they multiply and stimulate the growth of the nematode. Not all immunosuppression is protective to other organisms, however; that produced by *Plasmodium chabaudi* in mice did not affect resistance to reinfection with *Schistosoma mansoni* (Long et al., 1981).

Cross-immune responses among helminths are also well known (see review by Kazacos, 1975; also Au and Ko, 1979; Moqbel and Wakelin, 1979), as is expulsion due to nonspecific, but immunologically mediated, host responses. For example, the marked inflammatory response to *Trichinella spiralis* also expels *Trichuris muris* (Bruce and Wakelin, 1977), *Nippostrongylus brasiliensis* (M. W. Kennedy, 1980), and *Hymenolepis diminuta* in mice (Behnke et al., 1977); in rats, the inflammatory response stunted but did not expel *H. diminuta* (Christie et al., 1979), but it caused destrobilization of this tapeworm (Silver et al., 1960). Nonreciprocal cross-immunity has been suggested as a mechanism of interference competition by Schad (1966).

All these interactions involving various parasites and the immune system of the host illustrate the abundant possibilities for coevolution of co-occurring parasites and emphasize that an analysis of host–parasite coevolution cannot neglect the regular presence of other parasites.

SUMMARY

Parasitic helminths cannot survive without their hosts. This commitment to parasitism provides a strong selection pressure for adaptations to the hosts. Some helminths respond with generalized adapta-

184

tions and become host generalists; others respond with more specific adaptations and become host specialists; these patterns are not irreversible. Helminths are capable of rapid evolutionary change and track local conditions. As a result, individual species may have very different characteristics in separated localities, and coevolution should be looked for at the local population level.

Vertebrates exhibit many characteristics that suggest that pathogens, including helminths, exert important selection pressures. Various authors have built a strong case for the hypothesis that such pressures are responsible for much of the genetic variability in proteins characteristic of vertebrates, and even for the sexual reproduction that enhances that variability. The development of a highly complex, regulated immune system also evidences such selection pressure. Helminth counteradaptations to these systems, especially the immune system, provide the strongest evidence for reciprocal coevolution between helminths and their hosts.

Animal–helminth systems appear to be evolving along a variety of pathways, some leading to greater host specificity, some to less, some leading to high levels of pathogenicity, some to low pathogenicity or even to mutualism. Local conditions, including the presence of other species of parasites, appear to be very important. The complexity of such systems will provide challenging opportunities for investigation of coevolutionary principles for the foreseeable future.

CHAPTER 9

PARASITE-HOST COEVOLUTION

Robert M. May and Roy M. Anderson

INTRODUCTION

In this chapter, the term "parasite" is used broadly to include viruses, bacteria, protozoans, and fungi along with the more conventionally defined helminth parasites. The chapter thus meshes with those by Levin and Lenski, Barrett, and Holmes (Chapters 5, 7, and 8, respectively), chapters that give empirically oriented discussions of particular classes of such parasite–host associations, together with outlines of contending ideas about coevolutionary trends in these associations. Our chapter focuses mainly on the patterns of coevolutionary relationships between parasites and hosts that emerge from a diverse range of mathematical models.

The organization of the chapter is as follows. First, after some preliminary discussion, the various analytic approaches to modeling coevolution in parasite–host associations are indicated. The commonly held view that "successful parasites are harmless" is noted, and its theoretical and empirical bases examined. The various kinds of mathematical models for parasite–host coevolution are then surveyed, and the main conclusions are summarized. These conclusions are next confronted with the empirical patterns; in one particular case, we show how data for the epidemiological parameters characterizing the various grades of myxoma viruses infecting rabbits in Australia lead to the theoretical expectation that virulence may here be expected to evolve to an intermediate value (which appears to accord roughly with the observations). We conclude with brief remarks on "diffuse coevolution" and community patterns in parasite–host associations.

186

MACROEVOLUTION

Macroevolutionary trends in the interactions between hosts and parasites are extraordinarily difficult to assess, as the traces left in the fossil record by parasites are fragmentary at best. One attempt to chart the evolution of the number of parasitic infections (interpreting parasite in the wide sense defined above) over geologic time is by Moodie (1967). Mainly on the basis of known lesions on fossil bones, Moodie finds the first evidences of diseases approximately 300 million years ago (around the Carboniferous), with the number of parasitic species then rising steadily to attain a maximum of development among the dinosaurs, mosasaurs, crocodiles, plesiosaurs, and turtles in the Cretaceous. The curve of parasite diversity falls abruptly with the great dying at the end of the Cretaceous, and the fossil evidence suggests the mammals of the early Tertiary were less generally afflicted with disease than were the preceding groups of giant reptiles. Throughout the Tertiary and on into recent times, however, the number of parasite-induced diseases appears to rise steadily, indicating "that disease is much more prevalent at the present time than ever before in the history of the world."

Moodie and others emphasize the tentative nature of these conclusions, which may well be wrong; as is so often true of research in evolutionary biology, it is a lot easier to criticize the work than to improve on it. If one nevertheless accepts the view that "disease is, from the geological standpoint, of relatively recent origin" (Moodie, 1967, p. 39), the macroevolutionary status of parasite–host associations is somewhat different from that of competitive or predator–prey relations, whose genealogy would appear to extend back to the Cambrian and possibly beyond (Gould, 1981; Chapter 15 by Stanley).

TAXONOMY, BEHAVIOR, POPULATION BIOLOGY

In much of classical parasitology, the evolutionary emphasis is on using taxonomy to deduce parasite phylogeny, as Mitter and Brooks discuss in Chapter 4. From these genealogies, inferences may sometimes be drawn about the phylogenetic history of the host species. This is indeed the focus of much earlier research. Delyamure's treatise (1955) *Helminthofauna of Marine Mammals: Ecology and Phylogeny* is a good example of these rather different "coevolutionary" studies, in which parasite taxonomy is used to probe ancestral relations among host species.

As noted, with specific examples, in other chapters in this book, the evolutionary exigencies of parasite–host associations will often influence the behavior of individual parasites and hosts. Though less often discussed, it is also possible that the social organization and behavioral ecology of host populations may be influenced by factors having to do with parasitic infections. Thus, Freeland (1979b) has argued that one of the evolutionary determinants of the size of social groups of primates is pressure to avoid the high parasite burdens associated with relatively large and/or sedentary groups. In this general context, Wilson (1980) has emphasized the need to consider the spatial structure of populations in analyses of parasite–host coevolution; one serious oversimplification in the models dealt with below is their spatial homogeneity, which omits all such complications.

Although commonly overlooked in population biology texts, parasites may often be responsible—wholly or in part—for the regulation of host populations. Such dynamical aspects of the evolutionary relationship between parasites and hosts are reviewed elsewhere and may be responsible for phenomena ranging from the 5–12 year cycles in many forest insects (which may be governed by associations with baculovirus or microsporidian protozoan infections) to biogeographical patterns in the distribution and abundance of particular plants and animals (see, for example, Anderson and May, 1978, 1979, 1981; May, 1982a). Indeed, some of the broader patterns of human history may plausibly be regarded as having been shaped by the influence of parasitic infections on the dynamics of human populations (McNeill, 1976; Anderson and May, 1979).

APPROACHES TO MODELING PARASITE–HOST COEVOLUTION

Against this lightly sketched background of macroevolutionary, taxonomic, behavioral, and population dynamical aspects of parasite–host interactions, we turn now to the main theme of this chapter: how evolutionary forces affect the parameters characterizing parasite–host associations, in relatively simple mathematical models.

Ideally, the basic model for studying the coevolution of parasite and host populations should include both genetics and epidemiology in an explicit way. Such an approach is fraught with technical complications and—more important—a proliferation of parameters that makes it difficult to extract a clear message. Perhaps for these reasons, most of the existing studies tend to make one or another kind of simplification.

One class of models incorporates the population genetics of both parasite and host populations in an explicit way (usually in terms of the conventional diallelic loci, as discussed generally in Chapter 3 by Roughgarden) but ignores the density-dependent effects on fitness due to the interplay between epidemiological processes and population densities. Another class focuses on the genetics of the host population, with the genetical dynamics of the parasite subsumed in frequency-dependent fitnesses of the host genotypes. The frequency-dependent fitness functions may be chosen phenomenologically, or they may be derived from epidemiological assumptions. Yet other classes of models seek a relatively accurate account of the density-dependence and epidemiology of the interaction between hosts and different strains of parasites, without retaining the explicit genetics. This rough categorization serves to organize the material below.

THE CONVENTIONAL WISDOM

Before going further, we will pause to reappraise the conventional wisdom about the course of coevolution in parasite and host populations.

Most medical, parasitology, and older ecology texts assert with little or no reservation that "successful" or "well-adapted" parasitic species evolve to be harmless to their hosts. Although the arguments offered in support of this view are occasionally nakedly group selectionist ("Nature prefers that neither host nor parasite should be too hard on the other. For Nature, survival of the species is all that counts"; Burnet and White, 1972, p. 82), the basic position is not unreasonable: all else being equal, it is to the advantage of both host and parasite individuals for the parasite to inflict little damage.

A considerable amount of empirical evidence can be marshaled in support of this conventional view. Thus, in regions of Africa where trypanosomiasis is endemic, indigenous ruminants suffer mild infections with insignificant morbidity, whereas domestic ruminants that have been bred for a long time in the region suffer more severely, and recently imported exotic ruminants suffer virulent infections that are usually fatal if untreated. This is one of several such examples reviewed by Allison (1982). Wild rats captured in cities that have had recent plague epidemics show higher survival after infection with the plague bacillus, *Yersinia pestis*, than do rats from cities with no recent experience of plague; this resistance appears to be genetically based (Habbu, 1960; Levin et al., 1982). In a recent analysis of some 300 parasite-host associations (mainly among invertebrates), Dobson (1983) finds there is a general—though not invariable—tendency for the parasites that are "older" in evolutionary time to be less virulent. Approaching the same question from a different angle, Holmes (1982) has observed that parasitic infections appear to be more effective as regulatory agents (which broadly corresponds to being more virulent) among newly introduced species of plants and animals, or when the parasites are introduced into new regions.

The preceding theoretical argument rests on the assumption that the virulence of the parasite is uncoupled from transmissibility and duration of infectiousness. The damage inflicted on their hosts by viral, bacterial, protozoan and helminth parasites is, however, usually directly or indirectly associated with the production of transmission stages. As pointed out in Chapters 5, 7, and 8, once these complications are acknowledged many coevolutionary paths are possible, including even the possibility of escalating "arms races." We now proceed to show how the emerging patterns depend on the interplay between the virulence and the transmissibility of the parasite and on the costs to the host of evolving resistance.

MATHEMATICAL MODELS FOR PARASITE-HOST COEVOLUTION

Parasite-host genetics (no epidemiology)

There is a large and still-growing body of literature in which the fitnesses of various hosts and genotypes under exposure to various parasite genotypes are specified, and the ensuing dynamical behavior of host and parasite gene frequencies then studied. Much of this work is related, metaphorically or in detail, to crop breeding (Mode, 1958, 1961; Person, 1966; Yu, 1972; Leonard, 1977; Lewis, 1981a,b; Levin and Udovic, 1977; Fleming, 1980, 1982; biologically oriented reviews are given by Van der Plank, 1975; Day, 1974). As discussed by Barrett (Chapter 7), the usual models are based on the "gene-for-gene" assumption, in which a one locus/two allele system in a diploid host interacts with a diallelic locus in a haploid parasite; the fitness $\overline{w}_{Aa}$ of the Aa host genotype will be a weighted sum over the fitnesses $w_{Aa,i}$ for this host when attacked by parasites of the genotype i (weighted by the gene frequencies of the haploid parasite genotypes i), and so on. The equilibrium and stability properties of such systems can be studied, once explicit assumptions are made about the magnitude of the fitness constants, $w_{Aa,i}$ and the like.

Such gene-for-gene relationships have been documented for many parasite-host systems, especially in cultivated crops. For example, in one early and influential study, Flor (1955, 1956) found 27 genes distributed as multiple alleles at five loci in flax, *Linum usitatissimum*, for resistance to the rust *Melampsora lini*; virulence in the rust is governed by a complementary genetic system in which each gene in the host is identified, one-to-one, with a gene in the parasite. This example is typical both of the gene-for-gene interlocking and of the complexity of natural parasite-host systems (which possibly stems from the ability of parasites to overcome single stress factors; Pimentel, 1982).

Such systems are clearly capable of yielding polymorphisms in host and parasite gene frequencies. In the simplest models, such polymorphisms often have the peculiar neutral stability of the frictionless pendulum, but various kinds of realistic complications can lead to stable polymorphisms or stably cyclic oscillations between resistant and susceptible hosts and virulent and avirulent parasites; there can even be chaotic fluctuations in gene frequency. As several authors have emphasized (Lewis, 1981a; Levin, 1983; Anderson and May, 1982), such polymorphisms—whether as stable points, stable cycles, or chaos—do not automatically ensue: in the absence of costs associated with resistance or of constraints (due, for instance, to segregation) on its evolution, host populations should evolve toward being entirely resistant; conversely, in the absence of costs and constraints, parasites should evolve toward maximum virulence. Thus, the outcome of all such models depends on the assumptions made about the fitness constants, $w_{Aa,i}$. In general, as summarized by Lewis (1981a), polymorphic equilibria arise if the virulent and avirulent reactions are not too different.

This style of analysis has the merit that the genetics of both host and parasite are treated explicitly. Epidemiological factors, on the other hand, are not really considered: if two or more strains of parasite are present, it is usually assumed that all hosts are infected, with the fraction of hosts infected by each strain being in simple proportion to the relative abundance of that strain (which, for a haploid parasite, means in proportion to the parasite gene frequencies). At best, it may be assumed that a constant fraction x of the host population escape infection; this can introduce significant additional complications (Lewis, 1981a; Yu, 1972). Even here, x is treated as one more phenomenological constant rather than as a dynamic variable determined by the interaction between host and pathogen populations. We shall return to these points later.

For a more detailed account of this class of models, see the excellent review by Levin (1983).

Host genetics: phenomenological, frequency–dependent fitnesses

In most of the models just discussed, whether framed as difference or as differential equations, it is assumed that host and parasite generations tick over on much the same time scale. This is a reasonable approximation for many interesting plant–parasite systems and for some invertebrate–parasite associations. The parasites of many animals, however, cycle through many generations in a single generation of the host, and may thus evolve rapidly compared with their hosts. Under these circumstances, it is appropriate to focus attention on the genetic structure of the host population, subsuming the genetical dynamics of the parasites in frequency-dependent fitness functions for each genotype of host.

191

There have been several studies of this kind. The usual approach is to make some *ad hoc*, phenomenological assumption about the frequency dependence of the fitness functions of the various host genotypes—still without any explicit epidemiological analysis—and then to explore the possible range of dynamical behavior.

Clarke (1976), for instance, has argued that plausible biological mechanisms are likely to result in a rare genotype enjoying a selective advantage over commoner host genotypes in the presence of parasites or predators. Numerical exploration of models embodying these ideas showed that stable, or cyclic, or chaotic polymorphism could result, depending on the strength of the frequency-dependent advantage that accrued to rarer genotypes. Clarke (1976, 1975) has suggested that many protein polymorphisms may be maintained in this way. Later analyses have given formal explications of the way cycles and chaos arise in these nonlinear genetic systems (e.g., Rocklin and Oster, 1976; Oster et al., 1976; May, 1979), leading to complex and nonsteady genetic systems. The essential ideas here go back to Pimentel's (1968) "genetic feedback" and to Haldane (1949).

Jaenike (1978b) extended these ideas, making the tentative suggestion that frequency-dependent aspects of the parasite–host association may help explain the evolution and maintenance of sex in host populations. This idea has been developed further by Hamilton (1980, 1982), who shows that such models can generate complex cycles in host gene frequency, provided the frequency dependence associated with resistance and susceptibility to parasites is sufficiently intense. If the cycling is of sufficient amplitude, Hamilton finds that sexual species can obtain higher geometric mean fitness, over the long term, than any competing monotypic asexual species or mixture of species. Hamilton urges the bold notion that the widely discussed selective disadvantage of sex (Williams, 1975; Maynard Smith, 1978) may typically be outweighed by the advantage sexual recombination confers under the sort of strong, frequency-dependent, selective forces that the genetic interplay between host and parasites may often induce. Several testable hypotheses that follow from these ideas have been cataloged (Hamilton, 1982; Levin et al., 1982); in particular, gene frequencies at a polymorphic locus should exhibit systematic changes from generation to generation if the polymorphism is maintained by Hamilton's cycling.

Host genetics: epidemiologically derived, frequency-dependent fitnesses

An alternative to the approach adopted in the preceding section is to let conventional epidemiological assumptions, of one kind or another, dictate the form of the frequency-dependent fitnesses of the host genotypes. Such models represent the first realistic accounting for the density-dependent effects associated with transmission and maintenance of parasitic infections.

Gillespie's (1975) pioneering study initially assumes a population of haploid hosts and explores the usual metaphor of one locus with two alleles: individuals of the *a* genotype are resistant to some particular disease, but they pay a cost in having a lower fitness (by a factor $1 - s$) than the susceptible individuals who do not become infected; individuals of the *A* genotype are susceptible to the disease, and those individuals who actually contract the disease have their fitness decreased to $1 - t$, which is lower than the fitness of the resistant individuals (i.e., $t > s$). Gillespie assumes the disease spreads through each generation in an epidemic fashion, in a manner described by the standard equations of Kermack and McKendrick (1927; see also Kendall, 1956; Bailey, 1975):

$$dX/dt = -\beta XY \tag{1}$$

$$dY/dt = \beta XY - vY \tag{2}$$

$$dZ/dt = vY \tag{3}$$

Here X, Y, and Z are the numbers of hosts of the susceptible population in a given generation that are still uninfected, infected, and recovered (and thereby immune), respectively. New infections appear at a rate linearly proportional to the numbers of infected and still susceptible hosts, and the coefficient β measures the transmission rate (which may itself depend on the magnitude of the host population); v is the recovery rate. From these equations, it can be shown that the fraction I of the susceptible hosts that are affected by the disease is given implicitly, for these haploid hosts, by the relation:

$$I = 1 - \exp\left(-IpN/N_T\right) \tag{4}$$

Here N_T is the threshold host density, $N_T = v/\beta$, below which the disease cannot be maintained; N is the total host population density; and p is the frequency of the allele A (so that pN is the population of susceptible hosts). It follows that the fitnesses of resistant and of susceptible hosts, w_R and w_S, respectively, are

$$w_R = 1 - s \tag{5}$$

$$w_S = 1 - It \tag{6}$$

The fitness w_S is frequency dependent, in a way that depends explicitly on the epidemiological assumptions embodied in Equation (4). Gillespie (1975) also indicated how the analysis and broad conclusions ex-

193

tend to a diploid host population in which the heterozygotes Aa are equivalent either to the susceptible AA individuals (A dominant) or to the resistant aa individuals (a dominant).

Gillespie's analysis shows that a stable polymorphism, with both susceptible and resistant genotypes (both A and a alleles) present, will ensue for a particular range of values of the fitness values s and t, and of the magnitude of the host population in relation to the threshold population, N/N_T. If the fitness cost of resistance s is relatively small, then the population will usually evolve toward most individuals being resistant. Conversely, if the fitness cost of contracting the infection t is sufficiently small relative to s ($t \to s$), or if N is close to (or below!) N_T, resistance will not be present.

The study by Gillespie is for a disease that sweeps as an epidemic through each generation. The analysis can, however, readily be extended to embrace parasitic infections that are stably endemic within the host population. Thus, Kemper (1982) has given a thorough treatment of the case of endemic infections that do not confer immunity (roughly corresponding, for example, to gonorrhea or to many viral and protozoan infections of insects), and May and Anderson (1983) have extended this to the case of endemic infections that do induce lasting immunity. In these different circumstances, appropriate expressions [replacing Equation (4)] can be derived for the equilibrium fraction I of susceptible hosts that are infected. These expressions are, however, basically similar to Equation (4) in that I and thence [via Equation (6)] the fitness of susceptible hosts are both frequency dependent (by virtue of the parasite–host genetics) and density dependent (by virtue of the epidemiological assumptions); that is, I and w_S depend both on p and on N. As reviewed by Anderson and May (1982), the essentials of Gillespie's conclusions remain unchanged by these different epidemiological assumptions: the selective pressures exerted by the parasite lead to a stable polymorphic equilibrium, providing the fitness of the immunity-producing allele is neither too large nor too small compared with the fitness of a diseased individual.

Very generally, these studies show that if selective pressures always favored the evolution of "harmless" or "avirulent" parasites (so that $t \to 0$), we would *not* expect to find polymorphisms in host susceptibility (or resistance) associated with such infectious agents. In reality, inherited variability in host susceptibility to infection by a specific pathogen appears to be the rule rather than the exception.

Parasite–host population models (no genetics)

At the opposite end of the spectrum from the models with explicit genetics and no epidemiology are those with explicit epidemiology and no genetics. Such models give a relatively accurate account of the den-

194

sity dependence and epidemiology of the interaction between a host population and populations of different strains of a parasite; the genetics is crudely implicit in, for instance, the varying rates of transmission and virulence of different strains of the parasite.

Levin and Pimentel (1981) have used this approach to examine the coexistence or otherwise of a host population with two different strains of a pathogen: one of the strains is more virulent than the other and induces a mortality rate α_1 that is greater than that due to the less virulent strain α_2 ($\alpha_1 > \alpha_2$); both strains have identical transmissibility (susceptible individuals, on contact with an infected individual, acquire the infection at a "transmission rate" β in both cases), but the more virulent strains can "take over" individuals already infected with the less virulent strain (at a per capita transmission rate $\sigma\beta$). Denoting the populations of hosts that are susceptible, infected with strain 1, and infected with strain 2 by X, Y_1, and Y_2, respectively, Levin and Pimentel describe the dynamics of this system by the set of differential equations

$$dX/dt = aN - \beta XY_1 - \beta XY_2 - bX \qquad (7)$$

$$dY_1/dt = \beta XY_1 + \sigma\beta Y_1 Y_2 - (\alpha_1 + b)Y_1 \qquad (8)$$

$$dY_2/dt = \beta XY_2 - \sigma\beta Y_1 Y_2 - (\alpha_2 + b)Y_2 \qquad (9)$$

Here a is the per capita birth rate (assumed to be unaffected by infection); b is the per capita death rate in the absence of infection; and the total population is $N = X + Y_1 + Y_2$. It is assumed that both strains of the infection are lethal, so that, once infected, no hosts recover. This kind of model differs from conventional epidemiological ones in that the total host population is a dynamical variable, which may or may not be regulated to a stable equilibrium value by the infection; traditional studies (e.g., Bailey, 1975) assume N to be a constant, determined by other factors. Despite their simplicity, equations of this general type have been shown to give a good fit to data for endemic infections that regulate experimental populations of laboratory mice (Anderson and May, 1979), and to give plausible explanations for the population dynamics of associations between foxes and rabies in Europe (Anderson et al., 1981) and between various arthropods and viral or protozoan parasites (Anderson and May, 1981).

Analyzing the system of Equations (7)–(9), Levin and Pimentel show that if one strain has significantly greater virulence in relation to its transmission advantage, it will not persist. On the other hand, if the virulence α_1 is not significantly greater than α_2 although the transmission advantage is substantial (σ relatively large), the more virulent strain will win. For an intermediate range of the ratio α_1/α_2 in relation to the transmission advantage enjoyed by the more virulent strain, the two strains can coexist.

Levin and Pimentel's study can be generalized in a variety of ways, and the essentials of their conclusions remain intact. Thus, it is not necessary to assume that the more virulent strain can infect hosts bearing the less virulent strain but that the reciprocal process is impossible; it is only necessary to assume that $\sigma\beta$ is the net excess of infections $2 \rightarrow 1$ over $1 \rightarrow 2$. The actual data for myxomatosis (see, for example, the work of Saunders, 1980) suggest that individuals, once infected with one strain, do not acquire infection with another strain. Levin and Pimentel's general analysis can, however, be preserved by considering the more virulent strain (virulence α_1) to have a higher transmission rate (β_1) than that (β_2) of the less virulent strain (virulence α_2). In this case, if the disease is considered always to be lethal, the strain with the lower value of $(\alpha_i + b)/\beta_i$ will always win. But if recovery to a state of temporary or permanent immunity is possible, or if other realistic complications are admitted, a range of coexistence is possible, corresponding to the two strains having roughly comparable values of the overall ratio between virulence and transmissibility (May and Anderson, 1983).

Anderson and May (1981; May and Anderson, 1978) and Bremermann (1980) have also used models of this type to examine the general way in which epidemiological parameters, such as transmission rate β, virulence α, and recovery rate v, are likely to evolve in response to the selective pressures exerted on host and on parasite populations. As in all such studies of "evolutionarily stable strategies" (ESS; Maynard Smith and Price, 1973), the underlying genetics is brushed aside in pursuit of a more transparent but less rigorous analysis.

For many directly transmitted viral, bacterial, or protozoan infections, an appropriate set of population equations are (Anderson and May, 1979):

$$dX/dt = aN - bX - \beta XY + \gamma Z \tag{10}$$

$$dY/dt = \beta XY - (\alpha + b + v)Y \tag{11}$$

$$dZ/dt = vY - (b + \gamma)Z \tag{12}$$

Here X, Y, and Z are the number of susceptible, infected, and recovered-and-immune hosts, respectively; the total host population is $N = X + Y + Z$. The disease-induced mortality rate α, per capita birth rate a, disease-free mortality rate b, recovery rate v, and transmission rate β are all as defined previously, and γ is the rate of loss of immunity ($\gamma = 0$ if immunity is lifelong; $\gamma \rightarrow \infty$ if there is no immunity, so that recovered individuals are again susceptible). This population grows exponentially at the rate $(a - b)$ in the absence of the disease. Once the disease becomes established, it will regulate the

196

host population to a stable equilibrium if the virulence is sufficiently high (specifically, if $\alpha > [a - b][1 + v/(b + \gamma)]$), and will slow the rate of exponential growth otherwise (Anderson and May, 1979).

Bremermann (1980) and Anderson and May (1981) observe that the fitness of the parasite is increased by having large β, small α, and small v, whereas the host fitness is increased by having small β, small α, and large v. Therefore, were these parameters not inextricably linked by the biological processes whereby virulence, recovery rate, and production of transmission stages of the infective agent are intertwined, the ESS would clearly favor $\alpha \rightarrow 0$ (although the countervailing interests of hosts and parasites with regard to β and v would tend to drive these parameters to some intermediate value, or into cycles). This is the essential basis for the common view that "successful parasites and pathogens are harmless." But the biologically based interlinkage among α, β, and v will usually invalidate this simple argument.

Intrinsic reproductive rate of the parasite: R_0

The intrinsic reproductive rate R_0 is essentially the number of successful offspring a parasite or parasitic infection is capable of producing in the absence of density-dependent constraints. It is, in effect, Fisher's "net reproductive value" for the parasite. For the model discussed earlier in Equations (10)–(12), R_0 is the average number of secondary infections produced when one infected individual is introduced into a wholly susceptible host population:

$$R_0 = \beta N/(\alpha + b + v) \tag{13}$$

This relation may be derived formally from Equation (11), or in a more intuitive way by observing that the infectious individual will produce secondary infections at the rate βN per unit time and that the average duration of infectiousness is $1/(\alpha + b + v)$ before recovery or death occurs. In the particular system defined by Equations (10)–(12), the host population density N itself may depend on the epidemiological parameters; more generally, Equation (11) will give Equation (13) for R_0 even if N is determined by other ecological or environmental factors.

We see from Equation (13) that the intrinsic reproductive rate of the parasite would be maximized by $\alpha \rightarrow 0$ if α, v and β were not connected with each other. Conversely, if β was a faster than linear function of α, R_0 would be maximized by having α as large as possible; such pressures of individual selection on parasites would make for small populations of hosts and of parasites, providing a clear example where parasite group interests would be at odds with individual selection.

For helminth parasites, it is usually not sufficient simply to partition the host population into susceptible, infected, and immune individuals [as was done in the models underlying Equation (13)], but rather it is necessary to take account of the differing parasite burdens in individual hosts (Crofton, 1971; May and Anderson, 1978, 1979). The resulting expressions for the intrinsic reproductive rate R_0 of such helminth parasites differ in detail from Equation (13) but still have the property that R_0 is maximized by $\alpha \to 0$, provided that virulence and transmissibility (as measured by parasite egg output and other quantities) are unconnected—which they rarely are (May and Anderson, 1978). Likewise, the basic set of Equations (10)–(12) can be modified by including such realistic refinements as latent periods, vertical transmission, stress-related virulence, and free-living infective stages of the parasite; the consequent expressions for R_0 are reviewed elsewhere (Anderson and May, 1981), and they lead to conclusions that are generally in accord with those based on Equation (13).

In the next section, we will turn from these abstract models and will examine empirical evidence bearing on the relations among α, β, and v in some real parasite–host associations. Before doing this, some of the inadequacies of an approach centered on R_0 should be noted (Levin et al., 1982).

First, as mentioned at the outset, the genetic underpinning is missing.

Second, the evolutionary tendency for a parasite to maximize R_0 is most clearly seen when only a single clone or a single species of parasite infects the host population. Bremermann (1982; see also Levin, 1983; Levin et al., 1982) has shown that, in polyclonal situations where the genetic relatedness of parasite individuals within a given host is lessened, parasite evolution toward attenuation will in general be less marked than for monoclonal situations.

Third, this focus on the evolution of the intrinsic reproductive rate of the parasite ignores the genetic response of the host population. Such neglect is partly justified by the common circumstance where the generation time of the host greatly exceeds that of the parasite: the time taken for selection to produce significant evolutionary change scales roughly with the generation time of the organism, so that unilateral parasite evolution is a reasonable first approximation (Levin et al., 1982; Price, 1980; Hamilton, 1980). On a much slower time scale, there is often likely to be continued evolution of increased resistance by the host. But, as seen in the earlier genetic models, such resistance will usually have associated costs, which may put steady or oscillatory limits on this slower coevolutionary process.

EMPIRICAL PATTERNS IN PARASITE–HOST COEVOLUTION

Myxomatosis in rabbit populations

When first introduced into Australia in 1950, the myxoma virus was exceedingly virulent among populations of the rabbit *Oryctolagus*

198

cuniculus. Fenner and Ratcliffe (1965) have carefully documented how successively less virulent strains of the myxoma virus came to preponderate during the subsequent decade. At the same time, the rabbit populations in Australia exhibited increasing resistance, so that any description of the coevolutionary process in terms only of the Darwinian fitness of the parasite (along the lines of the preceding section) is a gross oversimplification. Even so, we pursue this crude approximation to see to what extent a discussion based on Equation (13) can give a consistent account of the observed facts. Our analysis is made possible by the meticulous experimental protocols adopted by the virologists working with the myxoma virus, in which the virulence of field strains of the virus was tested against standard laboratory strains of the rabbit and the susceptibility of wild rabbits was tested against defined laboratory strains of the virus.

Fenner and Ratcliffe (1965) divide the myxoma virus into six strains on the basis of grade of virulence. They provide basic data for the percentage mortality and for the mean survival time of those rabbits (a standard laboratory strain) that died rather than recovered, for each of these six strains of the virus. By ignoring the details of the statistical distribution of mortalities and recoveries and by assuming constant rates of disease-induced mortality α and recovery v, we can use these basic data to estimate the recovery rate v for the various strains.

The resulting values of α and v for the six strains of the myxoma virus are shown (as the solid dots) in Figure 1. Manifestly, α and v are inversely related in this system. The solid and dashed curves in Figure 1 represent empirical relations fitted to this data. The solid curve is the fit obtained using the functional form

$$v(\alpha) = c + d \ln \alpha \qquad (14)$$

With $c = -0.032$ and $d = -0.0129$, there is an excellent fit ($r^2 = 0.987$; $r = 0.993$). The dashed curve is the corresponding fit with the function

$$v(\alpha) = c \exp(-d\alpha) \qquad (15)$$

With $c = 0.088$ and $d = 65.0$, the agreement is good ($r^2 = 0.910$; $r = 0.965$), although less so than with Equation (14). Other functional forms give poorer fits (for details of this analysis of Fenner and Ratcliffe's data, see Anderson and May, 1982).

The results in Figure 1 concern only the parameters α and v in Equation (13) and provide no information about the transmission rate β as such. There is good qualitative evidence, however, that high

199

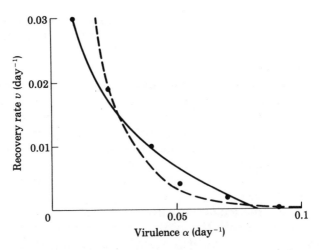

FIGURE 1. The empirical relationship between virulence α and recovery rate v for various strains of myxoma virus in wild populations of rabbits in Australia is shown. The six solid dots are the observed values; the solid and dashed curves are the best fits with the functional forms of Equations (14) and (15), respectively. (From Anderson and May, 1982.)

virulence (grades I–III) is typically associated with an abundance of open lesions and hence that mosquitoes (the vectors in Australia) or fleas (the vectors in Britain) can more easily bite infected wounds and acquire the virus. Low virulence (grades IV and V) is correspondingly associated with poor transmission. These tendencies for β to decrease with decreasing α [see, for example, Figures 12 and 13 and Table 30 in Fenner and Ratcliffe (1965)] are less pronounced than the inverse relationship between α and v and are more difficult to quantify. We therefore initially make the additional rough approximation of taking β to be a constant, independent of α, in Equation (13).

Using the empirical relationship between α and v shown in Figure 1, we can now use Equation (13) to plot the intrinsic reproductive rate R_0 of the various strains of myxoma virus as a function of virulence α for the Australian myxomatosis–rabbit system. This is done in Figure 2 [where we have put $b = 0.011$ day^{-1} (Fenner and Ratcliffe, 1965) and arbitrarily assigned $\beta N = 0.2$ day^{-1}]. R_0 is seen to attain its maximum value for an *intermediate* grade of virulence of the virus. Specifically, the virulence giving the maximum reproductive rate for the parasite is $\alpha = 0.013$ day^{-1} for the empirical relation of Equation (14) or $\alpha = 0.027$ day^{-1} for the less satisfactory empirical relation of Equation (15). It is understandable why this intermediate grade of virulence may be best: too high an α kills off hosts too fast, diminishing their capacity to transmit the infection; too low an α is associated with a very quick recovery time, so that again transmission is relatively weak.

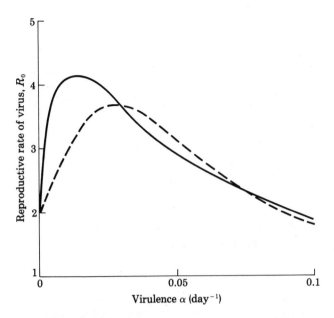

FIGURE 2. The relationship between intrinsic reproductive rate R_0 and virulence α for the various strains of myxoma virus in wild populations of rabbits in Australia is illustrated. Here R_0 is given by Equation (13); the disease-free mortality rate b is 0.011 day^{-1}, and βN is arbitrarily set constant at 0.2 day^{-1}. The relationship between recovery rate v and α is taken from Figure 1; the solid and dashed curves correspond to the empirical relationships of Equations (14) and (15), respectively. (From Anderson and May, 1982.)

These conclusions can be compared with the observed facts about the relative frequencies of the various strains of myxoma virus within field populations of rabbits. The virus was of virulence grade I when initially introduced into Australia in 1950–1951. At first there was rapid change, with increasing proportions of successively lower virulence grades. By the late 1950s and early 1960s, however, the relative frequencies of the different virulence grades had settled to roughly steady values (Fenner and Ratcliffe, 1965). These approximately steady proportions are set out in Table 1, which shows the virus apparently to have evolved to an intermediate degree of virulence centered around grades IIIA–IIIB–IV. This contrasts with our crude theoretical prediction, which on the basis of the empirical relation of Equation (14) suggests an equilibrium around grades IV–V [or an equilibrium around grade IV if the less accurate empirical relation of Equation (15) is used].

201

Table 1 also shows the distribution of frequencies of virulence grades to which the virus appears to have settled in wild populations of rabbits in Britain and France. The essential ecological difference among the myxomatosis–rabbit associations in the three countries is that mosquitoes are the vectors in Australia and France, and fleas in Britain. Transmission by fleas is thought to be less efficient because the longer and larger mouthparts of mosquitoes enable the latter vector to transmit lower virulence grades of the virus more effectively. This may explain the higher incidence of grade IV strains in Australia and France (Ross, 1982).

As discussed earlier, there is some evidence to suggest that β also changes with the virulence α of the virus (Mead-Briggs and Vaughan, 1975). There appears to be a curvilinear relationship between transmissibility and the average survival times of rabbits infected with the different virulence grades, with a maximum value of β at some intermediate value of α. This type of relationship would clearly have the effect of pushing the theoretically predicted grade of virulence closer to the observed grade around IIIB.

TABLE 1. Comparison of the virulence of field strains of myxoma virus in Great Britain, Australia, and France (from Ross, 1982), contrasting initial frequencies with those several years after introduction.

| | VIRULENCE GRADE (AS PERCENTAGE OCCURRENCE) | | | | | |
	I	II	IIIA	IIIB	IV	V
GREAT BRITAIN						
1953	100	—	—	—	—	—
1962–1967	3.0	15.1	48.4	22.7	10.3	0.7
1968–1970	0	0	78.0	22.0	0	0
1971–1973	0	3.3	36.7	56.7	3.3	0
1974–1976	1.3	23.3	55.0	11.8	8.6	0
1977–1980	0	30.4	56.5	8.7	4.3	0
AUSTRALIA						
1950–1951	100	—	—	—	—	—
1958–1959	0	25.0	29.0	27.0	14.0	5.0
1963–1964	0	0.3	26.0	34.0	31.3	8.3
FRANCE						
1953	100	—	—	—	—	—
1962	11.0	19.3	34.6	20.8	13.5	0.8
1968	2.0	4.1	14.4	20.7	58.8	4.3

So far, we have only considered the evolutionary pressures acting on the virus. The disease quite clearly also acts as a strong selective pressure on the rabbit. This is manifest in Australia and Britain by the evolution of increasing resistance to the myxoma virus; mean survival times of "resistant" rabbits infected with a grade IIIA strain (29–36 days) are significantly longer than those of fully susceptible laboratory rabbits (17–22 days). Such evolution of resistance among rabbits acts as a further selective pressure on the virus, tending to select for higher virulence grades (see, for example, Table 5 in Fenner and Myers, 1978). Unfortunately, quantitative data on the relative frequencies of "resistant" rabbit types are scanty.

Baculoviruses and insects

The interactions between nuclear polyhedrosis viruses and their lepidopteran hosts provide clear examples where parasite virulence is associated with enhanced transmission efficiency (Anderson and May, 1981). These baculoviruses are directly transmitted between hosts via inclusion bodies that contain virus particles; the replication cycle of the parasite within the insect is completed with the production of more occluded viruses, which are released into the environment when the caterpillar dies. The rate of release of infective stages of the parasite into the external environment is thus clearly dependent on the rate at which the virus kills its host.

Under the crudely simple assumption that the net rate of production of infective stages is directly proportional to the virulence of the parasite, the intrinsic reproductive rate R_0 takes the form

$$R_0 = \delta\lambda\alpha N/\mu(\alpha + b + v) \tag{16}$$

Here α, b, v, and N are as previously defined, $\lambda\alpha$ is the average number of infective stages released on the death of an infected host, δ is the transmission rate of infective stages to the subsequent insect host, and $1/\mu$ is the life expectancy of the inclusion body in the external environment. The intrinsic reproductive rate can be seen to increase from zero (at $\alpha = 0$) to a saturation value (at approximately $R_0 = \delta\lambda N/\mu$) as the virulence α becomes larger and larger. In reality, there will usually be a trade-off between the number of inclusion bodies produced by the virus and the life span of an infected insect; if the host is killed too fast, not enough time will have elapsed for production of the maximum number of transmission stages permitted by the resources available within the host. In general, it is therefore likely that $\lambda\alpha$ will rise to some asymptote as the life span of infected hosts increases. The up-

203

shot will tend to be that R_0 rises to a maximum and then decreases back to low values, as α increases. The pressures of evolutionary processes in such baculovirus–lepidopteran associations are thus apt to be toward intermediate levels of virulence, i.e. an intermediate time before the host is finally killed. [For a more mathematically detailed discussion, see Anderson and May (1981).]

A slightly different reproductive strategy has evolved for the cytoplasmic polyhedrosis viruses, which are a closely related group of insect viruses also belonging to the Baculoviridae. In contrast to the nuclear polyhedrosis viruses, they do not necessarily rely on the death of their host for the release of transmission stages, but produce and release inclusion bodies throughout the duration of the infection. Interestingly, cytoplasmic polyhedrosis viruses are in general less pathogenic to their insect hosts than are the nuclear polyhedrosis viruses.

Other patterns in parasite virulence

A variety of other circumstances in which evolutionary pressures are not necessarily toward avirulence may be identified. Some of these qualitative patterns are discussed in Chapter 8 by Holmes.

One broad class of examples comprises those in which the mechanisms for parasite transmission are directly associated with modification of host behavior, of a kind deleterious to the host. The preceding discussion of situations in which host death is necessary for dispersal of infective stages is an obvious limiting case, of which there are other instances. An important transmission route for rabies is through saliva-infected bites, and such transmission is greatly enhanced by the behavioral changes concomitant upon the final stages of viral infection of the nervous system; it is likely that the high virulence of rabies in many host species represents the evolutionary optimum for the parasite.

As noted by Holmes (Chapter 8), evolutionary pressures toward parasite avirulence will also be reduced in those situations where death of the host does not imply death of the parasite. Although we excluded them from our broad definition of parasites, the insect parasitoids constitute extreme examples in this category: these creatures constitute 10 percent of all metazoan species, and are distinguished by the characteristic that the adult females (only) search for hosts, which, when encountered, are oviposited on or in; one host is sufficient for development of the progeny, which kill the host as a developmental necessity. Such parasitoid–host systems deserve passing mention, as they form a connecting link between the parasites (considered here and in Chapters 5, 7, and 8) and the predators considered elsewhere in this book. For a systematic comparison of various features of parasites, parasitoids, and predators, see Table 1 in May (1982b).

204

Another dimension of complication enters when we pay proper attention to parasites whose transmission is indirect, involving arthropod, molluscan, or other kinds of intermediate vector hosts. One representative comment must suffice. For some viral infections of mice that ordinarily are vector borne, it has been shown that the virulence increases when the viruses are passed directly from mouse to mouse, avoiding the intermediate host; a plausible explanation is that this less efficient, direct transmission puts an evolutionary premium on more open lesions and other such correlates of increased virulence.

DIFFUSE COEVOLUTION AND COMMUNITY PATTERNS

The foregoing discussion has concentrated on tight associations between one species of parasite and one species of host. In actuality, host populations—and even host individuals—will usually be coping with a variety of co-occurring parasite species; many of these parasite species may, in turn, have an assortment of possible host species. It is presumably for this reason, indeed, that immune systems in vertebrates or chemical defenses in many plants represent generalized defenses against an array of possible parasites.

Holmes (Chapter 8) has already emphasized that parasite–host coevolution must ultimately embrace circumstances where many parasite species are present and has given an overview of some of the empirical patterns that have been observed for helminth parasites. In particular, Schad (1963, 1966), Rohde (1981), and other authors (referred to in Chapter 8) have tried to elucidate the extent to which site specificity or limits to niche overlap may govern the structure of multi-species communities of parasites. This subject even has medical implications, which have received very little attention: some of our tropical ecologist friends subscribe to the intuitive notion that it may be a good idea to have some of the relatively harmless tropical fauna in one's gut, as partial protection against the real nasties; conventional medical reactions are predictably to eradicate all "exotic" bacteria, amoebae, and protozoans. A clearer understanding of the interactions among the parasites, and their consequences for the host, would be useful.

Theoretical investigation of general situations of "diffuse coevolution" are due to Roughgarden (1979; and Chapters 3 and 17) and Rocklin and Oster (1976) for competitive interactions, and to Levin and Segel (1982) for predator–prey systems. As reviewed by Levin (1983), these coevolutionary studies can be broadened reasonably easily to include parasite–host interactions (and also plant–pollinator interactions; see Chapter 13 by Feinsinger). These mathematical models address such questions as when is coevolution likely to produce communities of generalized parasites, as opposed to highly specific parasite–host associations? That is, of the range of possibilities surveyed at the end of Holmes' Chapter 8, which circumstances favor specialization and which generalization? The preliminary studies of Levin and Segel (1982) showed that many different outcomes are possible, depending on the relative growth rates of the predator and prey (parasite and host) populations, on the "predation pressure" (virulence), and on other ecological and epidemiological parameters; for a review, see Levin (1983).

In short, real hosts usually harbor entire communities of parasite species. Both the data and the formal mathematical models tell us that a great variety of different patterns are possible, but neither empiricists nor theorists have yet produced a crisp and ineluctable codification of these patterns.

SUMMARY

In principle, mathematical models for parasite–host coevolution should take explicit account both of the population genetics of parasite and host species and of the nonlinear way epidemiological processes depend on population densities. Although the models reviewed in this chapter all tend to oversimplify either the genetics or the epidemiology, they tend to concur in finding that the coevolutionary trajectory followed by any particular parasite–host association will depend on the way the virulence and the production of transmission stages of the parasite are interlinked and on the costs to the host of evolving resistance.

With parasites defined broadly to include viruses, bacteria, protozoans, fungi, and helminths, at least two general conclusions emerge from the mathematical models. First, it does indeed seem likely that the genetic polymorphisms observed in natural populations of animals are often the result of coevolution between hosts and a wide variety of strains and species of parasites. Second, "well-balanced" associations are not necessarily ones in which the parasite does little harm to its host: depending on the relation between the virulence and the production of infective stages of the parasite, the coevolutionary course can be toward low virulence (as in many old-established associations of parasites with vertebrate hosts), or toward very high virulence (as for many viral and protozoan parasites of invertebrates), or toward some intermediate grade.

Too often, the real situation and the theoretical model bear a disquietingly metaphorical relation to each other. The myxomatosis-rabbit story is a nice counterexample, where the data collection is complete enough to permit comparison with an explicit, if excessively simple, model. More case studies of this kind are desirable.

CHAPTER 10

EVOLUTIONARY INTERACTIONS AMONG HERBIVOROUS INSECTS AND PLANTS

Douglas J. Futuyma

The interactions among herbivorous, or phytophagous, insects and their host plants are the subject of a vast literature. It includes work in plant breeding for resistance to crop pests (Painter, 1951; Sprague and Dahms, 1972; Gallun et al., 1975; P. H. Williams, 1975; Pathak and Saxena, 1976; Panda, 1979; Maxwell and Jennings, 1980), identification of chemical and physical features of plants that confer resistance to insects (Dethier, 1947; Kennedy, 1965; Whittaker and Feeny, 1971; D. A. Levin, 1973, 1976; Rosenthal and Janzen, 1979), toxicological (Dauterman and Hodgson, 1978; Brattsten, 1979) and behavioral/ neurophysiological (Schoonhoven and Dethier, 1966; Schoonhoven, 1968, 1972; Hedin et al., 1974) analysis of the responses of insects to plant chemistry, ecological analyses of plant-insect associations (Strong, 1979; Gilbert and Smiley, 1978; Lawton, 1978; Cates, 1980; Futuyma and Gould, 1979; Gilbert, 1977), genetic analyses of insect populations (Hatchett and Gallun, 1970; Mitter and Futuyma, 1979), and taxonomic studies of the relations between adaptive radiations of insects (Ehrlich and Raven, 1964; Benson et al., 1976) and the chemical features of the plants on which they feed (Gibbs, 1974). It is clearly impossible in a short review to touch on more than a few highlights of a subject to which many books and reviews (e.g., van Emden, 1973; Jermy, 1976a; Wallace and Mansell, 1976; Harborne, 1978; Rosenthal and Janzen, 1979; Edwards and Wratten, 1980) have been devoted.

PLANT-INSECT COEVOLUTION

Historically, the central problem of interest has been to account for the degree of host specificity of phytophagous insects. Verschaffelt (1910), Brues (1924), Dethier (1947, 1954), and Fraenkel (1959) were among the early authors who recognized that the chemical compounds of plants play a large role in determining host specificity, and this theme has been dominant in the literature in this field up to the present. Probably the most influential paper, at least for evolutionary biologists, was Ehrlich and Raven's (1964) description of the taxonomic relationships among butterflies and their larval food plants, in which they brought the term *coevolution* into general use. They observed that in many groups of butterflies, related species (constituting, for example, tribes or subfamilies) feed on plants that share common chemical compounds either by descent or by convergence, and they postulated that herbivory had imposed selection for divergence in defensive compounds among plant groups, thereby rendering certain plants relatively free of herbivores. These chemically diverse plants offered competition-free resources for butterflies that could overcome the defensive barriers; the butterfly species that evolved the capacity to use these plants gave rise by adaptive radiation to groups of species that inherited the same or similar host affiliation, using as cues for host recognition the very compounds that the plants had once elaborated as defensive features.

In what sense has coevolution actually occurred? Ehrlich and Raven (1964) did *not* postulate what we call, in the introduction to this volume, "parallel cladogenesis," or what Mitter and Brooks (this volume) call "association by descent." That is, unlike the situation in certain animal parasites, the divergence of lineages of insects is not temporally correlated with that of the plants on which they feed. Although there are some cases in which primitive insects feed on primitive plants, there is no overall correspondence, either for the Insecta as a whole or for any given order or family of insects, between insect and plant phylogeny (see Chapter 4 by Mitter and Brooks). And in no case of which I am aware has anyone postulated a correspondence between a given speciation event in insects and speciation of the host.

The pattern to which Ehrlich and Raven brought attention, rather, is one in which most of the members of a subfamily, for example, feed on related hosts, but related subfamilies feed on unrelated plant families. In the Pieridae, for example, most of the members of the subfamily Coliadinae (sulfur butterflies) feed on legumes, whereas most of the Pierinae (cabbage butterflies and relatives) feed on Cruciferae (mustards) and the closely related Capparidaceae. The legumes and mustards are not related to each other, but species within each plant family generally share compounds to which closely related butterflies respond.

208

The taxonomic pattern described by Ehrlich and Raven is by no means universal, as they pointed out. Within every major group of phytophagous insects, related species and genera often feed on phylogenetically and chemically very different plants. The majority of Pierinae may feed on crucifers, but some genera feed on unrelated families such as Berberidaceae (barberry family), Anacardiaceae (sumac family), and Leguminosae. The butterfly family Lycaenidae has a profusion of genera in which related species feed on utterly unrelated plants (Gilbert, 1979; Ehrlich and Raven, 1964). In the Psyllidae (relatives of aphids), related species feed on plants as different as willows, nettles, umbellifers, and composites; and among the aphids themselves, a single species usually alternates between a woody plant during the sexual generation and any of a great diversity of herbaceous plants during the asexual generations of the complex life cycle (Eastop, 1973a). In the fruit flies of the genus *Rhagoletis* (Bush, 1966), sibling species are limited to hosts as different as hawthorn (Rosaceae), dogwood (Cornaceae), and blueberry (Ericaceae).

These and many other such examples indicate that the major radiation of host–plant usage within most groups of insects did not accompany the diversification of the angiosperms, but followed it, and is continuing at the present time. Thus, if the "secondary compounds" that typify different plant groups were evolved as defenses, their diversification preceded the diversification of modern insects (Ehrlich and Raven, 1964; Benson et al., 1976; Vane-Wright, 1978). Jermy (1976b) has suggested that the evolution of insect responses to plant compounds is not coevolution, which he takes to imply simultaneity of genetic change, but "sequential evolution." The taxonomic relationships imply that the lag in response of insects and plants to each other has been long, at least when viewed in a macroevolutionary perspective: modern insects have evolved to use resources that evolved long before; and the fact that many plant compounds are taxonomically conservative (shared by most members of large groups) suggests that major changes in plant chemistry do not evolve very quickly in response to herbivory. That closely related insects can occupy very different hosts suggests, moreover, that host affiliation can be evolutionarily very labile; thus, the defensive barriers of many plants may be quite easily overcome by genetic change. Even within historical times, insects have evolved to attack new hosts. The codling moth *(Laspeyresia pomonella)* has evolved walnut- and plum-feeding populations from an apple-feeding ancestor within the past few decades (Phillips and Barnes, 1975); the Hessian fly, the brown planthopper, and many species of aphids have evolved the ability to attack previously resistant strains of crops; and five species of the Hawaiian moth *Hedylepta* have apparently evolved on banana from a palm-feeding ancestor within the past 1000 years (Zimmerman, 1960).

Although there is ample evidence (see later) that secondary compounds provide protection against herbivory, there is less evidence that they evolve rapidly in response to changes in insect fauna. This is surely not for lack of intense selection; insects clearly can have a severe impact on plant survival and reproduction, especially when, as is often the case, they attack meristems or reproductive structures (Kulman, 1971; Janzen, 1970, 1979). Nor is evolutionary conservatism due to a lack of genetic variation; crop plants can readily be selected for resistance to almost any particular species of insect (Maxwell and Jennings, 1980; Panda, 1979). It is more probable that in any plant that is attacked by dozens or hundreds of species of insects (as most plants are), only a very rare change in chemistry will provide protection against the full suite of enemies. Many of the slight changes in chemical profile are likely to improve resistance against some species, but lower resistance to others. For example, in the cocklebur *Xanthium strumarium*, a compound that seems to confer protection against a seed-eating moth apparently enhances susceptibility to a seed-eating fly (Hare and Futuyma, 1978). Thus, selection should be most effective on characteristics that provide a broad-spectrum effect, and most plant defenses appear to be of this kind—they are active as toxins or repellents against a wide variety of insects, as well as vertebrates and pathogenic microorganisms. Such broad-spectrum defenses, the product of "diffuse coevolution" (Janzen, 1980; Fox, 1981), will tend to be conservative in coevolution.

If, however, a plant is attacked by only a few species of insects, it is probably easier to respond to each of these several major sources of selection, by mounting a defense against each insect species' idiosyncrasies (Gilbert, 1977; Fox, 1981). True reciprocal coevolution is most likely in such circumstances. For example, vines of the genus *Passiflora* are attacked primarily by *Heliconius* butterflies and certain flea beetles; they have defenses against other insects, probably alkaloids. *Passiflora adenopoda* successfully reduces herbivory by the possession of hooked hairs that immobilize small *Heliconius* larvae (Gilbert, 1971), and many species of *Passiflora* have structures that mimic *Heliconius* eggs, which deter female butterflies from ovipositing (Williams and Gilbert, 1981; Chapter 12 by Gilbert). One of the clearest examples of reciprocal coevolution is in the carrot family Umbelliferae. Linear furanocoumarins, found throughout the family, are toxic to insects generally but not to a few umbellifer-feeders such as larvae of the black swallowtail butterfly. The most advanced tribes of umbellifers have not only linear, but angular, furanocoumarins, which are toxic to insects that are adapted to the linear compounds

210

(Berenbaum and Feeny, 1981). The phylogenetic analysis of the umbellifers is essential to showing that there has been stepwise coevolution in this case. It is interesting to note, parenthetically, that the advanced umbellifers have not replaced one defense by another but have added to their armamentaria. In *Acacia*, however, cyanogenic glycosides, which confer protection at least against mammals, have been lost in the species that harbor mutualistic stinging ants (Rehr et al., 1973).

PLANT DEFENSES

Many characteristics of plants confer at least some degree of protection against at least some species of insects. It would be rash, however, to conclude that all such characteristics were evolved to serve this function. Characteristics that deter some insects can stimulate feeding and oviposition in others, which suggests the possibility that some deterrent effects are accidental consequences of the behavioral idiosyncrasies of some species of insects. Moreover, it is possible, or even likely, that defense against insects is in some cases an incidental effect of plant chemicals that serve other metabolic roles, or that evolved as defenses against pathogens, or that are merely metabolic waste products (Robinson, 1974; Seigler and Price, 1976). Demonstrating that a plant feature evolved as a defense against insects is difficult because almost all plant populations are potentially subject to insect attack; therefore, it is hard to find plants that have evolved in the absence of herbivore pressure.

Among the characteristics that can reduce herbivory are physical features that reduce the likelihood of discovery. It is at least possible that the pebble-like coloration and form of the stone plants (*Lithops*) of South African deserts reduce discovery by herbivores. Gilbert (1975) has suggested that the great variety of leaf shapes in passionvines (Passifloraceae) is a consequence of selection by their major enemies, *Heliconius* butterflies, in which the search for oviposition sites is highly visual. Divergence in leaf shape may reduce the ability of female butterflies to form search images. Rausher (1980) has shown that females of the butterfly *Battus philenor* form search images based on the shape of their hosts' leaves. Other physical characteristics that deter herbivory include stipules, glands, and other structures of passionvines that are modified to look like *Heliconius* eggs; the female butterflies are less likely to oviposit on plants that already have butterfly eggs—or egg mimics (Williams and Gilbert, 1981).

The husks, pods, and seed coats of many plants deter some seed-eating insects; indeed, protection of the seeds is often cited as a possible reason for the evolution of the angiosperm carpel (Janzen, 1971a; Stebbins, 1974). The foliage and stems of many plants are invested with a pubescent coat of hair (trichomes), which is demonstrably effective in providing protection against many small insects (Levin, 1973; Gilbert, 1971; Norris and Kogan, 1980). Many plants have features, such as proteinaceous bodies and extrafloral nectaries, which attract ants that defend the plant against other insects (Janzen, 1966; Bentley, 1977). Species of *Acacia* that harbor ants lack the cyanogenic glycosides that non-ant acacias possess, a complementarity that argues for a defensive role as the adaptive raison d'être of both cyanogenesis and ant attraction (Rehr et al., 1973).

The diversity of chemical compounds in plants that are known or suspected to affect insects is staggering and is the subject of an almost equally staggering literature. Primary plant constituents such as sugars, amino acids, and proteins are important in stimulating feeding and oviposition and in supporting insect growth. Because of their universality, however, they are unlikely to be very important in host specialization. For this reason, much of the literature focuses on "secondary" compounds, of which many (e.g., pigments) play important roles in plant physiology; but many of these compounds, and others that are biosynthetically and structurally related, affect the responses of insects. Here is a brief description of the major classes of secondary compounds (following Whittaker and Feeny, 1971; Norris and Kogan, 1980; see also Rosenthal and Janzen, 1979). Members of most of these groups of chemicals have been shown to affect insects in some way.

The shikimic acid pathway leads to the synthesis of simple phenols, which in turn are the basis for lactones known as coumarins and for hydrolyzable tannins. This pathway also serves for the biosynthesis of the amino acid phenylalanine, the basis for many aromatic phenylpropanes. These include some of the volatile aromatic compounds in familiar spices and serve as one of the bases for the synthesis of quinones. Another pathway, based on acetate, provides one of the aromatic rings that make up flavonoid compounds, which figure in the synthesis of condensed tannins. Acetate groups also form the basis for isoprene units that in turn are the constituents of an enormously diverse group of terpenoid compounds. Some of these (monoterpenoids) include volatile "essential oils." The sesquiterpenoids may have toxic properties, as in the cucurbitacins, and may by dimerization form steroidal compounds. Some of the steroids act as mimics of insect hormones and can severely affect growth, development, and survival of insects. Others are modified into cardiac glycosides, which are toxic at least to vertebrates. Diterpenoids, when dimerized, form carotenoid pigments and certain alkaloids such as solanidine, which is the toxic principle in species of Solanaceae (e.g., potato).

212

Approximately 20% of the species of vascular plants contain alkaloids, which have in common a nitrogen-containing aromatic ring. Certain of the simple alkaloids (the pyrrholidine and piperidine alkaloids) are synthesized from aliphatic amino acids; others (e.g., the benzylisoquinoline alkaloids) are derived from aromatic amino acids. Many alkaloids are highly toxic.

Other classes of secondary compounds include sulfur-containing thiols and glucosinolates, especially the mustard oil glycosides in the Cruciferae and their relatives. The cyanogenic glycosides, which release hydrogen cyanide when hydrolyzed by enzymes, are broadly distributed among plants. Some plants, especially legumes, have toxic nonprotein amino acids, especially in the seeds. An important class of defensive compounds are small proteins, phenolics, and perhaps other compounds that inhibit proteolytic enzymes and thus may interfere with insect digestion of plant protein. These proteinase inhibitors typically are produced in response to damage of plant tissues by herbivores or pathogens (see Ryan, 1979), and damaged tissues are sometimes less palatable to insects (Haukioja and Niemalä, 1977).

The extent to which the taxonomic distribution of secondary compounds reflects phylogenetic relationships among plants varies greatly. As might be expected of compounds derived from ubiquitous biochemical pathways, many of them occur abundantly among vascular plants with little regard for taxonomic affinity and are the consequence of rampant convergent or parallel evolution. Others are typical of major taxonomic groups, but occur sporadically in unrelated species. For example, sesquiterpene lactones are abundant in the Compositae but also occur in some species of Magnoliaceae and Umbelliferae; the cyanogenic glycoside prunasin is abundant in the Rosaceae and the related Leguminosae but also occurs in unrelated families such as Caprifoliaceae, Scrophulariaceae, and even ferns in the Polypodiaceae. Glucosinolates are typical of the Cruciferae (mustards) and related families in the Capparales but are found in a few other unrelated groups such as the Caricaceae (papaya). Thus, some compounds can be used selectively to reinforce arguments about plant affinities based on other characteristics, but others are of decidedly limited value.

Phylogenetic relationships among major groups of angiosperms are the subject of very considerable disagreement; perhaps for this reason, as well as the extensive convergent evolution that apparently typifies many secondary compounds, no one to my knowledge has attempted a comprehensive phylogenetic analysis of the evolution of secondary compounds. Certainly many of the "advanced" angiosperm groups have fairly unusual secondary compounds: the Cruciferae with glucosinolates, the Asclepiadaceae with cardiac glycosides, the Compositae with sesquiterpene lactones, for example. But so do some "primitive" groups: for instance, pseudocyanogenic glycosides are restricted to the cycads (Conn, 1979). It is not clear whether "advanced" plant families tend to have more derived ("apomorphic," to use the terminology of cladistics) compounds, and "primitive" families more "primitive" (symplesiomorphic) compounds. Because of the prevalence of convergent evolution in secondary chemistry, it will often be difficult to tell whether a compound is a derived or a primitive character. Thus, it is hard to know whether, in vascular plants as a whole, there have been progressive trends in the elaboration of secondary chemicals.

MODES OF PLANT DEFENSE

At least some compounds in all of these chemical families have been shown to reduce attack by some insects. The mechanism of action is often obscure, however. A compound can act at the behavioral level by repelling an insect or inhibiting feeding (or oviposition); or it may act at a physiological level by (for example) poisoning the insect or reducing its ability to digest food. If the growth rate or survival of insects is used as a test for antiinsect activity of a compound, it is not possible to say whether the compound has merely repellent (behavioral) or antibiotic (physiological) activity, without monitoring feeding or behavior. In many cases, then, the mechanism of a compound's action, even at the very superficial level of behavior versus physiology, is unknown.

Clearly many compounds have a repellent effect; very often, however, some insect species are repelled or inhibited, whereas others are attracted to the same substance. Cucurbitacins, the sesquiterpenoids that occur in cucumbers and their relatives, are feeding inhibitors of the cucurbit-feeding beetle *Epilachna* but are attractive to another such beetle, *Acalymma* (Carroll and Hoffman, 1980). A sesquiterpene lactone of *Vernonia* (Compositae), glaucolide-A, is attractive to several *Vernonia*-feeding Lepidoptera, but repellent to others (Mabry and Gill, 1979). Dulcitol, a constituent of *Euonymus*, is an attractant to the *Euonymus*-feeding moth *Yponomeuta cagnagellus* but not to its apple-feeding sibling species *Y. malinellus*; phloridzin, a constituent of apple, is inhibitory to *Y. cagnagellus*, but not to *Y. malinellus* (van Drongelen, 1980). Many such examples could be given. It is commonly the case that a plant compound is repellent to polyphagous insects and to insects that do not feed on the plant but is attractive to some species that specialize on it. However, specialized insects are not necessarily attracted by all their hosts' compounds: the Colorado potato beetle (*Leptinotarsa decemlineata*), for example, is repelled by tomatine and other alkaloids of its solanaceous hosts (Hsiao, 1969). The list of plant compounds known to have a repellent effect is very long, and many

214

such compounds, such as nepetalactone, the volatile principle in catnip (Eisner, 1964), are known to repel a great variety of insects. It is likely, then, that many compounds evolved because of their broad-spectrum repellent effect, posing a defensive barrier that only certain insect species have overcome. In some instances an adapted insect is merely not repelled by a compound; in other cases, the insect is attracted to an otherwise repellent compound. Many insects that are specialized feeders on cruciferous plants, for example, are attracted to certain glucosinolates.

A great many compounds have also been shown to reduce insect growth and survival by toxic or other physiological effects. L-Canavanine, for example, is a toxic amino acid in certain legume seeds that is incorporated in place of arginine into insect proteins and that inhibits their function; a specialized beetle, *Caryedes brasiliensis*, can degrade the compound and use it for nitrogen metabolism (Rosenthal et al., 1978). Many plants contain compounds that are chemically similar to insect hormones, such as ecdysone and juvenile hormone, and that disrupt the development of nonadapted insects (Sláma, 1979). Toxic effects are also known for terpenoids, alkaloids, glucosinolates, and many other groups of secondary compounds. Many toxic compounds also have a repellent effect (Norris and Kogan, 1980). For example, sinigrin, a mustard oil glycoside in many cruciferous plants, is attractive to the cabbage aphid, the cabbage butterfly, and several other crucifer specialists but is repellent to the pea aphid. It does not appear to reduce growth in the cabbage butterfly but is toxic to a nonadapted insect, the black swallowtail butterfly, which feeds on umbellifers (Erickson and Feeny, 1974; Blau et al., 1978).

Some phenolic compounds, especially tannins, are believed to reduce insect growth not by directly affecting metabolic pathways but by forming complexes with dietary protein and thereby reducing its digestability (Feeny, 1970; Rhoades and Cates, 1976). Such compounds often occur in much higher concentrations than alkaloids, glucosinolates, and other compounds that presumably have a more directly toxic effect. Digestability-reducing compounds are especially prevalent in woody plants, almost independently of taxonomic affinity. A survey of the phytochemical literature suggests that among common plant families in the northeastern United States, trees, in which phenolics figure prominently, display a lower diversity of secondary chemicals than herbaceous plants (Futuyma, 1976). I suggested that in "fine-grained," species-rich communities of herbaceous plants in early succession, selection might more strongly favor divergence in defensive compounds, to reduce the likelihood of being attacked by in-

215

sects adapted to neighboring plant species. A similar view was expressed by Feeny (1976), who suggested that abundant, long-lived plants in communities of low diversity (e.g., trees in north temperate forests) are more "apparent" or "bound to be found" than ephemeral or "unapparent" plants in highly diverse communities. He suggested that insects would more rapidly adapt to "apparent" plants, which would achieve protection by massive quantities of compounds such as phenolics that have a dosage-dependent, "quantitative" effect on insect digestion, and would be difficult to counteract. Most insects would not be subject to strong selection for adaptation to "unapparent" plants, which therefore might achieve protection by deploying small quantities of highly toxic ("qualitative") compounds. In Feeny's view, most insects can more readily adapt to these compounds than to high doses of "quantitative" defenses, but most of them are not under strong selection to do so. Rhoades and Cates (1976) arrived at a similar rationalization of the differences between the chemistry of woody plants and herbs by a different argument: they postulated that iteroparous, long-lived species can "afford" to allocate a larger fraction of their energy budget to massive amounts of defensive compounds than can short-lived ("r-selected") species.

The "apparency" hypothesis on which these authors converged has been the best approximation so far of a general theory of the evolution of plant defenses, but it has not proved entirely satisfactory. It is clear that many insects that feed on woody plants are well adapted to deal with tannins and phenolics (Fox and Macauley, 1977; Bernays, 1978; Berenbaum, 1980), although no selection experiments have been done to determine whether insects can adapt to these compounds as rapidly as to others. The distinction between "qualitative" and "quantitative" defenses is fuzzy because most compounds display a dosage-dependent effect. And the role of phylogeny in determining the distribution of classes of compounds among plants has not been adequately explored. Herbaceous plants, for example, may have a phylogenetically more diverse ancestry. Fox (1981) has provided a critical evaluation of the apparency hypothesis.

INTERPRETATIONS OF SECONDARY COMPOUNDS

There is not the slightest doubt that one *effect* of many secondary compounds is to deter insects, but the evidence that reducing herbivory is the *function*, or adaptive raison d'être, of these compounds is scanty. The high turnover of alkaloids and other compounds suggests to some authors that their primary function might be to serve for storage of carbon and/or nitrogen, and in some cases they may play a role in regulating primary metabolic functions such as photosynthesis and respiration (Seigler and Price, 1976). Thus, it is hard to exclude the

216

possibility that some of these compounds have primary functions in plant physiology, although it is hard to imagine why such functions should require such a vast diversity of compounds that differ from one plant taxon to another.

Some of the evidence that secondary compounds have a primarily defensive function is entirely circumstantial and is cast in terms of adaptive plausibility. For example, McKey (1979) has amassed considerable evidence that alkaloids and other compounds are produced in, or translocated to, those plant parts that are most "valuable" or most in need of defense. Thus, young leaves, reproductive tissues, and seeds are commonly most heavily invested with toxic compounds. The fact that some compounds are induced by tissue damage or are translocated into damaged tissues argues strongly for a defensive role (Ryan, 1979; Carroll and Hoffman, 1980).

In a few cases, patterns of geographical variation in secondary compounds strongly suggest a defensive role. Cyanogenesis in *Lotus corniculatus* is more prevalent in southern than in northern populations and has been shown to confer protection against some herbivores. It is likely that the frequency of the trait is determined by a balance between herbivory, which favors cyanogenesis, and low temperature, which by causing cell damage can release cyanide into the tissues and cause self-toxification (Jones, 1973; Jones et al., 1978). Geographical variation in several monoterpenes in ponderosa pine is correlated with past outbreaks of the highly destructive mountain pine beetle; in particular, populations with a past history of attack have a high frequency of trees containing limonene, which is toxic to the beetle (Sturgeon, 1979). Sturgeon interprets the pattern as a consequence of selection for resistant trees during past outbreaks.

The variation in limonene in ponderosa pine illustrates the difficulty of inferring purely from patterns of geographical variation that plant compounds have evolved in response to herbivory. Suppose there were no historical records of beetle outbreaks and that there were no direct evidence that limonene is toxic to beetles. We observe at present that beetles are abundant in pine populations that have low limonene content and rare in high-limonene populations. A reasonable inference would be that limonene is advantageous in the *absence* of beetle attack, perhaps for other ecological reasons. Or we might hypothesize that limonene is disadvantageous where beetles are abundant; perhaps the beetles use it as a cue to find their host. Or we might hypothesize (as seems to be the case) that past infestations resulted in the evolution of high limonene content, which now confers protection. But suppose we made the opposite observation: that beetles were abundant in high-limonene stands and rare in low-limonene stands. Then we could argue that limonene is not an effective defense and did not evolve in response to beetle attack. Or we could argue that the presence of beetles has selected for high limonene content, to which the beetles have subsequently adapted. In the absence of historical and physiological information, then, it is possible to interpret diametrically opposed observations in exactly the same way—or in any of several ways. To show that a compound has evolved as a defense against a specific insect, it appears necessary to show that the compound is repellent or toxic to populations of the insect that have not had the opportunity to counterevolve (i.e., to coevolve) resistance to it. To show that coevolution by the insect has taken place (i.e., that the next step, evolution of resistance, has occurred), it appears necessary to show that populations of the insect that are exposed to the defense are more resistant to it than populations that have not been exposed. But if a third coevolutionary step occurs, that is, evolution by the plant of a defense against the insect's resistance, the insect may be eliminated from plant populations to which it was formerly adapted, so that the populations required to demonstrate step 2 may not exist.

It may therefore be extraordinarily difficult to demonstrate the stages of reciprocal coevolution, or even to show that it has occurred at all. It is even difficult to show that toxic plant compounds have evolved because of their effect on insects. Many studies have some, but not all, of the necessary ingredients. For example, Dolinger et al. (1973) showed that populations of lupines (*Lupinus*) that are phenologically protected from a lupine-feeding caterpillar (*Glaucopsyche lygdamus*) have lower alkaloid content than those that are heavily damaged by the caterpillars. One's first reaction is to suppose that herbivory by *Glaucopsyche* has favored the evolution of high alkaloid content. (Note, incidentally, that this is the reverse of the current association between limonene in ponderosa pine and the abundance of pine beetles.) But *Glaucopsyche* larvae show higher growth and survival when reared on lupines with high alkaloid levels than low levels! Perhaps lupines did evolve alkaloids in response to *Glaucopsyche*, but, if so, the insect has counterevolved, and the reasons for high alkaloid content are lost in evolutionary history.

THE DIVERSITY OF PLANT-ASSOCIATED INSECT FAUNAS

Plant–insect associations are likely to undergo reciprocal coevolution only in those special cases in which a plant harbors a few specially adapted insects. The majority of plant species, however, have a complex fauna of both specialized (oligophagous) and generalized (polyphagous) species (Lawton and Schröder, 1978; Futuyma and Gould, 1979). The number and composition of insect species on a plant is

218

partly determined by the plant's taxonomic (and presumably chemical) distinctness from other plant species in the community (Holloway and Hebert, 1979; Connor et al., 1980; Opler, 1974), but a larger fraction of the variance in species number is explained by the geographical range of the plant (Southwood, 1973; Opler, 1974; Strong, 1979) and its geometrical structure—complex plants such as trees harbor more species than simpler plants such as herbs (Lawton and Schröder, 1978; Strong, 1979; Southwood, 1973; Moran, 1980; Auerbach and Hendrix, 1980). Much of the increased insect diversity on widespread plants is made up of polyphagous insects that differ from one place to another (Lawton and Schröder, 1978). Thus, the fauna of most plants includes polyphagous species that show no evidence of special coevolved adaptation, and specialized species that appear specifically adapted to a particular group of plants. It appears, then, that specialization is not required for successful feeding on most plants, yet most plants harbor some specialists. Why, then, are so many insects specialized in diet?

GENETIC ASPECTS OF HOST UTILIZATION

A population faced with a spectrum of possible resources such as host plants may be monomorphic for a generalized (polyphagous) genotype if its fitness on each resource is not lower than that of specialized genotypes. The population can be stably polymorphic, consisting of a set of specialized genotypes (a "multiple-niche polymorphism") if there is marginal overdominance (heterozygotes have highest fitness, averaged over resources; Levene, 1953) or if there is frequency-dependent selection with each genotype being most fit on a specific resource (see Bulmer, 1974; Maynard Smith, 1966; Hedrick et al., 1976; Felsenstein, 1976). If the densities of the subpopulations are independently regulated on each resource, selection can favor the origin of reproductive isolation between the specialized subpopulations, yielding, under certain circumstances, sympatric origin of host-specific species (Mather, 1955; Maynard Smith, 1966; Bush, 1974; Futuyma and Mayer, 1980). If, however, a particular specialized genotype has highest fitness, the population becomes monomorphic and specialized, without speciation. Several authors have postulated that sympatric speciation is responsible for the origin of many host-specific insects (Thorpe, 1930; Bush, 1974; M. J. D. White, 1978). Other authors (Mayr, 1947; Felsenstein, 1981; Futuyma and Mayer, 1980) have emphasized, however, that such speciation will occur only under rather restrictive genetic conditions and do not find the evidence for sympatric speciation convincing. The entomological literature uses the term *host race* to refer to conspecific populations that differ genetically in host affiliation. In the agricultural literature, *biotype* is a frequently used, but poorly defined, term for insect samples that differ genetically in some trait of interest, such as their response to a particular species or strain of crop. The genetic analysis of biotypes is best developed for the Hessian fly (*Mayetiola destructor*), which is the one definite case in insects of a "gene-for-gene" relationship of the kind described for plants and pathogenic fungi (Flor, 1956; Chapter 7 by Barrett). For each of four nonallelic dominant genes (e.g., H_1, H_2, H_3, H_4) that confer resistance to Hessian fly in wheat, a recessive, nonallelic gene in the fly (*a, b, c,* or *d*) confers counteradaptation (Hatchett and Gallun, 1970; Gallun, 1977). Virulent fly genotypes have rapidly increased in North American wheat-growing areas within this century, stimulating the development and planting of new resistant wheat strains. Fly genotypes (e.g., *aa bb cc dd*) capable of attacking the most broadly resistant (H_1_H_2_H_3_H_4_) wheat strains have become abundant in some areas (Sosa, 1981). Adaptation of the fly to wheats with multiple resistance factors does not seem to preclude adaptation to wheats with fewer resistance factors.

Some populations of the brown planthopper (*Nilaparvata lugens*) similarly have responded to widespread planting of resistant strains of the host, rice, by rapidly evolving adapted biotypes (Panda, 1979; Pathak, 1977; Kaneda and Kisimoto, 1979). Biotypes are distinguished by their differential ability to attack strains of rice that differ in resistance by single genes. Each biotype appears to be a genetically heterogeneous entity, and the differences between biotypes are probably polygenically inherited (Claridge and den Hollander, 1980; den Hollander and Pathak, 1981). Again, adaptation to a more resistant rice variety does not reduce adaptation to other varieties (den Hollander and Pathak, 1981).

Biotypes are more prevalent in aphids (Easton, 1973b; Pathak, 1970), probably because specific combinations of genes are propagated for many generations by parthenogenesis (Mitter and Futuyma, 1982). Sympatric genotypes are often found on different cultivars [*Amphorophora rubi* on raspberry (Briggs, 1965); *Dysaphis devecta* on apple (Alston and Briggs, 1977); *Therioaphis maculata* (Nielson and Don, 1974) and *Acyrthosiphon pisum* (Cartier et al., 1965) on alfalfa; *Rhopalosiphum maidis* on sorghum (Cartier and Painter, 1956)] or on different species—*Acyrthosiphon pisum* (Frazer, 1972; Auclair, 1978; Müller, 1980) and *Aphis nasturtii* (Eastop, 1973b). Müller (1971, 1980) finds morphological differences and postmating isolation among host-specific forms of *Acyrthosiphon pisum*. There have been few genetic analyses of aphid biotypes and few mechanistic analyses of their host-specific adaptations. Each biotype clearly grows best on its own host, but this may be the consequence not of physiological adaptations to

220

different plant chemistry, but of differences in behavioral responses to repellent or stimulatory chemicals (Nielsen and Don, 1974).

Differences in host affiliation exist among parthenogenetic genotypes of the fall cankerworm (Mitter et al., 1979; Schneider, 1980). Pronounced differences in clone frequencies exist between neighboring stands of maple and oak, and larvae of a maple-associated genotype are more inclined to eat maple than those of an oak-associated genotype (Futuyma et al., unpublished). The "maple genotype" survives better on maple than the "oak genotype" does, but they survive equally well on oak (Futuyma et al., 1981). Our preliminary tests suggest that the major differences among genotypes are not in their efficiency of assimilation of either kind of foliage but in behavior and phenology, which may synchronize hatching with the phenology of the host species (Mitter et al., 1979).

There is much less evidence of sympatric host race formation in sexual than in parthenogenetic insects. Edmunds and Alstad (1978) have described local adaptation of scale insect demes to individual pine trees within a single stand; this differentiation is undoubtedly fostered by the extremely low mobility of these insects, which may persist for many generations on a single tree. Host-associated genetic variation seems not to exist in polyphagous species of geometrid moths (Mitter and Futuyma, 1979), fungus-feeding *Drosophila* (Jaenike and Selander, 1979), or wheat stem sawfly (Callenbach, 1951). Several putative cases of host races apparently are really sibling species, as in the fall webworm (Jaenike and Selander, 1980), the *Enchenopa binotata* complex of treehoppers (Wood, 1980; Guttman et al., 1981), and the ermine moths of the *Yponomeuta padellus* complex (Herrebout et al., 1976; Hendrikse, 1979). From the published evidence, it is not entirely clear whether or not the putative host races of the apple maggot *Rhagoletis pomonella* (Bush, 1974) are genetically distinct (Futuyma and Mayer, 1980).

Different geographical populations, however, are in some cases known to differ genetically in host preference and, of course, to encounter different hosts, to which they may or may not be differentially adapted (Fox and Morrow, 1981). Californian populations of the codling moth have adapted to walnut and plum by changes in phenology and oviposition preference (Phillips and Barnes, 1975) in regions where these crops are abundant and the ancestral host, apple, is rare. Populations of the Colorado potato beetle (*Leptinotarsa decemlineata*) from Arizona are behaviorally and physiologically better adapted to a locally abundant host, *Solanum elaeagnifolium*, than beetles from regions where this host is not abundant (Hsiao, 1978). The most widespread host, *Solanum rostratum*, is preferred by all populations; thus, Hsiao suggests that the shift to *S. elaegnifolium* required geographical isolation. Local populations of the butterfly *Euphydryas editha* also differ in host preference (Singer, 1971) and in physiological adaptation of each population to the locally abundant species of host plant (Rausher, 1982). Some local populations of swallowtail butterflies have higher growth efficiency on their usual hosts than on hosts used by other populations of the species (Scriber, 1983).

Remarkably little is known about the genetic basis of interspecific differences in host utilization. Huettel and Bush (1972) interpreted the oviposition preference of crosses between two host-specific tephritid fly species as being predominantly due to only a few genes, but the data permit a polygenic interpretation (Futuyma and Mayer, 1980). The sawfly complex *Neodiprion abietis* includes sympatric forms that differ in oviposition response to, and ability to survive on, balsam fir and spruce. These forms apparently do not interbreed in nature, despite seasonal overlap, and may be full species. Hybrids produced in the laboratory have oviposition preferences and feeding behavior compatible with the hypothesis that each trait in the fir form may be controlled by a single dominant gene (Knerer and Atwood, 1973).

The evidence to date suggests to me that most genetic differentiation in host plant utilization occurs in populations that are spatially isolated, although no doubt the isolation is often very local. Within populations, recombination and gene flow may prevent the sympatric formation of different gene complexes required for the several behavioral, physiological, and phenological adaptations that sympatric host race formation may often require. Linkage disequilibrium—the packaging of coadapted sets of polymorphic loci—appears rare in natural populations except in the case of chromosome inversions (Charlesworth and Charlesworth, 1973; Langley et al., 1978), and genetic correlations between host preference and physiological adaptation are not well documented within natural populations (but see Cavener, 1979, and Gelfand and McDonald, 1980, for a correlation between alcohol dehydrogenase activity and behavioral response to ethanol in *Drosophila*). However, when recombination is suppressed, as by parthenogenesis, sympatric "host races" can persist.

PHYSIOLOGICAL RESPONSES OF INSECTS TO PLANTS

Whether host shifts occur in allopatry or sympatry, the adaptive response to plant chemistry can, in principle, include evolution of both behavior and physiology. In many insects, the female lays eggs on the larval host; both the larva and especially the adult female must therefore be genetically programmed to seek and accept the plant. The feeding stage (often, as in Lepidoptera, the immature stage only) must not only accept the host, but also must be able to overcome its chemical and and physical defenses, and be able to obtain necessary nutrients [especially carbohydrates, sterols, and nitrogen, which is often limiting to growth (Slansky and Feeny, 1977; Scriber and Slansky, 1981)]. In some insects, specific chemical properties of the host are essential to yolk production and reproduction (Labeyrie, 1978). Different parasites, predators, and competitors are often associated with different species of host plants (Gilbert and Singer, 1975; Vinson, 1976; Price et al., 1980).

Plants do not generally differ substantially in the nutrients required by insects, and dietary imbalance is not the basis of host specificity (Beck and Reese, 1976; Fraenkel, 1969). In contrast, some oligophagous insects clearly possess special adaptations to the toxic properties of their host plants. *Drosophila pachea* is the only desert *Drosophila* that can survive the alkaloids peculiar to its host, the cactus *Lophocereus schotti*; it also requires a sterol provided by this cactus (Kircher et al., 1967). The allyl glucosinolate of cruciferous plants is toxic to noncrucifer feeders but not to the crucifer-feeding cabbage butterfly larva (Blau et al., 1978). Nicotine is highly toxic, except to both polyphagous and oligophagous insects that feed on tobacco; they avoid toxication by efficient excretion of the nicotine (Self et al., 1964; Brattsten, 1979). The bruchid beetle *Caryedes brasiliensis* feeds only on the seeds of *Dioclea*, and not only degrades, but uses for its own nitrogen metabolism, the abundant L-canavanine in the seeds, which is highly toxic to other insects (Rosenthal et al., 1978).

The major detoxifying mechanisms possessed by insects are generally, however, extraordinarily nonspecific, and it is likely that adaptation to specific plant toxins often requires very little genetic change, if any. The major detoxifying system is the mixed-function oxidase (MFO) system, consisting primarily of cytochrome P-450 and an associated reductase, that carries out the hydroxylation, epoxidation, and demethylation of an extraordinarily diverse array of lipophilic compounds (Agosin and Perry, 1974; Dauterman and Hodgson, 1978; Brattsten, 1979). The MFO system is ubiquitous among higher organisms and apparently shows no differential substrate specificity in different species. These enzymes can be induced by any of a vast number of toxic compounds and, when induced to higher levels, confer protection against chemically unrelated compounds (Brattsten et al., 1977). The evolution of insecticide resistance often entails the evolution of higher MFO activity, and insects selected for resistance to one insecticide are usually cross-resistant to many other, chemically unrelated, insecticides (Agosin and Perry, 1974), whether resistance is achieved by this or by other mechanisms. [Conversely, phytophagous mites selected for resistance to a toxic plant, cucumber, were cross-resistant to unrelated plants and to a variety of insecticides, although in this case the mechanism of resistance is unknown (Gould et al., 1982).] Polyphagous species of Lepidoptera have higher MFO activity than oligophagous species (Krieger et al., 1971), and it is quite possible that a difference in level, rather than kind, of enzyme is the major physiological adaptation to a broad spectrum of plant toxins.

Nonspecificity is characteristic not only of the MFO system but of many other detoxification systems as well. Conjugation of a wide variety of substrates is accomplished by glutathione S-transferases and other enzymes that are found in both specialized and generalized insects (Dauterman and Hodgson, 1978). Milkweed cardiac glycosides and other compounds are sequestered in the body of the monarch butterfly and some other milkweed-feeding insects (Roeske et al., 1976), but the ability to sequester any of a wide variety of compounds is typical of many insects, not only those that specialize on toxic plants (Rothschild and Marsh, 1978; Rothschild, 1973). It appears, therefore, that most insects are preadapted to deal with a great many compounds that they do not normally encounter and that this pre-adaptation is a by-product of their adaptation to degradation products of their own metabolism or to the harmful compounds that they naturally ingest. Genetic adaptation to the chemistry of a new plant must often entail minor quantitative changes in the levels of detoxifying or sequestering enzymes. Adaptation to "digestability-reducing compounds" such as tannins probably also entails slight genetic change in characteristics such as the pH of the midgut (Berenbaum, 1980; see also Bernays, 1978). If insects can so readily adapt to foreign compounds like insecticides, they probably can adapt to plant toxins with equal facility and rapidity.

It seems reasonable to suppose that the energetic "cost" of detoxifying a compound might be greater for polyphagous species than for specially adapted oligophagous species (Dethier, 1954). Feeding efficiency (growth in larval biomass/amount of food assimilated) appears to be more strongly correlated with factors such as water content of the food than with specialization of diet, however (Scriber and Feeny, 1979), and armyworms feed as efficiently on plants containing cyanogenic glycosides as on conspecific plants that lack them (Scriber, 1978). Addition of some toxic compounds, but not others, to the diet can lower feeding efficiency, presumably by inducing the energetically costly production of high levels of detoxifying enzymes (Schoonhoven and Meerman, 1978).

Whether or not specialists are more efficient than generalists seems to vary. Auerbach and Strong (1981) report higher feeding efficiency for specialists than for unrelated generalists feeding on *Heliconia*, a plant in the banana family. Congeneric species of generalists and specialists have been compared on a common host plant by Scriber (1979, 1982), who found that some specialized species of swallowtail butterflies utilize their hosts more efficiently than do generalists. In contrast, Futuyma and Wasserman (1981) found no difference in efficiency of the polyphagous tent caterpillar *Malacosoma disstria* and its oligophagous relative *M. americanum* when reared on cherry, the primary host of *M. americanum*. Smiley (1978) has also suggested that host specificity does not confer increased feeding efficiency, and entomologists are familiar with many cases in which a specialized species can grow quite well on plants utterly unrelated to the normal host (see, for example, Waldbauer and Fraenkel, 1961; Hsiao, 1969).

BEHAVIORAL RESPONSES TO PLANT COMPOUNDS

There is every reason to suppose that in many cases, the initial shift to a new host entails genetic change only in behavior; that genetic change in physiology is not immediately necessary for successful change of host but is a fine tuning of adaptation that occurs only after the species has already become specialized on a particular host. There is much to be said for Dethier's (1954, 1970) position, that host specificity in insects is very largely a consequence of behavioral responses to attractant and repellent chemicals, rather than specialized physiological adaptation.

The literature on behavioral responses to plant chemicals is enormous. Some summaries are provided by Schoonhoven (1968, 1972), Dethier (1976), and Mitchell (1981). Both oviposition and feeding are commonly governed by complex reactions to stimulatory and inhibitory chemicals. Feeding (Jermy, 1966) and oviposition (Jermy and Szentesi, 1978) in oligophagous insects can be inhibited by any of a wide variety of compounds, including some that occur in the hosts that the species uses (Hsiao, 1969). Many repellent chemicals are extremely general in effect, repelling even nonphytophagous insects (Eisner, 1964). It is likely that feeding by polyphagous insects is elicited by the absence of repellent stimuli, combined with nonspecific stimulants such as sugars (Fraenkel, 1969). For many oligophagous insects, feeding depends on a balance between repellents and specific compounds of the host that are required as attractants or stimulants. Thus, the glucosinolates of cruciferous plants stimulate feeding in crucifer-associated Lepidoptera (Gupta and Thorsteinson, 1960a; Schoonhoven, 1969) and Coleoptera (Nielsen, 1978), and hypericin in the Klamath weed *Hypericum* stimulates feeding in the host-specific

225

beetle *Chrysolina brunsvicensis* (Rees, 1969). Both kinds of compounds are toxic to nonadapted insects. Cucurbitacins stimulate feeding in some insects that specialize on Cucurbitaceae but are repellent and even toxic to other species (DaCosta and Jones, 1971; Nielsen, 1978; Howe et al., 1976). Although in some instances certain chemosensory cells are stimulated specifically by certain host compounds such as glucosinolates or hypericin, most chemosensory receptors are not compound-specific; feeding and oviposition are governed by the central nervous system, which integrates patterns of stimuli from the chemoreceptors (Dethier, 1976).

The sibling species of small ermine moths (*Yponomeuta*) illustrate the interplay of stimuli in the evolution of host choice. Among the European species are *Y. cagnagellus*, which feeds only on *Euonymus* (Celastraceae); *Y. evonymellus*, which feeds on *Prunus* (cherry, in the Rosaceae) but will accept *Euonymus* rather readily in the laboratory; and *Y. malinellus*, which feeds on *Malus* (apple, also Rosaceae) and will not accept *Euonymus* (Gerritts-Heybroek et al., 1978). *Euonymus* is thought to be the host of these moths' common ancestor. *Euonymus* contains dulcitol, *Prunus* prunasin, and *Malus* phloridzin. Dulcitol evokes a neurophysiological response in taste receptors and acts as a feeding stimulant for *Y. cagnagellus* and *Y. evonymellus*, but not for *Y. malinellus*. Phloridzin evokes a response in taste receptors but acts as a feeding deterrent for *Y. cagnagellus* and *Y. evonymellus*, but not for *Y. malinellus*. Prunasin evokes a sensory response in all the species and acts as a deterrent to *Y. cagnagellus*; it neither stimulates nor deters *Y. evonymellus* from feeding (van Drongelen, 1980). In the F_1 hybrid between *Y. cagnagellus* and *Y. malinellus*, the sensory response to dulcitol and prunasin is equal to that of *Y. cagnagellus* and so acts as if inherited by dominant genes; the response to phloridzin is intermediate between the parents (van Drongelen and van Loon, 1980). One could plausibly explain the shift from *Euonymus* (*Y. cagnagellus*) to *Prunus* (*Y. evonymellus*), therefore, by a few simple genetic changes: for example, the loss of sensitivity to the deterrent effect of prunasin in *Y. evonymellus* (a change in the central nervous system, because prunasin has the same sensory effect on *Y. evonymellus* and *Y. cagnagellus*) and the development of a positive response to an unidentified *Prunus*-specific compound. The shift to *Malus*, in *Y. malinellus*, would follow easily from a loss of the positive sensory response to dulcitol and loss of perception of the *Malus*-specific feeding inhibitor phloridzin.

SELECTIVE FACTORS FAVORING SPECIALIZATION

Clearly a plant compound that acts as a feeding deterrent or toxin to many insects and that has been evolved possibly as a defense against herbivory can serve as an attractant or stimulant to insects that become specialized on this very plant. Although some plant repellents are also toxic, many are not; and the nontoxic repellents seem to act merely as sign stimuli for insects that are physiologically preadapted to feed on a plant and that need merely to overcome the behavioral barrier of repellence or absence of positive stimuli (Dethier, 1970). To mount an effective defense against most potential herbivorous insects, a plant need not be toxic; it may suffice to be chemically repellent, or unrecognizable as food. Clearly, the chemical phenotype of most plants protects them against most specialized insects: composites, for example, are not attacked by cabbage butterflies simply because they do not have the glucosinolates that attract cabbage butterflies to crucifers. Defense against more broadly polyphagous species can similarly be effective at a purely behavioral level: the plant need merely clothe itself in a repellent. Such a defense may be a facade, masking a fully palatable diet, and some insects indeed break through such facades: many insects feed on hundreds of plant species in dozens of families. Given the ubiquity of broad-spectrum detoxifying mechanisms among insects and the ease with which insects adapt to both resistant crops and a great variety of toxins, the selective reasons for dietary specialization seem less likely to lie in toxic barriers than in more ecological factors. Gilbert and Smiley (1979) have similarly concluded that such "ecological monophagy" is a major kind of dietary specialization. There are several reasons why "ecological monophagy" may evolve.

First, many insects derive a positive advantage from sequestering in their bodies specific plant toxins that provide protection against predators and parasites (Campbell and Duffy, 1979; Gilbert and Singer, 1975; Brower et al., 1968; Rothschild, 1973; Roeske et al., 1976). This is a major reason for aposematism in many groups of insects. The best-known examples are in milkweed specialists such as the monarch butterfly. The cardiac glycosides of milkweeds are toxic to vertebrate predators but apparently not to milkweed-feeding insects, or to insects generally.

Second, many parasitoids are attracted to specific plant compounds, and they search for their insect hosts on some plants in preference to others. Parasitism and predation may therefore favor adaptive radiation of herbivorous insects into the "enemy-free space" (Gilbert and Singer, 1975) that many plants provide. Price et al. (1980) cite many examples of herbivorous insects that experience higher parasitism on some hosts than on others.

Third, in at least some cases (e.g., Whitham, 1978) populations of phytophagous insects experience density-dependent competition for resources. Interspecific competition may therefore also exist and foster specialization as it presumably does in other organisms. Patterns of resource utilization in communities of grasshoppers suggest that competition is reduced by specialization (Joern and Lawlor, 1980), but there is no such suggestion in the pattern of host-plant usage by stem-boring insects (Rathcke, 1976). Lawton and Strong (1981) have reviewed this topic and found little evidence for interspecific competition in phytophagous insects (but see Stiling, 1980, for a counterexample).

Fourth, the neurological sophistication of insects is probably insufficient to distinguish among all possible palatable and unpalatable plants. In at least some species, feeding preferences are not well correlated with the suitability of the plants for growth (Chew, 1980; Gerrits-Heybroek et al., 1978), and the correlation between the plants on which females oviposit and those on which the larvae grow best is far from perfect (Wiklund, 1975). Ovipositing females of the North American butterfly *Pieris napi macdunnoughii* do not distinguish between one of its suitable native hosts, *Descurainia*, and two introduced European crucifers, on which the larvae cannot survive; the plants share certain glucosinolates that probably act as oviposition stimuli (Rodman and Chew, 1980). In such a case, selection could favor genotypes that can survive on unsuitable hosts, but it could also favor genotypes that could distinguish the suitable from unsuitable plants or genotypes that abandoned all three and confined their oviposition to other native crucifers; the latter genotypes would have a more specialized diet. Levins and MacArthur (1969) have shown theoretically that the penalty of making mistakes and ovipositing on toxic plants that cannot be distinguished from palatable species can select for specialization on a common suitable host.

Fifth, optimal foraging theory (Pyke et al., 1977) predicts that given a spectrum of equally palatable food types, a species should accept only those for which the rate of acquisition of reward (e.g., energy consumed or eggs laid) is highest. The optimal food types will be those that are richest in energy (or other requisites), most abundant (minimizing search time), and least costly in terms of handling time. In vertebrates, the formation of a "specific searching image," whereby an individual focuses on the most abundant food and ignores less abundant items, increases the rate of acquisition (Dawkins, 1971; Murton, 1971), and a similar search image appears to increase the rate of plant discovery and oviposition by the butterfly *Battus philenor* (Rausher, 1980). The existence of the search image phenomenon implies that sensory recognition of prey is more efficient when the animal focuses on one prey type (or host) and ignores the rest of the complex, confusing world, even if that world harbors less abundant, though palatable,

228

items. A genotype with a genetically fixed search image for a particular host may therefore have a higher oviposition rate than a more generalized genotype, if the specific host is sufficiently abundant (Futuyma, 1983). Laboratory cultures of bruchid beetles, when reared for several generations on a single kind of bean, evolved a preference for this bean species; the base population oviposited on several bean species more indiscriminately (Wasserman and Futuyma, 1981). Thus, even in the absence of choice, a tendency toward monophagy can evolve—as it might in a allopatric population where a particular host is abundant (Hsiao, 1978). Crop pests such as the codling moth (Phillips and Barnes, 1975) and the many aphids referred to earlier show that preference can rapidly evolve for locally abundant species of plants.

It is sometimes supposed that a polyphagous species may become host-specific by acquiring the ability to detoxify a particular plant's secondary compounds and that a specialized genotype of this kind will replace the polyphagous genotype because it does not incur the energetic cost of detoxifying a broad range of compounds. There is, however, little support for this view, in that the energetic advantage of specialization appears to be slight, if it exists at all. Moreover, selection for a specialized detoxification ability will be strong only if the genotype is freed from selection for adaptation to other plants. This requires that the genotype be forced to occupy the host to which it is best adapted physiologically, either because only one host is commonly available or perhaps by virtue of a correlation between host preference and physiological adaptation. There is little evidence for such correlations in sexually reproducing species (although they seem to occur in parthenogenetic forms). Specialization is thus likely to evolve first at the behavioral level, for any of the several reasons described earlier. Once a population is specialized, selection will favor the host-specific detoxification abilities that are characteristic of many specialized species. Specialized species, once evolved, probably tend to give rise to other specialized species more often than they revert to polyphagy, for a dietary specialist doesn't recognize most plants as potential food.

THE NATURE OF COEVOLUTION BETWEEN PLANTS AND PHYTOPHAGOUS INSECTS

In the narrow sense, coevolution is an evolutionary change in one population in response to a second species, followed by an evolutionary response of the second species to the change in the first (Janzen, 1980a; Chapter 1 by Futuyma and Slatkin). Such reciprocity has not been demonstrated for any complex of plants and herbivorous insects. It is most likely to occur when an insect has few host species and its hosts harbor few insect enemies (Fox, 1981). In such "subcommunities," some plausible cases have been described in which plant properties have evolved in response to the properties of specific insects (e.g., egg mimics in *Passiflora*, terpenes in ponderosa pine), and other cases have been described in which insect properties have evolved in response to the characteristics of one or a few, related, plant species (e.g., adaptation of tobacco hornworm to nicotine, of cabbage butterfly to sinigrin, of *Chrysolina* to hypericin). In the vast majority of cases, however, the defensive properties of plants appear to be broad-spectrum adaptations to a very large suite of enemies, including not only insects, but vertebrate herbivores and pathogens. And many of the characteristics of insects (e.g., mixed function oxidases) that enable insects to deal with plant compounds confer resistance against a broad spectrum of chemicals. Specialized insects often have specific adaptations to the defensive properties of their hosts but seldom is there evidence that those defensive properties were evolved by the host in response to those particular species of specialized insects. Especially on the part of plants, coevolution—or, to be more exact, adaptation to herbivory—appears to be extremely diffuse and so perhaps does not merit the term *coevolution* at all.

As Janzen (1980a) has pointed out, Ehrlich and Raven (1964) did not define coevolution in their classic paper, "Butterflies and plants: a study in coevolution." Although they speak of the "stepwise" evolutionary responses of plants and butterflies to each other, Ehrlich and Raven clearly meant to describe these responses as a series of successive adaptive radiations. They supposed, first of all, that vascular plants acquired a diverse spectrum of secondary compounds early in their history, presumably in response to herbivores that existed at that time. The evolution of a new chemical defense that freed a plant from herbivory could permit it to enter, and diversify in, a new adaptive zone. The diversity of butterflies, which evolved subsequently, "has been elaborated against a dicotyledonous background." Ehrlich and Raven attributed the diversification of specialized feeding habits in butterflies primarily to avoidance of competition, but they noted that other factors could also favor specialization. They gave no reason to suppose that the adaptive radiation in the secondary compounds of angiosperms has occurred in response to, or been furthered by, the dietary diversification of the butterflies. Ehrlich and Raven merely attributed the adaptive radiation of the butterflies to the prior adaptive radiation of angiosperm chemistry, which presumably occurred in response to Cretaceous herbivores that probably weren't even related to butterflies. Thus, Ehrlich and Raven's use of the term *coevolution* did not refer to reciprocal adaptations of interacting lineages; nor did it

230

refer to cospeciation (parallel cladogenesis). It referred to the adaptive utilization by a group of insects of resources that had already diversified in response to selection pressures that had been similar in kind (herbivory) but not in identity (herbivore species). The evidence at present supports this view. Insects evolve adaptations to plants, and plants evolve adaptations to herbivory, but seldom if ever are the genetic changes in plants and insects highly coupled in an "arms race." Genetic changes in pairs of interacting species, each promoted by change in the other, have not yet been documented for plants and insects.

CHAPTER 11

DISPERSAL OF SEEDS BY VERTEBRATE GUTS

Daniel H. Janzen

The function of a ripe vertebrate-eaten fruit is to get the seeds into the right animals and keep them out of the wrong animals. This process occurs in the context of fruit and seed morphology, chemistry, and phenology that (1) protect the ripe fruit from the wrong animals and microbes, (2) protect the developing as well as mature seeds from climate and seed predators, (3) are an arena for parental rejection of offspring through fruit and seed abortion, (4) influence photosynthesis in the developing embryo and parental fruit tissue, and (5) suffer the expected economic and strategic restraints on offspring production. In the limit, the success of these functions can only be measured through the inclusive fitness of parents and offspring. The fitness of the parent plant may be either raised or lowered by the consumption of its fruits and seeds and, on occasion, even by the digestion of its offspring; there are great opportunities for evolutionary and ecological adjustments in this system of reciprocal parasitism ("mutualism" of other authors), and direct and diffuse coevolution between the animals and plants may occur as well. However, coevolution may not have occurred in any particular case (Janzen, 1980a).

To examine the evolutionary aspects of seed dispersal via vertebrate guts, I first offer a brief eclectic review of the literature and then briefly characterize the interaction and ask that you keep this portrayal in mind in reading the remainder of my chapter.

THE LITERATURE OF SEED DISPERSAL BY ANIMAL GUTS

The literature of dispersal of seeds by animal guts is widely scattered; prior to 1970 it was almost always imbedded in a study of something other than seed dispersal. Ridley (1930) did the world an enormous service in compiling almost everything that had been written that could be construed as seed dispersal. While we now have a few papers directed explicitly at the interaction of the animals and plants, with respect to the fitness of both (see later), for no species of plants has there ever been an intensive analysis of the interaction from flower to reproductive plant with the coterie of animals involved, integrated with the lives of the animals. The literature of animal-mediated seed dispersal is less developed than that of animal-mediated pollination. The causes are easy to identify (and see Wheelwright and Orians, 1982). Merely observing animals swallowing fruit and seed does not describe the process of dispersal, the end point of which is the location and fate of the seeds; this end point is far away in space and time. Pollination involves both removal of pollen (an essentially unstudied subject) and bringing pollen in; incoming pollen-bearing animals (or animals imagined to be pollen-bearing) are easy to observe. Second, understanding pollination and controlling it are of enormous economic importance; modern humans benefit from the harvest, rather than dispersal, of seeds.

The three essentials of seed dispersal by animals—who moves seeds, where seeds land, and what happens to a seed because it landed there—have never been the subject of a book-length review, and research on this sequence as a whole is only now beginning to receive attention. Although just as in floral biology there are many useful interspecific comparisons to be made among animal-mediated seed dispersal systems (e.g., Fleming, 1979, van der Pijl, 1957, 1966, 1972; Herrera 1982a,b; Foster, 1978; McKey, 1975; Wheelwright and Orians, 1982; Snow, 1981; Janzen and Martin, 1982; E. W. Stiles, 1980), we are now desperately in need of in-depth studies of just what *does* happen to the seeds for particular species, and who or what is responsible.

Who moves seeds

Recording who eats fruits (and sometimes, whether the seeds are swallowed) has been a favorite occupation of ornithologists (e.g., Leck, 1969, 1970a, 1971a,b, 1972a,b,c; Leck and Hilty, 1968; Greenberg, 1981; Frith et al., 1976; Kantak, 1979; Land, 1963; Baird, 1980; Robbins et al., 1975; Jordano, 1982; Herrera and Jordano, 1981; Racine and Downhower, 1974; Edson, 1918; Cruz, 1974, 1981; Howe and Vande Kerckove, 1979, 1981a,b; Foster, 1977a,b, 1978; Proctor, 1968; Noble, 1975a,b; McAtee, 1906, 1947; Temple, 1977; Willis, 1966; Phillips, 1910; Snow and Snow, 1971; Snow, 1962, 1976, 1981; Herrera, 1981b; Janzen, 1979a, 1981d; Grant and Grant, 1981) and of primate watchers (e.g., Jones, 1970, 1972b; Struhsaker, 1975; Lieberman and Lieberman, 1980; Lieberman et al., 1979; Gautier-Hion, 1980; Hladik, 1973; Hladik et al., 1971; Howe, 1980; Waser, 1977; Hladik and Hladik, 1967, 1969, 1972; Freeland, 1979b; Wickler and Seibt, 1976; Freese, 1977, 1978; Oppenheimer, 1977; Milton, 1980; Muskin and Fischgrund, 1981; Klein and Klein, 1973; Molez, 1976). Consumption of fruits by large terrestrial mammals is generally recorded as part of a larger list of their food (e.g., Sikes, 1971; Gautier-Hion et al., 1980; Anonymous, 1960; Hamilton et al., 1977; Jouventin, 1975; Hitchins, 1968; but see Alexandre, 1978; Janzen and Martin, 1982) as is generally the case with fishes (e.g., Hochreutiner, 1899; Goulding, 1980; Smith, 1981; Gottsberger, 1978). Bats have long been known as important movers of seeds (e.g., Jones 1972a; Greenhall, 1956; Fenton and Fleming, 1981; Gardner, 1977; Mutere, 1973; Vogel, 1969; Osmaston, 1965; Meijer, 1969; van der Pijl, 1957; Heithaus et al., 1975; Heithaus and Fleming, 1978; Fleming, 1981; Janzen et al., 1976) but probably because they can be "viewed" only by radio tracking, they have been better studied as dispersers than have been even the more readily visible birds (e.g., Fleming, 1981; Fleming and Heithaus, 1981; Morrison, 1975, 1978a,b,c,d, 1979, 1980; Daniel, 1976; Jimbo and Schwassman, 1967; August, 1981; Bonaccorso, 1979; Williams, 1968; Williams and Williams, 1967; Heithaus and Fleming, 1978; Jones, 1972a; Wickler and Seibt, 1976). Although not the subject of this chapter, seed dispersal by squirrels and other forest rodents (e.g., C. C. Smith, 1970, 1975, 1981; Smith and Balda, 1979; Morris, 1962; Fox, 1974; Stapanian and Smith, 1978; Sork and Boucher, 1977; Emmons, 1975, 1980; Abbott and Quink, 1970; Barnett, 1977; Calahane, 1942; Murie, 1977; Bonaccorso and Sanford, 1979; Bonaccorso et al., 1980; Shaw, 1936; Dennis, 1930; West, 1968; Rahm, 1972; Heaney and Thorington, 1978; Lloyd, 1968), seed predator corvids (e.g., Balda, 1980; Smith and Balda, 1979; Silvertown, 1980; Vander Wall and Balda, 1977, 1981; Lanner and Vander Wall, 1980; Darley-Hill and Johnson, 1981; Bossema, 1979; Tomback, 1978), arid land rodents, birds, and ants (Brown et al., 1979a,b; Hay and Fuller, 1981; O'Dowd and Hay, 1980; Lockard and Lockard, 1971; Clark and Comanor, 1973; Price, 1978; Reichman, 1979; Shaw, 1934; M'Closkey, 1980; Munger and Brown, 1981) and forest ants (Davidson and Morton, 1981; Heithaus, 1981; Culver and Beattie, 1978; Horvitz and Beattie, 1976; Beattie and Culver, 1981; Thompson, 1981) must be considered by anyone planning studies in seed dispersal through animal guts.

Quantitative studies of the entire set of animals removing fruits from a given fruit crop (e.g., Bonaccorso and Sanford, 1979; Salomonson, 1978; Howe, 1980; Bonaccorso et al., 1980) are almost nonexistent.

Although fruits are generally recognized as "bait" for the dis-

234

perser, until very recently (e.g., McDiarmid et al., 1977; Foster, 1977b; Herrera, 1981a,c, 1982a,b; Stapanian, 1980; S. C. White, 1974; Hladik et al., 1971) fruit tissue analyses have been from a human viewpoint (e.g., Chatfield and McLaughlin, 1928; Muell et al., 1949; Abdel-Rahman, 1977; Iwagaki and Kudo, 1977; Rathore, 1976; Soegeng, 1962; Meijer, 1969; Auda et al., 1976; Nagy and Shaw, 1980; Hulme, 1970, 1971; Munsell et al., 1949; Saimbhi et al., 1977; Cavalcante, 1976). Analysis from the disperser's viewpoint will not only help understand such things as why avocados have such oil-rich pulp (Cook, 1982) but will focus attention on the significance of the many spectacular secondary compounds in fruits. For example, ripe *Malpighia* fruits contain 19% dry weight vitamin C (Massieu et al., 1956); papain disappears from papaya on ripening and bromelin appears in pineapples on ripening (Czyhrinciw, 1969); drugs for animals occur in fruit pulp (Siegel, 1973; Janzen, 1978c); chili is eaten by birds and irreversibly avoided by rats (Rozin et al., 1979); essential oils occur in citrus rinds (Scora and Adams, 1973), antifeedants occur in ripe and green fruits (Kubo and Nakanishi, 1977); taste-modifying proteins occur in vertebrate-dispersed fruits (Higginbotham and Hough, 1977; Hough, 1978); resins occur in fruit walls (Kaikini, 1968; Gupta and Banerji, 1967); carcinogenic *N*-nitrosodimethylamine occurs in *Solanum incanum* fruits (Leete, 1979); bactericidal agents are apparently more common in winter fruits than in summer fruits (Stiles, 1980); ripe *Terminalia* fruits have such high tannin content that they were used in the tanning trade (Hathway, 1959); strychnine occurs in ripe bat-eaten *Strychnos* fruits (Bisset and Choudhury, 1974); and lethal doses of triterpenes occur in ripe cattle-eaten *Stryphnodendron* fruits (Tursch et al., 1963). Only very recently have fruit traits such as these been thought about as defenses of ripe fruits and mechanisms that deter the wrong "dispersers" (e.g., Herrera, 1982a;E. W. Stiles, 1980; Janzen, 1977a, 1979b, 1981e). There is an enormous unexplored domain in the integration of descriptive morphology of fruits (e.g., Roth, 1974, 1977; Roth and Lindorf, 1972) with the function of that morphology in relation to animals (e.g., Janzen, 1982e,g; Janzen and Martin, 1982; Prance and Mori, 1978).

The conversion from green inedible to ripe edible fruit is a most dramatic developmental process (Czyhrinciw, 1969; Goldstein and Swain, 1963; Jansen, 1965): fruits soften (e.g., ripe mangoes are one-tenth as hard as green ones; Krishnamurthy et al., 1960) and abruptly expand from tiny and dormant to large and edible (Janzen, 1982e,f,g); nutrient content changes (e.g., the starch in breadfruit only appears upon ripening; Czyhrinciw, 1969); protective tannins condense to an inactive state (Joslyn and Goldstein, 1964; Goldstein and Swain, 1963) but at times only incompletely, with the result that insoluble obstructive deposits form in the stomach of the frugivore (Allen, 1938); protein and amino acid content may decline (Auda et al., 1976); cyanide content declines 100-fold in *Passiflora* fruit (Gondwe, 1976); alkaloid content may *not* decline (Phillipson and Handa, 1976); oxalates disappear (Misra and Seshadri, 1968); pectin is lysed (Brecht et al., 1976; Abdel-Rahman, 1977); fluorescent materials increase in concentration (Magure and Haard, 1975); internally damaged fruits may "ripen" prematurely and be shed (Boucher and Sork, 1979). On the other hand, the fruit may not entirely change its chemistry but rather break open to expose an arillate seed (e.g., McDiarmid et al., 1977; Madison, 1979; Janzen, 1972; Howe, 1977, 1981b; Howe and De Steven, 1979; Howe and Primack, 1975; Howe and Vande Kerckhove, 1979, 1981a,b).

Frugivorous animals do not merely pluck fruits but rather display a variety of anatomical and behavioral traits for getting at them: Old World fruit pigeons have grasping, climbing squirrel- or parrot-like feet (Baker, 1913); coatis solve approach problems to bananas (Chapman, 1935); monkeys learn which artificial fruits are edible by observing parental choices (Jouventin et al., 1977); frugivorous bats have larger brains per unit body weight than do insectivorous bats (Eisenberg and Wilson, 1978); regular approach trails or flyways are used in arrival at fruit crops (Sikes, 1971; Morrison, 1978b, 1979; Heithaus and Fleming, 1978; Fleming, 1981; Howe, 1977); animals migrate long distances to track fruit seasons (Nelson, 1965); frugivores live a long time at one site (Snow and Lill, 1974; Wilson and Tyson, 1970); pouches are invented for carrying seeds (e.g., Long, 1976); large seeds are regurgitated to make room for more fruits (e.g., Snow, 1962; Howe, 1981b); seeds are spit out (Janzen, 1981a; Galdikas, 1982); complex stomachs separate fruit and seeds (e.g., Davies, 1978; Keast, 1958; Walsberg,, 1975); large incisors peel and scrape fruits (Hylander, 1975); bats lose gut complexity and flora (Klite, 1965); bats increase odor sensitivity (Fleming et al., 1977; Wolff, 1981); bats' time of reproduction is related to the fruiting season (Fleming, 1971; Wilson, 1978); bats fly as far as 0.6 to 10 km between day roosts and trees at which to feed at night (Morrison, 1978a,d; Heithaus and Fleming, 1978; Fleming, 1981); large herbivores harvest fruits dropped by other animals (Hamilton et al., 1977; Sharatchandra and Gadgil, 1975; Wrangham and Waterman, 1981); and so forth. Practically all frugivores eat a variety of things besides fruits and the seeds they sometimes digest (e.g., toucans and hornbills eat nestlings, eggs, lizards, insects; Swynnerton, 1908; Chasen, 1939; Murgatroyd, 1970; Skutch, 1944, 1971; Van Tyne, 1929), but the oilbirds, manakins, bell-birds, and cotingas of the Neotropics may be essentially frugivorous (e.g., Snow, 1962, 1971, 1976, 1981). Conversely, members of the Carnivora

236

are frequently highly frugivorous (e.g., Stuart, 1976; Gipson, 1974; Applegate et al., 1979; Gipson and Selander, 1976; Brunner et al., 1976). Flower-visiting birds are often frugivores as well (e.g., Gill, 1971; Leck, 1971a, 1972a; Snow and Snow, 1971). Seeds may also be consumed accidentally along with nonfruity foods (e.g., Balgooyen and Moe, 1972; Clark and Comanor, 1976) or leafy forage (e.g., Janzen, 1982i). Although it seems evident that seed shapes, sizes, specific gravity, surface contour, and other characteristics (Harper et al., 1970; Heiser, 1973; Janzen and Martin, 1982; Janzen, 1982e) will all influence the transit and survival of the seed in the animal, the great diversity in these parameters has hardly been thought about. Transit times are incredibly diverse and variable; small seeds may take as little as 10–20 minutes to pass through a small bird or bat (Docters van Leeuwen, 1954; Ziswiler and Farner, 1972; Klite, 1965; Fleming, 1981) or 24–72 hours to pass through a horse (Janzen, 1982g), whereas a large seed may take 24 hours to 2 months to pass through a horse-sized animal (Janzen, 1981b,c, 1982a) (rhinos may retain seeds for months as well; Ridley, 1930). Diet clearly influences the rate of passage of gut materials (e.g., Furuya et al., 1978), and there are numerous studies of the rates of passage of fine particles through gastrointestinal tracts (e.g., Furuya et al., 1978; Pickard and Stevens, 1972; Bailey, 1968; Clemens et al., 1975; Argenzio et al., 1974; Ziswiler and Farner, 1972; Dean, 1980; Bjornhag, 1972; Bjornhag and Sperber, 1977; Hoelzel, 1930), but these tell us almost nothing of the passage rates of seeds (or whether a trip of that duration is lethal). The trip is not a simple slide down a tube; for example, the large horse cecum (Janis, 1976) is apparently a retention site for large hard seeds (Janzen, 1981b,c), whereas the ruminant stomach of a cow passes seeds through rapidly (Janzen, 1982a). The presence of viable seeds in dung is often taken as evidence of a "safe trip through the animal," whereas it is clear that a very high percentage of the seeds may be digested (e.g., Gwynne, 1969; Janzen, 1981b,c; Krefting and Roe, 1949; Hladik and Hladik, 1967).

Seed dormancy, classically treated as a problem of interaction of the seed with the physical postdispersal environment, may range from none in bamboo (Janzen, 1976) and cacao (Barton, 1965) to 10,000 years in Arctic lupines (Porsild et al., 1967). Dormancy of animal-dispersed seeds is greatly confounded by the selection for seed coats to minimize germination inside the animal, which is commonly confused with the question of so-called enhanced germination by passage through the animal (e.g., Hladik and Hladik, 1967; Rick and Bowman, 1961; Oppenheimer, 1977; Krefting and Roe, 1949; Applegate et al., 1979; Noble,

1975a,b; Adams, 1927; Lamprey, 1967; Lamprey et al., 1974; Ng, 1975, 1978, 1980; Olson and Blum, 1968; Vasquez-Yanes, 1977). On the other hand, the fact that many dispersers also digest some seeds (Janzen, 1981b; Lamprey et al., 1974; Jarman, 1976; Davidge, 1977) is no excuse to grind up the seeds along with the fruits in doing nutrient analyses, as is often done (e.g., Milton and Dintzis, 1981; Hastings, 1966). Grinding and digestion of seeds is not only related to seed size and hardness (Gwynne, 1969; Janzen, 1981b,c), but seed spitting, which may be just as lethal to the seeds, is also related to seed size (Herrera, 1981a; Janzen, 1981a; Thom, 1937; Galdikas, 1982, Hunt, 1956).

The biology of birds that heavily use fruits has been under scrutiny for much longer and more broadly than the biology of other frugivores (e.g., Snow, 1971, 1980, 1981; Stapanian, 1980; Frost, 1980; E. W. Stiles, 1980; Sorenson, 1981; Howe, 1979; Howe and Estabrook, 1977; Wheelwright and Orians, 1982; McAtee, 1947; Foster, 1978; Dorst, 1947; Ridley, 1930). These studies tell us many, but far from all, of the traits that constitute the interaction. Fruit removal rate depends on location of the fruit (Thompson and Willson, 1978; Herrera and Jordano, 1981); brown and green fruits are least likely to be chosen (Turcek, 1963); consumption of rotting fruits may be dangerous (Janzen, 1977a, 1979b); fruit rewards range from hardly more than water and a bit of sugar to protein and oil-rich foods on which a bird can subsist entirely (McKey, 1975; Morton, 1973; Snow, 1962, 1971, 1976; Stapanian, 1980; Herrera, 1982b); seediness matters (Herrera, 1981a; Howe and Vande Kerckove, 1981a,b; Milton, 1980); frugivorous birds have food that is more predictable in detailed location and may therefore have more times for complex mating rituals (Snow, 1966, 1976; Foster, 1977b, 1978; Orenstein, 1973; and see Bradbury, 1977, for a chiropteran example); certain habitats of the wet lowland tropics have understory frugivorous bird faunas of strongly varying sizes (Karr, personal communication, 1976), and the amount of fruit present reflects this (Janzen, 1980b); failure of birds to harvest fruits may lead to the evolution of ant-dispersed seeds (Thompson, 1981); only a small fraction of the bird species to visit a large fruiting tree may be important dispersers of seeds (Howe and Vande Kerckove, 1979; Howe, 1977, 1980; Howe and Primack, 1975); birds may evolutionarily mediate the timing of fruit ripening (Leck, 1970a,b; Janzen, 1977b, 1978b; Thompson and Willson, 1979; McClure, 1966; Medway, 1972; McAtee, 1906; Stapanian, 1980; E. W. Stiles, 1980; Snow, 1965); the bird–fruit interaction has been around for many millions of years and has probably played a major role in plant speciation (Regal, 1977; Weigelt, 1930; Ridley, 1930; Voorhies and Thomasson, 1979; Givnish, 1980); tree fruit crop size (number of fruits, value of fruits, and combinations) influences removal rates of seeds and fruits (Howe, 1977, 1981a,b; Howe and De Steven, 1979; Howe and Primack, 1975; Howe and

Vande Kerckhove, 1979, 1981; Stapanian, 1980); bird arrival and exit from a fruit tree is a social as well as hunger phenomenon (Leck, 1971b; Howe, 1979); migrants may be more important than local birds in fruit removal and seed dispersal (Leck, 1972c; Herrera, 1982b); and so forth.

Where seeds land

Seed rain from animals can be recorded as a habitat-level phenomenon (e.g., Bews, 1917; Brunner et al., 1976; Foster, 1973; Vazquez-Yanes et al., 1975; Chrome, 1975; Guevara and Gomez-Pompa, 1972; C. C. Smith, 1975; Smythe, 1970; Lieberman et al., 1979; Thompson, 1980; Janzen, 1978a; Bullock, 1978; Burtt, 1929). However, without knowing the source of the seed, it is very difficult to relate the results to the fitness or productivity of the parent plant. Although a forester may not care about the source of the acorn that put a sapling in this or that tree fall, to the evolutionary questions about fruit morphology, acorn phenology, relatedness of neighbors, seed chemical composition, and so forth this information is indispensable. Following an animal that has eaten the seeds from a known plant until it defecates those seeds (e.g., Docters van Leeuwen, 1954; Darley-Hill and Johnson, 1981; Howe and Primack, 1975) is only slightly easier than following a long-distance pollinator until we know the fate of its contaminant pollen. With very few exceptions (Howe, 1977; Howe and Primack, 1975; Janzen et al., 1976; Fleming, 1981; Stapanian and Smith, 1978; Darley-Hill and Johnson, 1981; Vander Wall and Balda, 1977; Herrera and Jordano, 1981; Jordano, 1982), we do not know what animal-generated seed shadows of particular plants look like. Certainly gene flow via seeds dispersed by animals (Levin and Kerster, 1974) is totally unknown, though birds and large mammals presumably carry some seeds much further than pollinators carry pollen.

Animal-generated seed shadows are certainly not homogeneous, monotonically declining functions of distance from the parent. Heterogeneity is generated by such things as perches (Livingston, 1972; Salomonson, 1978; Howe and Primack, 1975; Cowles, 1936; Heumann, 1926; Heim de Balsac and Mayaud, 1930; Doctors van Leeuwen, 1954), water holes (Bews, 1917; Freese, 1978), repeated use of nest holes (Chasen, 1939; Kemp, 1975), lek sites (Snow, 1972), caves (Vazquez-Yanes, 1975; Snow, 1962), feeding roosts (Fleming, 1981; Morrison, 1975, 1980), cover from predators (Herrera and Jordano, 1981; Howe, 1979), favored defecation sites (Kakati and Rajkonwar, 1972; Shortridge, 1931), bouts of seed spitting (Thom, 1937), and practically anything else that one can think of that leads to heterogeneity of animal movement and rest. There are many cases that emphasize the need for explicitness in generalizations. For example, oilbird juveniles regurgitate the seeds below the nest (Snow, 1962) whereas the long-tailed manakin juveniles do not (apparently, they are removed by the parent; Foster, 1976); likewise hornbills discard seeds (and other trash) out the nest entrance whereas toucans carry them away. *Epophorus* fruit bats carry food to feeding roosts but not to the day roost (Wickler and Seibt, 1976); birds and bats may pass small seeds through and regurgitate or spit out large seeds (Snow, 1970; Wickler and Seibt, 1976). Seeds (or fruits) dropped or defecated by one dispersal agent may be picked up and killed or further dispersed by another (Morgan-Davies, 1960; Hamilton et al., 1977; Janzen, 1982c). The bearded bellbird in Trinidad is a deep forest bird, except when *Byrsonima* trees are in fruit; then it ventures into the open for its fruits (and presumably generates a quite different seed shadow) (Snow, 1970). Seeds deposited in caves (Snow, 1962; Vazquez-Yanes et al., 1975) may appear to be doomed, but if rainy season high water flushes out the cave, these seeds may be deposited in some of the very best sites in the area for seedling survival.

Seed shadow modeling and discussion of the consequences of various forms of dispersal have not escaped attention (Bullock, 1976; Hubbell, 1979; Janzen, 1970, 1971a; De Angelis et al., 1977; Fleming and Heithaus, 1981; Hamilton and May, 1977), but it is my opinion that these efforts are largely futile until we know a great deal more about which animals generate what kinds of seed shadows at what cost to the parent plants.

What happens to a seed because it landed there?

A discussion of the fate of dispersed seeds is beyond the scope of this chapter, but I would be remiss if I did not point out that we know even less about this than about seed shadows and who generates them. The seed and seedling survival depends in great part on the interplay between the duration of dormancy and the site. Those that land in dry caves or are picked up by migrating animals on the way out are in serious trouble. Seeds in the forest may have a very different mortality schedule than in adjacent grassland (Rees, 1963; Janzen, 1971b; 1982h). Seeds buried near logs or stumps may have a higher chance of being found by peccaries (Kiltie, 1981); recovery of cached seeds (Murie, 1977; Barnett, 1977; Shaw, 1936; Lockard and Lockard, 1971; Stapanian and Smith, 1978; Vander Wall and Balda, 1977; West, 1968; Abbott and Quink, 1970) depends on a plethora of site details. Seeds that fall in gaps may have different germination percentages than do those that fall in adjacent forest (Fleming, 1981); proximity of a defecated seed to an ant nest may be critical (Perry and Fleming, 1980;

240

O'Dowd and Hay, 1980; Davidson and Morton, 1981). Seeds left below the parent plant are in severe trouble both with respect to competitors and ease of location by animals (Sork and Boucher, 1977; Vandermeer, 1977; Janzen et al., 1976; Janzen, 1971b, 1972; Wilson and Janzen, 1972). Seeds in dung may be avoided by seed predators (E. W. Stiles, 1980) or be at higher risk from seed predators (Janzen, 1982c,h) or fungi (Bakshi et al., 1968). The nifty looking piles of seeds left in dung or under feeding perches in the forest appear to be the epitome of dispersal success; in fact, even if the seed predators don't find these piles, the seedlings are likely to suffer severe sibling competition (Howe, 1980; Foster, 1977b) or communicable diseases.

A CARICATURE OF SEED DISPERSAL BY ANIMAL GUTS

The interaction between fruits and frugivores is deceptively easy to visualize. Fruits ripen, a vertebrate goes to them and eats some (swallowing some seeds as contaminants); it moves away; the seeds pass through it; and sometime later, they germinate. This applies to plants from tiny epiphytes and herbaceous riverbank vines to enormous trees and lianes hundreds of meters in length. It occurs from Arctic cranberry bogs to melon patches in Saharan oases, from tropical swamps to the highest Andean paramo. It dates back at least as far as dinosaurs, if the sweet fleshy fruits on Chinese gingko trees and African podocarp trees are as old as we think they are. But although the basic pattern is simple, it bears an extraordinarily diverse overlay of details which, if modified even slightly, will have very visible and large effects on the population and evolutionary biology of the animals and plants involved. Above all, the system has the property that as partners are eliminated by ecological processes or evolution, seed dispersal is passed to the next animal and seed dispersers are passed to another plant. This continuity is promoted by the fact that the interaction benefits both members; vertebrates have many dietary needs in common, with the result that once a structure has been selected to be edible to one species of animal it is likely to be edible to at least some others, and once an animal has been selected to find and eat one species of fruit (or seed) it is likely to find and eat another. It generally does not proceed to a tightly coevolved one-on-one relationship because there are powerful forces acting against the evolution of the year-round high quality fruit production by one species that would be needed to support a completely monophagous vertebrate; in addition, a single vertebrate species is unlikely to reliably generate a seed shadow of as high quality as can certain kinds of disperser coteries

241

made up of several species of vertebrates and with a composition that changes over the geographical range of the plant. Finally, even with a tightly coevolved mutualism, the habitat occupied by a plant is normally of sufficient heterogeneity that in at least some part of the range the plant will be able to survive with no seed dispersal other than that that occurs by abiotic processes or sloppiness of seed predators.

Fruit ripening

Fruits do not merely hang on a plant until temperature-dependent physiological processes have run their course. Fruits and the seeds they contain mature synchronously yet are genetically and biochemically quite different objects; this alone is sufficient evidence that the time to maturation is not dictated by the chemistry of physiological processes. Rather, fruit and seed development time has been evolutionarily adjusted to some optimal duration in the context of seed disperser availability and hunger, photosynthate accumulation and dispensation, predispersal seed predation, postdispersal seed death, and optimal germination times. Fruits and seeds clearly do not ripen at random with respect to the calendar. There is no optimal date for a fruit to ripen within a plant's fruit crop, but rather there is an optimal distribution of times of fruit ripening. A crop of 5000 large indehiscent juicy fruits and a herd of horses may generate a seed shadow with very high fitness for the parent tree if 10 ripe fruits fall each day for 50 days in a tropical dry season; but if all 5000 fruits fall in a week of a tropical rainy season, disperser satiation, microbes, disperser attraction to succulent vegetation, and the five days needed for the horses to first locate the fruits will take a heavy toll of the tree's offspring.

Selective forces on a ripe fruit differ from those on the rest of the plant. Except for flower nectar and some pollen, all other parts of a plant have been under eons of selection for traits to reduce their digestibility and to poison consumers. For example, cellulose is the most widely distributed and abundant coevolved natural product, and almost no animal can digest it. It is very likely not to be an accident that among all the possible structural carbohydrates, an inedible one has become the dominant one; can you imagine the defenses a plant would have to have against herbivores if cellulose were easily digested?

It is not surprising that cellulose is not the structural carbohyhdrate in most edible fruit pulp (most exceptions are fiber-rich and woody fruits that are eaten by large browsing vertebrates, that is, animals whose guts contain microbial and protozoan specialists at degrading cellulose-rich plant parts). Ripe fruits have been under only slightly fewer eons of selection to be edible to a select subset of the vertebrates in the habitat and to be toxic or indigestible to the remaining animals and microbes.

242

Ripe fruit also differs from vegetative parts in a strategic sense. When a ripe fruit is removed by an animal or is shed, the parent has no feedback system to evaluate the cause or consequence of that removal; on the other hand, in many situations herbivory or other tissue loss can be evaluated by the plant as to cause, intensity, and kind, and repair and other facultative responses can be conducted.

Fruit ripening requires that a tissue that has been photosynthetic and intensely antibiotic to animals or pathogens abruptly become non-photosynthetic and highly edible to only a very small subset of the enormous array of herbivores and frugivores in the habitat. The intense nature of chemical and morphological defenses of green fruits is the consequence of the nutrient-rich and fitness-rich developing embryos and other tissues within. The need to convert all or part of a well-protected tissue to one that is highly edible puts numerous restraints on the kinds of defenses the immature fruit can have, the kinds of dispersal agents that can be in the disperser coterie, and the speed with which a fruit can ripen. However, in explicit disagreement with currently fashionable evolutionary theory (see Lewin, 1981), I would argue that these restraints are no different from any other environmental restraints against which a mutant is tried and are therefore a normal portion of the natural selection processes that drive evolution.

There are enough dietary needs in common among species that process fruit that when a tissue evolves to where it is edible to one species, the probability is high that its edibility will rise for some other species, ranging from the seed disperser's congenerics to microbes. A ripe fruit cannot therefore be expected to have the life span displayed by other parts of the parent. Fruit ripening is a game of positioning highly perishable items in space and time rather than simply a matter of putting out a bowl of cherries and letting the seed dispersers take their fill as their fancy and numbers dictate. Seed dispersers must locate and harvest a highly desirable and perishable object of capricious and short-term fixed location, and they do not clean it of its contaminants in the harvest. Humans have a very biased view of fruits; even though recognized as perishable, domestic fruits have been artificially selected to be less perishable and have been provided when ripe with the same chemical protection (pesticides) that was evolutionarily discarded in nature in order to get the seeds into the right animals.

Disperser arrival and feeding

Animals arrive at a plant's fruit crop in response to memory, location cues (odor, color, locality), instructions and cues from other animals, hunger, proximity of other fruit crops (allospecific as well as conspecific), fruit quality, random movement, and so forth. The wind—the other prominent disperser of seeds—arrives irrespective of a fruit crop's traits, except for fruiting phenology and plant (and fruit) location. The wind cannot be satiated. Animals may prey on seeds, eat fruit and discard or ignore seeds, eat fruit- and seed-eaters, socialize, eat fruit and cache seeds, and eat fruit and ingest seeds as contaminants. Although only the latter two categories are the focus of this chapter, the others have a strong influence on the composition of the disperser coterie (and hence seed shadow) of any particular plant or species bearing ripe fruit. (A plant's disperser coterie is that array of animal individuals and species that generate its seed shadow; when most narrowly defined, each member of a plant population may have a slightly different disperser coterie.) An individual animal often does several of the preceding acts on a given day, to a given fruit crop, during the year, or during its life cycle, and may do a different combination of these things to each of an array of species of plants. Most seed dispersers are simultaneously dispersing several species of seeds in a particular habitat. Within a time span of a few minutes, a Central American agouti *(Dasyprocta punctata)* may prey on seeds, cache seeds, defecate seeds, eat fruit pulp, fight over fruits, and eat smaller seed predators found in the fruit. The result of this complexity is that dispersal cannot be characterized by merely listing visitors to fruiting plants, by recording how many fruits are "eaten" by which animals, or by listing plants with fruits whose seeds appear to be dispersed by this or that vertebrate.

However, all is not chaos. Certain species and individuals are much more likely to treat a certain fruit crop in a certain manner than are others. It is safe to speak of parrots (Psittacidae) and peccaries (Tayassuidae) as usually being seed predators, and toucans (Ramphastidae) and horses are usually being seed dispersers. However, understanding any particular plant and its seed disperser coterie requires a detailed understanding of how different animals treat the fruit crop and where the seeds go. There is a large procedural difference between generating generalizations on dispersal and in manipulating or understanding the idiosyncracies of a particular plant and its disperser coterie. The statements about "average" individuals are likely to be very far out of focus unless the data are collected from individuals known to be representative of a particular distribution.

Animals at a fruit crop display many behaviors that strongly influence their subsequent dispersal of seeds. They select among fruits, become satiated, are sloppy, spit out seeds, fight and flee, defecate seeds swallowed long ago, and damage uneaten fruits. However, although seeds dropped, spit, regurgitated, or defecated below the parent plant may be abortions of the fruit's dispersal mission, in some habitats this seed placement is only the first step in a complex and

244

somewhat serendipitous further dispersal that may generate a portion of a seed shadow quite different from that anticipated by mere consideration of the fruit and vertebrate for which the fruit appears to have been evolutionarily designed. In addition, seeds below the parent are often picked up and carried off by a variety of seed predators who later lose them or never get around to eating them.

But the variation in the seed shadow is not generated solely by the variation in animal responses to fruits and seeds. Fruits and seeds are highly variable within and between species, and many of these variations can be easily interpreted in the context of influencing the ensuing seed shadow. Why do large seeds and nuts often have variously strong fibrous connections between them and the fruit pulp eaten by a vertebrate seed disperser? Why do fruits contain highly variable numbers of seeds within and between crops? Why does fruit seediness vary within and between seed crops? Why do seeds vary 2- to 3-fold in weight and volume within a fruit crop? Such questions have never received botanical attention equivalent to the attention bestowed on frugivorous and seed dispersing vertebrates.

Disperser departure from the fruit crop

Vertebrates move away from fruit crops to search for other kinds of food, to escape predators, to return to a nest or to distant feeding perches, to visit superior fruit crops elsewhere, and for many other reasons. Many of the details of a seed shadow will depend on when and how an animal leaves a fruit crop. The easiest detail to understand is that the longer an animal stays at the fruit crop, the more likely it is to spit, drop, defecate, or abandon the seeds directly below the parent plant. If there were ever an evolutionary prediction, it is that plants that bear fruits containing animal-dispersed seeds are likely to have chemical and morphological traits that make it unpleasant or difficult for seed dispersers to perch in them. Fruit-free seeds that end up below the parents plant not only have to compete with one of the worst competitors in the habitat but are also usually in an area of high concentration of conspecific seeds that is the focal area of search by seed predators and a hotbed of highly competitive siblings.

The patterns of movement of seed dispersers to and from a fruit crop are influenced by many fruit and seed traits such as size and other nutrient rewards of fruits, pattern of fruit ripening within the plant's crop, seediness of fruits, difficulty of getting at the fruit reward, conspicuousness of the crop, size of the crop, location of the crop within the plant crown, and habitat in general.

Disperser defecation and voiding of seed

Once a seed has been carried away from the fruit crop, its misery is by no means ended. If inside the animal, it may be ground up by molars or gizzard, digested anywhere from stomach to caecum to colon, scarified and thereby seduced into (probable) lethal germination (owing to the anaerobic conditions in animal guts), or just detained for the wrong amount of time. If outside the animal, it may be discarded in an intact fruit (often lethal owing to fruit decomposition or a fruit hull that the seedling cannot penetrate), chipped up, buried and later recovered, or buried in a microhabitat where it has little or no chance of survival as a seed or seedling even if forgotten or lost.

What happens to a seed in an animal (or in its paws) depends not only on the animal's traits, but very strongly on detailed seed traits such as seed coat texture, hardness, contour and toughness, volume, specific gravity, seed numbers per fruit and seed/fruit pulp ratios, weight, tenacity to fruit parts such as fibers and hard endocarp, elasticity, and permeability to germination cues. In addition, the fruit does more than just attract the disperser. Laxative chemicals may speed the seed's transit through the animal; hard items in the fruit may prevent molar occlusion on a soft seed; lubricating chemicals in the fruit pulp may render the seed more slippery and therefore difficult to sort out of the chewy fruit mass by the tongue (for spitting or more intense grinding); mildly astringent or otherwise annoying chemicals in the fruit may cause the animal to consume only small amounts at a particular feeding.

Even if the seed passes unharmed through the animal, many animals will defecate it in inappropriate sites (e.g., dry caves, cavities in trees, ponds or rivers, sun-baked soil). Sites that are good for the seeds of species A may be horrible for the seeds of species B; thus, desirable members of the seed disperser coterie will be different for different plants. There is no selection for seed dispersal to oceanic islands; most seeds that start the trip are defecated into the ocean, and those that make it do not report their success back home.

The high percentage of mortality often recorded for the seeds passing through an animal's gut may suggest that the animal is a low-quality disperser. However, virtually all dispersal systems kill most seeds by putting them in inappropriate habitats; highly directed seed flow (in animal dung) may in fact result in just as many surviving saplings as the much more gentle but blind generation of seed shadows by wind or water.

The most is not necessarily, or even likely, the best with respect to how far a disperser carries seeds and how many seeds a disperser carries away from the fruit crop. Beyond the immediate vicinity of the parent plant, the quality of sites isn't correlated with distance. A species of seed disperser that carries off the largest number of seeds

246

from a fruit crop will not necessarily distribute them over scattered suitable sites. Ten seeds defecated by a tapir (*Tapirus*) in a flowing creek may be worth 100,000 of the same species defecated by a horse in adjacent drought- and fire-prone grassland.

Seed shadows, just like subsequent germination regimes, have a temporal component. Animals with transit times of weeks or months may not only lay down a seed shadow in many different habitats (simultaneously with the seed shadows of seeds eaten at other times), but may also lay down a seed shadow in a different season from the one in which the fruits were eaten. This is especially apparent with large fruits eaten toward the end of a tropical dry season by large mammals that have long retention times for large seeds.

And sometime later, germination

Seeds are subjected not only to competitive and physical environments, but to severe postdispersal seed predation. Large genera of insects and vertebrates make their living digesting seeds, and no small number of fungi kill dispersed seeds. For many species of plants in mainland habitats, the majority of seeds in a crop are killed by animals and disease rather than by competition with other plants. The traits of the seed shadow, and especially the location and intensity of peaks in it, have a strong influence on the degree and kind of postdispersal seed predation. Seed predators search no more at random than do any other foragers. Even the timing of behavioral exit from the seed shadow (germination) may be determined by animals through the degree of scarification of the seed coat during passage through the disperser.

Germination needs a few words in passing. Perhaps one of the most poorly conceived and used concepts in plant biology is *germination rate*. First, it is usually not a rate but rather a percentage. Second, what is usually reported may matter to farmers, but it is quite irrelevant to the biology of wild plants. Third, the ungerminated seeds are often assumed to be dead. I am often asked, "But doesn't passage through the animal enhance germination?"; "Don't plants need dispersal to improve their germination success?" Or, "If germination is not enhanced by animals, why do plants need dispersers?" The implication is that somehow an increased percentage of immediate germination is somehow better for the plant, that ungerminated seeds are dead, and that the behavioral trait of germination is not subjected to the same evolutionary influence as are all other plant traits. Quite to the contrary, seed coats (and nut walls in drupes) are evolutionarily designed to withstand the rigors of a variety of animals' molars, gizzards, intestinal acids and enzymes, anaerobic atmospheres, and so forth and to pass out the other end with enough seed coat remaining to protect the embryo from weather, false germination cues, and seed predators until the appropriate germination conditions come along. The same applies to other fruit traits such as laxatives that speed seed passage through the animal and hard objects in the fruit that prevent molar occlusion. A seed coat that is too tough survives all conditions, but the embryo cannot know when to germinate. One that is too weak fails one of the many challenges inside or outside the animal. The extreme case is that where the parent plant pays offspring in the form of edible seeds (e.g., acorns, pine nuts, herbaceous legume seeds) for the dispersal of its surviving offspring. Here there is selection for a seed coat or nut wall that is penetrable enough that the animal remains interested but yet impenetrable enough not to make the seed available to all potential seed predators. An alternative solution is mass seeding, as in bamboos and many conifers. Here there is little protective seed coat or dispersal by animals, but escape occurs through seed predator satiation and dispersal occurs by wind or nothing.

Seeds that are sufficiently scarified by transit through the disperser to start germinating usually either die in the animal through digestion or in the dung outside of the animal through pathogens, dung processors, or seed predators searching in dung. Seedlings growing out of cattle dung are picturesque, but they are photographs of dead plants. Those who think it amazing that holly berry seeds or nutlets (*Ilex*) require either passage through a bird or treatment with strong acids to respond to germination cues seem unaware that the normal cycle of a holly seed *(Ilex)* is to survive a short trip through a hostile environment or a much longer stay in a more benign one (the litter). We certainly do not know enough about what yields maximum holly tree recruitment to know what pattern of germination in time is best for a holly tree seed crop.

ECOLOGY, EVOLUTION, AND COEVOLUTION

The evolution of these animal–host interactions, which exist in almost every terrestrial habitat, must be influenced by the processes in the following three scenarios.

1. A frugivorous animal new to the habitat arrives, feeds where it finds food and thereby remodels several species' seed shadows; a plant new to the habitat puts out its fruits and persists with the population structure produced by the seed shadow generated by the disperser coterie it acquires in ecological time. In either scenario, no evolutionary change occurs on the part of either plant or disperser coterie, though both may have been evolutionarily molded through previous encounters with other species.

248

2. A mutant fruit or seed, or seed disperser, persists because the ecological response by the partner to the new phenotype raises the fitness of the mutant higher than that of the parental type (and may lower or raise the fitness of the partner). This is ordinary evolutionary change.

3. The evolutionary change described in (2) occurs, followed by an evolutionary change in the partner; this latter change is directly associated with the change in the first mutant. Here direct coevolution has occurred. Diffuse coevolution occurs when the interaction occurs between suites of partners, such as between overlapping disperser coteries and one or all members of the array of plants whose seeds they disperse.

All three of these processes occur, and I suspect that direct coevolution of one seed disperser with one plant is the rarest. To bring these airy conceptualizations down to a level that I can understand, I will present a reconstruction that has been made as realistic as current knowledge of plants and dispersers allows and that traces the probable ecological change, evolution, and coevolution of a tree–disperser interaction over several millions of years. This reconstruction is representative of that for most plants whose seeds are dispersed by animal guts, given that the members of the disperser coterie feed heavily on food other than fruits. Space does not permit a similar reconstruction for the interaction between a narrowly frugivorous seed disperser, such as an oilbird (*Steatornis caripensis*), and its food plants, but the reader can provide that by adding to the following story the modifications expected when nearly all of the diet of the disperser is the fruit around large soft seeds of 20–100 species of rain forest trees.

A HISTORICAL RECONSTRUCTION OF GUANACASTE AND ITS DISPERSERS

At present, guanacaste trees (*Enterolobium cyclocarpum*, Leguminosae) are widespread, rare, but conspicuous, large plants in the lowland deciduous forest remnants and fencerows of Mexico to tropical South America. The large fruits of guanacaste (Figure 1) are eaten by livestock and the large seeds are frequently encountered in their dung. This species is currently the subject of an intense autecological and population study in Santa Rosa National Park, northwestern Guanacaste Province, northwestern Costa Rica (Janzen, 1969, 1981a,b,c, 1982a,b,c,d,h; Janzen and Higgins, 1979; Janzen and Martin, 1982).

Because the tree is a mimosaceous legume, I assume that at one

249

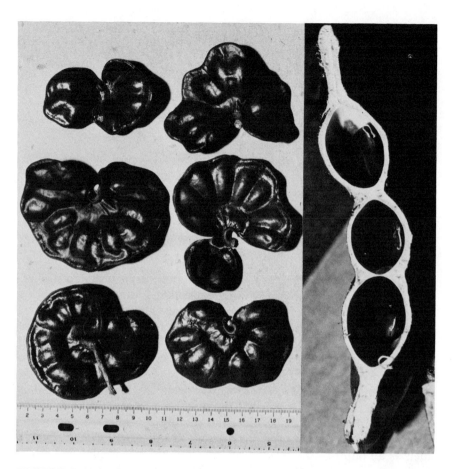

FIGURE 1. Left: Guanacaste tree fruits of various sizes. Right: Longitudinal section through a guanacaste fruit. Each cavity contains a single ovoid seed (Santa Rosa National Park, Guanacaste Province, Costa Rica).

time in its evolutionary lineage the fruits were strap-like flat pods. When ripe, the walls of these pods were fibrous and woody; the pods split apart when dried to drop or throw the flat, hard, and desiccation-resistant seeds into a xeric habitat, and they contained few nutrients. From this evolved a fleshy linear pod eaten by large Tertiary browsing mammals that dispersed those seeds that survived passage through the gut. Such a fruit type is found throughout the large genus *Pithecellobium*, which is very similar to *Enterolobium*. There are at least two reasonable pathways for this evolution, and which is more likely is in part determined by tree height. Various species of large mammals could have picked up the newly shed and largely seed-free dehisced pods as valuable forage at the end of the long dry season, a

250

time when green foliage is scarce or chemically well protected. The cellulose-rich pods would have been a useful dietary item for a ruminant or caecal digester. The pods would have been eaten as a minor item along with browse, and the consumption of a seed-free and therefore dead tissue would have led to no evolutionary change by the large mammals or tree. Assuming that the mammals were more likely to hit a local watering area with their dung than were wind-propelled seeds or seeds still attached to pod walls, any mutant protoguanacaste tree with tardily dehiscent pod valves would have been favored through large mammal dispersal of its seeds. Once this occurred, there would also have been selection for more molar- and moisture-resistant seed coats to delay initiation of germination until the seed had made the trip through the mammal.

Alternatively, the browsing mammal may have behaviorally sought the ripening pods of the plant as a superior dietary item in contrast with drying foliage at the end of the rainy season, or even in contrast with fully functioning foliage (owing to the high nutrient level of developing seeds, which are soft enough to be easily chewed up and digested). Here, any mutant plant that matured its seeds early, so that the seeds were hard enough to survive the molar mill and digestive tract, would be favored both through direct increased seed survival and more focused dispersal.

Once our protoguanacaste tree has its fruits in a state where they are sought as food and the seeds gain from this in enhanced survival through more accurate dispersal, the stage is set for a succession of small changes in the fruit, such as the evolution of indehiscence, increased nutrients in the fruit wall, a round flat fruit taken in as an easy mouthful, increased fruit to seed-volume ratio, seed streamlining to speed transit and increase survival during chewing, tough tissue layers to protect seeds from sharp molar cusps, fruit ripening at a time and pattern that optimizes fruit consumption by a disperser coterie whose numbers and dietary interests change with the season, and fruit crop sizes tailored to the expected consumption amounts. None of these changes select for direct evolutionary changes on the part of the large mammals. However, the mammals' density, patterns of movement, impact on other plants, and other demographic traits are probably influenced by the presence of a high quality fruit dietary input at a certain time of year, especially because some other trees in the habitat are probably evolving in the same direction. All of these ecological changes will select for a variety of evolutionary changes that are difficult, if not impossible, to blame only on our protoguanacaste tree. If there is diffuse coevolution, it would be in the evolution of molars that are more resistant to damage caused by biting rock-hard seeds, cud chewing directed at seed elimination, increased cleansing motions by the caecum to prevent obstructions by seeds, and so forth. However, these traits are of importance in the processing of many other dietary items as well. The coevolution is diffuse with respect to both the other trees that the mammals disperse and the mammals' other food items.

To date, perhaps the mid-Tertiary, our tree has a fruit type characteristic of many arid-land mimosaceous legumes in habitats rich in large mammals. Then our tree finds itself in a more moist habitat either through a general climatic change or by having been carried over a mountain range by a migrating mammal. More moist habitats select for larger seeds (to give the seedling the reserves it may need to compete with a lusher set of associates). Larger seeds are more easily noticed and more likely to be spit out or broken by large chewing mammals. Selection is therefore intensified for a higher fruit:seed ratio, which may be accomplished by a larger fruit and/or higher quality fruit of the same size (accomplished chemically or through timing of fruit availability). Simultaneously, the larger fruits are more attractive to the largest mammals present—ground sloths, proboscidians, large perissodactyls—than were the smaller ones before. The species richness of the disperser coterie declines as small animals drop out because the fruit and seed are larger, and because the large animals get a higher proportion of the fallen fruit through more attentive gathering. Although these very large mammals will swallow a larger seed, they also have very powerful jaw action, which again selects for a harder and tougher seed. Guanacaste trees have the largest mimosaceous legume seed in the dry Neotropics (300–1200 mg).

Throughout this fruit evolution, there have been at least two sources of intense pre- and postdispersal seed predation. Bruchid beetles (Bruchidae) and curculionid weevils (Curculionidae) oviposit directly on the fruits and the dispersed seeds, and the larvae mine into the seeds and consume the contents. One of the values of the large mammals to the plants is that they remove the fruits before the insects can prey on all the seeds. Also a variety of large to small rodents collect the seeds from the fallen fruits and the dung of the large mammals. They kill most of the seeds by chipping them up, but they cache some and lose some, which may produce viable seedlings.

In short, during the bulk of large mammal history up through the Pleistocene, the entire system appeared and evolved without any direct coevolution, but it is rampant with diffuse coevolution between the plant and suites of animals (and probably suites of similar plants).

Then enter the humans that were the Pleistocene big game hunters of 10,000–15,000 years ago. Over a period of only a few hundred years in any given habitat, their activities (coupled with the mop-up squad of large carnivores left short of food) remove nearly all the large mammal dispersers of guanacaste seeds (only the tapir—*Tapirus bairdii*—remains), but they hardly touch the smaller mammals. There are several immediate ecological consequences. The vertebrates that prey on guanacaste seeds in newly fallen fruits—peccaries and a variety of rodents—are suddenly confronted with large patches of rotting fruits from which the seeds are easily and thoroughly harvested. Seeds that were previously obtained rather haphazardly by many individual seed predators from far-flung dung are now obtained in large numbers by those few individuals that happen to live near or forage near large guanacaste trees, and by larger animals that make pilgrimages to fruiting guanacaste trees. Guanacaste seeds are, however, still a small part of the total diet for all these animals, which feed on a variety of similar seeds to which the animals are already physiologically, morphologically, and behaviorally well adapted.

The tree's demographic structure begins an immediate change in that its density falls and the number of habitats occupied declines. This is because most seeds are killed (because of their ease of location directly below the parent) and most survivors are below or very near a parent (seed-caching mice and agoutis do not carry them more than a few tens of meters). There is still the occasional more distant seed dispersal when a seed survives passage through a tapir. However, the demographic changes in the adult portion of the population are slow because the trees live 200–300 years once they are established. Their seed shadow is trivial in extent compared to what it used to be. The occasional seed that escapes from a rodent's seed cache or is dispersed by the occasional tapir is now the only dispersal that occurs and, therefore, relatively of very great importance. There is now a quite different equilibrium between demography and habitats occupied, and one that appears normal except for the telltale fruit type that has little or no disperser coterie. After hundreds of thousands of years, new fruit and seed traits would surely evolve because of the very high seed mortality; the evolved traits would either raise the amount and distance of seed dispersal by these animals or would incorporate other dispersal agents in the disperser coterie. The plant might go to lighter wind-dispersed fruits, to a seed so hard that it survives the molar mill of common and far-ranging peccaries, or to a seed so tough that small rodents with deep caches would not harvest it but larger rodents with shallow caches (e.g., agoutis) would. Such changes in the fruit and seed crop would be associated with evolutionary changes in the vegetative traits of the tree to match it better with its new proportional representation among the habitats occupied. There would be no diffuse co-

253

evolutionary change selected for in animals unless a number of other species of plants were simultaneously going through the same changes.

But after only 10,000 years of this (a mere 20–40 generations of the guanacaste tree), humans bring back the horse they extinguished and add the cow, a large mammal of a type present in the Pleistocene only over the northern portion of the geographic distribution of guanacaste trees (bison ranged as far south as Nicaragua). Turned loose in grassland–forest mix, both of these large mammals discover a wild fruit— the guanacaste fruit—evolutionarily designed for large herbivorous mammals. They consume the fruits, spit out some seeds, digest some seeds, carry some in their guts as long as 2 months, and disperse them in a widespread seed shadow. However, this seed shadow is of yet a third kind, similar to the Pleistocene one for guanacaste trees but different in that the horse and the cow are abruptly moved long distances at certain seasons, are restricted to certain pastures at other seasons, have high heterogeneity in freedom of movement on one ranch as compared with another, have densities and movements dictated in part by the capricious nature of the market system rather than through the fluctuating carrying capacity of adjacent habitats, and are not accompanied by other large mammals in dispersing guanacaste seeds. The guancaste tree population will again adjust to this new circumstance and will again appear "well adapted" to the habitat; the differences between this seed shadow and the Pleistocene one will be much harder to identify than was the case with the post-Pleistocene, pre-Spaniard seed shadow. We will never see the longer term results of this experiment because the tree will be extinguished by human harvest for lumber and fenceposts in the next 50–100 years except in a few parks and reserves.

In summary, the fruits and seeds of the guanacaste tree have clearly evolved in response to seed dispersal by an array of mammals. Although many traits of these mammals may be viewed as adaptive in harvesting guanacaste fruit and seeds, none can be said to be a specific adaptive response only to guanacaste trees. That is to say, there is no reason to postulate any kind of direct coevolution, but there has been ample evolution, diffuse coevolution, and ecological interaction.

QUESTIONS AND ANSWERS

In the preceding section, I went at biology by observing animals and plants and then puzzling over what they do and the way they are. However, reports of such puzzlings tend to evoke more generalized questions cast in a philosophy of ever-attempting to generalize the world, that is, to reduce it to a few rules. These questions deserve consideration by virtue of their omnipresence and by the fact that I cannot drag each of you through a year of Costa Rican fruits and dispersers; though often I feel such questions are of very limited use.

**What factors impede the evolution of ever more tightly coevolved
seed dispersal systems?**

This question takes many forms. Why don't all fruit types converge?
Why don't all frugivorous vertebrates become purely that (and seed
dispersers)? What is the cost of being a frugivore? Why doesn't each
plant have its own unique disperser? Why doesn't each disperser have
its own unique plant? What is the perfect fruit or the perfect disperser?
The answer to all these questions is approximately the same and is
complex.

The vertebrate that ecologically and evolutionarily moves into an
ever more frugivorous life form is immediately confronted with problems. Seeds and nuts are contaminants that break teeth, grind teeth,
impede chewing, obstruct passageways, poison when digested, fill
restricted spaces with inert useless bulk, slow digestion, and impact
sacs. The fruit tissue, although often rich in some particular nutrient(s)
is generally very poor in other nutrients. The fruits of a given species
are available only in particular seasons of the year; and worse, they are
completely absent in some years. The physiologies of different species
of animals have enough in common that the fruit the protoobligatory
mutualist specializes on will invariably be highly acceptable to other
animals, only some of which are seed dispersers.

All of these plant traits can of course be evolutionary modulated by
the plant in a diffuse or direct coevolutionary dance with some
animal(s), but as the plant eradicates these difficulties it makes itself
ever more susceptible to a host of others—smaller seeds produce more
incompetent seedlings, increased fruit nutrient balance attracts other
species of animals, and more continuous fruit production attracts more
yet. As the plant moves into ever-increasing dependence on a single
species of disperser, the seed shadow becomes restricted to fewer
habitats and reflects the capricious demographic changes all vertebrate populations experience. Various evolutionary solutions to the
attack by fruit and seed predators are no longer possible as they interfere with this particular dispersal agent's consumption of fruit.
Likewise, various evolutionary solutions to climatic changes will be
obstructed by the dispersal agent's failure to adjust its population
structure to match. The plant becomes ever more vulnerable to the appearance of other plants that seduce its disperser away.

A quite different impediment to tightly coevolved plant–disperser
relationships in species-rich tropical habitats (where so many other
kinds of tightly coevolved relationships have developed) is that each
individual conspecific tree in fruit is in a quite different and unpredictable microsociological relationship to the other fruiting plants, food

sources, potential members of a disperser coterie, and so forth. For example, a mutant that favored a particular seed disperser might well find itself nowhere near a usefully dense population of that species in its second generation. Because the disperser requires more than just fruit of the habitat, its presence cannot be molded by the plant with infinite flexibility. In extratropical species-poor forests (and hence, conspecific-rich forests), this would seem to be less of a problem and, indeed, it is there that some of the apparently most tightly directly coevolved plant–disperser relationships are found (e.g., squirrels and oaks).

I suspect that the largest stumbling block to very tightly coevolved plant–disperser interactions is the fact that the dispersers are vertebrates; and there are many difficulties in maintaining a year-round crop of fruits of sufficient quantity and quality to sustain a vertebrate population. Figs and fig bats (*Artibeus jamaicensis*) approximate this most closely in the Neotropics, but the system would probably be impossible without (and was probably generated by) the figs' very peculiar pollination system (see Chapter 13 by Feinsinger). There are also some plant–disperser relations that are very tightly coevolved with respect to seed predators that are also dispersers. Various extratropical birds and squirrels appear to live on a nearly pure diet of a particular seed species that they also disperse by caching. Here, the problem of year-round food is solved by the disperser rather than the plant, though the plant contributes by bearing chemically undefended seeds.

The comments in this section apply to plant–pollinator relationships as well as plant–disperser relationships, but there is an important difference. Many pollinators are small enough to get their entire food supply (or other needed material) from one or a very few species of plants. In addition, such animals are often inactive during the non-flowering part of the year (e.g., a bee underground in a cell), and therefore the periodicity of a plant's sexual reproduction is not nearly as disruptive to them as it is to a vertebrate (Chapter 13).

Finally, even if a one-on-one plant–disperser relationship coevolved, the opportunity for other animals and plants to evolutionarily move into the system is very great. As a new animal begins to intrude, for example, there is selection for plant traits that incorporate that animal in the disperser coterie just as well as for traits to keep it out. Seed dispersal is definitely one of the "if you can't beat 'em, join 'em" areas of nature.

How does one recognize coevolved plant–disperser relationships?

First, it is essential to be intimately familiar with the ways traits of animals and plants function in the field. Whenever the function of a trait is not clear, be very suspicious that either the field observations are incomplete or that some of the members of the habitat are not present. Second, one needs to show that the trait really does function in concert with the traits of the partner. However, the best one can still do is show that the trait functions in this manner at present. For example, if the agouti were introduced to Africa and it were to spread widely (there is no ecological analogue to the agouti in Africa), I suspect that it would very shortly be found to have many tight interactions with the fruiting and seeding biologies of a variety of forest trees. Once the demographies of these trees had adjusted to this new member of their disperser coteries, even very intense study of the system would probably lead to the conclusion that the agouti was one of the major members of the disperser coteries with which the trees had diffusely coevolved. This is because while there is no agouti analogue in Africa, the other members of the disperser coteries have, through diffuse coevolution, selected for an array of fruit and seed types quite similar to those that the agouti dealt with in the Neotropics, because it was for most of its evolutionary history a member of disperser coteries that contained large mammals similar to those that disperse seeds in Africa today. In short, African plants did not coevolve with the agouti (even though it will look like they did when it is introduced), but they did diffusely coevolve with disperser coteries that were similar to those that naturally contained the agouti.

Are fruits, seeds, and the animals that eat them highly anachronistic?

All organisms are collections of anachronisms, but it is clear that some organs and organ systems are more conservative than are others over evolutionary time. Fruits should be particularly conservative with respect to animal dispersal agent relationships for the simple reasons that (1) there is strong convergence among fruits as a consequence of strong agreement as to what is edible by large groups of vertebrates, (2) most plants have multispecies disperser coteries, so that when one animal becomes extinct, the remainder continue selection to maintain the fruit and seed phenotype (likewise, a newcomer cannot so easily evolutionarily tug the phenotype in its own direction because of the other species holding it in one place), and (3) the plant is doing its best to stay with the interaction, not escape from it as in predator/parasite-prey relationships.

Related to this is a very worrisome pragmatic aspect to understanding anachronistic aspects of fruit–seed–animal interactions. A glance at any tropical landscape with at least one passable road in the vicinity will show that there are survivors of plant–disperser interactions scattered about in a sea of extinct interactions. It may take 300–400 years for the population of animal-dispersed trees to die out of an area after all or the bulk of the disperser coterie has been shot, poisoned, or starved out of the habitat; trees have long life spans, and a dwindling population may dwindle yet more slowly by occasionally replacing its members through recruitment directly under the parent without dispersal.

Is there a best or optimal fruit?

Certainly there is no best fruit as long as disperser coteries are made up of many species of animals having to harvest other resources elsewhere. However, there is also probably no optimal fruit as well. There can only be an optimal distribution of fruit traits among a plant's fruit (and seed) crop. Normal distributions of fruit traits are certainly not to be expected, as the different likes and dislikes of animals in different proportions in a disperser coterie will distort the distribution in many ways. One cannot even design a theoretical optimal fruit crop without first having fixed the disperser coterie and its traits and decided on the type of seed shadow to be generated. Would you predict that intense seed predation by parrots might increase the fitness of a fig seed crop because their noisy chattering serves as a location cue for troops of spider monkeys? The optimal defecation pattern of guanacaste seeds by horses is probably generated by almost minimizing the number of fruits eaten by each horse, because the fewer seeds per dung pile the less likely seed predator mice are to find the seeds. As fruit quality rises, dispersal quality may well fall, if the rise just results in more animals sitting in the tree. These kinds of relationships cannot be predicted in any specific sense, though their presence can be predicted in general.

Do plants compete for dispersers?

Yes. All species of plants with animal-dispersed seeds share one or more dispersal agent with another species of plant; all conspecific plants have overlapping disperser coteries. Allospecific plants also support each other's disperser coteries; the interaction is ecological rather than evolutionary. However, dispersers, like other herbivores, may easily compete in evolutionary time without ever being in day-to-day contact. If a disperser mutant appears that does a higher quality job, selection may favor traits that favor that animal, and do it through decreasing the availability of fruits, nutrients, and seeds for another animal.

258

Do plants evolutionarily transform predatory activity into dispersal?

Yes. Many traits of fruits and seeds are easily and probably correctly interpreted as having appeared through selection for lowering the damage done to seeds as they pass through the hands and guts of animals. When a new fruit eater that is prone to grinding up seeds is added to a plant's disperser coterie, traits may appear that deter the newcomer from feeding on fruits or that make the seed less likely to be ground up as the fruit is eaten. (There may be no change if the animal kills some seeds, yet leaves the survivors in a very high-quality habitat.)

The dispersal of herb seeds by the guts of large browsing/grazing mammals may provide the best, but almost totally unexplored, example. Herbivory, whether by large or small animals, is generally viewed as detrimental to the plant. However, when the large mammals are also dispersers of the seeds (as is the case with many grasses and herbs that have no other dispersal than just falling to the ground beneath the parent), the game changes. When the vegetation was largely made up of established perennial plants, and herbs (especially annuals) made their living by moving among temporary disturbance sites (tree falls, landslides, new alluvial deposits), movement via the guts of the large mammals that grazed and browsed in these sites may have been a very important part of their biology (this has been hidden from us by the destruction of these plants' habitats and their movement into fields, roadsides, and so forth). Here, consumption of an herb along with its seeds is much like the consumption of a more conventional fruit with its seeds; the vegetation portion of the herb is the "fruit." Herbivory by the large mammal may then be far from detrimental to the plant. If such herbivory is beneficial and herbivory by insects detrimental, there are quite severe constraints placed on the possible defense options available through natural selection and on our attempts to categorize and understand the many patterns evident in herb defenses as compared to defenses of browse plants whose seeds are dispersed by other means. For example, the pungent mustard oils of the crucifers could not only be defenses against insects but also odoriferous cues to attract a large mammal that will consume the foliage and disperse the small hard seeds.

To what extent are seed dispersers also seed predators?

In most studies of seed dispersal via the guts of animals, the focus has been on the quality of what goes in rather than on what comes out. The simple presence of viable seeds in dung was generally regarded as the end of the investigation, probably because of the old attitude that plants make so many seeds that it doesn't matter if some are killed here or there. There is also no literature based on feeding animals known numbers of seeds and recording their fate during transit. I suspect that virtually all seed dispersers kill or even digest some living seeds, with the number rising the longer the transit time, the larger the seed, and the more the animal's gut normally digests hard food items. This kind of seed digestion is pragmatically seed predation but is usually best viewed as a by-product rather than something that selection has generated through evolutionary modification of digestive regimes to favor it. However, we simply don't know how important are the bits of nutrient so obtained in the general scheme of vertebrate diets.

Many plants clearly pay offspring in return for dispersal, and in this case the swallowed seeds are invariably dead and the dispersed seeds are a by-product of the caching behavior of the animal. However, there are a number of traits of the seed or nut that can be easily identified as increasing the probability that the disperser will (1) bother to cache the seed and (2) fail to recover it.

In the case of accidental seed digestion, there are many examples where it is clear that the plant has evolved traits that protect the seed, though such things as particular transit rates may derive their value from directing the placement of the seed in the habitat as much as from getting the seed out of the hostile gut. In the progression from animals that consume fruit pulp as a minor part of their diet (e.g., horses, elephants) to animals that live almost entirely on fruit pulp (e.g., manakins, quetzals, oilbirds), there is a general progression from those with no sign of traits to avoid digesting seeds to those with many traits that appear functional for this. In the latter case, the system may be accurately viewed as diffusely coevolved because even the most specialized dispersers are usually members of multi-species disperser coteries, and nearly all coteries are multispecific.

In the case of deliberate seed predation, not only has the plant evolved traits that temper the animal's impact, but it is possible that the animal may have solved some traits that minimize its impact. It is generally assumed that rodents scatterhoard seeds rather than cachehoard them because the rodent then need not protect a cache and because other animals cannot readily find all the solitary seeds. However, an agouti that turned to very accurate seed predation by virtue of caching and thorough recovery would simultaneously select strongly for chemical and other seed traits that rendered the seeds unavailable to it and available to other dispersal agents.

Are there local adaptations to specific dispersers?

If there is local adaptation to specific dispersers, it is occurring on a scale too fine to have been picked up by any study to date. However, virtually all studies of dispersal have been carried out in such disturbed systems that such tight evolutionary fine-tuning would have been severely disrupted. Furthermore, it is safe to say that *all* plant populations, tropical and temperate, have had their disperser coteries so badly distorted by recent human activity that this kind of question can probably never more be examined, except by inference.

Are adaptations for seed dispersal and protection mutually reinforcing or incompatible?

Both. Virtually all traits of seeds and fruits have probably been modified in response to selection for protection of the seed in passing through the guts of dispersing animals. Many of these traits will simultaneously increase protection from seed predators. It is not useful to worry about which animal did the selecting, but rather to note that dispersers and seed predators, for example, are just part of the suite of selective forces that produce the hard seed coat of a legume seed. The toxins in a large soft lauraceous seed that selected for the failure of bird gizzards to grind them up may also stop a rodent from eating them once dispersed. On the other hand, the amount of sugar in a blueberry is probably not a response to seed predation so much as to feeding by bears and birds.

Incompatibilities appear frequently in the area of fruit chemistry; a severely toxic fruit may well keep away the parrots, but it has to be nullified when the seeds are to be offered to a fruit pigeon. One solution is the fruit that splits when ripe, allowing the disperser to get at the aril-encased seeds without having to chew through the fruit wall rich in secondary compounds. Incompatibilities also occur in fruit crop ripening schedules. A conspicuous adaptation for seed dispersal is to spread the ripening period of a plant's fruit crop over a longer period, thereby depressing disperser satiation. Such plant behavior simultaneously depresses seed predator satiation. The crop of longer duration may also attract more dispersers, more seed predators, or both.

Incompatibilities also occur when the parent plant is paying offspring rather than fruit tissue for disperser services. A seed can clearly be evolutionarily designed to be so toxic that a rodent will ignore it—they ignore many species of forest-floor seeds. However, such a rodent will also not carry the seed off to a nest or cache. Yet if the seed is too edible, not only will the rodent eat it but so will many other animals in the forest. One solution has been to put a highly edible seed inside a very tough and hard seed or nut (bony fruit endocarp). Such a system also has the advantage that the nearly impenetrable object can be covered with a small and edible meal (fruit pulp), causing the rodent

261

to eat that and bury the harder-to-open seed for a meal when times are harder.

In all of these cases, generalizations are easy, but not very useful for specific cases. We desperately need case history studies before more specific generalizations are possible, yet in most parts of the world, the case histories are no longer possible. It is very difficult to even attempt to understand fruiting patterns of berries and nuts in the eastern United States, for example, without consideration of passenger pigeons.

IN CLOSING

The interaction of seed dispersers and plants is so rich in relevant detail and so idiosyncratic from one system to the next that it may appear so chaotic as to be uninteresting to study. But generalizations are possible: fruits containing seeds dispersed by small birds are *usually* red—and the proportion that are not is an interesting geographically varying trait of habitats. But the subject has been so thoroughly ignored, and the natural arena so thoroughly muddied, that we don't need a massive infusion of theory and generalizations but rather a massive infusion of detailed case studies of how things really work. There must be at least 100 papers on pollination biology for every one in seed dispersal biology. We don't even know if dispersers choose blemished fruit over perfect fruit.

The interaction between plants and their seed disperser coteries should be of special interest to evolutionists for the simple reason that much of the phenotype is spread out in the field to count and measure. When a mutation alters a leaf's temperature regime by 0.2 degree by increasing marginal crenulations, you know that change has many effects on the rest of the plant, but the changes are hidden in microscopic molecules, enzyme rates, and other parts of nature that cost billions to look at. When a mutation alters the color of a fruit from red to orange, ·the mutant's new seed shadow and disperser coterie is laid out for all to observe.

262

CHAPTER 12

COEVOLUTION AND MIMICRY

Lawrence E. Gilbert

INTRODUCTION

R. A. Fisher called mimicry theory "the greatest post-Darwinian application of natural selection." In his classic synthesis of Darwinian theory, Mendelian genetics, and mathematics, Fisher (1930) devoted an entire chapter to mimicry. Indeed, from A. R. Wallace (1889) to Ford (1971) to current evolution textbooks, the subject of mimicry has been given extensive attention, if not entire chapters.

There are several reasons for the continuing interest by evolutionary biologists in the subject of mimicry. First, mimetic traits of an organism include structures, patterns of color, behaviors, or other phenotypic attributes that set that organism apart from its close taxonomic relatives while promoting its resemblance to a more distantly related taxon. Thus, nonmimetic relatives provide an idea of the primitive phenotype, and the model phenotype is the standard toward which the mimetic species evolves by natural selection. Second, many mimetic species display both intra- and interpopulation variation with respect to mimetic attributes, and for some cases, the genetic basis of that variation is established. Finally, mimetic species are generally parasites of previously evolved systems of communication between other ("host") species, so that "host–parasite" coevolution is one possible consequence of mimicry.

This chapter concerns both the circumstances that might lead to detectable coevolution between organisms involved in mimicry systems and the extent to which mimicry itself might result from coevolution. Throughout this chapter the term *coevolution* is used in a restricted, microevolutionary sense, namely, evolution that occurs in populations of at least two species as the result of reciprocal selective influence that each has on the other. I am therefore using the narrowest definition described by the editors (see Chapter 1).

A BRIEF SYNOPSIS OF MIMICRY THEORY

The theory of mimicry was first developed by the great naturalist H. W. Bates (1862) as a means of explaining remarkable resemblances between certain South American butterfly species belonging to different families (Figure 1). According to Bates' friend Ralph Meldola, the theory occurred to Bates only after he was back in England working with his collections. He was struck by the geographical correlation of color pattern changes within entire sets of butterfly species that he had collected in the Amazon Basin. Having just read Darwin's *Origin of Species*, Bates proposed that edible species had evolved by natural selection to resemble warningly colored, noxious species and that predators such as birds were the likely selective agents.

Some twenty years after Bates' landmark contribution, another of the great nineteenth century naturalists, Fritz Müller (1879), identified another form of protective convergence, namely, that between different distasteful species. In this form of convergence, each species has different warning patterns initially but converges on the same pattern under selection from predators. Using a simple mathematical model, Müller argued that two warningly colored species in the same area should converge on a common appearance because such cooperative education of predators would require fewer deaths for each species. He proposed that mutants of species A bearing slight resemblances to species B would benefit from the umbrella of predator protection resulting from past predator experience in testing individuals of B and would have a selective advantage over pure A genotypes. Likewise, mutants of B that resemble A should be selected over pure B by the reciprocal argument. Müller proposed that the final mimetic pattern would be determined by the relative abundances of A and B, such that the rarer species should converge on the pattern of the common species faster than the common one converges on the rare pattern.

There is a vast literature on classical Batesian and Müllerian mimicry, most of which is concerned with butterflies and other insects. Reviews or monographs on this subject have appeared in practically each decade of the last 100 years. Those by Carpenter and Ford (1933), Brower (1963), Rettenmeyer (1970), Turner (1977), Rothschild (1979), and Vane-Wright (1980) are thoughtful analyses that also provide an index to the background literature dealing mainly with the Bates-Müller theory. A few highlights follow.

The geographical correlation in color pattern between model and mimic species, first noted by Bates and illustrated by Moulton (1909), has been further documented from a genetic perspective by Sheppard (1962) and Turner (1971). The potential for birds to act as selective agents on prey color pattern was given credibility by Swynnerton (Carpenter, 1942) and experimentally verified for captive birds by Brower (1958) and for wild birds by Jeffords et al. (1979). The genetics

264

Poulton.   LINN.SOC.JOURN.ZOOL.VOL.XXVI.Pl. 42.

1. Group as described by H W Bates in 1861.

2. Group as known in 1897.   EDWIN WILSON, CAMBRIDGE

RESEMBLANCES IN TROPICAL AMERICAN LEPIDOPTERA.

FIGURE 1. A plate from an early review of mimicry by E. B. Poulton. Bates was concerned with the resemblance between presumed edible pierid *Dismorphia* and the ithomiine *Methona*. Müller's theory was based on observations of the danine *Ituna* and the ithomiine *Thyridia*, both of which are distasteful. Neither Bates nor Müller had the modern view of systematics reflected in Poulton's arrangement of the plate.

265

of mimetic variation, which developed as a field of interest immediately after the rediscovery of Mendel's work, culminated in the classic work of Sheppard, Clarke, and their associates (see Turner, 1977). The unpalatibility to vertebrates assumed for model species by most students of mimicry has now been demonstrated for a few species such as the monarch butterfly (Brower and Glazier, 1975). It frequently results from the storage of host plant-derived chemicals (references in Rothschild, 1979). Ecological studies of model and mimic populations of burnet moths (Sbordoni et al., 1979) have verified Müller's assertion that the rare species of a distasteful pair should converge on the more common, and in this case, less noxious, species.

In addition to these contributions to the Bates–Müller tradition, other lines of biological research have revealed significant new categories of mimetic relationships in nature. Of particular interest are forms of mimicry not involving a predator as the selective agent. Many such examples emerged from the kinds of ethological studies pioneered by Tinbergen and Lorenz. Wickler (1968) reviews the entire subject of mimicry from this fresh perspective and develops a general theoretical framework that includes Batesian mimicry as a special case.

In Wickler's scheme, mimicry evolves only in the context of well-established communication systems, wherein organisms that send signals as well as organisms that receive and react to those signals have a strong mutual interest in clear, unambiguous transfer of messages. Both the unpalatable insect and its potential attacker benefit from a clearly transmitted warning signal. The insect does not sustain life-threatening damage; the predator does not waste time and energy in pursuit of unsuitable prey. Wickler defines mimicry as the sending of fake signals by a third organism, which derives some advantage (such as predator protection) in the deception of the signal receiver.

Late last century, Poulton (1898) pointed out that Müllerian mimicry probably should not be regarded as true mimicry. Wickler would agree because such convergence of warning signals does not involve deceit. Or does it? Another important feature of Wickler's theory is that phenotypic traits should only be categorized relative to the signal receiver. For example, a mantid that resembles a flower is an aggressive mimic from the standpoint of insects that are attracted to the imitated flower, but cryptic (imitation of nonsignals) to an insectivorous bird. Likewise, it must be true that as the degree of unpalatibility is variable within and among species of Müllerian complexes (Brower et al., 1963), some predators will treat the entire set as unacceptable whereas others, possibly specialists on noxious insects, will find part of the same complex edible and will lead us to view the system as Batesian mimicry. There are other reasons to treat Batesian and Müllerian mimicry as a continuum (Sbordoni et al., 1979, and references therein) that involves relative palatability and relative abundances—aspects of the problem largely anticipated by Müller himself.

266

COEVOLUTION AND MIMICRY

The Batesian–Müllerian gradient is a special version of the parasitism–commensalism–mutualism continuum well known in ecology. It is generally thought that relationships that begin as parasite–host relationships may evolve toward commensal or mutualistic relationships. For example, the relationship between pollinating insects and higher plants most likely began with insects as parasites of plant reproduction, and became mutualistic as plants and insects coevolved toward the specialized pollination systems of today. The tricky problem is how a parasite–host coevolutionary race becomes coevolved mutualism. We would expect Batesian models either not to evolve as mimicry evolves (in the case of a rare, and/or somewhat distasteful, *commensal* mimic) or to evolve away from the mimetic pattern (as with a common, edible, *parasitic* mimic). In the case of Müllerian models, either no coevolution occurs (with rare, and/or slightly less noxious, *commensal* mimic/comodel) or two warningly colored species coevolve toward an intermediate pattern (equally abundant, equally distasteful, Müllerian mimics). These relationships are shown in Figure 2.

From these simple considerations it is clear that coevolution need not result from the evolution of Batesian mimicry nor be the mechanism by which Müllerian species come to possess similar patterns. But, by the same token, certain instances of mimicry almost certainly involve coevolution and thus provide an unusual opportunity for developing direct evidence for the study and documentation of this elusive process.

MÜLLERIAN MIMICRY AS EVIDENCE FOR
COEVOLVED MUTUALISM

Coevolution is more likely to occur under circumstances of tight ecological association between species, as in host-specific parasitism. It is most likely to be detected if extensive comparative knowledge of the morphology, behavior, and genetics of the higher taxa involved can be used to establish a basis for picking out the reciprocally evolved traits of each species in a coevolving pair. For example, Janzen's (1966) study of obligately associated species of ant *Acacia* and *Pseudomyrmex* ants revealed numerous morphological and behavioral attributes of both the ant and the *Acacia* which, on the one hand, set them apart from congeners not participating in such mutualism, and on the other, are best explained as a consequence of a coevolutionary interaction.

267

Many Müllerian associates, although not involved in such a tight ecological linkage as *Pseudomyrmex* and *Acacia*, do appear to have evolved away from close relatives with respect to those attributes that function in the cooperative warning and education of predators. But as Müller anticipated and Figure 2 restates, strong coevolved mutualism is only expected where the mimics have similar abundance. Müller's comment on this point is, in my opinion, the first quantitative statement of coevolution: "If two or even several distasteful species are equally common, resemblance brings them a nearly equal advantage, and each step which the other takes in this direction is preserved by natural selection" (Müller, 1879, p. xxviii).

F. A. Dixey (1897, 1909) elaborated Müller's concept of mutual convergence—he called it diaposematism—based upon his studies of pierid butterflies. The Dixey papers touched off a debate that, in retrospect, appears to be a case of both sides being partly correct. Marshall (1908) argued against the possibility of diaposematism or coevolution. He reasoned that a rare species should converge on a common species, but the common species would not itself evolve toward the rare pattern because relatively few predators would have encountered and learned the pattern of the rare species. Therefore, mutants of the common species that resemble the rare form would have lower fitness than the standard individuals.

Modern experimental evidence from natural populations does support the notion that major alteration of a warning pattern results in significantly higher attack rates by birds. Benson (1972) blacked out the forewing red patch of wild *Heliconius erato* and established a control group visually identical to unaltered specimens (red, black forewing; yellow, black hindwing). He detected increased wing damage by birds and shorter longevity in the altered individuals. I have already mentioned other field studies that show convergence of rare species or the phenotype of the common species in distasteful moths (Sbordoni et al., 1979). Had these results been available in 1908, Marshall would have used them to support his case against coevolved Müllerian associations.

Fisher's (1927, 1930) analysis of this problem introduced a concept akin to what today would be called "punctuated equilibria" (for a discussion of this interpretation see Turner, 1977; 1981). Put briefly and simply, Fisher identified two classes of genes defined by the nature and degree of their impact on the phenotype: (1) Major genes, mutations of which cause major alterations of phenotype. (2) Modifier genes, mutations of which cause slighter alterations in color and pattern components and which account for gradual evolution of phenotype in the time period between major genetic changes.

Fisher's theory would predict that coevolution would occur between two distinct aposematic species only after a major gene mutation in the rare species has produced a phenotype that, though not exact, can be confused with the common species. Once rough resemblance is achieved (evolving unidirectionally as Marshall would have predicted), minor coevolved modifications of both species would lead to a more perfect resemblance, as Müller and Dixey had suggested.

The controversy over Müllerian convergence led to some very creative thinking (especially by R. A. Fisher), the influence of which is

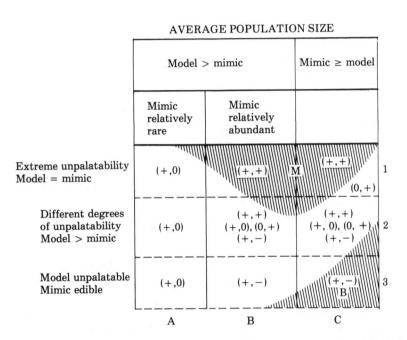

FIGURE 2. Ecological interactions and coevolution in Batesian–Müllerian mimicry. Hypothesized interaction between gradients of palatability (rows), relative abundance of model and mimic species (columns), and net ecological interaction coefficients between mimic and model (shown as negative, positive, or zero in parentheses in cells of chart) are related to potential for coevolution (hatched zone of table). The entry $(+,-)$ means that the relationship has a positive impact on the mimic but a negative impact on the model. Cells 2B and 2C show that the net effect of mimicry is basically uncertain without more knowledge of relative palatability (measured in terms of the most significant natural predators) and more precise population data. Depending upon variables such as toxicity of local host plants, local population sizes of each species through time, and seasonal effects on predator palatability thresholds, a given pair of mimetic species might occupy many cells of this chart through both ecological and evolutionary time. (M = zone of likely Müllerian coevolution; B = zone of likely Batesian coevolution.)

269

seen in the work of Ford, Sheppard, Turner, and others. The debate was, and is, more complex than my distillation would lead one to believe. Commonness and rarity are often treated as species phenotypes rather than ecological variables. Müllerian coevolution seems much more likely if the two species in question periodically switch rank in terms of numerical dominance, a circumstance that undoubtedly occurs within many Müllerian associations through time and space.

The best documented example of Müllerian convergence is the extensive parallel race formation of the neotropical *Heliconius erato*, *Heliconius melpomene*, and other members of the *H. erato* and *H. melpomene* groups (Figure 3). In Müller's day, too little was known of the taxonomic affinities within *Heliconius* for anyone to seriously suspect that such remarkable convergence reflected anything more than very close taxonomic affinities. But as careful biogeographical and morphological information on the genus developed it became clear that *H. erato* and *H. melpomene* represented distinctive radiations within the genus, and Eltringham (1916) presented what amounts to a modern view of the relationships between these species. We now know of extensive differences between these species groups of *Heliconius*. *H. erato* and its relatives have distinctive pupal and male clasper morphology. They often mate on pupae, and females are monogamous. They are generally associated with the Plectostemma subgenus of *Passiflora* (passionvines) as larvae (Gilbert, 1976; Benson et al., 1976; Brown, 1981). *H. melpomene* and its relatives have standard butterfly courtship, females mate repeatedly, and morphology and host relations are generally distinct from *H. erato* (see previous references).

H. erato and *H. melpomene* have been the subjects of extensive genetic and biogeographical analyses during the past 25 years. Highly distinctive races of these two species (Figure 3) are closely correlated in space (Turner, 1971) but a sample of genes not involved in color pattern determination is not differentiated geographically (Turner, 1979). Variation in a half dozen or so major genes and a host of minor modifying loci, most unlinked, account for the racial differentiation of each species (Sheppard et al., 1981). Thanks to the extent and quality of these data, it has been possible to demonstrate the parallel evolution of races in *H. erato* and *H. melpomene*. On the assumption that recessive alleles at a locus are replaced by dominant alleles and thus define ancestral color pattern components, it has been possible to construct

FIGURE 3. Parallel races in the genus *Heliconius* according to Eltringham ▶ (1916). *H. erato* (3, 7, 11, 15, 23, 25, 27) are matched against sympatric races of *H. melpomene* (4, 8, 12, 16, 24, 26, 28). Races of *H. erato* relative *sapho* (9, 13, 17, 21) are matched against sympatric varieties of *H. melpomene* relative *cydno* (10, 14, 18, 22). In one case, the mimic of *H. erato* is considered to be a *cydno* (19, 20). (This figure is reproduced in color on the cover.)

MODELS   MIMICS   MODELS   MIMICS

H. Eltringham, del   Andre, Sleigh & Anglo, Ltd

GENUS HELICONIUS, MODELS AND MIMICS.

ancestral patterns for both species. These appear to have been similar black and yellow insects (Sheppard et al., 1983). Cladograms based on gene substitutions and geographical relationships suggest periodic pulses of evolution for both species (Turner, 1981).

Unfortunately, parallel evolution of mimetic races does not prove coevolution (in its reciprocal, microevolutionary sense) because if one species is everywhere the more common, as indeed *H. erato* usually seems to be, the rarer species (*H. melpomene*) might be always converging on the common one (*H. erato*) as the latter differentiates. In the case of *H. erato* and *H. melpomene*, genetic evidence and common sense provide reason to reject the hypothesis that no coevolution has occurred, that is, it is simply not probable that all of the numerous unlinked "mimicry genes" that have undergone allelic substitutions during the parallel race formation would have evolved first in *H. erato*.

To my knowledge, the only strong evidence for mimetic coevolution in this system comes from analysis of clines in color pattern, which correlate with the absence of one of the species. Because *H. erato* is generally more abundant and more widespread geographically and elevationally (see maps in Turner, 1971), there is no doubt that its presence has strongly influenced the evolution of the *H. melpomene* pattern as it has, for example, been shown to affect *H. hermathena* (Brown and Benson, 1977). Thus in this case, to be certain that coevolution has occurred, it must only be shown that *H. melpomene* has some reciprocal impact on the evolution of *H. erato*. In extreme northern Mexico, *H. erato* is not accompanied by *H. melpomene*, and there the hindwing yellow bar is extremely thin (Figure 4). Where Central American *H. erato* is sympatric with *H. melpomene*, the part of the *H. erato* hindwing bar not shadowed out by the overlapping forewing closely matches that of *H. melpomene* in shape and width, being about twice the width of Mexican *erato* (see Figure 5). In my study site in Corcovado Park, Costa Rica, *erato* and *melpomene* are indistinguishable not only in flight, but also when visiting flowers under strong vertical light, which has the visual effect of cropping the *erato* yellow bar to the length of *melpomene's*. The extensive geographical range of this species pair provides a rich assortment of distributional relationships (overlap versus nonoverlap situations) between the two species which now should be studied in detail to establish Müllerian coevolution as a fact of life for *Heliconius*.

BATESIAN MIMICRY: HOST–PARASITE COEVOLUTION?

As defined in this chapter, coevolution between a Batesian mimic and its model occurs only if the model species evolves in response to the mimic. The conditions under which a model's color pattern might evolve are rather stringent because there is a cost to altering a warning

272

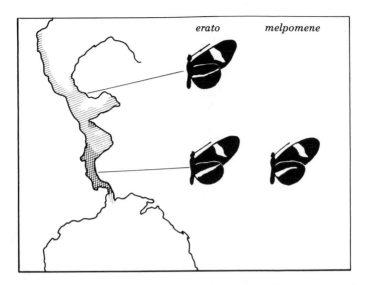

FIGURE 4. Evidence for effect of *H. melpomene* on color pattern evolution in *H. erato*. *H. erato*'s hindwing bar widens in zones of sympatry with *H. melpomene* (stippled area). (See Figure 5 and its legend.)

signal. Benson's altered *Heliconius* were still black and yellow and were unaltered in shape and flight behavior, yet they suffered increased mortality from birds. Even though most birds might generalize from their bad experience with bright-colored prey by ignoring all remotely similar insects, a few species of specialists on conspicuous insects, such as the neotropical jacamar or African bee eaters, may be sufficient to act as major selective agents in this system, because their success in feeding depends upon fine distinctions between similar prey that differ in degree of unpalatability.

Nicholson (1927), Fisher (1930), and later Charlesworth and Charlesworth (1975) explored the consequences of Batesian mimicry for both model and mimic species. Using the realistic assumption that the population sizes are determined by density-dependent factors acting on larval stages, the latter authors show (consistent with Fisher's earlier conclusions) that the evolution of a gene for mimetic pattern in an edible species will result in little or no change in its population size but will cause substantial reduction in the model species' population size. Consideration of this problem had earlier led Fisher (1930) to postulate that models should, in response to Batesian parasitism, evolve new patterns that predators can distinguish from the mimetic patterns. Charlesworth

273

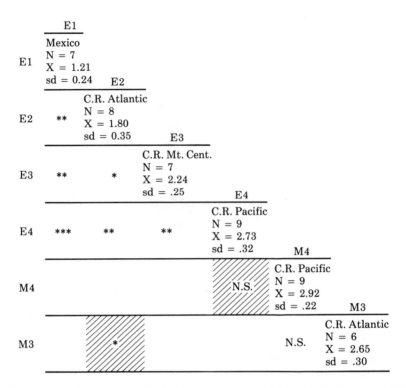

FIGURE 5. Preliminary analysis of variation in width of hindwing bar in *H. erato.* In the chart, E1, E2, E3 and E4 represent four geographically separate samples of *H. erato* and M3 and M4 represent *H. melpomene* sympatric with E2 and E4. The diagonal cells give location, sample size, mean bar width (in mm), and standard deviation for each sample. Asterisks in the cells of the chart indicate probability from one-way ANOVA, that differences are due to chance (∗, $p < 0.01$; ∗∗, $p < 0.001$; ∗∗∗, $p < 0.0001$). Thus, all *H. erato* samples are different, with a continuous increase in bar width from Mexico to the Pacific side of Costa Rica, at which point there is no significant difference between *H. erato* and *H. melpomene* bar width (hatched cell, M4/E4). Note that Atlantic side *H. erato* differ significantly from sympatric *H. melpomene* (hatched cell, M3 E2). Thus, gene flow from the north may be preventing accurate mimicry in the Atlantic lowlands, whereas mountains protect the Pacific lowland *H. erato* populations from this influence.

and Charlesworth (1975) disagree: "However, a new form of the model will lose the protection from which all members of the model species share, through the fact that the predator learns best to avoid the form of the model which is most common, so that the model species may well be unable to evolve a new pattern, despite the evolution of mimicry in another species living in the same area."

Are we then to conclude that Batesian mimicry involves only convergence by edible parasitic species toward the predator warning signals of a distasteful model or host species and that coevolution, *sensu stricto*, does not occur in Batesian systems? As tempting as this conclusion appears to be, it would be made in ignorance of the evolutionary options that are available to the model in nature.

Simple theories of coevolution necessarily consider evolution in parallel traits of model and mimic, such as the components of color pattern visible at a distance. However, a model's evolutionary response to "mimic pressure" could take the form of behavioral avoidance of mimics in time or space, or the development of nonvisual signals (such as bad-tasting wings or strong odors) and the addition of subtle pattern elements (such as the white thoracic spots of monarch butterflies) that would improve predator ability to distinguish model from mimic (after capture but before killing prey). Such traits should vary within an aposematic, distasteful species between geographically separate populations exposed to different degrees of Batesian parasitism. However, much more study of natural predators that are important as selective agents will be needed before we will be able to adequately understand the coevolutionary options of Batesian models.

Returning to the possibility of coevolutionary changes in model patterns resulting from the evolution of Batesian mimicry, it should be pointed out that not all mutant patterns of the model species would necessarily have lower fitness than the original type. Most authors have ignored the possibility that a model species might escape a Batesian mimic by switching to another sympatric mimicry system containing more comodels and fewer edible Batesian mimics. As with the standard argument for the evolution of Batesian mimicry itself, this mechanism works only if the initial mutation has a major effect on the model's phenotype in the direction of its new Müllerian model.

One of the frequently recited principles of mimicry is that distasteful, warningly colored species should be monomorphic for aposematic traits. Yet we know of several polymorphic, yet distasteful, butterflies, including African *Acraea encedon* (Owen and Chanter, 1968) and American *Heliconius ethilla* (Turner, 1968) and *H. numata* (Brown and Benson, 1974). Brown and Benson hypothesize that *H. numata* has a Batesian relationship to more numerous and more distasteful ithomiine butterflies and argue that color pattern switch genes have evolved in *H. numata* in the context of temporal and spatial patchiness of different ithomiine species. But why are there so many patterns to copy among the ithomiines (Papageorgis, 1975)?

One case of polymorphism in a distasteful model appears to be consistent with a coevolutionary escape hypothesis. D. A. S. Smith (1979) sampled the African danaid *Danaus (Limnas) chrysippus* and its Batesian mimic *Hypolimnas misippus* for 11 months near Dar es Salaam, Tanzania. Four major morphs of *D. chrysippus* are paralleled closely by four morphs of *H. misippus* (see Smith, 1973). Smith recorded bird beak marks on the two most common forms of *D. chrysippus* to determine whether different levels of predator pressure were experienced by different morphs of *D. chrysippus* in relation to the abundance of corresponding mimetic morphs of *H. misippus* (see Figure 6).

As classical theory would have predicted, the *D. chrysippus* morph most frequently attacked by birds had the larger following of mimics. *D. chrysippus* form *aegyptius* showed a more significant incidence of bird damage than did *D. chrysippus* form *dorippus* (12.7% to 6.0%; D. A. S. Smith, 1979). During that period (1974–1975) *D. chrysippus* form *aegyptius* was 5 times as abundant as its mimic *H. misippus* form *misippus*, whereas *D. chrysippus* form *dorippus* was 32–41 times as abundant as its mimic *H. misippus* form *inaria* (D. A. S. Smith, 1976, Table 2). *D. chrysippus* form *dorippus* also participates in a large Müllerian complex containing several distasteful *Acraea* and relatively few Batesian mimics.

Genetic evidence is consistent with the idea that *D. chrysippus* form *dorippus* evolved from *D. chrysippus* form *aegyptius*. The latter

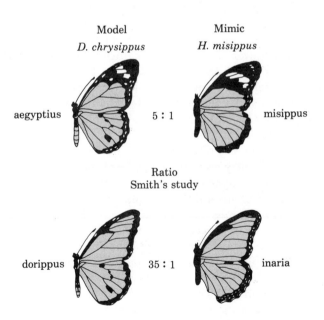

FIGURE 6. Smith's illustration of parallel polymorphic forms of *D. chrysippus* and Batesian mimic *H. misippus* in Tanzania.

276

is converted to *dorippus* by a dominant allele at the *C* locus (Smith, 1975). So, if the East African populations of *D. chrysippus* are composed mainly of toxic individuals, the evidence is in general support of the idea that was put forward by Owen and Chanter (1968) that genetic polymorphism in this species results from an overloading of Batesian mimics. However, it should be emphasized that evolutionary escape from Batesian mimics is but one of several possible explanations for the polymorphism in *D. chrysippus* (Smith, 1980). It should also be pointed out that in some areas of West Africa, *D. chrysippus* populations consist of a majority of individuals that are edible because they feed on non-toxic hosts (Brower et al., 1975). Thus until African *D. chrysippus* is better studied, any explanation for its polymorphic status must be tentative.

MIMETIC COEVOLUTION IN NON-BATESIAN SYSTEMS

This chapter has focused on the classic forms of mimicry in butterflies because it is here that the most extensive taxonomic, geographical, and genetic information exists. However, it is in other forms of mimicry that coevolution is more likely to be a constantly ongoing process. In cases of intraspecific mimicry such as egg mimicry by male cichlid *Haplochromis burtoni* (Wickler, 1968), wherein a mimetic egg on the male's anal fin organizes oral insemination in this mouth breeder, coevolution between morphology and behavior of the sexes seems certain. But putting aside the special case of intragenomic coevolution, there remain many cases where coevolution is made likely by the nature of relations between signal-sending and signal-receiving species.

In the case of cleaner wrasses (*Labroides*), whose colors and behaviors elicit passive behavior in larger predaceous fish, strong mutual benefits prevent dissolution of the relationship under the pressure of mimicry by the saber-toothed blennies (*Aspidontus*), which parasitize rather than clean the larger fish (Wickler, 1968). The highly specific and obligate nature of this mutualism, coupled with the predaceous abilities of the signal receiver, promotes coevolved mimicry. Apparently, regular removal of ectoparasites by cleaner fish is crucial to the survival of many larger marine fish. The relationship between the cleaner wrasse and its "host" involves communication of cleaner status both through color pattern and ritualized dances by the wrasse. The entry of mimics into this system has apparently created selective pressures that favor improved visual discrimination by the "cleaned fishes" and, at the same time, select strongly for a more complex or sophisticated mode of signaling by the cleaner.

Other mimicry systems, within which ongoing coevolution is virtually certain, also possess the elements of strong mutualism coupled with signal receivers that have intimate relationships with model and mimic. The most striking cases are those in which signal senders and receivers are either juveniles and adults or males and females of the same species. Reproduction, sex, and death are the stakes in this most obligate and specific of signal transfers. Mimetic eggs or young of brood parasitic birds (Payne, 1977), false courtship signals in male-eating female fireflies (Lloyd, 1965), mimetic floral signals (Wiens, 1978), and dummy butterfly eggs on passionvines (Williams and Gilbert, 1981) can all be included in this category.

The case of plants mimicking insect eggs illustrates the indirect and tortuous sequence of coevolutionary events that culminated in this mimicry between trophic levels. The first steps that must be considered involve the evolution by ancestral Passifloraceae of allelochemicals in response to a diffuse background of herbivores. According to one view (Gilbert, 1971), each advance in chemical defense would have eliminated all but those herbivores able to evolve counter defense. This process would lead to specialization on Passifloraceae by a few insect groups. Once the fauna using Passifloraceae had been reduced to a few types such as heliconiine butterflies, it is hypothesized that successful advances in herbivore defense would have been increasingly aimed toward heliconiine and other specialists and would have been less chemical in nature because those insects are in the business, as it were, of circumventing the chemical defenses of Passifloraceae.

Mechanical defenses such as hooked trichomes represent the type of specialized nonchemical defense that would appear to be most likely to arise after the plant fauna had been reduced during diffuse coevolution (Gilbert, 1971). Egg mimics also represent a mechanical device, but one that manipulates the behavior of ovipositing insects.

Explaining why a butterfly would avoid mimetic eggs further emphasizes the complex web of relationships that must be understood in order to trace the probable course of coevolution. It appears that relatively recently the genus *Heliconius* arose from other heliconiines and underwent spectacular adaptive radiation (Brown, 1981). Two steps seem to have been involved. One of these is the expanded reproductive life span and the increased adult investment in reproductive effort allowed by the habit of feeding on the pollen of certain curcurbit vines (Gilbert, 1975). The second innovation of ancestral *Heliconius* was their specialization on the very youngest shoots of *Passiflora* and other Passifloraceae (Benson et al., 1976). Both of these evolutionary developments set the stage for further selection favoring sophisticated foraging behavior by adult *Heliconius* (Gilbert, 1975), one aspect of which is the careful choice of oviposition site by females. Because the larvae exist on small parcels of suitable host, the likelihood of larval

278

crowding and competition is high, and aggressive, cannibalistic larvae have evolved among many of the *Heliconius*. Females ignoring the presence of eggs already on shoots risk leaving no offspring.

In contrast to most other heliconiines that lay relatively inconspicuous eggs, *Heliconius* lay bright yellow eggs, the color of which constitutes a signal used by many species in the genus. This signal would have both an intra- and an interspecific function, because *Heliconius* species often share *Passiflora* species. Several species appear to respond strongly to the presence of eggs on the host plant and the role of egg color has been implicated by experiments on discrimination by ovipositing females of *H. cydno* (Williams and Gilbert, 1981).

When the butterflies evolved discrimination against ovipositing near previous eggs, the way was open for the coevolution of egg mimics by the *Passiflora*. Based on current taxonomic and geographical occurrence of the trait, the evolution of egg mimicry seems dynamic and ongoing. Only approximately 2% of the more than 500 species of *Passiflora* are known to possess these structures. Also, several independent origins of the trait are evidenced by the use of different structures (nectaries, stipules, buds) to form the fake egg and by the independent occurrence of egg mimicry in several subgenera. Furthermore, the occurrence of the trait is geographically variable among populations of the same or closely related species (Williams and Gilbert, 1981). Shapiro's (1981) discovery of a similar phenomenon in crucifer hosts of pierids in California suggests that such mimicry may be widespread.

The coevolutionary responses by *Heliconius* to evolving egg mimicry in *Passiflora* are not yet identified, but careful study of species whose geographically separate populations are differentially exposed to egg mimicry by their host plants should provide some clues.

In both documented cases (Williams and Gilbert, 1981; Shapiro, 1981), egg mimicry has apparently evolved in the absence of conspicuous demographic impact by the butterflies on the plants. Thus, the arguments that butterflies are minor components of plant faunas and therefore are not likely to select for plant defensive traits are similar to arguments against the role of birds as selective agents in butterfly mimicry (on the grounds that most mortality occurs to eggs and larvae).

These and other "Wicklerian" mimicry systems constitute many spectacular cases of mimicry that are, no doubt, dynamically coevolving systems deserving of the detailed attention that has been given to Batesian and Müllerian systems. It is tempting to propose egg mimicry in plants as a model case for intertrophic level coevolution.

279

MIMICRY AND ITS ECOLOGICAL CONSEQUENCES

It should now be clear that systems of Batesian or Müllerian mimicry do not, given our present knowledge, provide many unambiguous examples of coevolution, in the strict sense of the word. This is in part because much past research has been overly occupied with proving facets of mimicry theory that, thanks to a vast amount of systematic, genetic, geographical, and natural historical data, have been virtually certain for most of this century. Given limited resources and a possible time limit on the survival of intact natural ecosystems, it is now time to begin using mimicry systems as probes for more general ecological and evolutionary problems (see Vane-Wright, 1980) rather than working to convince the last skeptic that mimicry exists.

We should also not stall too long on the questions of whether coevolution occurs or how pervasive it is. Rather, we should soon proceed with testing predictions of an emerging theory of coevolution as best we can. Mimicry systems will prove valuable in part because sets of species involved in mimicry will have a much greater likelihood of having coevolved with respect to particular attributes than similar sets of nonmimetic relatives. Mimetic groups can therefore be compared to nonmimetic relatives to answer such questions as the impact of coevolution on rates of differentiation in particular parts of the genome, as well as on other micro- and macroevolutionary phenoma.

Evolutionary theory and its corollary—mimicry theory—were spawned in the minds of nineteenth century biologists who had the good fortune to observe and study tropical faunas and floras. We now view tropical rain forests as the most species-rich, highly organized, and perhaps most energetically dynamic ecosystems on our planet. Such forests are characterized by an overwhelming variety of mimicry systems and other coevolving subsystems (Gilbert, 1980). Practically every diurnally active and conspicuous arthropod participates in some form of mimicry, and the obvious cases are just a hint of what might be found using sensitive assays of auditory, visual, and chemical signals being sent, received, and imitated in the system. Mimicry *reduces* the apparent phenotypic variety in a system by increasing the similarity of unrelated species, but it *increases* the phenotypic diversity among closely related species that segregate ecologically by habitat and by mimicry complex (see Gilbert, 1983; Gilbert and Smiley, 1978). Polymorphic mimicry and the differentiation of mimetic races increases within- and between-habitat diversity, respectively.

In any ecosystem, exchanges within and between trophic levels of food webs usually are based on interacting pairs of individuals. Likewise, the persistence of populations within the system involves courtship, mating, oviposition, brood care, pollination, and other key behavioral interactions (involving communication) between individual or-

280

ganisms. Building on Wickler's theory (i.e., mimicry of communication signals), we can view the incidence of mimicry as a rough index to the degree of specificity of behavioral interaction in an ecosystem. The incidence of mimicry should thus reflect the long-term mildness and predictability of the climate within which a particular ecological system has been evolved. This is because mimicry often represents a parasitic specialization on small parcels of materials and energy being exchanged in the system, the constant and predictable availability of which is mandatory for the evolution and persistence of the mimicry.

It could be argued that the apparent increase in mimetic relationships in tropical rain forest is a trivial outcome of the fact that such habitats possess more species. However, I suggest that involvement in mimicry may account for the local persistence of many rare species (Gilbert, 1982) so that mimicry actually promotes local species richness. Indeed, when our knowledge of the geographical incidence of mimicry within taxonomic groups of organisms is more complete (along the lines of Hespenheide, 1973b), we may find evidence that this phenomenon correlates with the general importance of deterministic relationships between species and with the likelihood of community coevolution. Mimicry is possibly the most compelling evidence that community patterns are more than random noise.

CHAPTER 13

COEVOLUTION
AND POLLINATION

Peter Feinsinger

INTRODUCTION

Many angiosperm plants rely on animals for pollination. Many insects and some vertebrates rely on flowers for food or other requisites. Some of the more striking relationships are well known. For example, yucca-moths actively pollinate yucca flowers and use the developing fruits for brood chambers (Powell and Mackie, 1966); *Pedicularis groenlandica* has a flower that resembles a "little pink elephant" and bumble-bees straddle the flower's "trunk" to vibrate out pollen (Macior 1968); sword-billed hummingbirds *(Ensifera ensifera)* pollinate passionflowers that are over 10 cm long (Snow and Snow, 1980); euglossine bees are manipulated by scent-mongering orchids (Dodson et al., 1969); tiny fig wasps live and die within figs (Wiebes, 1979). Although most interactions between plants and pollinators are less spectacular, nearly all appear to involve some degree of mutual adaptation between flower and animal (Baker, 1961; Faegri and van der Pijl, 1979). To what degree is evolution in each group—plants and pollinators—influenced by evolution in the other?

Broadly speaking, plants and pollinators undoubtedly have contributed to each other's evolution. For example, Regal (1977) attributed much of the evolution of Cretaceous angiosperms to the influence of insect pollinators, insect herbivores, and vertebrate seed dispersers evolving concurrently. Mulcahy (1979) suggested that flower-visiting Cretaceous insects, by depositing many pollen grains simultaneously on receptive stigmas, created opportunities for the best-adapted, rather than the first-arriving, pollen tubes (male gametophytes) to fertilize ovules. The result, Mulcahy argued, was rapid adaptation of plants to specific microhabitats and proliferation of angiosperm species. Certainly the radiation of angiosperms in the Mesozoic and Cenozoic created opportunities for flower-visiting animals to diversify as well, leading to the broad interdependence between angiosperms and flower visitors that we see today. However, fossil evidence for coevolutionary interactions between particular plant and animal species is sparse [although see discoveries reported by Crepet (1979) and Heinrich (1979a, p. 180)]. Therefore, evolutionary pathways must be inferred from present-day interactions.

In this chapter, I shall consider coevolution between plants and pollinators in the context of the ecological community, following the route charted by four earlier reviews: Grant and Grant (1965), Baker and Hurd (1968), Heinrich and Raven (1972), and Faegri and van der Pijl (1979). I shall examine selective forces affecting specialization by plant or pollinator, describe three case histories of well-studied interactions, and conclude with a series of generalizations suggested by the case histories and with suggestions for future research.

Because of lack of space, I shall restrict myself to discussing plants that have hermaphroditic flowers adapted for outbreeding, a restriction that ignores the diversity of sexual systems in plants and the roles pollinators play in the evolution of sexual systems (see reviews by Jain, 1976; Janzen, 1977c; Charnov, 1979; Lloyd, 1979; Willson, 1979; Bawa, 1980; Givnish, 1980; Bawa and Beach, 1981). I shall ignore asexual propagation (White, 1979) and somatic mutations, which may occur within clones or single large plants (Whitham and Slobodchikoff, 1981) and which could affect the evolution of pollination systems. Although neighboring plants might be genetically similar either because of cloning or because of restricted dispersal and inbreeding (e.g., see Price and Waser, 1979, Linhart et al., 1981, Waser and Price, 1982), the role of kin selection in plant evolution (Nakamura, 1980) has been overlooked by most pollination biologists and so will not be considered in this chapter. Hopefully, future research will deal with these considerations.

SPECIALIZATION, CONVERGENCE, AND COEVOLUTION

Conflict between "mutualists"

The ecological mutualism between plants and pollinators actually involves an element of conflict (Baker and Hurd, 1968; Heinrich and Raven, 1972; Feinsinger, 1978; Waser and Price, 1982). Selection acting on plants favors traits that ensure reception of numerous, genetically compatible pollen grains on flower stigmas and that ensure dispersal of pollen to numerous stigmas on other conspecifics (Janzen, 1977c; Charnov, 1979; Willson, 1979; Waser and Price, 1982). The optimal pollen vector would contact anthers and stigma, move rapidly among plants, and remain constant to flowers of that species even in the presence of other flowering plants. To force the animal to visit many individuals, selection might favor the secretion of small quantities of nectar per flower or per plant (Heinrich and Raven, 1972; Heinrich, 1975b,c). Thus, a harried, underfed, yet constant pollinator is ideal for the plant.

According to foraging theory, however, pollinators seek to remain fat and sedentary. Animals that visit flowers for carbohydrates in nectar may attempt to maximize energy intake while minimizing energy expenditure, that is, to obtain the most carbohydrates with the least movement (e.g., see Waddington and Heinrich, 1981), even though their foraging route might include flowers of several species. The contrast between the foraging behavior optimal for the plant and that optimal for the pollinator shapes the evolution of most plant–pollinator relationships.

When will selection favor specialization in pollinators?

The typical animal pollinator forages among an array of flowers having different shapes, sizes, colors, and rewards. It must visit several plants during a given foraging bout and numerous plants over its lifetime. Which plants will the animal choose? (1) A forager that has few morphological or behavioral links to particular flower species and is incapable of associative learning may forage rather indiscriminately among many flowers. If the animal happens to encounter a dense clump of one flower species, it becomes a temporary, facultative specialist. Some flies and some beetles may fit into this category (Faegri and van der Pijl, 1979). (2) An animal capable of associative learning may form temporary search images for especially rewarding flowers and may learn to use those flowers as adeptly as an obligate specialist (Motten et al., 1981; Tepedino, 1981), while remaining capable of switching to more profitable flowers (see Heinrich, 1976a). (3) A forager may be genetically predisposed to visit one flower species, responding innately to cues from that flower, with little trial and error. Foraging in such species is often precise and efficient (see Linsley, 1958; Strickler, 1979).

Where the growing season is quite long and flowering in the plant community is spread out, the active foraging life of flower-visiting animals often exceeds the flowering season of any single plant species. This situation confronts social insects (see Heinrich, 1976a, 1979a; Eickwort and Ginsberg, 1980) and vertebrates (see Heithaus et al.,

1975; Stiles, 1978a). Absolute commitment to one plant species is unlikely to evolve even if commitment would confer greater foraging efficiency (Strickler, 1979). If the flowering peaks of potential food plants are asynchronous, selection is likely to maintain diverse behavioral alternatives in the pollinator population (Levins, 1968; Real, 1980). Experience with particular flowers can decrease search or handling times so that such pollinators must experience intense selection for associative learning abilities. Even though they often specialize temporarily on abundant flowers (Heinrich, 1976a, 1979a), such pollinators are unlikely to be influenced evolutionarily by a single plant species.

If most plant species flower simultaneously each year in a short, intense burst, however, the foraging lifetime of the pollinator must be equally short. Where flowers of the average plant species are around for the animal's entire foraging lifetime, selection on the pollinators may favor commitment to particular flowers. The life cycles of animals could then respond to those environmental cues that stimulate plants to flower, with the result that adult foragers emerge at the time of the flowering burst. Such foragers could locate and exploit appropriate resources efficiently (Strickler, 1979), without wasting time on trials at several different plant species (see Laverty, 1980). If these specialized foragers happened to carry pollen more consistently than other visitors to the same blossoms, then selection might favor reciprocal specialization of plants. Thus, environments with short, synchronous flowering seasons could display a number of discrete, coevolving pairs of plants and pollinators.

Oligolectic bees, their pollen plants, and specialization

Warm deserts, semideserts, and sites with Mediterranean climates exemplify environments having brief favorable seasons. Many plant species flower more or less simultaneously following heavy rains. Often their most frequent visitors are not eusocial insects, but are solitary or semisocial bees, which are especially diverse in these regions. Females of many such bees are oligolectic, that is, they collect pollen from only one or a few of the many plant species available to them. Often females show intricate behavioral or morphological adaptations for exploiting the pollen of a single plant species (Linsley, 1958; Linsley and MacSwain, 1958; Thorp, 1969). Have oligolectic bees, then, coevolved with their pollen plants?

Earlier research suggested that specialization to particular pollen plants was genetically determined in univoltine oligolectic bees. But more recent studies imply that the "Hopkins host selection principle" (Fox and Morrow, 1981) operates among oligolectic bees: females are conditioned by larval food to forage for the same pollen their mother used to provision the nest (Linsley, 1978; Eickwort and Ginsberg, 1980). During times of food shortage, females often switch to pollen of other plant species (Thorp, 1969; Cruden, 1972a; Linsley, 1978; N. M. Waser, personal communication). Females and males of most species obtain nectar, at least, from a variety of plants (Baker and Hurd, 1968; Thorp, 1969). The geographical ranges of oligolectic bees and pollen plants rarely coincide precisely (Linsley, 1958). Where the "typical" pollen plant is absent, the bee uses other species; and where the oligolectic bee is absent, entirely different insects may pollinate the plant effectively (Cruden, 1972a). Moreover, not all solitary and semisocial bees in seasonal habitats are oligolectic. Few if any of the numerous colletid, megachilid, and halictid bees in Australian habitats specialize on single pollen plants (Michener, 1965). Few solitary or semisocial bees in the Neotropics are oligolectic (Linsley, 1958), even those in highly seasonal habitats (Heithaus, 1979).

There is little evidence that particular bee species exert significant selective pressures on their pollen plants (Eickwort and Ginsberg, 1980). Many of the plants that attract oligolectic bees have open flowers with exposed anthers. A single such plant may attract numerous oligolectic and polylectic species of bees (Linsley and Hurd, 1959; Linsley and MacSwain, 1959; Thorp, 1969; Cruden, 1972a; Estes and Thorp, 1975). Many of these visitors might be effective pollinators, or none might be. Tepedino (1981) showed that the oligolectic squash bee *Peponapis pruinosa* pollinated squash no more effectively than the unspecialized, recently introduced honeybee. Motten et al. (1981) found that specialized, oligolectic solitary bees (*Andrena erigeniae*) pollinated the spring beauty (*Claytonia virginica*) no more effectively than unspecialized bee-flies (*Bombylius major*). Other flowers supporting oligolectic bees, such as many evening primroses (*Oenothera*), are pollinated much better by entirely different visitors, such as hawkmoths (Gregory, 1963–64). In this case, oligolectic bees may even act as parasites rather than mutualists, stealing pollen before more effective visitors have had a chance to disperse it. Thus, it seems likely that few plants coevolve exclusively with particular oligolectic bees.

When will selection favor specialization in plants?

Plants offering nectar or pollen rewards are likely to attract many flower visitors (Baker et al., 1971; Carpenter, 1979; Faegri and van der Pijl, 1979). According to the argument above, however, if a plant has a flowering season shorter than the active foraging life of its animal visitors it may have a constant set of visitors over its entire flowering season. Furthermore, only a subset of the visitors might transfer pollen effectively. The "most effective pollinator principle" of Stebbins (1970) holds that selection should favor traits that attract and maintain only those visitors that provide the best pollination service. Therefore, we might expect plants to be more specialized in their choice of animals than flower-visiting animals in their choice of plants.

There are several reasons why flowers might specialize for the most effective animals, even though relatively ineffective visitors still perform occasional pollinations. Some visitors might transfer pollen so erratically that both the male function (pollen dispersal to receptive stigmas) and female function (pollen deposition on the plant's own stigmas) suffer instead of benefiting (Janzen, 1977c; Willson, 1979; Price and Waser, 1979). Whether or not they contact reproductive parts, ineffective visitors may lower the nectar volumes in flowers, which discourages potentially better pollinators from forming search images for those flowers (Lyon and Chadek, 1971; McDade and Kinsman, 1980) and which might lower the quality of those visits that the potentially effective visitors do make. Thomson and Plowright (1980) showed that longer flower visits by foraging bumblebees result in greater pollen deposition and that the length of visit is a positive function of nectar volume encountered. When the same effective pollinators arrive consistently generation after generation, then, selection favors traits that discourage less effective visitors.

Still, a plant species need not become dependent on a single taxon of effective visitor. Several animal species could forage in ways indistinguishable to the plant. Different animals may be more numerous at some times or places than others, so selection on the plant can vary. Thus, many plants exploit a set of "acceptable alternatives" among animals, rather than a single animal species (see Grant and Grant, 1965; Levins, 1968; Heinrich, 1979a; Real, 1980; Waser and Price, 1982). Although a plant's effective pollinators often have similar morphologies, behaviors, and metabolic needs (Faegri and van der Pijl, 1979), very different animals sometimes are acceptable alternatives. In western North America, for example, a single plant species may receive effective pollen transfer from hawkmoths, hummingbirds, and bees (Chase and Raven, 1975; Miller, 1978; Waser, 1978a; Carpenter, 1979). In short, it seems possible that few plant species become specialized to a single species of animal pollinator, and even fewer pollinator species become specialized to a single species of food plant.

Rarity, specialization, and convergence

If a plant population is quite densely distributed, nearest neighbors are likely to be conspecific. Nearly any visitor, no matter how uncommitted, is likely to bring useful pollen to a plant and to disperse the plant's own pollen to conspecific stigmas. Selection on plants to specialize is relaxed. Consider a population of widely dispersed plants with few flowers each, however. If these flowers invite all comers, then the pollinators may not distinguish the rare species from more common ones. Indeed, some terrestrial orchids, whose devices for pollen transfer are so precise that inconstancy among their visitors rarely results in pollen loss or stigma clogging, appear to take advantage of indiscriminate foragers: their floral signals closely resemble those of sympatric plants in other families, and their flowers secrete no nectar whatever. Thus, they act as "aggressive mimics" of the more common plants they resemble (Heinrich, 1975c; Dafni and Ivri, 1981a,b; Bierzychudek, 1981).

Plants with less exact pollen transfer systems, however, suffer from indiscriminate foragers (see Waser, 1978b). Incoming visitors deposit foreign pollen, and consign to oblivion the pollen they pick up from the rare plant's anthers. Most plants growing in sparse populations, then, face strong selective pressure to diverge in signal, morphology, and reward from more common flowers, so as to appeal to foragers that will ignore more common plant neighbors if given sufficient incentive. Such species tend to protect their bonanzas from indiscriminate foragers with a complex floral morphology (Heinrich, 1975c; Ostler and Harper, 1978; Laverty, 1980; Pleasants, 1980) that also provides the effective pollinator with a discrete search image. According to this argument, the typical rare, outbreeding plant will have more highly specialized flowers, which appeal to a narrower range of pollinators and reward them more generously, than the comparable plant that typically finds itself in dense monospecific clumps.

The most effective pollinator for one rare plant may be an animal already flying to other rare plants. There is evidence that foragers that feed at rare plants and travel long distances between flowers include intervening plants in their diet if a net energy benefit is possible and if they recognize the cues presented (see Janzen, 1971c; Feinsinger and Chaplin, 1975). Therefore, rare plants, if they make use of long-distance pollinators, might converge on one another in signal, morphology, and reward (Macior, 1971; Heinrich, 1975c), effectively achieving an increased density by forming a "mimicry ring" (Macior, 1971; see also Grant and Grant, 1968; Brown and Kodric-Brown, 1979). Although mimicry increases the number of pollinator visits to each species (Thomson, 1981), pollen could be lost to each other's stigmas, and stigmas could be clogged by foreign pollen (Waser, 1978b). Therefore, selection on rare plants might favor divergence in the placement of their reproductive parts (Waser, 1982). Some rare plants share pollinators despite the potential for pollen loss and stigma clogging, however. For example, two Panamanian gingers

(*Costus*), pollinated by the bee *Euglossa imperialis*, exist in sparse, intermingled populations (Schemske, 1981). Only one flower matures each day in each ginger inflorescence. The two species differ greatly in vegetative characters (which suggests they are not close relatives), but they have nearly identical habitat preferences, flower colors, floral morphologies, patterns of nectar secretion, and placement of reproductive parts. Individual bees forage at both species indiscriminately and carry mixed pollen loads. Schemske (1981) concluded that the advantages of increased visitation outweigh the problems of mixed pollen.

Thus, rare and common plants illustrate a coevolutionary paradox. Plant species that typically grow in high densities are apt to have quite open flowers that appeal to a variety of visitors, and are unlikely to coevolve with particular species of pollinators. Nevertheless, most movements by their visitors are bound to be among conspecific flowers, so that each visitor's diet is narrow, at least at any one time. In contrast, rare plant species may be expected to have highly specialized and rewarding flowers, more prone to selective pressure from a few pollinator species, and therefore likely to evolve quite fast. Nevertheless, each visitor's diet may be quite broad!

Evolutionary responses of plants to pollinators

A plant species that invades a new habitat confronts flower visitors and other plants that are somewhat different from those experienced by the parent population. In many cases the new flower visitors forage like the old ones, and selection favors little or no change in floral traits. Sometimes, however, the group of animals that formerly acted as effective pollinators is absent, or it is already thoroughly exploited by extant plant species, or it is ineffective for reasons of climate (Cruden, 1972b). Nectar (and pollen) accumulates unused in the invader's flowers until the attention of other foragers is attracted (Grant and Grant, 1965; Cruden, 1972a). Even if they are inefficient at locating and exploiting the invader's flowers, the new visitors may transfer some pollen at least (e.g., Valentine, 1978). The invading plants may experience strong directional selection for any traits that increase these visitors' effectiveness (Stebbins, 1970). For example, a lowland plant species, pollinated by hawkmoths, might invade a cool montane habitat where hawkmoths are less effective (Cruden et al., 1976). If enough nectar or pollen builds up to attract the local hummingbirds or bumblebees, and if genetic variation exists, the invader may rapidly adapt to bee and/or bird pollination. For example, high montane populations of columbine (*Aquilegia caerulea*) have mostly blue flowers that are attractive to bees, whereas populations at lower elevations have whitish flowers attractive to their hawkmoth pollinators (Miller, 1981).

Occasionally, adaptation to a new pollinator group arises not through invasion of new habitats but through the spontaneous origin, *in situ*, of new traits that happen to attract animals who ignored the parent plants. Straw (1955) proposed that purple, wasp-pollinated *Penstemon spectabilis* in California arose from natural hybrids between red, hummingbird-pollinated *P. centranthifolius* and blue, bee-pollinated *P. grinnellii*. Hybrid flowers may be too narrow to entice bees, the wrong color to entice hummingbirds, but they are attractive to wasps, whose foraging activities and pollen transfer would then act to refine floral traits over time. Of course, new floral traits may also originate through mutation. Many plant populations are polymorphic for flower color (Kay, 1978; Mogford, 1978). In one polymorphic population of larkspur (*Delphinium nelsonii*), Waser and Price (1981) showed that pollinators discriminated against the rare white morph, perhaps because, unlike the blue morph, the former did not resemble other popular flowers in the vicinity. White-flowered plants set only 30–45% as many seeds as blue-flowered plants. Waser and Price suggested that spontaneous mutations, not pollinators, were responsible for maintaining the rare morph in the larkspur population. In other situations, though, a new color morph might be favored. The classic study by Grant and Grant (1965) on pollination in the phlox family gives several examples of geographical variation, in which the floral traits of different populations of a species correspond to the behavior and morphology of the potentially most effective visitor at each location. Geographical variation in floral traits and nectar volumes, presumably related to a changing spectrum of pollinators, also occurs in bumblebee-pollinated plants (Brink, 1980; Brink and DeWet, 1980) and hawkmoth-pollinated plants (Miller, 1981). These are among the few careful studies that combine geographical variation in plants with changing effectiveness in pollinators.

Geographical variation in hawkmoth-pollinated plants suggests one interesting possibility for coevolution between plant and pollinator. Proboscis ("tongue") length varies considerably among hawkmoths, even among conspecifics (Gregory, 1963–64). Miller (1981) showed that some moths foraging at the columbine *Aquilegia caerulea* had tongues so long that their head and body never made contact with anthers and stigma. Shorter-tongued moths, though, could not reach nectar without landing on the flower or at least brushing against its sexual organs with their, heads. Hawkmoths (*Hyles lineata*) that carried columbine pollen had significantly shorter tongues than those without pollen. *Aquilegia* species in the southwestern deserts of the United States, where many long-tongued moths exist, have deeper nectar spurs than populations pollinated primarily by *Hyles lineata* (Miller, 1981). Thus, moths and flowers could evolve through reciprocal selection: frequent visits from long-tongued moths select for long-tubed plants, which select in turn for longer-tongued moths capable of reaching the nectar. Evolution in moths and flowers is probably not that simple: long-tongued moths may have broad diets that include many short-tubed flowers; some of the variation in length among hawkmoth tongues is developmental, not genetic (Gregory, 1963–64); population densities of long-tongued moths are especially erratic, so that they are not likely to exert consistent selective pressures; during lapses in long-tongued moth populations, shorter-tongued species such as the cosmopolitan *H. lineata* feed on the nectar welling up in long-tubed flowers (Gregory, 1963–64; Miller, 1981). Nevertheless, selection for traits that force the visitor to thrust its entire face, rather than just its tongue or bill, into the flower may drive the evolution of deep nectaries in many plants, such as those visited by long-tongued bees or those visited by hummingbirds.

Certainly, selection does not inevitably lead toward a deep corolla and a narrow range of "acceptable alternatives." Especially in species that frequently invade new patches of habitat (D. A. Levin, 1976; Pickett, 1976) and face a new assortment of animals and other plants every few generations, selection is likely to favor a flower that can attract some animal or another regardless of the particular community (see Feinsinger, 1978; Parrish and Bazzaz, 1979)—or a breeding system that does not require pollinators (Baker, 1965; V. Grant, 1975; Jain, 1976).

Evolutionary responses of pollinators to plants

With the exceptions noted later, extraordinarily little evidence exists on the selective impact plants might have on pollinator populations or on geographical variation in pollinators coupled to changes in the array of flowers available. Apparently, foraging behavior in most pollinators is predicated on the distribution of food rewards over available flowers (Pyke, 1978; Heinrich, 1979a). Often this distribution is influenced more by other flower visitors than by the traits of particular plant species. Furthermore, unlike plants, flower-visiting animals are free to roam among different plant species, even among different habitats, in order to locate the most profitable resources. Therefore, not only is coevolution between most plants and pollinators diffuse, but it is also skewed: pollinators are even less apt to respond to selec-

291

tive pressures from particular plants than plants to selective pressures from particular pollinators.

The special cases

In some cases, however, obligate mutualism between one plant and one pollinator has arisen, perhaps from floral parasitism. The preceding premise was that an individual animal must forage at many plants to collect sufficient food for itself or its offspring. If the flower itself is the larval food, then a female animal has to visit only one flower, for oviposition. Specialization for the quirks of a single host plant species need not incur a risk, because only one new host must be located over the female's reproductive lifetime. Selection on some parasites might favor behaviors that make the host flower a better nursery: one such activity could be pollination—at first accidental, later active. Pollination ensures that the developing fruit continues to draw nutrients from the parent plant. Viewed in this light, the complex behaviors for active pollination that some flower parasites display should be no more astonishing to us than other complex, innate behaviors exhibited by insect parasites specialized on arthropod hosts. Finally, confronted by numerous, pollen-carrying flower parasites, the plant might adapt to, and benefit from, the specificity and high population density of the parasite.

The wind-pollinated ancestors of figs (*Ficus*) undoubtedly suffered from tiny chalcidoid wasp parasites, just as many other plants, and arthropods, suffer today. Even today's obligate "mutualists" among the fig wasps are more properly termed "seed parasites" (Janzen, 1979c). Female fig wasps make only one flight, from a male-phase inflorescence (actually a highly specialized, fused inflorescence termed a "synconium") to a synconium in female phase. Because wasps are tiny and abundant, every female-phase synconium on a tree is likely to be pollinated by females arriving from various male sources. As long as fig trees in a population are out of phase with one another (Janzen, 1979c), specialization by wasps does not lead to extinction of wasp or fig, and figs always experience an abundance of wasps. Many figs and fig wasps are coevolved in the most restrictive sense of the term, in that speciation in one apparently leads to, or occurs along with, speciation in the other. Often, each fig has a single effective wasp pollinator, and each wasp a single fig host, although exceptions occur (Wiebes, 1979). Routes for switching host or pollinator exist, however, through the choice of "wrong" hosts by wasps or through the normally non-pollinating wasp parasites that occur in most figs (Janzen, 1979c). The presence in the community of plants other than figs and of animal pollinators other than wasps has no effect on the system, in contrast to

292

interactions between more typical plants and pollinators. The wasp–fig system is unique; apparently it evolved, and evolves, independently of other plants and their pollinators.

In contrast, another widely cited mutualism, that between yuccas and yucca-moths (*Tegeticula*), probably originated in a more generalized animal-pollinated plant whose flowers were parasitized consistently by ovary-boring moth larvae. At present, *Yucca* is pollinated by adult female *Tegeticula*, which oviposit in one to three of the carpels in the flower but actively pollinate all three (Powell and Mackie, 1966; Aker and Udovic, 1981). Adaptations of *Yucca* to the *Tegeticula* life cycle (including the sacrifice of some seeds) and the complex pollen-carrying behavior of the female moth suggest a lengthy coevolutionary history. Yet vestiges of parasitism remain. When flowers are scarce, female moths oviposit in developing seed pods (Aker and Udovic, 1981), which does not benefit the plant. Other moth species (*Prodoxus*) closely related to *Tegeticula* oviposit in *Yucca* flowers without pollinating them, acting purely as parasites (Powell and Mackie, 1966). The cost of obligate dependence on *Tegeticula* is evident in the frequent occurrence of low seed set, resulting from low population densities of moths (Powell and Mackie, 1966; Cruden et al., 1976; Udovic and Aker, 1981). Perhaps if yuccas could emerge from the cul-de-sac of moth specialization, they would profit from inviting a wider range of pollinators.

Thus, coevolution of one plant and one pollinator into an exclusive relationship appears to begin only when the one flight made by parasitic, reproductively profligate insects happens to transfer pollen more consistently than other means. Surely this condition applies infrequently. Let us examine three more widespread cases. These represent a biased sample, too, because all three involve fairly large, long-lived, and "intelligent" animals rather than the shorter-lived and less adept insects that may contribute to the majority of pollination events (see Faegri and van der Pijl, 1979). Nevertheless, some of the features described below are presumably common to less complex, less highly visible interactions as well.

CASE HISTORIES

Orchids and euglossines

Many orchid species in lowland neotropical forests depend entirely on male euglossine bees (tribe Euglossini, in the honeybee family) for pollination (van der Pijl and Dodson, 1966). The orchids, which have no nectar, emit fragrances from secretory cells on the lip of the flower. Male euglossines land on the secretory area, "mop up" fragrance compounds with brushes on their front legs (Dodson et al., 1969; N. H. Williams, personal communication), and transfer the fragrances to chambers in the enlarged hind tibiae, for storage and perhaps chemical transformation. Recent studies suggest that the fragrances serve as biochemical precursors to more complex pheromones secreted from the male bees' huge mandibular glands (N. H. Williams, personal communication; see Williams, 1982b). These pheromones appear to attract far-ranging females into males' territories (Kinsey, 1980). Male euglossines obtain no food from these orchids, and females ignore the orchids entirely. Instead, bees of both sexes forage widely for nectar (Dodson et al., 1969). Some females travel especially far, along extensive routes that include flowers of many plant species (Janzen, 1971).

While entering or leaving an orchid flower, the male euglossine may brush against the pollinarium (a unit consisting of two clumps of pollen, or pollinia, and an attachment device), which adheres to it. Or the bee may brush against the stigma and deposit pollinia it has carried from another flower. Euglossine orchids display intricate mechanisms that serve to manipulate the bee into the correct position for these events (van der Pijl and Dodson, 1966; Williams, 1982b). Pollinaria of a given species attach to a precise, specific location on the bee's body (Dressler, 1981). Some bees may be too large or too small to fit the orchid's precise morphology and so fail to contact its sexual organs. A proper-sized bee, though, has little choice but to be an effective pollinator once it enters the flower (van der Pijl and Dodson, 1966; Dressler, 1981) (Figure 1).

Different orchid species produce different sets of chemical fragrances. Different euglossines prefer different combinations of scents. Therefore, sympatric orchids tend to attract different bee species. In what sense has coevolution produced this relationship?

Apparently, the evolution of euglossine–orchid relationships is guided by two phenomena discussed earlier: (1) the relationship between plant rarity and convergence, and (2) the spontaneous origin of floral traits that attract new flower visitors. Ackerman (1981) showed that population density and recruitment of euglossines on Barro Colorado Island, Panamá, were independent of the density of orchids, apparently for two reasons. First, euglossines of both sexes obtain food from entirely different plants. Second, the male bees can obtain scents from many sources besides the orchids they pollinate, including orchids not adapted to manipulate them, flowers in several other plant families, rotting fruits, decaying logs, wet leaf litter, and exposed tree roots (Ackerman, 1981; Williams, 1982a,b; N. H. Williams, personal communication). This combination of food plants and scent sources

FIGURE 1. A male euglossine bee entering an orchid flower from which one petal has been removed. (Courtesy of Norris H. Williams.)

supports an abundant, diverse pool of euglossine bees that is used only sporadically by orchids. One-third of the euglossine species on Barro Colorado Island have never been observed to visit scent-mongering orchids there or to carry pollinaria, although some of these species visit other orchids elsewhere (Ackerman, 1981). Only 1.8–6.8% of the individual males Ackerman attracted to scent baits in a given month were carrying pollinaria at the time, although most of them belonged to the species known to visit orchids.

Because populations of many orchid species are sparse, a single orchid species is unlikely to supply enough fragrances for an entire euglossine population. An orchid whose scents fortuitously appealed to preexisting fragrance needs among some of the euglossines in the pool, however, would find itself attracting occasional visitors. Through chance convergence on other fragrance sources, an orchid may obtain effective pollination regardless of its rarity. Because the pollinaria, which remain viable for days or weeks, are firmly attached to species-specific sites on the bee, they are unlikely to slough off onto foreign plants even if days pass before the bee visits a receptive, con-

295

specific flower (Williams and Dodson, 1972). Euglossine–orchid relationships, like less showy plant–pollinator relationships, are rarely exclusive (Table 1). The true redundancy in bee species per orchid species, and orchid species per bee species, is undoubtedly even greater than that reported in Table 1. N. H. Williams (personal communication) has recorded individual male euglossines carrying pollinaria from as many as five different orchid species (see also Dressler, 1981). Use of the same individual bees by different orchid species effectively increases the density of each plant population and the probability of a pollinating visit to each flower, at little or no risk of pollinarium loss.

Do orchids affect evolution in euglossines? Probably not. The existence of a euglossine pool largely independent of orchids suggests that bees evolve against the entire set of scent sources and food plants, not just against orchids (Ackerman, 1981). Speciation events among bees may involve the use of different fragrance sets as reproductive isolating mechanisms, but they probably entail normal allopatric divergence of bee populations rather than divergence mediated by orchids (see Dressler, 1968). Do euglossines affect evolution in orchids? Undoubtedly. The identity of fragrance compounds and the relative importance of each vary widely from orchid to orchid, even in the same species (M. Whitten, unpublished data). Most of the variation within an orchid population may fall within the preference range of the pollinator species. Occasionally, though, an unusual fragrance spectrum, due to *in situ* mutation, might appeal to a different set of euglossine species (Dressler, 1981). Viable interspecific or intergeneric hybrids can occur among orchids, so that new fragrance spectra could often arise through hybridization as well. Dressler (1981) observed that a hybrid between the orchids *Gongora gibba* and *G. quinquenervis* attracted a euglossine species that did not visit either parent species. Moreover, an orchid invading a new habitat might appeal to new euglossines and experience a shift in selection for fragrances (see Hills et al., 1972; Dressler, 1981). If new species of bees differ in size, behavior, or visual orientation from the previous effective pollinators, they must exert strong selection on other floral traits, leading to morphological features that best manipulate them past the flowers' sexual organs (Dressler, 1981).

In short, orchids may possess more lability (in fragrance) than many other plants. New fragrance profiles may arise frequently and result in orchid speciation. New species persist, however, only if they appeal to euglossines already present, supported by other plant species in the community. Little reciprocal evolution is likely to occur between particular orchids and bees: an orchid with a new scent profile may attract a new bee species entirely, and a bee developing new fragrance preferences is likely to switch to a new orchid entirely (or none at all). Therefore, coevolution in this system remains diffuse

296

TABLE 1. Redundancy in the interactions between euglossine bees and orchids on Barro Colorado Island, Panamá.[a]

A. Use of orchids by bee species

Number of orchid species used per bee species	NUMBER OF BEE SPECIES IN CATEGORY	
	From pollinaria recorded on bees[b]	From all records[c]
1	2	10
2	9	8
3	3	5
4	3	4
5		
6	2	3
7		
8	1	
9		1

B. Use of bees by orchid species

Number of bee species visiting an orchid species	NUMBER OF ORCHID SPECIES IN CATEGORY	
	From pollinaria recorded on bees[b]	From all records[c]
1	5	2
2	7	9
3	6	5
4	2	3
5		
6		1
7		1
8	1	
9		
10		1
11	1	
12		
13		
14		1

[a]From Ackerman (1981).
[b]Ackerman (1981) found pollinaria on 978 of 24,818 male euglossines attracted to chemical bait traps over a 15-month sampling period.
[c]Ackerman's (1981) pollinaria records, pollinaria records Ackerman and others obtained at other times, and bees observed to visit flowers.

(Ackerman, 1981; Dressler, 1981) and unilateral, with evolution among the pollinators apparently uncoupled from evolution in plants. Except for the unique consequences of fragrance-collecting, relationships between euglossines and orchids do not differ greatly from those in "typical" pollination systems.

Bumblebees and their flowers

Only at first glance do bumblebees (*Bombus*, in the honeybee family) seem more prosaic than their euglossine relatives. Heinrich (1979a) describes the life of a temperate-zone bumblebee colony, from its founding in the spring to its death in the autumn. Throughout its life, the colony dispatches a continuous stream of foragers to fields of wildflowers, beginning with the queen and continuing through many shorter-lived workers. Plants exploited by bumblebees tend to occur in clumps (Heinrich, 1976a). Although each forager has access to several such plant species at one time, some have complex flowers. These may require precision foraging (see Figure 2) that may take some time for a naive bee to master (Heinrich, 1979b; Laverty, 1980). Thus, each individual bee usually specializes temporarily on a single flower species,

FIGURE 2. A bumblebee foraging for nectar at flowers. (Courtesy of Bernd Heinrich.)

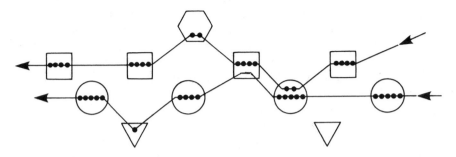

FIGURE 3. Schematic diagram of majoring and minoring by two bumblebees (after Heinrich, 1976a). Geometric shapes represent different plant species; lines and dots represent bee foraging paths.

its "major," and occasionally samples one or more "minor" flower species (Figure 3). Heinrich (1976a, 1979b) has shown that depletion of the major's nectar, a decline in its flower density, or an increase in the minor's nectar rewards or density will motivate the bee to switch preferences. Therefore, each bee, and the colony as a whole, has the capacity to track changes in the floral resources available.

New foragers often locate a plant species underexploited by other bees, learn quickly to associate that species' particular floral cues with a net reward (Heinrich, 1976a, 1979b), and learn an efficient foraging technique (Laverty, 1980). Thus, different bees from the same nest may have different majors (Figure 3), so that at any one time the colony has a broad diet but each forager is a "sequential specialist" (Colwell, 1973), majoring on a single plant species as long as that is the best option available. Tongue length, which varies among bumblebee species, affects the efficiency with which a bee extracts nectar from shallow or deep flowers (Inouye, 1980) and thus affects the choice of majors. The result is that coexisting colonies of different *Bombus* species have somewhat different patterns of resource use (Inouye, 1978; Heinrich, 1976b, 1979a).

It is possible that the behavior of bumblebees can produce patterns in the flowering of their food plants. If individual bees major on two similar flower species at once, interspecific transfer of pollen would select for character displacement among plant species in cues and required foraging techniques, to promote species-specific majoring (see Waser, 1978b). If plants that flower sequentially provide similar cues to bees and require similar foraging techniques, however, each except the first may capitalize on the search images bees have formed for its

299

predecessors, thereby acquiring its predecessors' majors. Heinrich (1975a) found that simultaneously flowering, bumblebee-pollinated plants displayed quite different floral traits from one another and required distinct foraging techniques, whereas sequentially flowering species formed "replacement sets" with similar floral signals (see also Macior, 1971). The earliest species in such replacement sets, which have no predecessor to mimic, have flowering phenologies strongly skewed to the right (Figure 4), possibly in part a result of selection presures for initially profuse, synchronous flowering that establishes search images among inquisitive bees (Thomson, 1980). Each replacement set of plants, then, experiences a mix of competition for pollinators while flowering seasons overlap, and indirect mutualism through the maintenance of long-term search images (Figure 4).

Admittedly, the direct evidence for competition and mutualism among these plants is scanty, and it is risky to assume that communities are structured by such forces (Connor and Simberloff, 1979; Strong et al., 1979). Nevertheless, theory and those data available both suggest that each plant flowers when it is most likely to receive a "bonus" of foragers left over from a similar predecessor, and when the competitive effect of the predecessor is minimized (Heinrich, 1975a; Waser, 1978a,b; Waser and Real, 1979; Pleasants, 1980; Thomson, 1980, 1981; see also conflicting evidence presented by Rabinowitz et al., 1981).

If coexisting plants influence one another's reproductive success and evolution through their bumblebee pollinators, then does each community begin with a chaotic hodgepodge of plant species, which natural.selection sorts out through character displacement in timing or morphology? It is more likely that a suite of plant species evolves gradually over an entire regional landscape that supports numerous

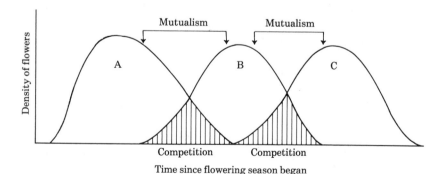

FIGURE 4. Phenological relationships within a "replacement set" of three plant species, A, B, and C, pollinated by bumblebees. (After Heinrich, 1975a; Waser and Real, 1979; Thomson, 1980.)

Bombus populations (Heinrich, 1979a). Except for a geographical cline in floral shape and nectar production reported for one species (*Aconitum*; Brink and DeWet, 1980), most bumblebee-pollinated flower species appear to show few changes in floral traits and flowering phenologies over broad areas (Macior, 1971, 1978). Natural selection within a single community may not have time to alter greatly the traits and phenologies of bee-pollinated plant populations, which are reshuffled into new communities every few generations. Rather, each community is built from plants and bees colonizing from the surrounding region, according to the traits each possesses and according to chance colonization events.

If bumblebees and their food plants engage in reciprocal evolution over the landscape, they do so only in the broadest sense. Little correspondence exists between the range of particular *Bombus* species and particular plants. A single plant population might attract effective foragers from ten different *Bombus* species (Macior, 1978; Miller, 1978; Kwak, 1979; Inouye, 1980), just as a single bee colony uses numerous flower species. What matters to the plants is that the individual bees visiting them remain constant for a short time. Constancy is affected more by the surrounding plant species than by the species identity of the bee. Therefore, plants probably coevolve much more closely with one another than with particular bees. The delicate energy balance between bee and plant and the precision with which bees are guided to rewards and manipulated to contact sexual organs (Heinrich, 1979a) are evolutionary responses to traits shared by bumblebees in general and often by other animal visitors as well. Evolution of foraging behavior in a *Bombus* species is exempt from the influence of particular plant species (Heinrich, 1979c); it is influenced somewhat by competition from other *Bombus* species or other flower-visitors (Inouye, 1978; Heinrich, 1979a); however, it is influenced primarily by the intraspecific demands of metabolism, social behavior, and colony life. Plants adapted for pollination by bumblebees form a single guild (Root, 1967) or a series of guilds (Pleasants, 1980) that evolve with *Bombus* only in diffuse ways.

Tropical hummingbirds and their flowers

Far from the bogs and fields dominated by bumblebees and bee-flowers, long-lived vertebrates pollinate many plant species. For example, at Monteverde, Costa Rica, nearly 100 of the approximately 600 species of angiosperms use 22 hummingbird species as pollinators. Hummingbirds (family Trochilidae) and plants at Monteverde and

301

other neotropical sites tend to fall into two groups (Feinsinger and Colwell, 1978; Feinsinger et al., 1979; Snow and Snow, 1980). Hummingbirds with long, often curved, bills forage at long, often curved flowers, which secrete copious nectar and occur on widely scattered plants (termed "rich flowers"). At low elevations, most such hummingbirds belong to the subfamily Phaethorninae, the hermit hummingbirds. At high elevations, some species from the other hummingbird subfamily, Trochilinae, have adopted the hermit form. Other tropical hummingbirds, including Trochilinae and a few Phaethorninae, have fairly short, straight bills (Figure 5). These birds forage at fairly short, straight flowers, including many that are better suited for insect visitors. Flowers that are best pollinated by short-billed hummingbirds (termed "moderate flowers") secrete considerably less nectar than hermit-pollinated flowers.

Although their bills allow access to long and short flowers alike, hermits and hermit-like hummingbirds apparently prefer the long, rich flowers. Flowers in clumps usually support bellicose, short-billed birds, which chase hermits away (Colwell, 1973; Linhart, 1973). Hermits fly speedily, however; like euglossine bees, they can travel great distances to "trapline" scattered rich flowers whose nectar is relatively inaccessible to short-billed species other than nectar-robbers (Stiles, 1978a). Instead of majoring on single plant species, then, hermits obtain a high energy intake per unit flight cost by visiting a variety of rich flowers on their foraging routes. Therefore, as in other rare plant species, selection may operate on plants with rich flowers to capitalize on birds' capacities for associative learning, to increase the frequency of hermit arrivals, and effectively to increase the population density by favoring convergence in general appearance on other rare,

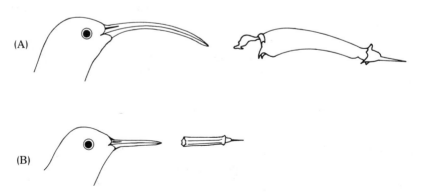

(A)

(B)

FIGURE 5. The two groups of hummingbirds and bird-plants. (A) A representative "hermit" hummingbird and rich flower. (B) A representative short-billed hummingbird and moderate flower.

hermit-pollinated plants. Hermit flowers of different species also secrete similar amounts of nectar, perhaps because hermits would ignore those individual plants that had failed to reinforce their efforts. Hermit plants that flower simultaneously tend to place pollen at different spots on the bird's long bill, or on the forehead, crown, nape, or chin (Stiles, 1975; see also Brown and Kodric-Brown, 1979). These differences are much less distinct than those among orchid pollinaria on euglossine bees, however, and many hermits carry highly mixed pollen loads (Feinsinger et al., 1982). In marked contrast to bumblebees, individual hermits tend to have fairly similar, broad diets (see Table 2).

Some hermit plants may coevolve with hermits. Sicklebill hummingbirds of the hermit genus *Eutoxeres* use certain plants with

TABLE 2. Comparison of individual and population diets in the long-billed hermit hummingbird, *Glaucis hirsuta*, and the short-billed territorial hummingbird, *Amazilia tobaci*, from the same site on Trinidad, West Indies. Individual *G. hirsuta* carry pollen of more plant species (mean = 2.1) than individual *A. tobaci* (mean = 1.6), but the diet of the entire *A. tobaci* population is broader.

A. Diversity of pollen carried by individual hummingbirds[a]

No. of plant species in pollen load[b]	No. of *Amazilia tobaci*	No. of *Glaucis hirsuta*
1	5	7
2	8	7
3	0	8
4	0	1

B. Diversity of flowers used by entire hummingbird populations[c]

Species	Diet diversity (monthly average)[d]
Amazilia tobaci	2.885
Glaucis hirsuta	1.990

[a]Mann-Whitney U test results: $z = 1.685$, p (one tail) < 0.05.
[b]Pollen loads collected from mist-netted hummingbirds (see Feinsinger et al., 1982).
[c]Results of Wilcoxon matched-pairs signed ranks test for $n = 13$ months: $T = 5$, $p \ll 0.01$.
[d]Calculated following the procedure outlined in Feinsinger (1976). The diversity index is $1/\Sigma p_i^2$, where p_i is the proportion represented by the ith plant species in the population's diet. The index ranges upward from 1.0, the value for a population using flowers from one plant species only.

flowers uniquely adapted to their strongly curved bills (personal observation; see Stiles, 1978a; Feinsinger et al., 1979), although *Eutoxeres* also feeds at less strongly decurved flowers (Stiles, 1975). *Lafresnaya lafresnayi* is the only hermit-like hummingbird that Snow and Snow (1980) found at temperate elevations in the Colombian Andes; three of four plant species from which Snow and Snow saw *Lafresnaya* feed had flowers accessible to no other bird present, suggesting that these had evolved primarily under the selective pressure exerted by *Lafresnaya*. Many habitats, however, support several hermits or hermit-like hummingbirds. From the plant's point of view, one far-flying, opportunistic hermit might be as effective as another. For example, each of the nine wild plantain (*Heliconia*) species that F. G. Stiles (1975) studied at La Selva, Costa Rica, had at least two species of hummingbird visitors, and some had as many as nine, all probably acting as effective pollinators. Therefore, most plants pollinated by hermits must evolve against a shifting background of several, interchangeable hermits or hermit-like hummingbird species.

In a similar way, many plant species have most likely been interchangeable during hermit evolution. Stiles (1975) argued that the Phaethorninae have been closely tied to the diverse plant genus *Heliconia* throughout their evolution, but today's hermits forage on many other flowers as well. Stiles and Wolf (1979) report 12 plant species besides *Heliconia* in the diet of *Phaethornis superciliosus* at La Selva. High-elevation, hermit-like hummingbirds have no *Heliconia* to exploit and tend to visit many plant species. At least one such hummingbird may coevolve with a particular plant, however. Snow and Snow (1980) reported a mean length of 83.1 mm for bills of Colombian sword-billed hummingbirds, *Ensifera ensifera*, but they noted that bills were highly variable, some being as long as 105 mm. This *Ensifera* population fed exclusively from flowers of a passionflower vine, *Passiflora mixta*, whose pollen apparently traveled on *Ensifera* foreheads. *Passiflora mixta* flowers also varied considerably in length, some being as long as 114 mm (Snow and Snow, 1980). *Ensifera* is primarily a *Passiflora* specialist throughout its range. Perhaps *Ensifera* and *Passiflora* are caught in an elongation race like the one proposed earlier for long-tongued hawkmoths and their food plants. Except for these cases, though, most hermits, hermit-like hummingbirds, and their food plants exemplify diffuse coevolution between two diverse groups of species.

Hummingbirds influence one another's choice of flowers, which may have evolutionary consequences. The extent of interspecific competition among hermits has not been determined (see F. G. Stiles, 1975, 1980; Stiles and Wolf, 1979). Hermits certainly suffer, though, from the presence of shorter-billed hummingbirds that exclude them from clumps of moderate and rich flowers alike (Linhart, 1973; Stiles,

304

1975; Feinsinger and Colwell, 1978). Short-billed species often compete intensely with one another; this competition results in partitioning of flowers among different bird species (Snow and Snow, 1972, 1980; Feinsinger, 1976, 1980; Wolf et al., 1976). Some short-billed hummingbirds, behaviorally subordinate to others, visit diverse species of low-quality flowers that are snubbed by more aggressive species and by hermits (Feinsinger and Chaplin, 1975). The diet breadths of these individuals suggest those of hermits, but morphologies of the two "trapliner" groups and the species of plants they choose differ strikingly (Feinsinger and Colwell, 1978). In contrast bellicose, short-billed hummingbirds defend feeding territories that often consist of a single plant or plant species. If several such plant species exist, however, the diet of an entire population of territorial birds can be quite broad (see Table 2).

In what sense have short-billed hummingbirds and the moderate flowers they visit coevolved? Unlike hermits, short-billed species visit flowers of many colors and shapes. Many of these flowers can be pollinated by nearly any species of hummingbird or even by butterflies or other insects. Although short-billed hummingbirds and plants with moderate flowers occur in most neotropical habitats (Stiles, 1978a), they reach especially high densities in ephemeral, early successional communities, perhaps because their opportunistic natures equip them for facing nearly any set of competitors and mutualists they encounter (Feinsinger, 1978; Feinsinger and Colwell, 1978). Many short-billed hummingbirds wander far and wide during the year (Feinsinger, 1980), in contrast to hermits, whose stationary mating leks prohibit extensive seasonal migrations (Stiles and Wolf, 1979; F. G. Stiles, 1980). In nearly any habitat, a short-billed bird can locate suitably shaped flowers. Likewise, because assemblages of short-billed hummingbird species are often limited by the nectar resources available, any moderate flower is likely to receive attention from one bird or another (Feinsinger, 1976, 1978). The chance for restricted coevolution is considerably less than that between sedentary hermits and plants with rich flowers.

Do hummingbird-pollinated plants affect one another's evolution, aside from those that converge to use hermits? Competition for attention from hummingbirds may not affect plants with moderate flowers (as defined earlier), which often luxuriate amidst hordes of hungry hummingbirds. Furthermore, hummingbirds foraging at those moderate flowers that occur in clumps will necessarily be quite constant, so that strong selection for character displacement between such plant species is unlikely to occur. Small or rare plants with moderate

305

flowers, however, attract meek hummingbirds with broad diets. If several such plants place pollen at the same sites on birds, those with unique flowering seasons should receive better pollination than those with highly overlapping seasons. Thus, flowering phenology could influence the ability of a plant species to invade a community. Perhaps as a result, blooming seasons of coexisting neotropical plant species, whether pollinated by hermits or by short-billed hummingbirds, tend to be scattered over the year instead of aggregated into a single season (Feinsinger, 1978; Stiles, 1977, 1978b; Cole, 1981).

In summary, most present-day interactions between plants and hummingbirds probably arose through diffuse coevolution at best. Only a few possibilities exist for reciprocal evolution between particular plant and bird species. Evolution of foraging behavior in most hummingbirds is influenced more by interactions with other hummingbirds than by a particular set of plant species (Feinsinger and Colwell, 1978). Floral traits in some plant species undoubtedly have evolved in response to foraging by hummingbirds and to the presence of other plant species that influence hummingbird foraging, but seldom do they respond exclusively to a particular hummingbird species.

CONCLUSIONS AND SUGGESTIONS

Surprisingly little quantitative research has specifically addressed the question, "Do plants and pollinators coevolve, and if so, how?" The gaps in our present knowledge suggest promising directions for future research.

It may be useful to compare the evolution of plant–pollinator interactions with evolution between plants and other animals. Although the fruits of many plants are adapted for dispersal by animals and although some animals display adaptations for frugivory, the evolution of most frugivores seems even less tightly linked to particular plant species than the evolution of pollinators. Most seed dispersers, aside from ants, are long-lived vertebrates, few of which can subsist solely on a diet of one fruit species. Thus, specialization is unlikely to evolve, and exclusive coevolution between one plant and one frugivore is even more unlikely (see Chapter 11 by Janzen). Many animals are only facultatively frugivorous, just as some flower-visiting birds (Snow and Snow, 1971) and bats (Heithaus et al., 1975) are only facultatively nectarivorous. Likewise, selection rarely leads to specialization of plants for particular frugivores; because of the many random events that occur between the act of frugivory and an adult plant in the next generation, differences among the seed shadows that different frugivores create may not have distinguishable selective effects (Wheelwright and Orians, 1982). Thus, plants generally evolve to exploit an interchangeable set of animals as seed dispersers, although there are

306

exceptions (e.g., see Snow, 1966; Temple, 1975), and animals are likely to evolve independently, exploiting an even wider range of fruits and other items as foods, although again there are exceptions (e.g., see Snow, 1961, 1962, 1966).

In contrast, as Futuyma (Chapter 10) indicates, many insect herbivores appear to evolve in response to traits of particular host plant species, whereas plant traits evolve in response to selective pressures coming from entire suites of herbivore species. Food requirements for many herbivorous insects can be met by the tissues of a single plant, so that selection leads to traits that enable insects to make the best use of specific tissues. Furthermore, in contrast to evolution in plant–pollinator systems, selection acts on plants to diminish, not increase, the spectrum and effects of specialist ("constant") herbivores. Plants confronted by herbivores are thought to evolve toward increased, not decreased, effective separation from one another in time and space and toward decreased "apparency" to insects.

In a sense, then, the degree of refinement in coevolution of plants and pollinators seems to be halfway between that of plants and frugivores on the one hand and that of insect herbivores and plants on the other. Coevolution of plants and pollinators occurs, or fails to occur, simply because plants benefit from short-term constancy in animals and animals benefit from short-term exclusive access to food in flowers. These benefits may lead to temporary specialization of pollinators on plants and to the evolution of features that appear to be coadapted. Nevertheless, the species identity of the animals, the animals' activities at other times of the year, or the activities of the animals' conspecifics may have no direct impact on the fitness of the plant (Baker, 1961; Stiles, 1978a). An obligate, exclusive, coevolving relationship does not necessarily increase the fitness of plant or animal; in fact, the opposite may often be true.

Most plants and pollinators move independently over the landscape, not in matched pairs. Even at a single site, most plant populations experience simultaneous selection from a suite of pollinators, and vice versa. Each plant population and each pollinator population is caught in a complex web of ecological interactions. Particular plant populations, like those pollinated by bumblebees, may indeed affect one another's evolution through their mutual use of pollinators. Particular pollinator populations, such as euglossine bees, may strongly influence the evolution of traits in plants even if there is little reciprocal effect. Particular pollinators, like hummingbirds, may compete strongly with one another and influence one another's evolution. Figure 6 diagrams the relative intensity of various selective pressures among populations, for the three case histories.

Competition among animals may benefit plants, such as those pollinated by short-billed hummingbirds (Feinsinger, 1978), by providing a visitor for every flower. Competition among plants may benefit pollinators, if temporal displacement is truly a result of competition, by providing flowers for every possible season in which a pollinator might be active (see Figure 4). This complex mutualism provides opportunities for the evolutionary diversification of both plants and pollinators. Nevertheless, the community-level mutualism is a result, not a cause, of selection acting in each plant or animal population (see Waser and Real, 1979).

One major stumbling block to understanding the nature of plant–pollinator coevolution is that we know almost nothing about the evolution of flower choice in flower-visiting animals, except perhaps for fig wasps (Janzen, 1979c; Wiebes, 1979). We know little about the extent of heritable variation in traits that influence flower choice, such as the length of hawkmoth tongues or the pollen preferences of solitary bees. Likewise, despite a tremendous body of research on plant evolution and ecological genetics, little information exists on microevolution in plants in response to foraging by pollinators. Is evolution of floral traits gradual or episodic? The interaction between orchids and euglossines implies that new, spontaneously produced floral traits might attract an entirely new pollinator species, which then alters the selective pressures on other floral traits until a new equilibrium is reached. Similar punctuational events could occur in isolated populations at the edge of species' ranges, where the "typical" effective pollinators, which exert stabilizing effects on central populations, are absent (e.g., see Baker, 1961; Grant and Grant, 1965; Cruden, 1972a,b;

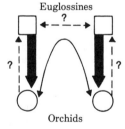 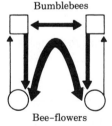 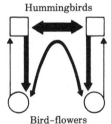

FIGURE 6. The relative intensity of selective effects among populations in the three case histories. Width of arrows indicates relative intensity; admittedly, some arrows might be wider were more known about the interaction. Two pollinator populations (squares) and two plant populations (circles) are shown for each case. The loop joining the plants indicates the strength of their interaction mediated through the mutual use of pollinators.

308

Levin and Kerster, 1974; Cruden et al., 1976; D. A. Levin, 1978). At this time, however, there is no evidence that the floral traits in most plant populations evolve suddenly rather than gradually (e.g., see Brink and DeWet, 1980).

At an even finer level of resolution, with few exceptions (e.g., Waser and Price, 1981) we don't know the extent or selective consequences of heritable variation in those plant traits used to entice and reward pollinators, even though research is very thorough on other aspects of floral morphology and coloration. Now that techniques for quantifying pollinator effectiveness are available (see following), more research on plant populations polymorphic for floral traits should be undertaken.

The most promising situations for investigating plant–pollinator coevolution may be those situations most unusual for the organisms concerned. If central populations, in normal years, are likely to experience stabilizing selection, then investigators should go to the edge of the range of a selected plant or pollinator species, to a new habitat just being invaded, or to a climate other than that in which the central populations exist. Furthermore, studies lasting for many years are needed, not only to document unusual events in flowering phenologies (see Stiles, 1977) but also to discover changes in the diversity and density of effective pollinators. Occasional, drastic, temporary shifts in the pollinator assemblage might be the most important events in the evolution of floral traits or flowering phenologies (see Wiens, 1977). The same consideration applies to flower-visiting animals, the evolution of whose foraging behavior might be influenced by occasional events, not only usual conditions (Feinsinger and Swarm, 1982).

The entire framework for discussing coevolution of plants and pollinators is based on the "most effective pollinator principle" (Stebbins, 1970), in which different flower visitors exert different selective forces due to the relative effectiveness with which they transport pollen. Nevertheless, the "most effective pollinator" is rarely determined accurately. One technique for quantification has been available for several years (Primack and Silander, 1975), but only very recently have researchers actually examined the effects of different pollinators on the seed set from the particular flowers they visit (e.g., Motten et al., 1981; Tepedino, 1981) or on the distance pollen flows (Price and Waser, 1979; Schmitt, 1980). These studies have already revealed some surprises that counter accepted dogma, for example, the discovery that oligolectic bees may be no more effective than "generalized" flower visitors. Indeed, many older studies inferred a bee's potential as a pollinator by examining the pollen in its corbiculae (pollen baskets), pollen that is no longer available to stigmas. Only pollen on the bee's body, missed during its grooming, is truly available for fertilization, and this pollen has seldom been examined. Careful studies on the effectiveness of different flower visitors, using both observational and experimental approaches, are crucial to further progress in understanding coevolution. It is fun and intellectually stimulating to speculate on the selective pressures leading to traits of plants and pollinators, but we really do need more data!

CHAPTER 14

INTIMATE
ASSOCIATIONS AND
COEVOLUTION
IN THE SEA

Geerat J. Vermeij

INTRODUCTION

Biological processes such as predation, competition, and parasitism have unquestionably played a leading role in the evolution of marine organisms. If the term *coevolution* is applied to all evolution resulting from biological interactions (see Chapter 1), as it was by Vermeij (1978), the impact of coevolution has been profound; but if the term is restricted to reciprocal adaptation involving the heritable traits of two or more species, the role of coevolution in the history of marine life is more difficult to evaluate. Although coevolution is often a plausible explanation for known historical correlations between the traits of interacting organisms, the available evidence is rarely sufficient to rule out the simpler hypothesis of unilateral adaptation, that is, evolution of one of the interacting members of an association without specific counteradaptation by the other.

In this review, I consider the following questions: (1) What evidence exists for reciprocal adaptation in cases of intimate association between marine species? (2) Which agents of selection favor the evolution and maintenance of intimate associations, and which additional conditions are required for coevolution? (3) How is the intensity of this selection distributed geographically, ecologically, and temporally? (4) What factors prevent or limit coevolution?

311

EVIDENCE FOR RECIPROCITY

Reciprocal coevolution entails adaptations in both species in an association. The adaptation of one member of the association must be specific to the other member, and vice versa. In the fossil record, the adaptive change in one member must be closely correlated in time with the adaptive change in the other member. That these conditions are difficult to satisfy is evident in the examples below.

Reciprocal adaptations in prey–predator associations

Stanley and coauthors (Chapter 15) outline the replacement of solenoporacean algae, which were apparently susceptible to grazing, by the much more resistant encrusting calcareous Corallinaceae during Jurassic time. This event, which involved adaptive changes of several kinds, was well correlated in time with the diversification of various herbivorous animals whose capacity to excavate calcareous crusts was greater than that of earlier grazers. Although the evolution of the grazers apparently influenced the proliferation of the corallines, the paleontological evidence does not indicate whether coevolutionary change in any of the morphological characteristics that mediate the interaction took place.

Balanomorph barnacles during the Cenozoic era show a tendency in several shallow-water lineages toward a thickening of the test wall by the formation of tubules and toward a reduction in plate number from 8 to 6 to 4 (Newman and Ross, 1976). Palmer (1982) showed that the sutures between plates are the chief sites where predaceous muricacean gastropods are able to drill holes successfully. Drilling through the plate walls often proved to be unsuccessful. He therefore interpreted the morphological trends in barnacles through time as adaptive responses to the evolution and diversification of drilling gastropods. It is not known whether the drilling apparatus or capacity of muricaceans changed in response to increasing armor in barnacles; therefore, a coevolutionary interpretation of the snail–barnacle interaction seems premature.

Shallow-water shelled gastropods evolved narrow and toothed apertures, tight coiling, and other defenses after the Triassic period in the Mesozoic era (Figure 1). At about the same time, shell-breaking fishes, reptiles, and crustaceans with specialized dentition evolved and diversified (Papp et al., 1947; Vermeij, 1977; Vermeij et al., 1981). An increase in the incidence of sublethal shell injuries between Late Triassic and Late Cretaceous time suggests that selection in favor of resistance to breakage became more important. Moreover, the relationship between the incidence of sublethal injury and shell size suggests that post-Triassic predators were more powerful than their earlier counterparts. In shells from the Late Paleozoic (Pennsylvanian

312

period) and Early Mesozoic (Late Triassic), the frequency of sublethal injury was lower for large shells than for small ones; but in shells from Late Cretaceous, Miocene, and Recent age, the frequency of repaired damage rises with increasing shell size. These observations suggest that agents of shell breakage (chiefly predators) became increasingly capable of attacking larger shells, and hence became stronger with time (Vermeij et al., 1981). Whether and how the predators responded to evolution in gastropods and in other types of shelled prey is not known. Diffuse coevolution may have taken place between shell-bearing animals and their shell-breaking predators, but the available evidence does not prove that it did.

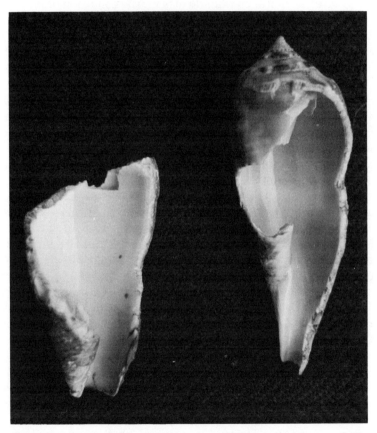

FIGURE 1. *Conus gladiator* (left) from Naos Island, Panama; and *C. tulipa* (right) from Merizo, Guam, found broken in the field. Both animals probably died by being crushed by a fish. Broken shells like these are abundant on tropical shores and point to the importance of breakage as a cause of selection.

313

In contrast to the preceding examples, in which evolutionary correlations exist between large numbers of prey and large numbers of predators, many marine associations whose origins could have been co-evolutionary involve a single host and guest or a small number of closely related species. Here again, however, evidence for reciprocal adaptation is often weak, and when it does exist, nothing is known about the coevolutionary dynamics of the association.

Consider the associations of paguridean hermit crabs with other marine organisms. Many workers have shown that hermit crabs are limited in their growth, reproduction, population density, competitive ability, and resistance to predators by the size and geometry of the gastropod shells or other hollow objects they inhabit (Vance, 1972; Bertness, 1981a,b,c,d). Because the shells of some gastropod species confer greater advantages than do others, the condition of hermit crab individuals and populations depends on the supply of gastropod shells, which in turn depends on the predators and reproductive characteristics of gastropods. It is not known whether gastropods derive any benefit from the presence of hermit crabs, but this possibility cannot be ruled out without experiments.

Hermit crabs are frequently associated, facultatively or obligately, with hydroids, sea anemones, or sponges (Wells, 1969; Cutress et al., 1970; Ross, 1970, 1971, 1974; Provenzano, 1971; Hand, 1975; and references therein). Sea anemones protect hermit crabs of the genus *Dardanus* against certain predators such as *Octopus* (Ross, 1971). The benefit to the anemone is less obvious and has not received attention from experimenters. The suggestion that shells provide attachment surfaces that would otherwise be unavailable should be examined by observing settlement or movement of anemones on to shells in comparison to other available substrata. A possible nutritional benefit to the anemone could be evaluated by studying uptake of particles by the anemone with and without hermit crabs in the shell. In many cases, anemones of the genus *Calliactis* have specialized behavior to mount shells occupied by hermit crabs, and some hermit crabs pick up and transfer anemones from another shell to their own. Some deep-sea parapagurids and temperate pagurids live in shells that are constructed or enlarged by sea anemones (*Stylobates*), hydroids (*Hydrissa, Hydractinia*), or sponges (*Xenospongia, Suberites,* and others) (Wells, 1979; Provenzano, 1971; Miyake, 1978; Dunn et al., 1981). The probable course of events is that a small gastropod shell or other object with a juvenile hermit crab occupant is colonized externally by a cnidarian or sponge. Using the original shell as a template, the cnidarian or sponge extends its basal membrane beyond the original shell's outer lip in a spiral direction, forming a "false shell" or carcinoecium in which the growing hermit crab continues to live. Experiments to substantiate these interpretations are needed, and the specificity of the association must be verified, but coevolution between hermit crabs and shell-generating cnidarians and sponges seems plausible.

The obligate mutualism in California between hydroids of the genus *Zanclea* and bryozoans of the genus *Celleporaria* may have arisen by coevolution, but the evidence is not yet compelling (Osman and Haugsness, 1981). With its stinging nematocysts, the hydroid reduces the impact of one of the bryozoan's predators, the nemertean worm *Notoplana californica*. The hydroid also prevents other animals from overgrowing the bryozoan's colony. *Celleporaria* has specialized areas between the zooids where the hydroid can attach and grow. The benefit to the hydroid is purportedly similar to that which anemones receive from hermit crabs, but, as in that case, the evidence is incomplete. It is also unclear whether the unique features of *Zanclea* are interpretable as adaptations to life with *Celleporaria*.

Many intimate associations whose origin might be coevolutionary are characterized by a phenotypic response by the host to the presence of the guest. The coevolutionary interpretation then rests on whether the phenotypic response is to the guest in particular or to any foreign intruder in general. Corals of the genus *Porites*, for example, produce membranes that protect various nudibranch gastropods, which feed on coral polyps, from the harmful effects of the coral's nematocysts (Rudman, 1981). The presence of the crab *Echinoecus pentagonus* in the sea urchin *Echinothrix calamaris* in Hawaii evokes the formation of a gall (Castro, 1971). Formation of galls is also known in corals with guest hapalocarcinid crabs and seastars with guest eulimid gastropods (Patton, 1967; Castro, 1976; Waren, 1980a,b). When the rhizocephalan barnacle *Lernaeodiscus porcellanae* infects the anomuran decapod crustacean *Petrolisthes cabrilloi* in southern California, the guest "takes over" the reproductive functions of the host and thus alters the host's behavior (Rotchie and Höeg, 1981). Modifications of host behavior and morphology by parasitic guests have been interpreted as possible adaptations for the transmission of the guests to primary hosts that prey on the modified host. They have been documented for parasitic worms in gastropods and in other animals (Feare, 1971; Holmes and Bethel, 1972). These effects may result entirely from the guest's interference rather than from reciprocal adaptation between host and guest.

Many other intimate associations have an equally obscure origin. Clownfishes of the genera *Amphiprion* and *Premnas* (family Pomacen-

315

tridae) live in close association with the large anemones of the family Stichodactylidae (see Roughgarden, 1975, and Dunn, 1981, for recent discussions). The anemones appear to protect the fish against predators, whereas the fish possibly enhance the anemone's food supply, although this claim requires confirmation. Mucus on the skin of the fish appears to protect the guest against the anemone's stinging nematocysts (Mariscal, 1971), but it is unclear which, if any, features of the anemones specifically function as adaptations to the presence of the fish. Pearlfishes of the family Carapidae that inhabit holothurians (sea cucumbers) are, in the more specialized forms, equipped with special entry behavior and with other adaptations suitable for life in these wormlike echinoderms (Trott, 1970); but the holothurians have no behavioral attributes that could be interpreted as adaptations to the presence of pearlfishes, so that a coevolutionary origin of this association has not yet been demonstrated.

Several small xanthid crabs and pontoniine alpheid shrimps live in an apparently mutualistic association with branching scleractinian corals. In the tropical Pacific and Indian Oceans, the coral provides space and nutrition for the crustaceans, and the crabs and shrimp protect the corals against attack by the crown-of-thorns seastar *Acanthaster planci* (Glynn, 1976, 1977, 1980). Although the adaptations of the guest to the host are obvious, it is unclear whether the coral has adapted to its guest.

Many associations between snapping shrimps (Alphaeidae) and fishes (usually of the family Gobiidae) are known. Experiments by Harada (1969) suggest that the gobies warn the shrimps against predators by touching the shrimps as the fish retreat into the shrimp's sand burrow. The burrow provides protection and brood space for the fish, and the shrimp evidently does not attack the goby as it would other intruders. The role of coevolution in the development of this association remains obscure, but reciprocal adaptation is likely to have been important.

Several fishes and shrimps have evolved the behavior of gleaning parasites from the bodies of other fishes (for reviews and data, see Feder, 1966; Losey, 1974). An adaptation on the part of the host is the behavior of establishing specific sites where the cleaning is performed. These associations, then, would seem to be good candidates for a coevolutionary interpretation.

Vance (1978) interpreted the occurrence of tunicates and other sessile invertebrates on the rough shells of living chamid bivalves in California as evidence of a facultative mutualism in which one or both parties gained protection from predaceous sea stars, and the epibionts benefited by being spared the intense grazing by the sea urchin *Centrostephanus* that organisms on bare rock experience. Although the rough surface of the chamid shell may promote settlement and sur-

316

vival of epibiont larvae, I believe that the association is little more than a happy coincidence for which coevolution need not be invoked.

Many shallow-water stony corals (Scleractinia), sea whips (Gorgonacea), sea anemones (Actinaria), zoanthideans, giant clams (family Tridacnidae), opisthobranch gastropods, didemnid tunicates, and other marine animals contain photosynthesizing organisms (for reviews, see McLaughlin and Zahl, 1966; Taylor, 1973). The hosts often have specialized areas or tissues in which the symbionts grow. Whether and how the symbiotic algae are adapted for life in particular hosts is not known. With the exception of the peculiar prokaryote-like organism *Prochloron* growing in the test and cloacal cavities of didemnids (Newcomb and Pugh, 1975), symbiotic algae are morphologically indistinguishable from free-living forms (Taylor, 1973). Schoenberg and Trench (1980a,b,c) have now shown that the dinoflagellate *Symbiodinium microadriaticum* occurs as many enzymatically distinct strains, only one of which is found in a given host species at any one site. Moreover, although a given strain can be introduced artificially into other hosts, the density of the cells of the naturally occurring strain is higher than that of symbionts in a foreign host whose own symbionts had been removed previously. These results suggest that the strains are adapted to particular hosts and that coevolution may have contributed to the establishment of the mutualism, but the nature of the adaptations remains a mystery.

The examples in this section show that coevolution is a plausible explanation for some marine associations between predators and prey and between hosts and guests that mutually benefit one another. The available evidence, however, is usually weak, and in no case can any conclusions be drawn about the timing, rate, or mechanism of the purported coevolution. In the next section, I shall argue that coevolution should often not be expected and that an appearance of coevolution could arise from parallel rather than from reciprocal adaptation.

AGENTS OF SELECTION

In his cost–benefit model for the origin of marine intimate associations, Roughgarden (1975) points out that the host should be persistent enough to provide to the guest a protective advantage that exceeds any disadvantage incurred during searching for the host. The host should, in other words, be a relatively secure place to live. In an environment where predation is an important cause of death and agent of selection, predator-resistant hosts would be especially susceptible targets for potential guests. Once the two are associated in a commensal or parasitic relationship, the association could be refined through coevolution.

The hypothesis that predator-resistant hosts have relatively more species associated with them than do prey that are more vulnerable to predators has not received systematic attention, but preliminary evidence is consistent with it. Presumably because of the presence of stinging nematocysts, corals and other cnidarians are usually avoided by predaceous crustaceans, fishes, and mollusks (Ross, 1971). Numerous animals, including mollusks (Robertson, 1970, 1980; Harris, 1973; Rudman, 1981), crabs (Castro, 1976), pontoniine shrimps (Bruce, 1976, 1977), and cyclopoid copepods (Humes, 1979) are obligately associated with stony corals. Crinoids in the Red Sea are rarely assaulted by fishes and support a rich assemblage of specialized associates (Fishelson, 1974). Many barnacles living on sea turtles, whales, manatees, and crabs, where they do not experience the predation by drilling gastropods and scraping animals that barnacles on shallow-water rocky surfaces must endure (Newman, 1960; Palmer, 1982) are characterized by large opercular openings, by thin test walls, and by the presence of six or eight plates instead of four. These features are, according to Palmer (1982), found principally in barnacles that do not experience heavy predation.

Dawkins and Krebs (1979) embody in their "life–dinner principle" the expectation that selection on one member of an association is frequently much stronger than on the other. According to their interpretation, failure by a predator in an encounter between predator and prey merely results in the loss of a meal, whereas failure by the prey results in death. In the predator–prey interaction, antipredatory selection takes place when the predator is less than 100% successful in detecting, chasing, or subduing its prey (Vermeij, 1982b). Association with a persistent host is one form of response by the prey to such selection. Counteradaptation by the persistent host may result from strong selection by the guest if the latter harms its host. Selection imposed by the prey on a predator, however, is probably often weak and indirect, except if the prey poses a potential threat to the predator. I believe that the principal agents of selection favoring improvement in predatory efficiency are not the prey, but other predators (Vermeij, 1982b). Structures that function in the predator as feeding organs also frequently serve as weapons for defense against other predators or against competitors. Although the evolutionary result might look like counteradaptation to the prey's defenses, in fact the cause could be selection that is imposed by the predator's own enemies. An explanation somewhat more consistent with a coevolutionary interpretation is that predators respond to the weak selection imposed by their prey only when selection is reinforced by the predator's enemies. These speculations lead to the postulate that predator–prey coevolution is most likely in associations in which the prey or guest are potentially lethal to the predator or host. They further suggest that coevolution by reciprocal adaptation may be the end point in a graded series of types of evolution and that the search for such pure "coevolution" is less interesting than the documentation of patterns in the importance of biological agents of selection.

In his cost–benefit model, Roughgarden (1975) makes the interesting prediction that mutualisms evolve only when the guest can substantially improve its host's persistence. As he points out, this can occur without coevolution, but the host is likely to adapt to the presence of a potentially beneficial guest. One important way in which the guest could enhance the host's survival is to make the host a better competitor. Reef-dwelling animals with symbiotic algae are, for example, typically much larger and grow faster than their asymbiotic relatives. These differences may explain why tropical alga-free epifaunal animals are often confined to interstices and undersurfaces in reefs, whereas the larger, competitively superior plant-containing forms occupy well-lit open surfaces (Jackson, 1977; Kott, 1980). The presence of nitrogen-fixing cyanobacteria in some sponges may confer a similar growth advantage over species that lack these symbionts (Wilkinson, 1978; Wilkinson and Fay, 1979; Wilkinson and Vacelet, 1979). The competitive improvement resulting from the presence of symbiotic microorganisms may be particularly advantageous on reefs, where standing stocks of nutrients (phosphates and nitrates) are typically low. In areas of upwelling or periodic plankton blooms, on the other hand, animals with photosynthetic or other food-producing symbionts may not have a distinct growth advantage over those lacking these associates.

The deep sea is another environment with chronically low food levels. Accumulating evidence shows that the gutless animals in the deep sea and in sulfur-rich sediments and deep-sea vents are associated with chemoautotrophic prokaryotic cells of the gram-negative bacterial type (Cavanaugh et al., 1981; Felbeck et al., 1981). Examples includes several vestimentiferan Pogonophora and bivalves of the genus *Solemya*. Whether these associations are in any sense coevolutionary remains to be seen.

Many environments exist in which nutrients are difficult for metazoans to extract without the aid of bacteria or other symbionts. Seilacher (1977), for example, has suggested that the complex burrow systems of some crustaceans and other animals found far below the marine sediment surface might serve as sites where bacteria capable of breaking down refractory organic compounds could be "gardened" by

319

the occupant. Wood-borers might have symbiotic bacteria capable of breaking down lignins and cellulose. Work is needed to establish whether these mutualisms exist, and if so, whether they are the result of reciprocal adaptation.

These examples and speculations suggest that low availability of nutrients favors the evolution, unilaterally or by reciprocal adaptation, of associations between a guest and a host. I therefore predict that the incidence of such mutualisms is greater in nutrient-poor habitats than in areas where nutrients are abundant for part or all of the year.

THE GEOGRAPHY OF INTIMATE ASSOCIATIONS

If predation and low nutrient availability are important conditions for the evolution of intimate associations and mutualisms, respectively, the incidence of these associations should vary geographically and ecologically. Evidence from shell architecture, frequencies of sublethal shell injury, and experiments with tethered shells suggests that antipredatory selection with respect to armor increases from the temperate zone to the tropics, from fresh-water to marine habitats, from the deep sea to shallow water, and within the tropics from the Atlantic to the eastern Pacific to the Indo–West Pacific (Vermeij, 1978; Vermeij and Covich, 1978; Vermeij and Currey, 1980; Vermeij et al., 1980; Bertness et al., 1981). Similar gradients may exist for predation by drilling gastropods, but comparisons between the various tropical regions have not yet been made (Dudley and Vermeij, 1978; Palmer, 1982). Patterns in the toxicity of sponges and holothurians further support the view that antipredatory selection intensifies toward the equator (Bakus, 1974, 1981; Bakus and Green, 1974). Intimate associations, whether they originated by unilateral or by reciprocal adaptation, should increase in frequency in the same directions. The distribution of mutualisms should be broadly similar to that of intimate associations generally, but there are some points of discrepancy. Within the tropics, nutrient levels (as measured by standing stocks of phytoplankton) are generally lower in the Indo–West Pacific than in the western Atlantic (Highsmith, 1980b). Although upwelling and continental runoff occur in parts of both these regions, these sources of nutrients are truly characteristic of the eastern Pacific and eastern Atlantic. Levels of nutrients are as much as an order of magnitude lower in the deep sea than in typical coastal waters (Rex, 1976; Grassle, 1977). Accordingly, the two regions in which predictions about the frequencies of mutualisms and intimate associations should differ are the tropical eastern Pacific and the deep sea.

Latitudinal comparisons of intimate associations are scarce in the literature. Symbiotic associations between animals and photosyn-

320

thesizing organisms seem to be relatively much more common in the tropics than in the temperate zones. Only one sea anemone in Europe and one in western North America are photosynthetic, and there are no temperate corals with oxygen-producing symbionts. Numerous anemones, corals, and other animals are associated with various algae in the tropics (Taylor, 1973; Schoenberg and Trench, 1980c).

Other associations also are better developed in the tropics. In his survey of shrimps, Bruce (1976) found no obligate commensals in Great Britain, whereas more than 50% of tropical East African shrimps are obligate commensals. All tropical pearlfishes (family Carapidae) are associated with holothurians; the three free-living species of the family have a temperate distribution (Trott, 1970). The lack of commensalism among shrimps and carapids in the temperate zones is not explained by the absence of appropriate hosts, because sessile invertebrates, in general, and holothurians, in particular, are abundant in both temperate and tropical waters.

Rohde (1978a,b) has studied the host specificity of parasitic marine trematodes in fish. He found an increasing host specificity toward the tropics among the Digenea, but a constant and high specificity in the Monogenea. However, whereas most temperate fish rarely have more than one monogenean parasite species, tropical fish may harbor as many as nine monogenean species per individual.

Symbioses in which shrimps and fishes remove ectoparasites from other fishes may not conform to the predicted latitudinal pattern, although data are still inconclusive. In his compilation of these symbioses, Feder (1966) lists 43 marine fishes as cleaners, of which 9 (6 in California and 3 in the Mediterranean) have a strictly warm-temperate distribution, and 32 (12 Caribbean, 12 eastern Pacific, 6 Indo–West Pacific, and 2 Pacific–Atlantic species) are strictly tropical. Of six cleaning shrimps, one is temperate Californian and the other five are tropical. These increases in the number of cleaners toward the tropics are no greater than the trend seen in the diversity of fishes and shrimps generally. Temperate cleaners, however, are not as specialized to this mode of life as are their tropical counterparts (Feder, 1966). Moreover, there is reason to believe that Feder's compilation reflects the distribution of observers more than it does the geography of cleaning. His data were assembled before long-term observation under water became commonplace in remote parts of regions such as the tropical Pacific and Indian Oceans, where cleaning associations would be most expected. California and the West Indies, on the other hand, were biologically already quite well known.

A more plausible counterexample to the latitudinal gradient in in-

321

timate associations is provided by acmaeid and patellid limpets that are obligately associated with kelps (large phaeophytes) and sea grasses. These limpets feed on their hosts but may in some cases promote or stimulate the host's branching and reproduction (Black, 1976). Kelps and their associated limpets are confined to the temperate zones where water temperature is below 20°C (Vermeij, 1978). Curiously, limpets living with sea grasses are also confined to the temperate zones. This is surprising in view of the great abundance of sea grasses in the tropics. Smaragdine neritid gastropods possibly replace limpets on tropical grasses, but little is known about their relationships to the hosts.

Very few data are available on the occurrence of intimate associations along a depth gradient. Associations between hermit crabs and cnidarians seem to be most frequent in deeper water. A survey of hermit crabs from Sagami Bay, Japan (Miyake, 1978), reveals that none of the seven intertidal species is associated with anemones, whereas 14 of the 54 strictly subtidal species (26%) have shells adorned with cnidarians. Whether the absence of anemones on the shells of intertidal hermit crabs is related to desiccation stress is uncertain, but many free-living anemones are characteristically intertidal in distribution.

Data on the occurrence of intimate associations in general and of mutualisms in particular in relation to nutrients are also inconclusive. Kropp and Birkeland (1982) found that obligate commensals of *Pocillopora verrucosa* make up a smaller proportion of the crustacean species that are found among the branches of that coral at a productive site in the lagoon of a high island (Moorea) than at a nutritionally poorer site in the lagoon of Takapoto Atoll. Abele's (1976) data on the crustacean associates of *P. damicornis* in Pacific Panama show a similar trend, although they were interpreted by Abele as supporting the hypothesis that obligate associations are more frequent in less fluctuating environments. In this case, the less productive site (Gulf of Chiriquí) is also more constant in temperature than is the more productive site (Bay of Panama).

In the tropics, the incidence and degree of specialization of intimate associations appear to be much greater in the Indo-Pacific than in the western Atlantic. This difference would be expected on purely statistical grounds because the Indo–West Pacific fauna is locally and globally richer than is that of the Caribbean; but I believe that the difference usually exceeds in magnitude the expected factor of 2 to 5. For example, all nine bivalve species that settle on living corals and subsequently burrow into the skeleton are known from the Indo–West Pacific (Highsmith, 1980a). At least one of these clam–coral associations may be obligate. Highsmith (1980a) presents data suggesting that the larva of *Montipora berryi* preferentially settles near the empty cavities

322

of *Lithophaga curta*, a mytilid bivalve that in Enewetak Atoll is restricted to *M. berryi* and a few related species. In a recent compilation, Bruce (1977) found that 37 pontoniine shrimps are obligately associated with scleractinian corals in the Indo-West Pacific, whereas in the Caribbean no obligate pontoniine-coral associations are known. Castro (1976) lists 59 Indo-West Pacific brachyuran crabs associated obligately with corals, as compared to only 8 in the eastern Pacific and 3 in the tropical Atlantic. Barnacles obligately associated with scleractinians are most diversified and most specialized morphologically in the Indo-West Pacific (Ross and Newman, 1973; Newman and Ladd, 1974). In the balanid subfamily Pyrgomatinae, for example, four or five species in two genera are known from the western Atlantic, as compared to eight genera in the Indo-West Pacific and none in the eastern Pacific. Clownfishes of the family Pomacentridae are exclusively Indo-West Pacific; their host sea anemones (chiefly in the family Stichodactylidae), with one free-living Caribbean exception, are also confined to the Pacific and Indian Oceans (Dunn, 1981). Although at least ten fishes live facultatively with anemones in the Caribbean (Hanlon and Kauffman, 1976), obligate associations of the type between clownfishes and anemones are unknown there. Facultative fish-anemone associations are also known from warm-temperate regions (Japan and the Mediterranean). Pearlfishes (family Carapidae) live in the coelomic cavity and sometimes the respiratory trees of sea cucumbers. Only one species is known from the Caribbean, whereas 17 are known from the tropical Indo-West Pacific (Trott, 1970). The only pearlfishes that feed on the gonads and respiratory trees of their hosts are two species that belong to the Pacific genera *Enchyliophis* and *Jordanicus*. The Caribbean pearlfish (*Carapus bermudensis*), like the other tropical species, feeds on crustaceans (Trott, 1970). Ponder and Gooding (1978) found three genera of eulimid gastropods obligately associated with diadematid sea urchins in the Indo-West Pacific, whereas in the tropical Atlantic no obligate commensals reside in these abundant spiny animals. Waren's (1980a,b) continuing studies of eulimids also suggest that Indo-West Pacific seastars and sea cucumbers have more parasitic gastropods per species than do Caribbean forms. A commensal gastropod (*Caledoniella montrouzieri*) is known from stomatopods in the Indo-West Pacific but not from any Atlantic species (see Reaka, 1978).

Symbiotic associations of animals with algae are far more numerous in the tropical western Pacific and Indian Oceans than in the Atlantic. Giant clams of the family Tridacnidae and the cardiid genus *Corculum* are found in the Indo-West Pacific but are absent in the modern western Atlantic fauna. Although *Prochloron*-bearing didemnid tunicates are now known in the Caribbean (Lewin et al., 1980), the algae are found in the test rather than in specialized common cloacal canals as in many Indo–West Pacific species (Kott, 1980).

Cyclopoid copepods living with scleractinian corals may provide an exception to this interoceanic pattern. Humes (1979) has published a summary table of cyclopoids and their host corals in several widely scattered tropical localities. At Curaçao in the southern Caribbean, 14 copepods have been described from a total of 22 host species (host to copepod ratio, 1.57:1). The area around Nosy-Be in northwestern Madagascar supports 45 copepods on 84 host corals (ratio, 1.87:1), and the Moluccas in Indonesia have 29 species on 40 hosts (ratio, 1.34:1). Although these data are subject to criticisms about sampling and completeness, they suggest that host specificity is broadly similar in the Caribbean and Indo–West Pacific despite the much higher local species number of corals in the latter region.

The compilation above suggests that the incidence of intimate associations, as expected, increases from high to low latitudes and is greater in the Indo–West Pacific than elsewhere in the tropics. Data on the relationship between intimate association, mutualism, and nutrient availability are insufficient to permit conclusions to be drawn. It must be emphasized that our sampling of marine intimate associations is both incomplete and ecologically biased. As a result of the preoccupation of biologists with reefs, very little is known about associations in muddy and sandy habitats, where the intensities of antipredatory and competition-related selection may be less than on hard substrates (Vermeij, 1978; Peterson, 1979).

The predictions that led to the compilation in this section are derived from empirical observations on geographical and ecological patterns in the intensity of antipredatory selection and nutrient availability. How did these patterns in selection come about? Within the tropics, the inferred intensity of antipredatory selection for armor is greatest where the diversity of species is highest and least where it is lowest; but high diversity in such temperate areas as California does not guarantee strong selection in favor of armor (Vermeij, 1978). High diversity may be a necessary condition for strong selection in favor of armor and for the widespread development of intimate and perhaps coevolutionary associations, but it is not a sufficient condition. It is also noteworthy that intimate associations are most prevalent in areas where late Tertiary and Pleistocene extinction was relatively minor. Extinctions of mollusks were particularly numerous in the western Atlantic, less in the eastern Pacific, and still less in the Indo–West Pacific (Vermeij, 1978; Stanley et al., 1980; Stanley and Campbell, 1981). Whether predation was a more important agent of selection in the Atlantic before the rash of extinctions is under investigation.

If the increase in antipredatory selection in shelled mollusks (Vermeij et al., 1981) applies to other types of predation in the sea, the interesting possibility arises that intimate associations and conditions for reciprocal adaptation have become more common over the course of geologic time, especially after the Triassic period. Evidence on this point has not been systematically compiled. Clearly, there were intimate associations in the Paleozoic; they include platyceratid gastropods living on crinoids (Bowsher, 1955) and various Permian brachiopods possibly associated with zooxanthellae (Cowen, 1970). I believe, however, that a systematic survey will show that mutualisms and other associations are more frequent in today's ocean than they were in the distant past.

LIMITATIONS ON COEVOLUTION

Even if the selectional environment favors the evolution of intimate associations between marine species, such evolution may be constrained by genetic mixing. The degree to which genes that promote intimate association can spread through and become fixed in a population depends on the scale of gene dispersal relative to the scale of selectionally meaningful environmental variation (Levins, 1968; Vermeij, 1978, 1982a; Stanley, 1979). Most studies of predation, competition, and parasitism reveal substantial variation in the ecological and selectional impact of these relationships on different populations of the same species. Mortality due to drilling naticid gastropods, for example, varies by a factor of 5 between populations of the bivalve *Ctena bella* in Guam (Vermeij, 1980). Frequencies of predator-induced sublethal shell injury vary by factors of ≤ 12 among local populations of *Neritina virginea* (Andrews, 1935), *Terebra affinis* (Vermeij et al., 1980), *Littorina rudis* (Elner and Raffaelli, 1980), *L. littorea* (Vermeij, 1982a), and other gastropods. Other examples of great local variation have been described for predation by *Nucella lapillus* on the mussel *Mytilus edulis* (Menge, 1978a,b), parasitism of *Littorina* and *Hydrobia* species by trematodes (Honer, 1961; Pohley, 1976), and abundance of epibionts on shells inhabited by hermit crabs (Rasmussen, 1973; Conover, 1979). If biologically different conditions are distributed haphazardly on a scale smaller than that of dispersal, genetic mixing may prevent the establishment of genetic variants that are beneficial under particular circumstances. After the green crab (*Carcinus maenas*) spread north of Cape Cod early in the twentieth century, the periwinkle *Littorina littorea*, which itself had begun to colonize shores from Nova Scotia to Cape Cod between 1850 and 1872, did not increase

325

in shell thickness despite a demonstrably higher incidence of sublethal shell injury after the arrival of the predaceous crab (Vermeij, 1982a).

In general, I would expect intimate biological associations, local antipredatory adaptation, and reciprocal adaptation to arise more often in species with a limited dispersal phase (e.g., no planktonic larvae) than in those whose larvae spend a long time in the plankton. This hypothesis, like so many others considered in this chapter, has not yet been tested on a large scale. The gastropod *Nucella lapillus*, however, conforms to it. This species is known to vary greatly in shell form, with the most crush-resistant morphs occurring in environments where predaceous crabs are most abundant (Hughes and Elner, 1979). The same appears to be true for *Littorina rudis* (Elner and Raffaelli, 1980). These species have lost the planktonic dispersal part of their life cycle. Their principal predators, however, have substantial planktonic dispersibility, so that coevolution between crabs and snails is unlikely in these cases.

Many reef-associated species that are widely distributed are intimately associated with wide-ranging corals (Garth, 1974; Bruce, 1976; Vermeij, 1978; Rudman, 1981). Several of these associations, such as those between corals and some crabs, nudibranchs, and pontoniine shrimps, even show evidence of coevolution, as discussed in the section on reciprocal adaptation. These examples provide compelling counterexamples to the expectation evoked in the last paragraph and raise important questions about when species can adapt to each other. Can they do so when populations are large and widely dispersing, or is adaptation restricted to the time when populations are fragmented by temporary barriers? How and when did such separation by barriers take place? Must isolation of guest populations be simultaneous with that of the host in order for the two parties to coevolve?

Finally, although knowledge of marine associations is still very superficial, I have the strong impression that intimate associations and coevolution by reciprocal adaptation are far more prevalent on land than they are in the sea. Some authors (Wiegert and Owen, 1971; Schiel and Choat, 1980) have suggested that large marine plants and corals do not have the three-dimensional complexity that is characteristic of large land plants, but this suggestion begs the question of why such differences might exist. Moreover, no reliable measurements of physical complexity have been made. Insects and flowering plants are the first groups that come to mind when coevolution on land is considered. Both groups are perplexingly poorly represented in the world ocean; but even in fresh waters, where the two groups are diverse, intimate associations are seemingly much rarer than on land or in the sea. Are populations of land plants and insects more easily isolated than those of marine species, so that speciation and reciprocal adaptation become more probable? Has animal-mediated dispersal of pollen

326

and seeds contributed to speciation on land (Mulcahy, 1979) and, if so, why has such dispersal not evolved to a greater extent in the ocean or in fresh water?

These questions illustrate the extent of ignorance that still pervades the subject of coevolution. I believe that one of the most important questions about coevolution in the sea remains: under which ecological, geographical, and population-genetic conditions can reciprocal adaptation, in particular, and intimate associations, in general, take place?

CONCLUSIONS

Knowledge of marine intimate associations and coevolution is still so scanty and so anecdotal that none of the four questions posed in the introduction can be answered satisfactorily with presently available evidence. Evolution by reciprocal adaptation remains a plausible interpretation for many marine intimate associations and predator–prey interactions, but the alternative interpretation of unilateral adaptation cannot be ruled out in most instances. Moreover, the coevolutionary interpretation is complicated by the likelihood that reciprocal adaptation can appear to have taken place when in fact adaptation was induced by species not directly involved in the interaction in question. Predation is seen as an important cause of selection favoring the evolution of intimate associations, whether the latter arise by unilateral or by reciprocal evolution. Mutualisms may be especially favored in environments where availability of nutrients is low. In the Indo–West Pacific, where antipredatory selection appears to be more intense than elsewhere in the tropics, the incidence of intimate associations is higher than elsewhere. The latitudinal decline in the frequency of these associations from the equator to the poles is also correlated with an apparent decrease in antipredatory selection in marine invertebrates. The geographical, ecological, and population-genetic conditions under which intimate associations, in general, and coevolved relationships, in particular, can arise are for the most part shrouded in mystery. Wide dispersal of larvae across biologically heterogeneous environments would seem not to favor coevolution, yet many examples of intimate associations, some of which may be coevolutionary, are known in widely dispersing reef invertebrates.

CHAPTER 15

COEVOLUTION AND THE FOSSIL RECORD

Steven M. Stanley, Blaire Van Valkenburgh, and Robert S. Steneck

INTRODUCTION

However the word *coevolution* may be formally defined, to many biologists it has come to signify the gradual evolution of two inter-acting species—gradual evolution that in at least one of the species represents a response to evolution in the other (Leppik, 1957; Janzen, 1966; Southwood, 1973). The original, broad definition of the word *coevolution*, introduced by Ehrlich and Raven and discussed in the introduction to this volume, embraces all evolution resulting from biological interactions. It does not connote only gradualism but also includes relationships that have evolved rapidly during the origins of species. In order to examine the pattern of this process, we look to the fossil record, the only source of documentation of evolution within interacting groups over long spans of time.

Specifically, we review evidence that species that coexisted within ancient communities did not evolve toward optimal adaptive mor-phologies with regard to community structure. What we find instead is that the species' imperfect initial adaptations persisted with little change until the communities disappeared from the earth. More generally, we observe (as does Bakker; see Chapter 16) that the fossil record shows established species of all kinds to be relatively stable en-tities, changing little in the course of 10^5, 10^6, or 10^7 generations. This general stability implies that the evolutionary response of fully established species to changes in the physical and biotic environment is weak.

328

Coevolved relationships are nonetheless present in the world and must be explained. We conclude that coevolution operates effectively at two levels. At the *micro*evolutionary level, intimate coevolutionary relationships between species must be set up rapidly in localized areas, when species first form. This is the only reasonable explanation for the fact that neither the origins of such relationships nor changes in these relationships are generally documented in the fossil record.

At the *macro*evolutionary level, there is evidence of general changes in the compositions of higher taxa in response to changes in others. We attribute these large-scale changes to species selection or to the tendency of certain kinds of species to survive longer and/or speciate at higher rates than others. At this level, coevolution is a coarse-textured, imprecise phenomenon.

STABILITY OF AN ECOSYSTEM IN GEOLOGIC TIME

One way to test the traditional idea that gradual coevolution is commonplace is to identify ecological imperfections within fossil communities and to trace these imperfections through millions of years, in order to determine whether they were substantially diminished by adaptive change. We will describe an example of this kind of analysis involving carnivorous animals. Here we find virtually no gradual coevolution of the kind that the gradualistic model of coevolution would predict.

Within the context of the fossil record, the guild is a particularly useful but neglected unit of study. Root (1967) defined a guild as a group of species that exploit the same class of resources in a similar way. As such, a guild is a functional unit not bound by taxonomic considerations. Guilds are the building blocks of communities, both extinct and extant, and they represent useful units for the analysis of coevolution (see Chapter 19 by Orians and Paine). Coevolution within guilds is a response to competition for shared resources; members evolve behavioral and morphological differences that favor coexistence.

To examine the effect of coevolution on guild structure, we consider here a guild in which competition appears to be significant. This is the guild of large (> 10 kg) terrestrial, mammalian predators. In this guild, interference interactions are relatively frequent, hunting success is low, and prey represent sizable, discrete, defendable packages (Bertram, 1978). Ecological separation among Recent sympatric predators is effected through differences in diet, habitat, and escape behavior; consequently, predator guilds from tropical, temperate, and montane regions share structural similarities. Our question is, how does this structure develop? To answer this, we examine the evolutionary history of an ancient example of this guild, which when first seen in the fossil record exhibits gaps in its structure and similarities among member species. In the traditional, gradualistic view of coevolution, we would expect the species of this guild to diverge from one another and capitalize on weakly exploited and unexploited resources. A more detailed description of this work appears in Van Valkenburgh (1982a,b,c).

Figure 1 is a diagrammatic representation of predator guild hypervolume. Three separate, but ecologically equivalent, guilds are shown: two are extant—the East African Serengeti guild and the North American Yellowstone guild—and one is extinct—the Late Chadron–Orellan guild, which existed in western North America approximately 30 million years ago. The axes of the three graphs are the same: habitat preference/escape behavior are represented on the x-axis, body weight/prey size on the y-axis, and diet on the z-axis. Although these axes are labeled in terms of behavior, they also reflect quantified aspects of morphology. Habitat preference and escape behavior are a function of claw size, claw shape, and intermembral proportions of the axial skeleton. Position on the diet axis reflects quantified dental characters. Body weight is closely correlated with head–body length.

Group hunting behavior cannot at present be deduced from the fossil data. However, all extant large, running predators are social (e.g. wolf, hyena, wild dog, dhole), and the only ambushers with some tendency toward group hunting are the lion and the cheetah, two cats that have converged more closely with cursors than any other living forms (Kleiman and Eisenberg, 1973; McVittie, 1979). In accord with this pattern, prey size for group hunting is indicated in the Orellan for only the two most cursorial species. For extant species, prey sizes were estimated from behavioral literature. For fossil species, prey size was estimated by analogy with extant animals of similar size and morphology.

Comparison of two Recent guilds

A comparison of two recent guilds shown in Figure 1A and B reveals interesting similarities and differences. The guilds contain similar numbers of predatory species, seven in North America and eight in Africa. Both have three obligate terrestrial animals and four potential climbers. In each guild, species that are similar in habitat preference and escape behavior usually differ in either prey size or diet. Examples are lynx versus coyote, wild dog versus hyena, and lion versus cheetah. Similarly, predators with similar diets, such as grizzly and black bears, spotted and striped hyenas, or lion and leopard, differ in body size or escape behavior.

These similarities in structure are surprising in view of the climatic

330

and vegetational differences between the two areas. East Africa is warm, seasonally arid, and characterized by extensive grasslands (Sinclair, 1979). The North American Yellowstone is forested, moist year round, and snow-covered in winter (Murie, 1940). These environmental factors are not without effect: seven of East Africa's eight predators are meat or meat–bone specialists, whereas in North America these categories contain only four of seven species. Furthermore, the largest predators in East Africa are carnivorous (lions, hyenas), whereas in North America omnivores (bears) are the largest. Presumably, this greater emphasis on carnivory in Africa is a result of the higher diversity and year-round availability of prey in a tropical area relative to a temperate area (Eisenberg, 1981).

Similarly, carcass availability may determine the size and number of bone-eaters. In a habitat such as the Serengeti, with high ungulate densities and an open sky in which circling vultures are visible, carcasses must be easier to discover than in a forested, less densely populated area (Houston, 1979). Hence, East Africa has two large bone crushers, the hyenas, whereas North America has only one of moderate size, the wolverine.

The comparison of the two Recent guilds reveals a shared similarity in overall structure with differences that are due to interrelated environmental factors, including climate, prey diversity, and prey distribution. The fact that two geographically and historically separate guilds share a common pattern of adaptive divergence suggests that processes intrinsic to the guild, like competition, have dominated over extrinsic variables, like climate. Thus, intraguild coevolution seems to be a significant component of guild development. However, a second intrinsic factor, that of heritage, may also play a major role. To examine the effect of history or heritage on guild structure and development, it is necessary to compare guilds with widely disparate histories.

Comparison of Recent and Ancient guilds

In this section we will compare the modern guilds described in the preceding section with a paleoguild of the North American Oligocene (Late Chadron–Orellan). The relevant fossil data are derived from the extensively collected White River Formation of South Dakota, Wyoming, and Nebraska (Scott and Jepsen, 1941; Clark et al., 1967). It is significant that this fauna persisted with little change for approximately 2.5 million years, from 31.5 to 29 million years ago. The Orellan guild members (listed in Table 1) are all members of extinct orders, families, or subfamilies and hence are quite distinct phylo-

USUAL MAXIMUM
PREY SIZE (kg)

> 350

100–350

20–100

< 20

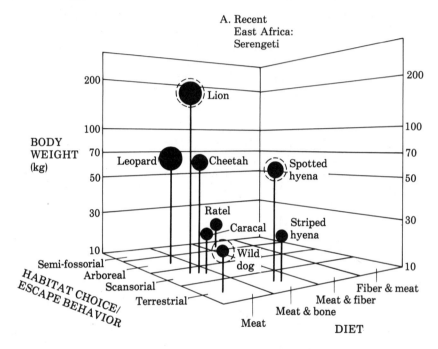

FIGURE 1. Predator guild hypervolumes. (A) East African Serengeti, Recent. (B) North American Yellowstone, Recent. (C) North American Nebraska–Dakota, Late Chadron-Orellan, 31.5 to 29 million years ago. The diet and habitat preference/escape behavior axes are each divided into four categories. Any animals lying within the same box on the floor of the volume are considered roughly equivalent in habitat preference. Body weight is plotted on the y-axis and the estimated maximum prey size is indicated by the size of the dot representing each species. The solid dot indicates the maximum size of prey taken by an individual, and the circle surrounding the dot is the maximum size of prey taken during group hunting. (After Van Valkenburgh, 1982b.)

332

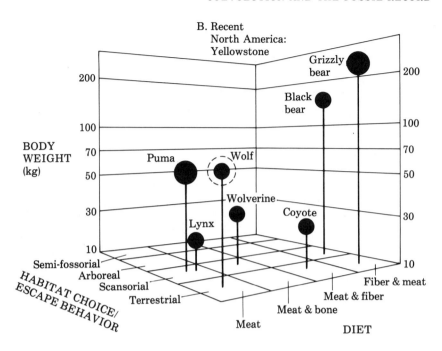

B. Recent
North America:
Yellowstone

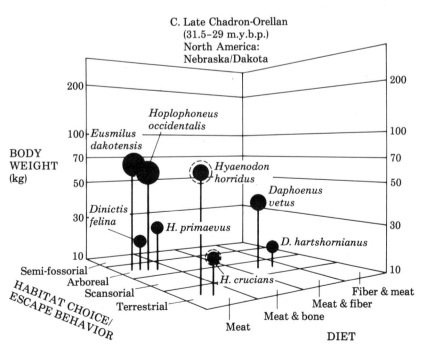

C. Late Chadron-Orellan
(31.5–29 m.y.b.p.)
North America:
Nebraska/Dakota

TABLE 1. Three guilds of large (> 10 kg) terrestrial predators.

Guild site	Order	Family	Species	Body wt estimate[a] (kg)
East Africa: Serengeti (approximately one million years ago to the present)	Carnivora	Canidae	Wild dog (*Lycaon pictus*)	22
		Felidae	Lion (*Panthera leo*)	163
			Leopard (*Panthera pardus*)	50
			Cheetah (*Acinonyx jubatus*)	58
			Caracal (*Caracal caracal*)	16.5
		Hyaenidae	Spotted hyena (*Crocuta crocuta*)	52
			Striped hyena (*Hyaena hyaena*)	32
		Mustelidae	Ratel (*Mellivora capensis*)	10
North America: Yellowstone (approximately 7500 years ago to the present)	Carnivora	Canidae	Gray wolf (*Canis lupus*)	45
			Coyote (*Canis latrans*)	15
		Felidae	Puma (*Puma concolor*)	60
			Lynx (*Lynx canadensis*)	11
		Mustelidae	Wolverine (*Gulo gulo*)	24
		Ursidae	Grizzly bear (*Ursus arctos*)	264
North America (Late Chadron-Orellan, 31.5–29 millions years ago)	Carnivora	Amphicyonidae	*Daphoenus vetus*	40
			D. hartshornianus	10
		Felidae	*Eusmilus dakotensis*	55
			Hoplophoneus occidentalis	58
			H. primaevus	12
			Dinictis felina	20
	Creodonta	Hyaenodontidae	*Hyaenodon horridus*	60
			H. crucians	20

[a]Body weights for fossil and extant animals were estimated as described in Van Valkenburgh (1982b).

genetically from the modern predators in Figure 1A and B. Because of the large number and high quality of the fossil specimens, it is unlikely that any large predatory species have been overlooked. Because the predators occur uniformly both vertically and horizontally within the formation, they are assumed to have lived sympatrically (Clark et al., 1967).

At first glance, the Orellan guild (shown in Figure 1C) resembles the two modern guilds. The number of species (eight) is similar, and the pattern of nonoverlap between species of like diet or size is maintained. Closer inspection shows that the dispersion of species within the guild is less than that within either Recent example. There are no bone crushers, and none of the predators is larger than 60 kg (African leopard size). The Orellan resembles East Africa in the emphasis on carnivory but is comparable to North America in the size of its nonomnivorous predators. It differs from both in (1) having a greater proportion of climbing animals and (2) lacking a specialized bone crusher of any kind.

The dominance of small-sized meat-eaters in the Late Chadron–Orellan can be explained as a response to the year-round availability of relatively small prey. The emphasis on climbing ability is a bit misleading because physical characteristics used to place an animal in this category may be equally useful in short-distance ambush predation. Predator locomotor morphology reflects prey escape behavior. The typical Orellan ungulates were not cursorial, and probably escaped pursuers by rapid bounds into holes or dense brush. Hence, the Orellan predators, as ambushers, were adapted to the available prey.

The absence of a bone-eater is less easily explained. Although the Late Chadron–Orellan paleoenvironment was probably not as open as the modern Serengeti, there were open spaces. In North America the entire Oligocene was a time of transition from warm forest to increasingly arid savannah conditions (Clark et al., 1967; Singler and Picard, 1981). As this transition occurred, bone-eating should have been favored. The ability to crush and consume large bones seems to be an eminently advantageous adaptation. By eating bone, a predator utilizes a carcass more fully and gains access to calories unavailable to fellow guild members. In the late Chadron–Orellan, there is a preponderance of meat-eaters, some of which resemble others in body size (and, thus, prey choice): *Hoplophoneus occidentalis* resembles *Hyaenodon horridus*, and *Dinictis felina* resembles *Hoplophoneus primaevus*. Competition among these animals must have been intense, and selection should have favored divergence, including trends toward bone-eating. It might be argued that a hyena analogue did not appear

335

because there was no potential ancestor. However, *Hyaenodon horridus* appears to be exactly that, a carnivore with a tendency toward bone-crushing abilities; yet it, like most of the fauna, shows no appreciable evolutionary change.

Evolution was demonstrably sluggish within the Orellan fauna, and a potentially profitable mode of life remained unexploited within a highly competitive system. Such stasis seems incompatible with gradualistic models of evolution, which predict that natural selection should act persistently. Such models predict that, barring extinctions, a guild tracked through 2.5 million years should exhibit adaptive divergence toward a stable, saturated structure.

The structural similarities of the fossil and extant guilds suggest that intraguild coevolution normally acts at some point to produce a *rough* spacing of species within the guild hypervolume. However, the lack of continued adaptive divergence shown by the Orellan predators and the conspicuous underutilization of resources over a long span of time indicates that guild coevolution is not universal and perpetual.

THE STABILITY OF SPECIES

Paleontology offers broader testimony than is provided by the fossil record of the carnivorous animals described earlier. Evidence showing that within many different kinds of higher taxa (barring extinction) species typically persist almost unchanged for more than a million years continues to accumulate. By "almost unchanged" we mean that they experience so little evolutionary change that, according to competent taxonomists, they do not deserve new names. For many groups, mean longevity exceeds 10 million years (Stanley, 1979, 1982). All Cenozoic taxa for which relevant fossil data are available support this generalization. The relevant data represent (1) taxa with remarkably complete fossil records or (2) taxa for which the incompleteness of the record can be circumvented by evaluating survivorship from past intervals to the present (as an example, approximately one-half of all species of marine bivalve mollusks 7 million years old have survived to the present).

The marine taxa for which excellent data on species longevity are available include the Bivalvia, Gastropoda, Echinoidea, Foraminifera (both benthic and planktonic), and diatoms; the freshwater taxa include the fishes (teleosts) and gastropods; and the terrestrial taxa include the mammals, beetles, and higher plants. Taken together, these groups not only represent great anatomical and ecological diversity, they include a large numerical sample of the species existing in the modern world. Species concepts are not at issue here. Sibling species may be unrecognized in the fossil record, but sibling groups encompass so little change that their unnoticed presence has no bearing on

336

the evaluation of substantial macroevolutionary transitions. Further-more, any group of fossil populations assigned to a single species displays very little total morphological variation and can embrace very little net evolutionary change.

Thus, it can be concluded that (barring extinction) a typical well-established species of animal or plant will survive for 10^5–10^7 generations without undergoing enough evolution to be regarded as a new species (Stanley, 1982).

TWO LEVELS OF COEVOLUTION

If established species are generally stable entities, then most micro-evolutionary change must take place rapidly and within localized populations. The implication is that most intimate coevolutionary relationships have become established rapidly, whereas during most of its existence a typical species experiences little coevolutionary change.

Coevolution is now commonly assumed normally to be a gradualistic process, or one entailing the long-term transformation of established species (Leppik, 1957; Janzen, 1966; Southwood, 1973). The term *coevolution* was originally introduced to describe "a stepwise pattern of co-evolutionary stages" (Ehrlich and Raven, 1964, p. 605). In another semi-nal study, Baker (1963) concluded that the involvement of a single species of bee, *Euglossa cordata*, in two elaborate pollination mech-anisms represents strong evidence that at least one of the interactions "is the product of recent, sudden evolution." He also observed that some floral features are under simple genetic control. A single gene, for example, determines the presence of the nectariferous spur of the colum-bine genus *Aquilegia*. In mimetic butterfly species of the genus *Papilio*, single genes produce color patterns that approximate those of the im-itated species (Sheppard, 1975). These examples suggest that the fossil record is not demanding the impossible in attesting to the episodic for-mation of coevolutionary relationships.

Viewing phylogeny from a greater distance, we can recognize co-evolution at a second, higher level of biological organization. This en-tails change in the biological composition of a higher taxon. Coevolu-tionary relationships between species often become raw material for change at a macroevolutionary level, where the species is the basic unit. Once a new feature having any degree of coevolutionary significance is present within a higher taxon, species that possess the feature may proliferate to the point where the composition of the higher taxon changes drastically, *with or without the formation of ad-ditional kinds of coevolutionary relationships*. Such changes that occur in the composition of the higher taxon represent the process of *species selection*: differential rates of extinction and multiplication among component species. Species selection may be driven by internal factors, such as traits that endow a particular kind of species with a propensity to speciate, or it may be driven by external agents. The external agents of species selection are ecological limiting factors, the biotic varieties of which are predation (including parasitism), competition, and provision of food or substratum.

When a higher taxon changes its composition under the influence of biotic interactions, we can assume that these are seldom specific, one-to-one interactions, except in the case of narrowly adapted parasites, symbionts, or epibionts. Few species of a taxon will serve as predator, prey, or competitor for only one other species. The result is that coevolution at a macroevolutionary level will generally be of a diffuse nature. What this means is that an individual species is not driven to extinction or prevented from speciating by one other species in particular. Rather, more than one species of one group affects each species of the other, coevolving group. Often the set of interactions will vary in time and space.

If the effect of the interaction is negative (as in competition, predation, or parasitism), it may represent a condition in which certain kinds of species elevate the rates of extinction of certain other kinds. The increased extinction rate may result from destabilization and contraction of populations through a web of biotic interactions, and although they contribute in this way, the interactions may not represent the sole, or even the final, cause of extinction of "victim" species. Species selection may also result from a suppression of speciation (by preventing certain kinds of divergent isolates from expanding to become full-fledged species). An elevated extinction rate and a depressed speciation rate could act in concert to bring about the decline of certain kinds of species. If the coevolutionary effect of the interaction is positive (as in provision of food or substratum), it may reduce the extinction rate by increasing or stabilizing population size or it may increase the speciation rate by improving opportunities for small populations to expand into new species. The diffuse nature of most species selection must be understood in this light. Ecologists are accustomed to contemplating competitive and predatory interactions between species pairs, but certainly many interactions follow complex spatial and temporal patterns involving many species. A slight elevation of mean extinction rate and/or reduction of mean speciation rate can tip the balance against a particular taxonomic or ecological group.

Coevolution resulting from species selection—what might somewhat awkwardly be termed *macrocoevolution*—may be unilateral or reciprocal. In other words, one higher taxon may change in composition in response to the presence of another, or both may change in a reciprocal fashion. The operation of coevolution at this level is not always easy to document. One of the classic alleged examples, the pattern of increase in brain size of mammalian carnivores and their ungulate prey during the course of the Cenozoic era (Jerison, 1973), remains unsubstantiated. Stratigraphic discrepancies, biased taxonomic sampling, and large variability in the index used to represent brain size leave the case for coevolution unproved (Radinsky, 1978).

Inadequate data represent only one potential problem. Difficulty in establishing causal relationships is another. Even where a pattern is consistent with the operation of coevolution, the process may remain largely hypothetical. The examples that follow rely on circumstantial evidence. It remains possible, though we believe unlikely, that each of the observed trends that we will discuss developed without influence from other taxa. One of the strongest forms of evidence that we can seek in support of a coevolutionary hypothesis is evidence that living representatives of the taxa in question are experiencing in the modern world the interaction believed to have dictated coevolution in the past.

EXAMPLES OF COEVOLUTIONARY SPECIES SELECTION

Coevolution at the level of species selection must often fail to provide raw material for effective species selection within small groups. Also, within small groups, species selection may be indistinguishable from phylogenetic drift (random changes in the species composition of a segment of phylogeny). The examples that follow represent large taxa, in which trends almost certainly result from the influences of biological agents on rates of speciation and extinction.

Predation and the Bivalvia

The great Mesozoic expansion of marine predators, including crabs, teleost fishes, and gastropods, had a major effect on marine biotas (Stanley, 1974, 1977, 1979; Vermeij, 1977). Siphonate bivalve mollusks are among the prey of these predators but have nonetheless undergone a major adaptive radiation since Paleozoic time. In the modern world, they outnumber nonsiphonate species approximately 10 to 1 (Figure 2). Siphonate taxa apparently owe their success to their ability to burrow rapidly and, in some cases, deeply (Stanley, 1968). Nonsiphonate burrowers, being for the most part sluggish animals that lie close to the sediment surface, are highly vulnerable to predation.

The macroevolutionary effects of predation on nonsiphonate bivalves seem to be expressed in rates of extinction. Late Cenozoic non-

339

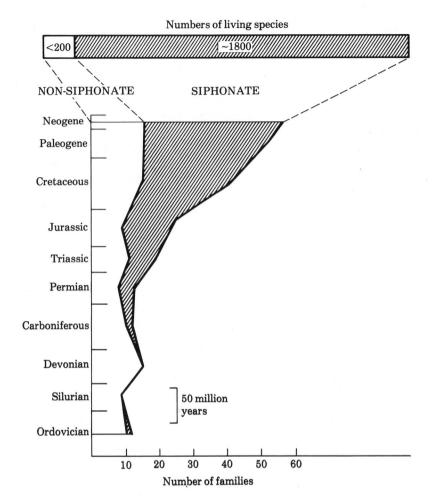

FIGURE 2. Diversity of nonsiphonate and siphonate Bivalvia. The lower part of the diagram shows the temporal increase in siphonate families. The numerical disparity between the two groups today is even greater at the species level (upper bar). (After Stanley, 1977; with data on species diversity from Boss, 1971.)

siphonate bivalves have experienced dramatically higher rates of extinction than siphonate bivalves: in areas that did not experience Pleistocene mass extinction, approximately 78% of Pliocene siphonate species survive to the present, compared to a survival rate of only 41% for nonsiphonate species (Stanley, 1981). The implication is that the great numerical expansion of siphonate species is largely due to their low rates of extinction. Probably, advanced predators have reduced the population sizes of species within the vulnerable nonsiphonate

340

taxa to the point where rates of extinction have been elevated.

The scallops (Pectinacea) exhibit a temporal trend that illustrates a similar phenomenon. Most species of scallops lie on the surface of the substratum, exposed to predators. The scallops' survival to the present in moderate diversity probably results from a special adaptation that provides them with an edge against predation: they can swim, even if only awkwardly. Sedentary epifaunal bivalves and brachiopods that, like the scallops, lack thick shells or spines are rare in subtidal settings today; in contrast, these forms were quite diverse in Paleozoic and Mesozoic seas. Despite the scallops' greater success, few of their species more than 5 million years old are alive today, which shows that the mean rate of extinction of scallops is extremely high compared to rates for other bivalve groups (Stanley, 1981). What is more, an average scallop species of Jurassic and Cretaceous age survived much longer—more than 10 million years (A.L.A. Johnson and A. Dhondt, personal communication, 1981). It seems likely that the expansion of modern predators has elevated rates of extinction considerably for scallops since some time in the Cretaceous period.

Calcareous red algae: crusts and herbivores

The fossil record of the calcareous red algal crusts reveals the dramatic expansion of one taxonomic subgroup (the coralline algae) during the Jurassic period and the simultaneous decline of another subgroup (the solenopores). The temporal pattern is so striking that a causal connection between the two events must be suspected a priori, and studies of the ecology and anatomy of the highly diverse and successful living corallines support the idea of a connection. The connection, however, is linked to the sudden appearance of herbivores capable of excavating calcium carbonate substrata.

Ecological studies have shown that calcareous crusts formed by coralline algae are competitively inferior to most other algal forms (Paine and Vadas, 1969; Paine, 1980; Adey, 1973; Vine, 1974; Wanders, 1977; Van den Hoek et al., 1978; Steneck, 1982a,b). The competitive deficiencies of coralline species probably result from their very low photosynthetic efficiency (Marsh, 1970), attributable to the fact that the photosynthetic cells of corallines are encased in calcium carbonate, and the fact that one side of the coralline thallus rests against the substratum.

How, then, have corallines succeeded in radiating in the presence of competitively superior fleshy algae? There is much evidence that corallines thrive today only because herbivores remove epiphytes that com-

341

pete with corallines for light and nutrients. Because calcareous algae are not known to possess effective allelopathic chemicals, the presence of herbivores is a necessity for survival of many species.

It is only because corallines possess certain anatomical traits that they can tolerate the grazing activity that removes their competitors. These traits, illustrated in Figure 3, have evolved polyphyletically and occur in a variety of genera and subfamilies (Steneck, in preparation). One of these traits is the presence of an *epithallus*, which is a layer of cells that overlies the *meristem* (the region of growth), and that protects the thallus from deeply grazing herbivores (Figure 4; Steneck, 1977, 1982a,b). Another important feature is the *hypothallus*, which is a layer of cells that underlies the coralline crust and, by elongation, causes the crust to grow laterally. Among coralline taxa, rate of lateral growth is positively correlated with the thickness of the hypothallus. The hypothallial cells generally lack photosynthetic plastids, and their

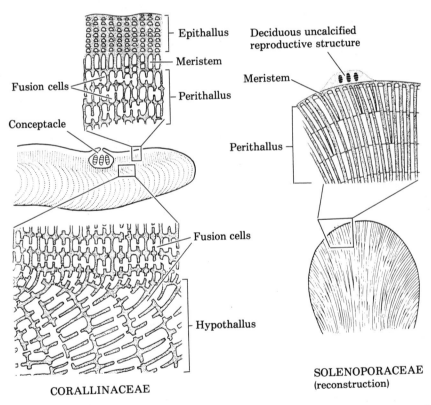

FIGURE 3. A comparison of the anatomy and morphology of coralline and solenopore algal crusts. (From Steneck, 1982b, in preparation.)

342

(A)

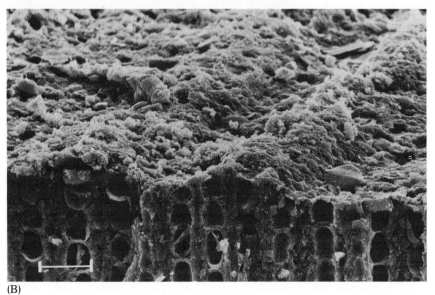

(B)

FIGURE 4. The intensity and frequency of grazing on Recent coralline algal crusts by excavating herbivores. (A) Chiton grazing marks on *Clathromorphum* in the Gulf of Maine. ×8.3 (B) Single grazing stroke of a limpet on the epithallus of *Clathromorphum*, also from Maine. ×22,500. Bar represents 10μm.

343

nutrition and gas exchange are provided for by intercellular conduits called *fusion cells*. *Conceptacles* are enclosed cavities within calcareous thalli that contain both sexual and asexual reproductive structures. Conceptacles protect their contents against deep-grazing herbivores (Steneck, 1977, 1982a).

The extinct solenopores, which became well established during the Paleozoic, differed from corallines in lacking an epithallus, hypothallus, fusion cells, and conceptacles (Poignant, 1977; Johnson, 1960). Just prior to the solenopores' extinction, some of their early Tertiary representatives possessed a hypothallus and perhaps also protected reproductive structures (Elliott, 1964). Solenopores reached their peak abundance and diversity during the Jurassic period (Figure 5), but even in their prime, they were not well suited for encrusting and were less common on organic reefs than corallines are today (Elliott, 1964; Wray and Playford, 1970; Wray, 1977; Flugel, 1977). Because no definite reproductive structures have been found within their skeletons, pre-Tertiary solenopores are thought to have had raised, uncalcified reproductive structures (Figure 3). Such structures could not have sustained severe grazing pressures.

The slowly growing, massive solenopores probably required grazers to remove fouling algal epiphytes, just as corallines do today. Throughout the Paleozoic era, the predominant grazers were probably small, generalized herbivores (malacostracans, rhipidoglossan mollusks, and primitive regular echinoids). As documented for mollusks by Steneck and Watling (1982), these forms were probably incapable of excavating calcareous substrata.

During Mesozoic time, the origins and radiations of herbivorous invertebrates with advanced feeding apparatuses radically altered ecological conditions for calcareous algae (Figure 5). The new, more effective excavating grazers included docoglossan limpets, which radiated after evolving from rhipidoglossan archaeogastropods, and echinoids with new kinds of mouth parts (Kier, 1974; Bromley, 1975). The oldest known grazing tracks on calcareous substrata are of Mesozoic age (Voight, 1970; Bromley, 1975). The increase in grazing pressure seems to have accelerated during the Cretaceous period (Bromley, 1975; Vermeij, 1977; Steneck, 1982b).

It seems reasonable to attribute the post-Jurassic decline of the solenopores to the intensification of grazing. Solenopores presumably benefited from moderate grazing pressure before Jurassic time but were unable to withstand the impact of grazers that excavated calcareous substrata and thus damaged solenopore tissues. The corallines, in contrast, were specifically equipped to deal with advanced grazers by virtue of their protective epithallus, hypothallus, and the conceptacles that enclosed the reproductive structures. In fact, the Jurassic-to-Recent radiation of the corallines may be attributed to the

344

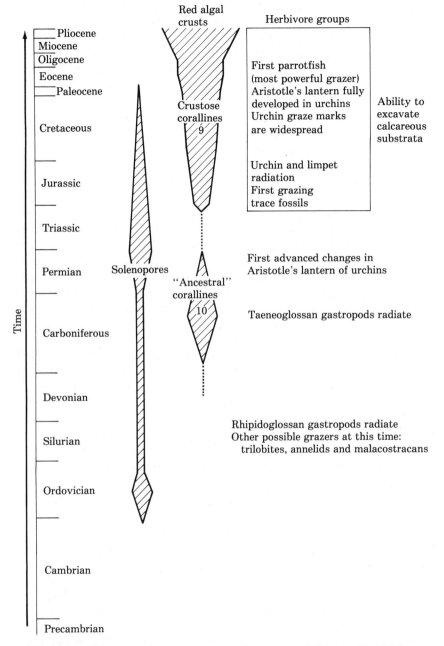

FIGURE 5. Patterns in the evolution of calcareous red algae and herbivores. (References supplied in Steneck, 1982b.)

intensification of grazing. Although modern corallines owe their success to grazers, which remove competitively superior fleshy epiphytes, the balance between beneficial and detrimental grazing is a delicate one. Beginning in Jurassic time, it seems to have shifted against the vulnerable solenopores and in favor of the resistant corallines.

Several more specific points deserve mention. One is that the coralline algae originated in Carboniferous time, long before their great adaptive radiation began. Yet their radiation since Jurassic time has not altered their basic anatomy with regard to the four key characters described earlier: epithallus, hypothallus, fusion cells, and conceptacles. Before the Jurassic period, these features may have offered the same advantages that they do today, but their adaptive significance was not fully exploited in an ecological sense until advanced grazers enhanced their value (Steneck, 1982b).

A second point is that the enclosure of reproductive structures in conceptacles may entail an energetic cost. Raised reproductive structures release their contents more effectively than do sunken ones (Steneck, 1982b). Zygotes or spores are commonly released by a breaking open of the entire roof of the conceptacles, and the contents of some conceptacles become trapped, never to be released (Adey, 1965).

Finally, although the success of corallines may have come in part through a release from competition with solenopores, this was probably a minor factor. Most solenopores grew as nodular masses. Because they had no hypothallus, they were not effective at growing laterally and encrusting calcareous substrata in the manner of modern coralline algae. The interactions of primary macroevolutionary significance have not been between these two groups but between each of these groups and two others: fleshy algae and excavating grazers.

Barnacles against barnacles

The preceding examples focus upon predation as an agent of macroevolutionary change. An example in which competition seems to have caused the decline of a particular group is the ascendancy of the balanoid barnacles over the chthamaloid barnacles, as described by Stanley and Newman (1980) and elaborated upon by Newman and Stanley (1981) in response to the skepticism expressed by Paine (1981). The case rests in part on circumstantial evidence that the chthamaloids have declined in diversity. Many of their component higher taxa are represented by very small numbers of relict and often geographically disjunct species. The balanoids, in contrast, have undergone a large radiation since evolving less than 50 million years ago. Three kinds of evidence support the idea that the balanoids have been competitively displacing the chthamaloids. One is that living balanoids characteristically defeat chthamaloids in competition by virtue of

346

strong fixation to the substratum and rapid growth (Hatton, 1938; Barnes and Powell, 1953). Another is that the competitive advantages of the advanced balanoids can be shown to relate to an inherent feature of the group—a tubiferous wall structure (Figure 6) that by its porous nature enables rapid growth and firm basal attachment. (An exception tending to prove the rule is *Notobalanus flosculus*, which is a South American balanoid species that, like the earliest fossil balanoids, lacks a tubiferous wall and loses in competition to the chthamaloid *Jehlius*.) The third line of evidence is that the bathymetric concentration of balanoid species lies between the two zones occupied by chthamaloids. Chthamaloids are concentrated in deep subtidal settings and in a high intertidal refuge but are virtually absent from the intermediate (low intertidal and subtidal) region where balanoids have their peak diversity. The chthamaloids have failed to evolve morphologically in any obvious way in response to the expansion of the balanoids. Had some speciation event within this group yielded a tubiferous wall structure, we might predict that the resulting clade would have expanded at a rate comparable to that of the radiating balanoids.

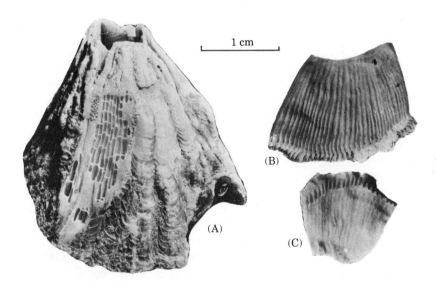

FIGURE 6. Tubiferous skeletal structure of a typical balanoid barnacle, *Balanus regalis* Pilsbry, from southern California and Baja California. (A) External view, with outer surface of rostral plate removed to expose longitudinal tubes and transverse septa. Also figured are interior surfaces and articulating margins of the wall (B) and basis (C). (From Stanley and Newman, 1980.)

ALTERNATIVE CHARACTER STATES

The examples that we have presented all entail selection between alternative character states—between the siphonate and nonsiphonate conditions in bivalves, between grazer-resistant and non-grazer-resistant morphologies in coralline algae, and between tubiferous and solid wall structures in barnacles. In all three cases, there is evidence of polyphyletic origins of the advantageous features, but there is little evidence of selection across a spectrum of character states—a graded series from one end-member condition to another. The analogue at the microevolutionary level is selection between two kinds of individuals within a polymorphic species.

Coevolutionary species selection operating across a spectrum of character states has not yet been well documented in the fossil record. Perhaps the best opportunity to analyze this kind of pattern will be found within the phylogeny of the Cenozoic Mammalia, which are represented by a fossil record of especially high quality and which include many taxa with skeletal features that lend themselves to adaptive analysis. When we are dealing instead with alternative character states that coincide with higher taxa, and especially when diffuse interactions are involved, there is some justification for viewing the process of selection as one that occurs between higher taxa rather than between species.

THE COARSE FABRIC OF MACROEVOLUTION

In contemplating ineffective coevolution within the well-established species of the Orellan guild of predators and the large-scale coevolutionary changes that are observable in sizable segments of phylogeny, we are impressed with the coarse fabric of coevolution. We suggest that coevolution is not, in general, the precise, effective, and persistent phenomenon that it has often been portrayed to be. When microevolution sets up coevolutionary relationships rapidly, it must do so haphazardly and unpredictably, as implied by Baker's previously cited example of the bee forming a unique, complex pollination interaction with each of two species of plants. This is not to say that natural selection does not usually govern the origin of the relationship. The point is that the timing and nature of the events leading to new relationships are highly unpredictable. For a given lineage, we do not know where, when, or in what subpopulation the next speciation event will occur. This means that there is a strong random component in the production of the variability upon which species selection can operate. When

348

coevolutionary change takes place on a macroevolutionary level, by species selection, it operates at the mercy of such vagaries of speciation. Persistent gaps, then, are to be expected within ecosystems. Thus, slow perfecting processes are not working efficiently in the world to yield precise coevolutionary trends that result in the constant restructuring of species and the rapid filling of ecological vacuums.

CHAPTER 16

THE DEER FLEES, THE WOLF PURSUES: Incongruencies in Predator- Prey Coevolution

Robert T. Bakker

Let us take the case of a wolf, which preys on various animals, securing some by craft, some by strength, and some by fleetness; and let us suppose that the fleetest prey, a deer for instance, had from any change in the country increased in numbers, or that other prey had decreased in numbers, during that season of the year when the wolf was hardest pressed for food. Under such circumstances the swiftest and slimmest wolves would have the best chance of surviving and so be preserved or selected—provided always that they retained strength to master their prey at this or some other period of the year, when they were compelled to prey on other animals. I can see no more reason to doubt that this would be the result, than that man should be able to improve the fleetness of his greyhounds by careful and methodical selection, or by that kind of unconscious selection which follows from each man trying to keep the best dogs without any thought of modifying the breed.

—Charles Darwin, 1896, *On the Origin of Species,*
6th Edition, pp. 110-111

CONGRUENCY IN DIRECTION

A notion of continuity rules Darwin's description of coevolution be- tween running predator and fleeing prey. Just as the breeder of hounds selects the slimmest and swiftest from each generation, so natural selection continuously favors the survivorship and reproductive success of the swifter deviants. Both natural and artificial selection drive progressive change in body form until an optimum is reached, namely, the swiftest predator possible within the constraints imposed by the canine morphogenetic pattern and by the competing functional demands of intraspecific competition, reproduction, and the other disparate components of fitness.

Large cursorial predators (those that pursue prey for long distances over flat terrain, Figure 1) and their cursorial prey seem to make an ideally simple couplet for the production of unreversed, lock-step coevolution. Most of the prey species pursued by these cursors are ungulates (that is, hoofed mammals) which respond to attack by fleeing at high speed. Despite eighteenth century beliefs that wolves and hunting dogs exhaust their ungulate prey through many miles of pursuit, modern observations have shown that if the predators do not catch their quarry in the initial rush of a kilometer or less, greater sustained speed saves the ungulate (Kruuk, 1972; Van Lawick-Goodall, 1971; Mech, 1970; Allen, 1979). Some ungulates are exceptions. Large, social species—water buffalo and muskoxen—counterattack aggressively; however, most ungulate species are smaller and rely nearly entirely upon flight. Warthogs and wild boar are double exceptions. Little heavier than the hyenas and wolves that hunt them, these formidable suids repel attacks by brandishing their tusks and by possessing an additional escape mechanism virtually unknown among deer and antelope: retreat down a burrow or into a lair within the thickets (Grundlach, 1968; Eisenberg and Lockhart, 1972). Coevolutionary responses of water buffalo or warthog to their predators should be a complex interaction among the competing functional advantages of flight, counterattack, and burrowing. But response from the ungulate majority (those with deer-like or antelope-like body forms) should be far simpler and more predictable: the pursued prey should evolve toward the longest, swiftest limb form possible. Likewise, the predator component of the coevolving system is complicated by divergent locomotor needs, but not so much that a parallel trend for increasing speed should be obscured. Burrows or dens of some sort are necessary for pup-raising among hyenas, wolves, and hunting dogs, and burrows give sanctuary to adults harassed by other predators (Kruuk, 1972; Mech, 1970; Van Lawick-Goodall, 1971). Burrows are inherited from warthogs, badgers, or marmots. However, some modification by the predators often is required, and wolves do dig their own dens when none are available for occupancy (Mech, 1970; Allen, 1979). Thus, the paws of the cursors must have a certain minimum strength and breadth for digging—requirements that prevent the acquisition of the elegantly tapered, elongated, and slender distal extremities of ungulates. Nevertheless, most of the locomotor moments in these predators' lives are spent walking, trotting, and running after prey, and selection for sustained speed should produce long-term trends congruent with those of the ungulate prey.

Followed backward through biostratigraphic time, the lineages of ungulates and all predators are remarkably congruent and converge to one common ancestral body type at the Cretaceous–Tertiary boundary, that is, at the dawn of the Age of Mammals (Matthew, 1937). Best described as a mechanical hybrid of opossum-like and raccoon-like features, this mammalian archetype had five spreading, clawed digits fore and aft and short, highly flexed limbs with great mobility at all the joints; the characteristics of the limbs permit a wide locomotor repertoire of climbing, digging, holding prey objects with the forepaws, and rearing up backward on the hindlimbs and thick, muscular tail. Directly from this Early Paleocene archetype came the successive waves of ungulates (Figure 1): first, the raccoon-like arctocyonids; and then from among these, the more progressive phenacodontids, which in turn produced the spectacularly diverse Perissodactyla (the rhino–tapir–horse league, now in its waning period). The Artiodactyla (pigs, hippos, deer, antelope, cattle, giraffes, and camels) probably issued from a separate arctocyonid stock and paralleled in a remarkably close way the Perissodactyla in the rapid evolution of elongated, compact, and stiff limbs for running. First among the cursorial predators were the mesonychids (also arctocyonid descendants), which evolved elongated, compact, and stiff cursorial limbs as early as the Middle Paleocene and which achieved a degree of cursorial locomotor advancement fully equal in many features to that of Recent hyenas and dogs by the Late Eocene (Figures 1–3). Thus, the arctocyonids produced both of the major cursorial ungulate orders and the earliest ungulate-pursuer. Other, less diverse and shorter-lived ungulate groups issued from among Paleocene arctocyonids in North America and Eurasia (Van Valen, 1978). On the island continent of South America, an endemic radiation of arctocyonid–phenacodontid descendants produced a splendid array of ungulates with elongated deer-like or horse-like limbs (Patterson and Pasqual, 1972). Mammals failed to supply distance-pursuit predators in South America until wolves arrived from North America in the Pleistocene. Giant, flightless ground birds were the only pre-Pleistocene pursuit predators on that continent. After the mesonychids, the second wave of running predators in North America–Eurasia–Africa were the hyaenodonts, which were descendants of primitive, tenrec-like Paleocene insectivores of the Family Palaeoryctidae. Hyaenodonts filled the wolf role in the

352

Oligocene and Early Miocene. From a different line of palaeoryctids issued the order Carnivora, which replaced the hyaenodonts as running predators in the Miocene and Pliocene. Primitive Carnivora differed little from the Paleocene mammal archetype in their short, sharply flexed limbs and spreading paws; among advanced Carnivora, the long-limbed, swift-footed configuration evolved independently in hyenas, dogs, and the running bear dogs (*Daphaenodon-Pliocyon*). Within the other island continent of the Cenozoic (Australia), the

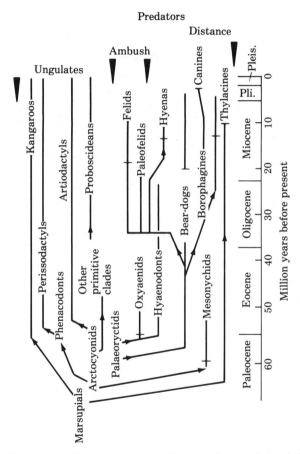

FIGURE 1. Phylogeny of ungulates and their predators; line of descent indicated by solid lines and arrows. The earliest occurrence of predator species with body weight ≥ 30 kg is marked by a bar. Vertical scale in millions of years.

353

marsupial equivalents of antelope and zebra are the kangaroos, which, though they hop instead of gallop, show distal elongation of the limbs and compaction of the hind toes not unlike that of the placental ungulate orders. Coevolving with these high-speed hoppers were the thylacine marsupials, wolf-like predators that were descendants of a Paleocene opossum stock.

When the direction of appendicular evolution is examined in detail, the intimate parallelism of ungulate and distance-pursuit predator is extraordinary and, at first glance, seems to corroborate the view that continuous coevolution maintained optimal phenotypes in both predator and prey. In every independent line of ungulate and mammalian pursuit predator, the evolutionary transformation from the primitive archetype followed the same morphological pathways (Figures 2 and 3): (1) angles at all the limb joints became less acute; (2) distal extremities of limb bone shafts became attenuated; (3) side toes, especially the first, became reduced or eliminated entirely (the dewclaws of some domestic dogs are remnants); (4) ball-and-socket joints became hinge-like, cylinder-in-groove joints; (5) the long bones of the paws (metapodials) became elongated, the phalanges became shortened; (6) claws became hoof-like. So close are the transformation pathways in predator and prey that primitive artiodactyl ungulate feet have been misidentified as those of advanced mesonychid predators (Scott, 1892). Certainly this striking case of iterative parallelism and convergence among six separate lines of mammalian pursuit predator and their ungulate prey is a powerful argument that observed long-term changes in the fossil record are the result of directional natural selection, not random walk through genetic drift.

But congruence of evolutionary direction does not necessarily imply that natural selection was working rapidly enough to maintain the best predator morphology at each instant of time. In other words, fossils show that pursuit predators were evolving in the correct direction to keep up with their prey, but the question remains whether the predators were evolving fast enough to keep constant the relative advantages of pursuer and pursued.

CONTINUITY VERSUS DISCONTINUITY OF RATE

Current imbroglios between advocates of punctuated equilibrium and phyletic gradualism reveal two extreme points of view about the fine-grain continuity of evolution—differences in opinion that have a long history.

My own opinion has long been—and I have many times given reasons for it—that there is always an ample amount of variation in all directions to allow any useful modification to be produced, very rapidly, as compared with the rate of those secular changes (climate and geography) which necessitate adap-

354

tation; hence no guidance of variation in certain lines is necessary. For proof of this I would ask you to look at the diagrams in Chapter III. of my "Darwinism," reading the explanation in the text. The proof of such constant indefinite variability has been much increased of late years, and if you consider that instead of tens or hundreds of individuals, Nature has as many thousands or millions to be selected from, every year or two, it will be clear that the materials for adaptation are ample [A. R. Wallace, 1898; in *Letters of Alfred Russell Wallace*, edited by J. Marchant, 1916].

Wallace shared Darwin's experience as naturalist–collector and his belief in the ubiquity of selection: all phenotypic properties varied within local populations; all properties had significant heritability; therefore, all properties were continuously fine-tuned by natural selection. Useless organs became eliminated; suboptimal organs became optimized. Cain and co-workers are recent eloquent proponents for the ubiquity, comprehensiveness, and efficacy of natural selection; optimization is seen as nearly instantaneous; suboptimal stages last but a fleeting moment (Cain, 1964). Quite as venerable a tradition, however, is the notion that increases in adaptive grade are not continuous in time and space. From the Cenozoic fossil record Deperet and Osborn described sudden intercontinental immigrations of ungulates and predators, which introduced exotic clades and grades far superior to the local resident lineages (Osborn, 1909; Deperet, 1905). Extinctions among endemic taxa on one continent were ascribed to competition and predation from the immigrant taxa.

Such biogeographical events suggest horizontal discontinuity in evolutionary rates; some continents produce faster progressive evolution than others, and therefore the evolutionary products of some continents displace endemics on other continents. Immigration of progressive exotics is widely accepted by paleontologists as a mechanism for stepwise advancement of adaptive grade within any one biogeographical province.

Generating far more controversy is the notion that the species is a tightly cohesive morphogenetic unit with stubbornly static boundaries, and that large, well-established species cannot produce significant adaptive change even if the species are far from the optimum theoretically possible for their bauplan (body design) within the local habitat. According to the view known as punctuated equilibrium or the rectangular model (Stanley 1979; Gould and Eldredge, 1977), the unit of selection is the species or the isolated population that is a potential species. Predator–prey coevolution provides excellent tests for several corollaries of punctuated theory. If significant genetic deviations are severely concentrated in small, isolated populations, then, assuming similar selection intensities, rates of long-term directional evolution should be diversity-dependent. Clades with many species, semispecies, and nascent species provide more units of deviation than clades with few species. Therefore, everything else being equal, ungulates should evolve more rapidly than their predators because the standing diversity of ungulates is now and always has been an order of magnitude higher than that of large predators. Only two cursorial predators—the spotted hyena and the hunting dog—occur today in the Serengeti amid thirty ungulate species. Second, rates of progressive evolution should be faster in large continents than in small continents because large biogeographical provinces support more species and more isolated populations. Therefore, when a barrier breaks and two formerly isolated continents exchange evolutionary products, the more progressive predators from the larger continent should displace part of the endemics in the small continent. To be sure, similar predictions have been made in traditional neo-Darwinian theory. Third, because the supply of deviant populations is limited for guilds with few species (such as large predators), the long-term rate of directional adaptive change should be very low and not congruent with Wallace's belief in rapid continuous optimization.

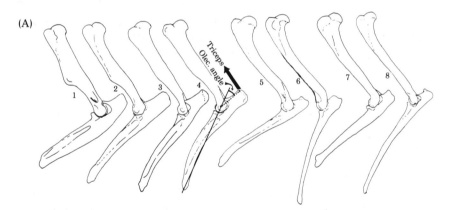

(A)

FIGURE 2. From flexible flatfoot to long-distance runner: the universal pattern of transformation from primitive mammal to specialized distance-pursuit predator, as illustrated by mesonychids. (A) Change in flexure at elbow. External view of left humerus and ulna, with the line of action of the triceps (arrow) shown as parallel to the humerus and perpendicular to the olecranon. *1, Arctocyon primaevus* (Late Paleocene; Early Paleocene mesonychids were similar); *2, Ancalagon saurognathus* (Mid Paleocene); *Pachyaena gracilis* (Early Eocene); *4, Dromocyon vorax* (Mid Eocene). In subsequent parts of this figure, the numbers refer to these species. For comparison, four extant predators are shown. *5, Speothos venaticus* (Brazilian bush dog); *6, Canis lupus* (wolf); *7, Panthera pardus* (leopard); *8, Acinonyx jubatus* (cheetah). (B) Anterior view of

ADAPTIVE GAPS, ADAPTIVE LAG

Speed indices

Fine-grain analysis of cursorial predators and their coevolving ungulate prey is possible because the functional anatomy of running on the flat is well understood, fossils from these guilds are common, and the structure of the ungulates permits reconstruction of the local habitat. Two indices are especially useful in assessing grade of distance-running adaptations (Figures 2 and 3):

1. Grooving of the proximal ankle joint (astragulo-tibia articulation). In primitive, generalized mammals, the astragulus presents a cylindrical surface that fits loosely into the distal end of the shank (tibia and fibula) and can rotate relative to the shank axis; these characteristics enable inversion and eversion of the sole, which are useful motions for climbing or moving over uneven substrata. In all fast-running ungulates and predators, the ankle joint limits motion to one plane; the astragular cylinder is deeply grooved and pulley-like, and

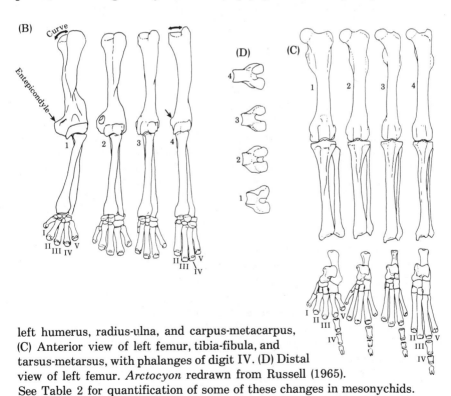

left humerus, radius-ulna, and carpus-metacarpus,
(C) Anterior view of left femur, tibia-fibula, and
tarsus-metarsus, with phalanges of digit IV. (D) Distal
view of left femur. *Arctocyon* redrawn from Russell (1965).
See Table 2 for quantification of some of these changes in mesonychids.

357

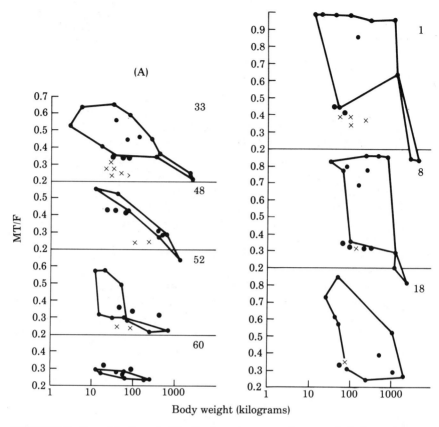

FIGURE 3. Speed indices in North American ungulates and their predators. (A) Body weight and metatarsal ÷ femur in successive faunas. Faunal references are given in the Appendices. Each dot or × represents one species: ●, distance-pursuit predators; ×, ambushers; •, ungulates. Number in upper right corner gives age, in millions of years. (B) Mean metatarsal/femur index. (C) Mean astragular groove index (depth of groove/width of trochlea) in successive fauna. For each index in B and C the mean value was computed for all

the tibia fits tightly around and into the astragulus. Restriction of movement to one plane also characterizes all of the other appendicular joints among cursors, flexibility being sacrificed to minimize the dangers of dislocation in high-speed runs.

2. Elongation of the metatarsus (long bones of the foot) relative to the thigh (femur). Generalized mammals have a relatively short, spreading metatarsus. Elongation of distal segments increases the stride length during the quick strokes of high-speed runs, although distal elongation reduces the mechanical advantage of the limb muscles for digging. Among extant mammals, the highest distance running speeds occur in species with deeply grooved astraguli and long extremities (Figure 4).

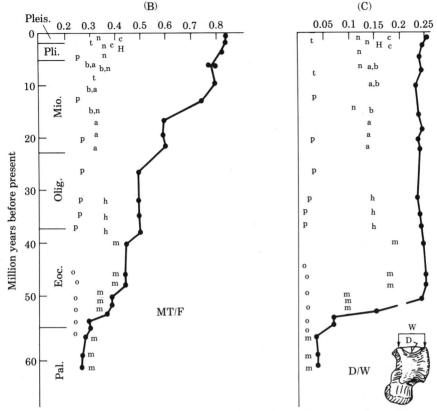

specimens from all species with a body weight 30 to 300 kg (the range that includes most extant ungulates and their predators). Fauna citations given in the Appendices. Predators are indicated by abbreviations: m, mesonychid; o, oxyaenid; p, paleofelid; h, hyaenodontid; n, neofelid; a, amphicyonid (*Daphaenodon-Pliocyon* linage); b, borophagine; c, canine; H, hyaenid; t, Tasmanian wolf (given for comparison with North American fauna). Ungulates are shown as dots connected by the solid line.

Predators seem to enjoy one scaling advantage—among fast species, with top speeds exceeding 40 $km \cdot h^{-1}$, predators can achieve a given top speed with a lower metatarsal femur index than can ungulates (Figure 4). Thus coyotes appear faster than elk, though elk have more elongated metatarsals. Quite probably this advantage is the result of the higher percentage of body weight allotted to limb muscle in predators, which do not require large, heavy, complex guts. However, the scaling advantage decays as the length of the chase increases. Wolves may overtake moose in the initial rush, but in long-distance pursuit, the moose usually escapes (Allen, 1979). A similar pattern can be observed among hyenas and zebra (Kruuk, 1972). It could be argued that the hunting success of hyenas and wolves would

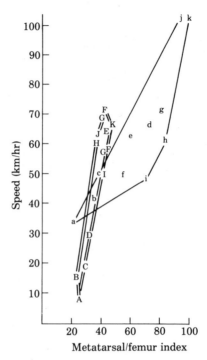

FIGURE 4. Running speed as a function of the metatarsal/femur index. Each point represents an individual animal charging or chased by a car and driven to maximum speed. A, Virginia oppossum; B, striped skunk; C, zorilla; D, European badger; E, grey fox; F, red fox; G, coyote; H, Mongolian wolf; I, brown hyena; J, spotted hyena; K, servalina; a, African elephant; b, white rhino; c, black rhino; d, Mongolian wild ass; e, Asiatic water buffalo; f, American bison; g, American elk; h, mule deer; i, desert bighorn; j, prongbuck; k, Mongolian desert gazelle. Predators enclosed by double line, ungulates by single line. Speed sources: Andrews (1922, 1939) H, a, d, j, k; Neal (1977) D; Burrows (1968) e; Kruuk (1972) J; Guggisberg (1966) b, c; Sweeney (1969) C; Sweeney (1973) I; Lee Rue (1978) h; Cottam (1937) E; Cottam and Williams (1943) f, g, i; Terres (1941) and Burrows (1968) F; Layne and Benton (1956) A; Verts (1967) B. Citations are listed in Appendix III.

increase markedly if they were equipped with limbs giving a higher sustained speed for long-distance pursuit.

Habitat index

Success and failure of pursuit and escape depends not only upon the speed adaptations of ungulate and predator but also upon the vegetative architecture. Dense woodland limits prey detection and long

360

pursuits and favors instead short ambushes by predators and escape by quick dashes into thickets or water by prey. Cursorial predators make up a progressively smaller proportion of the guild as cover density increases; a diversity of cats, not dogs, characterizes tropical woodlands (Lekagul and McNeely, 1977). Fortunately, one simple index gives a useful guide to the habitat selection of ungulate and predator. Among dense-habitat predators and ungulates, limb joints operate from sharply flexed modal positions and give quick acceleration and a capacity for bounds and leaps. Open-habitat species tend to have straighter, less-flexed limbs that give longer strides during sustained runs. All joint surfaces reflect this contrast in flexure, but the elbow is easiest to quantify (Figures 2 and 3): the elbow extensor muscles (the triceps group) pull upward on the olecranon process of the ulna; therefore, in highly flexed limbs the olecranon is bent backward less, relative to the ulnar shaft, than in less highly flexed limbs. Olecranon reflection angles indicate habitat use quite precisely. Leopards occur on the Serengeti, but hunt by short ambush from cover; cheetahs make much longer pursuits and are less dependent upon cover (Bertram, 1979). Cheetahs have much higher olecranon reflection angles than leopards (Figure 2). Brazilian bush dogs, which are hunters in woodlands and jungles, have less reflected olecranons than those of wolves, which are hunters of more open habitat (Figure 2).

Gaps and breaks

Startling gaps and incongruencies mark the 60 million year history of these three indices among pursuit predators and their prey (Figures 3 and 5). Within the first wave of ungulate—predator coevolution from 60 to 40 million years before the present (m.y.b.p.), both mesonychid predators and prey show increases in locomotor grade and a shift away from dependence upon cover. But the rate of change in the predatory cursors lags behind that of their prey. Mid Paleocene mesonychids were as fast as or faster than any prey of comparable body size. Mid Eocene mesonychids lost much of this initial advantage. This increasing morphological distance between predator and prey is the observed *adaptive gap*. A panselectionist view could explain away adaptive gap by invoking the principle of optimal adaptive compromise. According to this principle, the Mid Eocene mesonychids were the fastest possible configuration that still retained the paw strength necessary for digging burrows and performing the other nonrunning locomotor functions. Adaptive compromise is rendered less probable by the observation that modern wolves, which depend

361

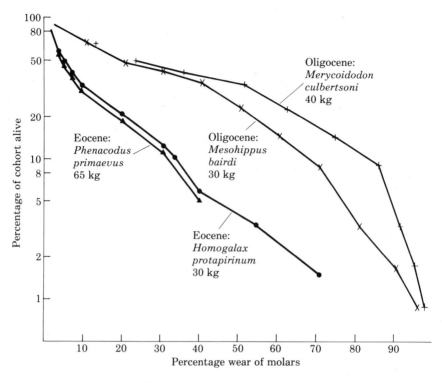

FIGURE 5. Tooth-wear survivorship curves for intermediate-size ungulates in the Early Eocene (▲,●) and the Mid Oligocene (×,+). Each curve was derived by measuring the percentage of the crown area worn through with dentine exposed in one hundred specimens from one narrow stratigraphic interval. Note that the Eocene curves are nearly linear, with far fewer dentally senile individuals than in the Oligocene curves.

upon burrows, exceed Eocene mesonychids in distal elongation of limb segments. *Adaptive lag* is here defined as the morphological distance between the observed structure and the optimal compromise possible within the given bauplan and habitat. Thus defined, the adaptive lag is that part of the adaptive gap that should be closed if natural selection were free to drive the evolutionary transformation toward the optimum phenotype. Mesonychid evolutionary patterns certainly raise suspicions that adaptive gap and adaptive lag did increase from Paleocene to Mid Eocene.

Fine grain resolution is possible for one segment of mesonychid evolution—the genus *Pachyaena*—which was the only distance pursuit predator for the first 3 million years of the Eocene. Adaptive gap was great between *Pachyaena gracilis* and medium-sized ungulates at the beginning of this interval. Nevertheless, *Pachyaena gracilis* shows no

362

significant change in astragular grooving for its entire history—an interval equivalent to half a million generations or more. Increase in adaptive grade occurs only when *Pachyaena* disappears and new, more advanced mesonychids—*Dromocyon* and *Mesonyx*—suddenly appear in late Early Eocene. Therefore, it becomes difficult to resist the conclusion that *Pachyaena* was not the best possible mesonychid phenotype and that the stasis in *Pachyaena* astragular grooving represents 3 million years of suboptimal stasis with no observed trend toward improvement.

MASS EXTINCTION AND SLOW REBOUND

A spectacular discontinuity marks the total extinction of mesonychids suddenly at the Eocene–Oligocene boundary, a time of intercontinental immigration. Many taxa of ungulates, rodents, and small predators dispersed from Eurasia to North America and vice versa (Osborn, 1909). Mass extinctions occurred at 10- to 30-million-year intervals all through the land vertebrate record (Bakker, 1975; Fischer and Arthur, 1977); most occurred when sea level fell and continental faunas were able to disperse widely. Sudden immigration of exotic species during a sea level fall not only can introduce new competitors and macropredators, but also new parasites and diseases of all types. Complete extermination of the mesonychids in North America, an event that emptied the large running predator guild, occurred *before* any other group appeared to fill this trophic role. Thus, direct competition seems unlikely as an agent of extinction. Because introduced diseases have wrought havoc among Recent African mammals (Sinclair, 1979), mass extinction by exotic pestilence must be considered seriously as an agent for mesonychid destruction in the New World.

A popular tenet of paleontological theory is that an empty guild invites rapid filling by evolution, and the mesonychids were in fact replaced by the hyaenodontids, which achieved large size and compact paws by the Early Oligocene (Mellett, 1977)—only a few million years after the last mesonychid disappeared in North America. Among ungulates, the trend toward increasing speed continued uninterrupted through the Eocene–Oligocene transition. Therefore, to retain the same degree of advantage in catching ungulate prey, the hyaenodonts that replaced mesonychids in the Oligocene would be required to have a locomotor grade higher than that of Eocene mesonychids. Large Oligocene hyaenodonts evolved from small, mongoose-like predators of the Mid Eocene. As they filled the vacancy left by mesonychid departure, large hyaenodonts did evolve more elongated, more compact limbs, but hyaenodonts never equaled the locomotor grade of Eocene mesonychids. The most advanced species of *Neohyaenodon* in the Mid Oligocene represents an adaptive grade inferior to that of Mid Eocene *Dromocyon*. Thus, the adaptive gap increased in the Oligocene when hyaenodonts replaced mesonychids. Ad hoc hypotheses could be invoked to explain away the slow evolution and low adaptive grade of hyaenodonts; for example, perhaps advances in hyaenodont social behavior and group hunting compensated for the gap in limb equipment. Or perhaps some special configuration of Oligocene topography favored ambush by hyaenodonts. I find such ad hoc explanations unsatisfactory. However, slow evolution of hyaenodonts conforms to punctuational theory because large hyaenodonts constituted a low-diversity clade that would produce few deviant populations.

After the large advanced hyaenodonts disappeared in North America in the Late Oligocene, two groups of Carnivora produced running predators; the bear dogs (amphicyonids) and true dogs (canids). First among the large predaceous canids were the hyena dogs (borophagines), which were big-headed creatures with massive jaws and dentition specialized both for slicing meat and crushing bones. Though highly modified for dismembering prey after they caught it, borophagines score very low on the metatarsal elongation index and had short-limbed proportions like those seen today in the Brazilian bush dog (*Speothos*). *Speothos* is a small hunter of dense tropical bush (Langguth, 1975), but the Borophaginae include species of wolf size that were the most common predators in woodland and woodland-savannah habitats throughout North America in the Miocene and Pliocene. Predaceous pursuit predators among the bear dogs all belong to the *Daphaenodon–Pliocyon* lineage, which emphasized the meat-slicing properties of the dentition, in contrast to most other amphicyonids. Predaceous bear dogs, ranging in bulk from the size of a wolf to that of a grizzly bear, were common through the Miocene and Early Pliocene, but none achieved the elongated, compact foot construction of Mid Eocene mesonychids. Because contemporaneous ungulates were but little inferior to Recent New World woodland species in locomotor grade, a literal reading of the fossil record would lead to the conclusion that borophagines and bear dogs suffered a greater disadvantage in catching prey than did the advanced mesonychids of the Eocene (Figures 2 and 3). Evolution of modern wolves and hyenas finally closed part of the gap in the last few million years of the Cenozoic. Long-limbed hyenas were widely distributed in North America in the Pleistocene but disappeared at the end of the Ice Age (Kurtén, 1963).

Rather than corroborating the notion of quick optimization and congruent coevolution, the fossil record of distance-pursuit predators shows that the predators fell behind their prey in the arms race of locomotor advancement (see also Chapter 15 by Stanley et al.). All the predator groups evolve in the predicted direction, toward greater speed and less dependence upon cover, but at rates that were far too slow to track the progress of their prey. Even among extant species, the observed adaptive gap between predator and ungulate is still large; it should not be assumed that Recent predators are the optimum grade possible within the constraints of habitat and ontogeny. Recent intensive study of the Serengeti ecosystem has produced a consensus that predation is not the immediate limiting control of wildebeest, buffalo, and zebra populations (Sinclair, 1977, 1979). Culling of ungulates by predators does not prevent prey populations from expanding to sizes where both inter- and intraspecific competition for grass and browse is severe. This apparent inefficacy of Serengeti predators could simply be the result of the inherent constraints placed upon predator optimization.

EVOLUTIONARY BALANCE OF PAYMENTS: THE LARGE CONTINENT ADVANTAGE

Throughout the Cenozoic, North America was intermittently isolated from Eurasia and Africa; faunas in the latter two provinces are very similar from the Miocene onward. Export of new predator clades clearly favored the larger, Old World landmass (Table 1). Most of the new predator types appearing in North America, both ambusher and cursor, seem to have an Old World origin. Wolves are the notable exception and are a probable North American export that has spread throughout Eurasia; extinct species invaded Africa and produced the hunting dog.

A second case of large-continent advantage is provided by South America and its great hunting birds. South America remained isolated from Holarctica until the late Pliocene, when a flood of immigrants from North America arrived (Patterson and Pasqual, 1972). Endemic South American mammalian predators were all marsupials; none were cursorial. Filling the large running predator role were the giant, flightless birds—the phororhacoids. These birds ranged in size from jackal-size to lion-size, and all had long, very compact, ratite-like hind limbs. Bird and mammalian functional anatomies are so fundamentally different that phororhacoid adaptive grade is hard to compare to that of wolves or hyenas. However, the results of the Late Pliocene dispersal strongly indicate that the South American forms were inferior to their Holarctic analogues. Within a short time interval after

365

TABLE 1. Immigration of predator taxa between Old World and New World.[a]

Time	Taxon	Guild	Direction
Late Eocene	Pterodontines	Distance pursuit	Old → New
Early Oligocene	*Hyaenodon*	Distance pursuit	Old → New
Mid Miocene	Pseudaelurine cats	Ambusher	Old → New
Late Miocene	Hemicyonines ⎫	Bear-like	Old → New
Early Miocene	Amphicyonines ⎭	Omnivores	Old → New
Late Miocene	Smilodontine cats	Ambusher: Sabre-tooth	Old → New
Pliocene	Homotherine cats	Ambusher: Sabre-tooth	Old → New
Pleistocene	Pantherine cats	Lion-like ambushers	Old → New
Pleistocene	*Chasmaporthetes*-type hyenas	Distance pursuit	Old → New
Pleistocene	Wolves	Distance pursuit	New → Old

[a]Data from Webb (1977); Kurtén (1957, 1963); Mellett (1977).

dogs arrived in South America, the phororhacoids were completely extinct.

Australia, isolated nearly completely for the entire Cenozoic, provides the third case. Marsupial foot architecture resembles that of placentals, and the endemic pursuit predator—the Recent Tasmanian wolf—shows an adaptive grade no higher than *Pachyaena gracilis* of the American earliest Eocene. Wolves, in the form of the domesticated dingo, arrived on the Australian mainland approximately 3000 y.b.p. (MacIntosh, 1975)—probably introduced by human colonists. At just about the same time, Tasmanian wolves became extinct on the mainland and were restricted to Tasmania (Keast, 1977). Feral dingos surely would have competed with marsupial wolves, for the two have widely overlapping ranges of body size and very similar dental–cranial adaptations.

The consistent pattern of large-continent advantage, especially the rapid displacement of the endemic South American and Australian predators by Holarctic invaders, demonstrates that the larger areas with more diverse faunas produce a certain type of adaptive superiority, as predicted by the diversity-dependent model of progressive evolution.

ADAPTIVE LEAPS AMONG AMBUSHERS

Ambush predators attack suddenly at close range from cover, sometimes stalking slowly and painstakingly before the final rush. Coevolution of ambush predators and their prey should be far less narrowly predictable than that of distance-pursuit predators, because ambush locomotor adaptations are not so completely concentrated for simple running over flat terrain. Leopards, cheetahs, and lions are sympatric ambushers in East Africa and have strikingly disparate adaptations (Schaller, 1972; Bertram, 1979). Leopards climb with ease, cache carcasses in trees, are solitary, and have a very short attack distance. Cheetahs overlap leopards widely in prey species, but have much longer, much straighter, more dog-like limbs, climb less easily, sometimes hunt socially, have much longer attack distances, and have spectacularly high speeds for several hundred meters. The average weight of lions is two or three times greater than those of cheetahs and leopards; lions climb only to escape heat or flies, hunt socially, have short attack distances, are relatively slow, but have immense forelimb strength for bringing down large prey. Cheetahs have the most elongated, least flexed limbs; leopards the least elongated, most flexed. The range of locomotor divergence among these three cats is greater than that between hunting dogs and hyenas, which represent two very different families of Carnivora.

Nevertheless, despite these complications, the fossil record of ambushers demonstrates significant overall trends in adaptive grade (Figure 3). Earliest among the specialized ambush clades were the oxyaenids, a sister group of hyaenodonts which appears at the Paleocene–Eocene boundary and disappears by the end of the epoch— victim of the same terminal Eocene mass extinctions which eliminated mesonychids. Oxyaenids were strictly terrestrial, with hoof-like claws and elbows incapable of the supination necessary for inverting the palm during climbing. Oxyaenid adaptive grade in distal elongation and astragular grooving was much lower than that of any modern cat. Replacing oxyaenids in the Oligocene and remaining common through the Miocene were the paleofelids, which were sabre-toothed predators with cat-like teeth, retractile claws, and great powers of supination that conferred excellent climbing ability. But Oligocene paleofelids score no higher in speed indices than their ecological predecessors, the oxyaenids. Because distal limb elongation and astragular grooving show little improvement over the oxyaenid condition, Oligocene paleofelids show a greater gap in locomotor grade behind the contemporary ungulates than existed between oxyaenids and their prey (Figure 3). Modern large cats with leopard and lion-grade limbs become common in the Late Miocene, and the observed gap between ambusher and ungulate decreases a bit.

SURVIVORSHIP CURVES AND ADAPTIVE GAPS

Can this hypothesis of large adaptive lag in the Oligocene be tested in some quantitative ecological way? Survivorship curves provide one test. If predator grade is high compared to that of the prey, then one would expect that all age classes of prey would be vulnerable to predators, even the vigorous young adults. If predator grade is relatively low, then one would expect the young adults to be underrepresented among predator kills, and the more vulnerable classes—the very young and very old—to be overrepresented. Survivorship curves can be constructed for fossil samples by using toothwear classes to construct age categories if the processes that produced the fossil sample from the live community are known. For Eocene and Oligocene samples in North America, most specimens come from flood-plain deposits and seem to represent cumulative mortality (Bakker and Bickert, 1983; Bown and Kraus, 1981). The best gauge for age estimation among the Eocene and Oligocene ungulates with low-crowned teeth is the area of dentine exposed by wear. Data from extant mammals suggest that the area of wear increases approximately linearly with elapsed time (Bakker and Bickert, 1983). Because juvenile specimens suffer disproportionate destruction from predators and scavengers, Eocene and Oligocene survivorship curves can be constructed only for adults. Curves for Oligocene ungulates show a much higher young adult survivorship than that seen in Early Eocene ungulates (Figure 5). Oligocene curves are convex-up, like that of Serengeti buffalo (Sinclair, 1977); Eocene curves are steep and linear and demonstrate a high, nearly constant probability of mortality throughout adult life. Furthermore, far fewer of the Eocene ungulates reach dental senility—when 70% or more of the crown enamel is removed by wear. These patterns strongly suggest that Eocene predators did have less difficulty pursuing and catching young, healthy adult ungulates than did Oligocene predators. Miocene survivorship curves resemble those of the Oligocene (Voorhies, 1969, Kurtén, 1953). Comparison of survivorship curves certainly does not corroborate the notion that predators and their ungulate prey have coevolved tightly and continuously, with the relative advantages of pursuer and pursued constant through geological time.

ABSOLUTE RATES OF WHOLE-BODY EVOLUTION

The intensity of selection depends, in turn, on the very complex relationships between the organism and its environment. Selection will tend to produce modal change in population only if something in this relationship disturbs existing adaptation. In most well-recorded sequences there is reason to believe that the rates of evolution are far from a maximum, although this is

368

evidently not true of all evolutionary events. This is indirect evidence that in such sequences adaptation keeps up with the environment [G. G. Simpson, 1953; *Major Features of Evolution*, p. 147].

Simpson observed that the usual rate of evolution in the fossil record was far slower than the theoretical maximum indicated by experimental and theoretical genetics. Therefore he concluded that most fossil lineages were evolving fast enough to keep up with their changing environments. As shown above, in selected limb indices, large predators were not keeping up with their prey. If the total amount of selective mortality among these predators was very small, then the predator's failure to track their prey's evolution is paradoxical. But perhaps the analysis of single limb indices grossly underestimates the total amount of evolution within predators. Perhaps predators were, in fact, evolving as fast as they could under very heavy directional selection which was acting over thousands of separate biomechanical functions, each controlled by a separate set of genes. In such a case the total number of selective deaths necessary to produce the total phenotypic change could be large, even though for any one feature, taken in isolation, the selection would appear weak.

How fast and how complicated was the total phenotypic change in these predator–prey systems? During the entire transmutation of each distant pursuit clade from opossum-like archetype to hyena-like predators, every limb segment, every joint surface was remodeled. Hundreds of ligaments and muscle attachments were rearranged. At first glance, an enormous amount of phenotypic change requiring many generations of intensive selection would seem necessary to produce this total musculoskeletal transformation. If each joint surface were under the control of a separate morphogenetic command, then the total musculoskeletal change indeed was complex and required much selective death. However, nearly all the changes fall into one or another of a very small number of simple ontogenetic patterns. Elongation of distal shafts occurs late in mammalian ontogeny; thus, puppies, horse embryos, and human infants seem stubby-limbed (Figures 2 and 6). Elongation of all distal limb shafts in the evolution of mesonychids or any other cursorial lineage would require only a simple heterochronic acceleration of this basic mammalian ontogeny. Puppies and horse embryos seem flexed-limbed and flexible-footed compared to adults because ontogeny produces both compaction of the joint action (thereby concentrating motion in one plane) and reduction of the flexure of all the joint angles (Figure 6). Thus, the evolutionary trend toward compact, hinge-like joints with reduced flexure can repre-

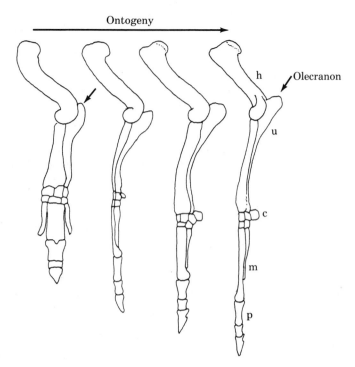

Ontogeny

h   Olecranon

u

c

m

p

FIGURE 6. Ontogenetic change *in utero* in the domestic horse. External view of left forelimb, redrawn from Ewart (1894). Limbs drawn to unit humeral length; far left, 20 mm embryo; far right, 350 mm embryo. H, Humerus; U, ulna; C, carpus; M, metacarpus; P, phalanges. Note the progressive ontogenetic lengthening of distal shafts, the mediolateral compression and reduction of side toes, the reflection of the olecranon, and the transformation of the last phalanx from a claw-like to a hoof-like core. These ontogenetic changes are congruent with the evolutionary transformation of adult shape from primitive arctocyonids to their advanced mesonychid descendants, as illustrated in Figure 2. (J. C. Ewart, 1894. *Anat. Physiol.* 28: 236–256; 273–369)

sent another simple result of ontogenetic acceleration. Such ontogenetic compaction can eliminate a toe position. The first toe, fore and aft, is slightly divergent in primitive mammals and embryos; developmental compaction tends to squeeze out these toes, and they are reduced or even missing in advanced dogs, hyenas, and mesonychids and nearly all ungulates. Because a single heterochronic change will effect mechanical transmutation of many joint surfaces and muscle attachments from shoulder to paw, hip to toe, only a dozen or fewer separate changes are needed to complete the transformation of the primitive, flexible-limbed, flatfooted archetype into the advanced cursor (Table 2).

If this minimalist morphogenetic model for whole-body transmutation is accepted, and if we assume a constant rate of selection for the

TABLE 2. Ontogenetic categories of phylogenetic transmutation.

	63 m.y.b.p. *Eoconodon* stage	59 m.y.b.p. *Ancalagon* stage	53 m.y.b.p. *Pachyaena gracilis* stage	49 m.y.b.p. *Dromocyon vorax* stage	Total change in Mesonychids 63–49 m.y.b.p. (Final ÷ initial)	Wolf (Canis lupus) s.d. ÷ mean for this shape index (n = 25)	b = distance from Z to truncation point B in Figure 7 for total change in mesonychids, using wolf s.d. for calculation
Distal shaft elongation + attenuation							
A. Metatarsal ÷ femur	0.227	0.309	0.336	0.400	1.76	0.06	4.70
B. (Tarsus + metatarsus) ÷ femur	0.459	0.518	0.564	0.664	1.45	0.06	4.79
C. Ulna length ÷ ulna width above distal radio-ulnar joint	7.10	13.0	13.0	15.6	2.20	0.05	4.59
Mediolateral compaction							
D. Distal femur breadth ÷ ant.-post. depth	0.885	0.885	1.08	1.29	1.46	0.04	4.70
E. Humerus shaft breadth ÷ entepicondyle breadth	1.92	2.85	4.50	6.00	3.13	0.04	4.46
F. Femur shaft breadth ÷ proximal width metatarsals	0.440	0.579	0.600	0.582	1.32	0.06	4.85
Reduction in flexure							
G. Angle between ulna shaft and olecranon	1°	5°	12°	15°	15	0.30	4.71
Change in joint shape from convex, ball-in-socket, to concave, grooved cylinder							
H. Degrees of arc in humeral head curvature in mediolateral direction as seen in posterior view.	104°	50°	20°	10°	0.097	0.20	4.66
I. Depth of astragulus groove ÷ astragulus trochlea breadth	0.04	0.04	0.12	0.18	4.5	0.10	4.40

371

entire history of mesonychids, then the total amount of selective death required to produce an advanced cursor is astonishingly small. Lande (1976) developed an elegantly simple expression for calculating selective deaths needed to produce observed skeletal changes, assuming severely directional, truncating selection (Figure 7):

$$b = \pm \ (-2 \ln\{\sqrt{2\pi}[|\ln(x_f/x_i)|/\sigma](h^2t)^{-1}\})^{1/2}$$

where x_i is the initial and x_f the final dimension; b is the distance from the truncation point to the mean x, measured in units of the standard deviation of x; σ is the standard deviation of $\ln x$; h^2 is the heritability of the property measured; t is the elapsed time in generations. For large mammals the coefficient of variation is relatively constant at approximately 5% of the mean for any length or breadth measurement of bones and teeth (Ognev, 1963), and heritability is high—approximately 0.5 (Lande, 1979). Thus, Lande's expression can be simplified to

$$b = \pm \ \{-2 \ln[2.507(|z|/0.0275t)]\}^{1/2}$$

where z is $\ln (x_f/x_i)$.

Shape change in mesonychid evolution can be expressed as the change in one measurement relative to another, for example, increase in metapodial length compared to femoral length. In Figure 2 I have illustrated shape change for mesonychids, because their fossil record preserves the most complete history of the transformation of the archetypal flexible-flatfooted omnivore into a slim-limbed cursor. The four species illustrated were chosen because they represent stages in this transmutation spaced approximately 5 million years apart and because they have approximately the same body size (equivalent to that of a large wolf) thus eliminating allometric complication. *Arctocyon primaevus*, representing the first stage, is actually a Late Paleocene arctocyonoid, but its skeleton differs little from the more fragmentary remains of the Early Paleocene mesonychids (*Eoconodon* spp.). Whole limb reconstructions for the first three stages are all composites, that is, they are based on several specimens for each species, but the observed range of shape variation within each species is small. Total change in shape in each limb proportion, (x_f/x_i), is only approximately 1.5, which amounts to an evolutionary change of only eight standard deviations, assuming a coefficient of variation of 5%. For simple truncation selection (Figure 7), only 1 death per 10^5 per generation is needed to produce this change, over 15 million years, assuming a generation time of 5 years. If the separate morphogenetic categories of shape change are under totally separate genetic control, then the total selective deaths necessary for whole skeleton transformation is the sum of the deaths for each category. If only a dozen or so categories are necessary for the total change, then only a dozen selective deaths per 10^5 per generation are required to explain all musculoskeletal evolution (Table 2).

372

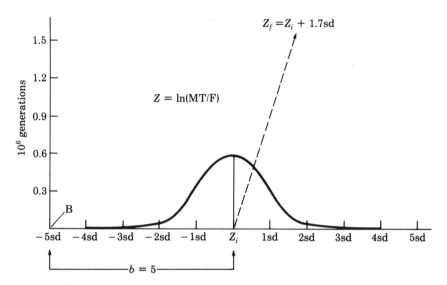

FIGURE 7. Simplified model of continuous, gradual evolution of relative metatarsal length in mesonychid evolution, assuming (1) a constant coefficient of variation in metatarsal/femur ratio (= 0.05); (2) a constant generation interval of 5 years; (3) truncation selection; (4) polygenic control of phenotype; (5) panmixia; (6) large population size. Simplified from the model of Lande (1976). The ln(metatarsal/femur) = Z. At any one time, the distribution of Z is normal; the fitness of all individuals to the left of B is zero and that of all individuals to the right of B is unity. The distance from the population mean to B is b and is measured in units of the standard deviation of Z. To illustrate the method, B is located for change in Z between 60 m.y.b.p. and 52 m.y.b.p., an interval of 1.6 million generations. Here the truncation point is five standard deviations from the population mean. Therefore, if the assumptions are correct, only one death per 10^5 per generation need be selective to produce the observed change in Z.

Surely total phenotypic change in the history of mesonychids or any other cursorial line was more complex than this; behavior, physiology, coat color, and other phenotypic attributes must have suffered changes. Nevertheless, the musculoskeletal system is approximately 60% of total body weight in cursorial mammals (Crile and Quiring, 1940) and is the most complex system in gross anatomy. It is sobering to consider that a model for total phenotypic change that requires even as many as 10^2 separate developmental categories would still need only one selective death per 10^2 or 10^3 per generation during mesonychid history—very weak selection by the standards of animal breeders.

373

Even if this minimalist model underestimates the necessary selective deaths by an order of magnitude, the results still are paradoxical. Mesonychids evolved rapidly compared to other mammalian lineages, and mammals evolved rapidly compared to most other organisms (Stanley, 1979; Simpson, 1944; Lyell, 1830). Yet the intensity of selection necessary to explain mesonychid evolution, assuming constant gradualism, is so low that it is difficult to believe that random genetic events would not swamp directional selection (Lande, 1975). Of course, the relief from this paradox is obvious—the assumption of constant rate must be incorrect. The fossil record for mesonychids does demonstrate remarkable cases of stasis—the primitive genus *Dissacus* persists without significant change in limbs from 60 million years b.p. to 50 million years b.p.; *Pachyaena ossifraga* and *P. gracilis* show no change in limbs or craniodental structure between 53 and 50 million years b.p. Such stasis is the rule for Early Cenozoic mammals with good fossil records (Schankler, 1982; Bakker and Schankler, 1983), but for mesonychids stasis is an especially intriguing pattern because it can be argued that *Dissacus* and *Pachyaena* were suboptimal stages in the long-term evolution of cursorial adaptations.

PARENT-DAUGHTER OVERLAP AND STEPWISE OPTIMIZATION: CUTTING TEETH OF OXYAENIDS

Mesonychids show chronic stasis and offer an example of progressive evolution stalled at a suboptimum for 3 million years. Unfortunately the fossil record has not yielded a sample demonstrating the coexistence of the last population of *Pachyaena* and early populations of the more advanced *Mesonyx* or *Dromocyon*, and thus the case for displacement of parent taxon by daughter taxon is not yet firm. An ideal case of a punctuated progression would show a long-lived, suboptimal parent species suffering final extinction as a new, more progressive daugher species, product of allopatric speciation, invades the parental range. Oxyaenid predators from the Early Eocene Willwood Formation provide just such a paradigmatic case history (Figure 8).

Limbs change little in oxyaenid history, but the flesh-cutting teeth display a remarkable transformation over 5 million years from a primitive, insectivore-grade mechanical design to an optimized blade identical in form and function to that of modern cats (Figure 9). In the earliest oxyaenids, upper and lower molars occluded in the double-wedge embrasure pattern typical of all primitive placentals. As the wedges shear past one another, the sharp edges act like a pinking shears and cut a zigzag pattern across the food item, and finally the internal upper cusp crushes the item against the basin of the lower tooth. Such multiple action is well suited for insectivores and omnivores, which must deal with relatively small food items, but severe and ubi-

374

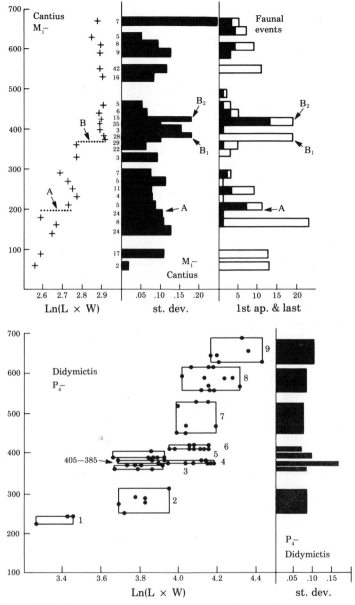

FIGURE 8. Displacement of parent species by sudden immigration of daughter. Vertical scale in meters of the Willwood Formation. First appearances of species among all genera indicated by white bars; last appearances by black bars; sample interval = 20 m. Size of *Cantius* and *Didymictis* expressed by Ln($L \times W$) of a central cheek tooth. For *Didymictis*, each specimen is shown as a dot. For the much larger *Cantius* sample, crosses show means; sample size is given by small numbers. Note sudden increase in standard deviation (st. dev.) during the 370–420 m events and the net size increase produced during these events. An earlier event may be recorded for *Cantius* at 200 m (A). From Bakker and Schankler (in preparation).

375

quitous competition among large predators drives dental evolution away from the double-wedge occlusion and toward a much simpler, single-action mechanism for cutting large chunks of flesh off a prey carcass. Large predators that kill large prey are especially vulnerable to carcass theft by a host of competitors of many sizes. A single zebra carcass represents an extraordinary local concentration of protein and calories; if the predator does not consume the carcass immediately, vultures, lions, hyenas, and three kinds of jackals begin to congregate to steal. Intraspecific competition for carcasses can be unrelenting even among littermates of lions and hyenas, although hunting dogs display more kin-selected altruism (Bertram, 1979; Kruuk, 1972; Schaller, 1972; Van Lawick-Goodall, 1971; Allen, 1979).

One simple mechanism to maximize the rate of carcass consumption is to minimize the time spent cutting the flesh into chunks small enough to swallow. Zigzag cuts with the "pinking shears" of primitive teeth are not optimal for this task. All the most specialized, extant, purely carnivorous large predators have an identical cusp pattern for meat-slicing: upper and lower cutting teeth have only two major cusps each; both upper and lower cusp pairs are aligned parallel to the side of the jaw and form opposing, guillotine-like blades for cutting large chunks off big carcasses. All the primitive complexity of lower molar structure is sacrificed to produce these blades: the internal-posterior cusp (metaconid) of the lower anterior wedge is lost; the entire 3-cusped, basined, posterior wedge is reduced to a tiny, featureless lump (Figure 9). Iterative convergence toward this simple blade function is excellent proof of optimization: from a great variety of insectivorous and omnivorous ancestral patterns, this dental form has evolved in Tasmanian wolves, oxyaenids, hyaenodontids, limnocyonids (a sister group of the hyaenodontids), paleofelids, true cats, hyenas, some bear dogs, weasels, wolverines, the Malagasy cat-civet (*Cryptoprocta*), and others. A few large predators, such as wolves, retain a metaconid and a basin in the lower molar, but

FIGURE 9. Suboptimal stasis in the evolution of oxyaenid cutting teeth. (A) ▶ Occlusion in an intermediate (*Oxyaena forcipata*, left) and an advanced species (*Patriofelis ulta*, right); left external view of the last two molars separated and occluded. (B) Occlusal view of right lower and left upper molars. Upper second molar is lost in *Patriofelis*. (C) Outline of uppers and lowers in occlusion, uppers in heavy outline. m, metaconid; b, crushing basin. (D) Stratigraphic record of the metaconid reduction in a parent–daughter species pair. *O. forcipata* and *O. ultima*, Willwood formation, Early Eocene. Vertical scale in meters of sediment; 1 m = 5 × 10^3 years. Note interval of parent–daughter temporal overlap. *O. intermedia* is a very early, very primitive species known from only a few specimens.

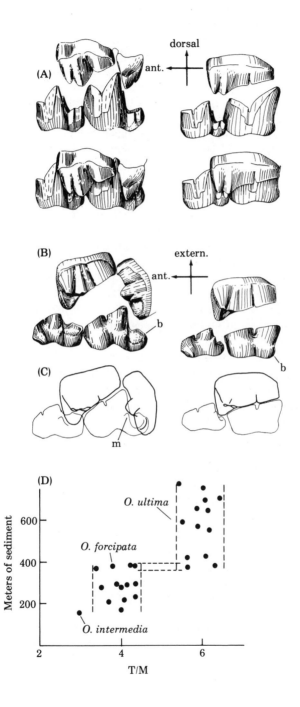

these species do consume small prey and fruit at times, food items demanding the retention of some crushing function (Allen, 1979).

Oxyaenids achieved the two-cusped, cat-like molar by the latest Early Eocene, 5 million years after the debut of the family in the Late Paleocene. Three or four successive waves of species show progressive decrease of metaconid and basin. Several of these species seem to last a million years with little net change, and rich collections from the Willwood Formation show the temporospatial overlap of the last sample of *Oxyaena forcipata* and the first sample of a more advanced daughter species, *Oxyaena ultima*. Metaconids were fully 50% more reduced in the daughter species, and parent and daughter co-occur in the same localities for 10 m of sediment (representing some 50,000 years), after which the parent disappears. Parent and daughter must have overlapped in prey species, because the two *Oxyaenas* were virtually identical except for the more advanced meat-slicing blades in the daughter. *Oxyaena forcipata* had existed for 10^6 years with no observed decrease in metaconid size. Retention of relatively large metaconids in *Oxyaena forcipata* cannot be explained away as adaptation for seasonal omnivory, because the lower molar basins of all oxyaenids are delicate and sharp edged and lack the strong, low cusps seen in wolves and other predators that crush fruit. The *Oxyaena* story seems to be a paradigmatic example of chronic suboptimal stasis and sudden replacement of a parent species by its more progressive daughter.

It has been claimed that the best mammalian fossil records show the gradual transformation of skull and teeth from primitive parent to more progressive daughter species. However, a recent review of the superb Willwood sample has revealed several cases like that of *Oxyaena*, where sudden immigration of a progressive daughter species coincides with extinction of a conservative parent. Two such daughter-parent displacement events occur in the same 380–410 m level of the Willwood which records the *Oxyaena* event (Figure 8); the wolverine-like predator *Didymictis* and the lemur-like primate *Cantius* both show a sudden increase in size variation, indicating coexistence of resident parent and immigrant daughter of larger size, and then a sharp increase in size with a decrease in variation, indicating extinction of the smaller parent and complete replacement by the larger daughter (Events B_1 and B_2 in Figure 8). In both *Cantius* and *Didymictis* increase in size is interpreted as a progressive adaptation which confers superiority during interspecific interaction, because among frugivores, such as *Cantius*, and predators, such as *Didymictis*, a 5 or 10% size increment can give a marked advantage in direct confrontations over food, territory or refuges. All three events, in *Oxyaena*, *Cantius*, and *Didymictis*, occurred at a time when a wave of immigration, followed by a wave of extinction, restructured the entire mammalian fauna. The

378

Willwood record leads me to suspect stasis, immigration and daughter-parent displacement as the three principal components in long term evolution.

PARADOX OF SLOW, STUTTERING COEVOLUTION

Coevolutionary patterns among large mammalian predators and their prey are paradoxical: (1) Predators evolve in the predicted direction, paralleling the locomotor adaptations of ungulates, but at such low rates that the adaptive gap between predator and prey increased for long segments of geological time. (2) Survivorship curves strongly suggest that when this adaptive gap was large, young, healthy adult ungulates were relatively immune from predation. (3) When mass extinction suddenly removed long-lived predator lineages, other predator groups evolved in the predicted direction to fill the void but at such low rates that gaps between predator and prey increased. (4) On large continents the rate of progressive evolution, although slow, is faster than on small continents. (5) A minimalist–gradualist model for total phenotypic change requires only a few selective deaths per thousand per generation to explain long-term evolutionary rates in the predators. (6) Well-sampled predator lineages show long periods of stasis at suboptimal morphology. (7) Long-term directional trends among predators were built up by successive stepwise replacements; in the best-sampled lineages steps occur when a more advanced daughter species suddenly appears and replaces the parent species.

Ubiquity and continuity are two of the most important properties of evolutionary models. As outlined above, simple neo-Darwinian models of ubiquity and continuity are falsified by both overall pattern and detailed replacement sequences in the fossil record. To preserve the idea that predators evolve rapidly (in 10^2–10^4 generations) to track changes in prey escape adaptations and to maintain optimal predatory efficacy, a series of ad hoc explanations must be generated to deal with the observed adaptive gaps between predator and ungulate in the fossil record. Because I am uncomfortable with such hypotheses, I choose the simpler course of concluding that many, perhaps most, of the extinct predators were not evolving fast enough to track their prey closely. A corollary of this conclusion is that most species, most of the time, are not fine tuned, tightly integrated, and optimized in all components by an omnipotent and omnipresent natural selection. Rather, most species are adequate, sometimes barely adequate, compromises of heritage and adaptation to local conditions, compromises that survive only until a better compromise appears.

I do not wish to imply that all or even most evolutionary biologists over the last century have advocated views of instantaneous optimization and taut coevolution of predator and prey. However, the quotes from Darwin, Wallace, and Simpson (given earlier) are fair representatives of a major theme in paleontology and biology, namely that changes in the environment regulate evolutionary rates and that enough variation exists in all traits to permit the evolving population to track environmental change with little adaptive lag. Most current students of evolution would accept some mechanisms that inhibit the evolution of optimal phenotypic compromises. The fossil record of mammals shows that stasis for 10^5 or 10^6 generations is the rule, not the exception, for species (Schankler, 1981, 1982; Bakker and Schankler, 1983) and that in many cases chronic stasis preserved phenotypes probably very far from the optimal compromise possible within the given habitat. Ever since Darwin, most evolutionary theory has concentrated on providing explanations of why populations should evolve. The fossil record demands more emphasis on explaining why populations do not evolve, even when considerations of functional morphology argue they should. Perhaps, as a complement to our Society for the Study of Evolution, we need a Society for the Study of the Prevention of Evolution, to explore explanations of the apparent rarity of major adaptive change.

APPENDIX I

The arguments presented in this chapter require precise estimates of body weight in extinct mammals. Estimates were derived from the relationship in extant species of weight and head-body length. W, body weight; L_{HB}, length of head and vertebral column to base of tail. The weight of mammals can be described by the function $W = k(L_{HB})^3$. The clusters of parallel lines shown in the figures have a slope of 3 and illustrate some values of k.

For fossil species, W was calculated using the following k values: Mesonychids, *Hyaenodon*, bear-dogs, borophagines, extinct wolves: 25. Oxyaenids, paleofelids, neofelids: 20. Arctocyonids, phenacodonts: 40. Horses and tapirs: 30. Titanotheres, rhinos, proboscideans, anthracotheres, entelodonts: 60. Other artiodactyls: 30.

Data from the following: W. Ansell, *The Puku* 2, 10 (1964); 2, 14 (1964); 3, 1 (1965); 5, 1 (1969). D. Davis, *Field Zool. Mem.* 3, 1 (1964). R. Meinertzhegen, *Proc. Zool. Soc. Lond.* A, 108, 433 (1938). W. L. Robinette, *The Puku* 1, 207 (1963). W. L. Robinette and G. Child, *The Puku* 2, 84 (1964). R. Sachs, *E. Afr. Wild. J.* 5, 24 (1967). S. P. Young and E. Goldman, *The Wolves of North America*, New York, Dover (1964); *The Puma*, New York, Dover (1964).

380

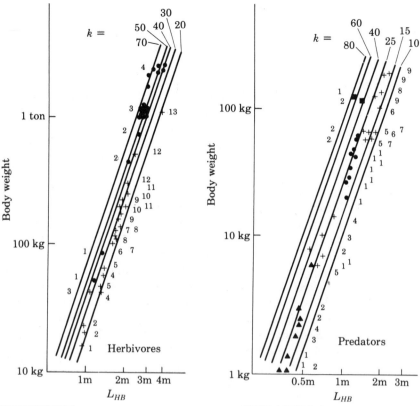

HERBIVORES

+ Gracile forms

1. *Gazella thomsoni*
2. *Cephalophus silvicultor*
3. *Tragelaphus scriptus*
4. *Aepyceros melampus*
5. *Gazella granti*
6. *Kobus kob*
7. *Damaliscus korrigum*
8. *Alcelaphus buselaphus*
9. *Connochaetes gnou*
10. *Kobus ellipsiprymnus*
11. *Equus burchelli*
12. *Taurotragus oryx*
13. *Giraffa camelopardus*

● Heavy ungulates

1. *Phacochoerus aethiopicus*
2. *Syncerus caffer*
3. *Diceros bicornis*
4. *Hippopotamus amphibius*

PREDATORS

▲ Primitive; generalized skeleton

1. *Bassariscus astutus*
2. *Martes flavigula*
3. *Bdeogale crassicauda*
4. *Genetta angolensis*
5. *Martes pennanti*

+ Cats

1. *Lynx rufus*
2. *Felis lybica*
3. *Felis serval*
4. *Lynx caracal*
5. *Acinonyx jubatus*
6. *Felis concolor*
7. *Panthera pardus*
8. *Panthera onca*
9. *Panthera leo*

● Long-distance running predators

1. *Canis lupus*
2. *Crocuta crocuta*

■ Heavy omnivores

1. *Ailuropoda melanoleuca*
2. *Euarctos americanus*

381

APPENDIX II

Data for limb proportions in all fossil samples are original. Data for abundance of fossil ungulates and predators are restricted to faunas from the High Plains and Rocky Mountains. Sources:

Mid Paleocene: K. Rose, *Univ. Mich. Pap. Paleont.* 26 (1981). W. D. Matthew, *Bull. Amer. Mus. Nat. His.* 9, 259 (1897). **Early Eocene:** M. McKenna, *Univ. Calif. Pub. Geol. Sci.* 37, 1 (1960). F. B. Loomis, *Amer. J. Sci.* 23, 356 (1907). D. A. Guthrie, *Mem. So. Calif. Acad. Sci.* 5, 1 (1967). D. A. Guthrie, *Annals Carn. Mus.* 43, 47 (1971). R. M. West, *Fieldiana Geol.* 29, 1 (1973); C. L. Gazin, *Smith., Misc. Coll.* 117, 1 (1952). P. Robinson, *Bull. Peab. Mus. Nat. Hist. Yale Univ.* 21, 1 (1966). Original census, Johns Hopkins University and USNM Collections. **Mid Eocene:** C. L. Gazin, *Smith. Cont. to Paleobiology* 26, 1 (1976). **Late Eocene:** Original census, USNM, Carnegie, Mus., Princeton Collections. **Oligocene:** J. Clark, J. Beerbower, K. Kietzke, *Fieldiana Geol.* 5, 1 (1967). W. D. Matthew, *Mem. Amer. Mus. Nat. Hist.* 1, 355 (1901). Original census, Johns Hopkins University and USNM Collections. **Miocene:** J. MacDonald, *Bull. Amer. Mus. Nat. Hist.* 125, 143 (1963). M. Voorhies, *Contrib. Geol. Univ. Wyo. Spec. Pap.* 2, 1 (1969). S. D. Webb, *Univ. Calif. Pub. Geol. Sci.* 78, 1 (1969). **Pliocene:** J. A. Shotwell, *Ecology* 36, 327 (1955). **Pleistocene:** C. W. Hibbard in W. Dort and J. K. Jones, *Pleistocene and Recent Environments of the Central Great Plains*, Univ. Kansas, Dept. Geol. Spec. Pub. 3 (1970) (specific faunal papers cited here).

APPENDIX III

Figure 4 sources: R. C. Andrews, *Nat. Hist.*, 39, 212 (1939); *On the Trail of Ancient Man*, Garden City Pub., Garden City (1922); R. Burrows, *Wild Fox*, Taplinger, New York (1968); C. Cottam, *J. Mamm.* 18, 240 (1937); C. Cottam and C. W. Williams, *J. Mamm.*, 24, 262 (1943); L. Guggisberg, *S.O.S. Rhino*, Survival Books, New York (1967); H. Kruuk, *The Spotted Hyena*, Univ. Chicago Press, Chicago (1972); J. Layne and A. H. Benton, *J. Mamm.*, 35, 103 (1969); L. Lee Rue, *The Deer of North America*, Times-Mirror, New York (1978); E. Neal, *Badgers*, Blanford Press, Poole (1977); M. Perkins, *Mutual of Omaha's Wild Kingdom*, Australian Roundup, aired 1981; C. Sweeney, *Grappling with a Griffon*, Pantheon, New York (1969); *Naturalist in the Sudan*, Taplinger, New York (1973); J. K. Terres, *J. Mamm.* 22, 453 (1941); B. Verts, *Biology of the Striped Skunk*, Univ. Ill. Press, Urbana (1967); S. P. Young and S. Jackson, *The Clever Coyote*, Univ. Oklahoma Press, Norman (1951).

CHAPTER 17

COEVOLUTION BETWEEN COMPETITORS

Jonathan Roughgarden

Competition is one of the important interactions between populations. It may cause each of the competing species to evolve in response to the other, thus producing coevolution.

THE HALCYON DAYS

The literature about the possible evolutionary consequences of interspecific competition is extensive. The modern literature begins with the work of Lack (1947) on Darwin's finches in the Galápagos islands. Lack noticed that these birds had bill sizes that depend on whether they co-occur with a presumed competitor. For example, there is a genus of finches that feed primarily on the ground. On islands where there is only one species of this genus, the bill depth is approximately 10 mm. But on islands where there are two (or more) species of this genus, the bill depth for the smallest species is approximately 8 mm whereas that for the next larger species is approximately 12 mm; there is no 10 mm form. This observation was interpreted as an evolutionary consequence of competition. The bill depth of a finch was presumed to be related to the size and thickness of the seeds that it eats. If so, a difference in bill size between two species indicates that, to some degree, the species use different resources. Lack's interpretation of the bill size data in the Galápagos was that competition caused the evolution of a difference in bill size, with the result that interspecific competition between the species was lowered.

Almost ten years later Brown and Wilson (1956) offered a similar observation. They brought to general attention Vaurie's (1951) description of two congeneric species of insectivorous birds (nuthatches), whose ranges overlap in central Asia. These species have different bill size and coloration where they coexist as compared to their bill size and coloration at locations where their ranges do not overlap. Brown and Wilson termed this phenomenon "character displacement," that is, the characters of bill size and coloration in each species are displaced away from those of the other species at places where their ranges overlap. Brown and Wilson suggested that character displacement could result from two evolutionary mechanisms. One is interspecific competition which, as Lack postulated, causes the species to evolve differences in bill size so as to use different food resources. As the differences in bill size evolve, the strength of the competition continually diminishes, thereby also lowering the strength of the selection pressure causing the divergence. The second mechanism is hybridization. If hybrids between the species are infertile, or poor in survivorship, then Brown and Wilson suggest that evolution will lead to an enhancement of the differences, like coloration, that promote species recognition. They interpreted the character displacement in color as due primarily to this mechanism.

Another issue related to the coevolution of competitors was raised by Van Valen (1965). This study established that there was more variation in bill size *within* a population of birds on islands than at adjacent continental locations. The comparison is between populations of the *same* species; on islands, a given species is more variable in bill size than that species is on the adjacent mainland. Because islands have fewer species overall than comparable mainland locations, Van Valen suggested that an island population would have fewer interspecific competitors than a mainland population. He interpreted the increased variation in bill size on islands as an indication that the island populations had evolved to use a wider variety of resources on islands than they do on the mainland. Presumably the individual birds with different beak sizes to some extent feed on different food items. This phenomenon now is often called "ecological release" or "competitive release."

So by the mid 1960s the stage was set with two main ideas about the coevolution of competitors. First, coevolution leads to an enhancement of the differences between competing species so as to bring about a partitioning of the resources and a consequent reduction of competition. Second, the removal of interspecific competitors leads to an expansion of the variety of resources used by the remaining species.

384

THE TEMPEST

During the late 1960s and throughout the 1970s, the literature related to these ideas exploded, and literally hundreds of theoretical and empirical studies were published relating, from almost every conceivable angle, to the coevolution of competing species. The dust is beginning to settle and the main questions addressed by the literature from the mid-1960s until the present can be listed as follows:

1. Do the ideas about the evolutionary consequences of competition make theoretical sense? The early theoretical papers treat the evolution of character displacement separately from the evolution of ecological release. Early papers on conditions for the evolution of character displacement include Bulmer (1974), Lawlor and Maynard Smith (1976), and Roughgarden (1976). Early papers on the conditions for the evolution of ecological release are Roughgarden (1972) and Matessi and Jayakar (1976). More recent papers treat the simultaneous evolution of character displacement with ecological release (and/or compression). Such papers include Fenchel and Christiansen (1977), Slatkin (1980), and Case (1982). Very recent theoretical work is also concerned with the invasibility of a coevolved community as discussed in more detail later in the chapter.

    These theoretical papers show that character displacement is a possible outcome of the coevolution of two competitors. But character displacement is not an inevitable outcome. According to present theory, character displacement results from coevolution provided the populations are initially quite similar to one another, the competition between the two populations is roughly symmetric, and the variance within each species remains approximately constant. Furthermore, the theory suggests that the rate at which the variance within a species changes is slower than the rate at which the character mean shifts during coevolution. An illustration of how to model the coevolution of character displacement appears in Chapter 3 by Roughgarden.

2. Is it true that interspecific differences in traits like bill depth actually correspond with interspecific differences in resource use? Papers relating to this include Ivlev (1961), Willson (1972), Hespenheide (1973a), Schoener (1974b), Pulliam (1974), Feinsinger (1976), Abbott et al. (1977), Leviton (1978), and Roughgarden et al. (1981). These and a great many other studies confirm a

relation between particular traits and the kinds of resources, including both food and space resources, that an individual characteristically uses. But evidence is required anew in each case; there is no automatic relation between a trait and resource use (occasionally a relation is not found), and the relation itself may vary with season, habitat, and food abundance.

Is it true that, within a species, differences in traits like bill size actually correspond to differences among individuals in their resource use? Key papers here include Soulé and Stewart (1970), Van Valen and Grant (1970), Grant (1971), Roughgarden (1974), Hespenheide (1975), Lister (1976), Bernstein (1979), Grant (1979), and Ebenman and Nilsson (1982). The answer is generally yes. However, the level of variation of traits like bill size is not generally a valid indication of the breadth of resources used by a population. The reason is that the breadth of resources used by a typical individual itself varies among populations. Hence, a given variance in bill size may have a different meaning for each population. The way to avoid this difficulty is to calculate the so-called between-phenotype component (BPC) and within-phenotype component (WPC) of the population's niche breadth (or niche width). These quantities allow comparison of the niche breadth of an individual with that of the whole population to which it belongs. (See Roughgarden, 1979, p. 529, for details on how to do the calculation.) When these quantities are used, it has generally emerged that the variety of resources taken by an entire population is not much greater than the variety taken by a typical individual in the population. Furthermore, when the niche breadth of a population is greater than that of a typical individual in it, the explanation lies mostly in phenotypic differences between sexes and ages and not in phenotypic differences among adults of the same sex. Finally, there is some empirical support that ecological release occurs more slowly than the evolution of ecological shifts. This point is an important assumption in models for the evolution of character displacement, but the data so far are too preliminary to be accepted without reservation.

3. Is competition present? This question has been extremely controversial because of the paucity, until recently, of experimental demonstrations of competition under field conditions (see Wiens, 1977; Connell, 1980; and Schoener, 1982, 1983). Schoener reviewed over 100 experimental studies of competition and pointed out that competition was almost always detected. Many other studies over the last decade have presented nonexperimental evidence of competition as well. But the successful detection of competition cannot stand alone; it must be accompanied

by evidence that other considerations are not more important than competition even though competition has been detected.

4. When competition is present, what is the relationship between its strength and the degree of overlap between the species in their use of resources? There is now increasing support for the assumption that the strength of competition is positively correlated with the degree of overlap in resource use provided the species are in fact competing and that the resources whose overlap is being examined are in fact limiting (see Fenchel and Kofoed, 1976; Pacala and Roughgarden, 1982b).

5. When the strength of competition is related to the degree of overlap, how can the overlap data be quantified to predict the strength of competition? At this time several indices are widely used for quantifying overlap data (see Schoener, 1974b; Abrams, 1975, 1980). Some are automatically symmetric, others allow estimates of asymmetrical competition between species. Still, nothing is known empirically about the relation between the numerical values of these indices and the numerical values of the competition coefficients in the Lotka-Volterra competition equations.

6. Are differences in bill size and in other characters related to resource use heritable? Boag and Grant (1978) and van Noordwijk et al. (1980) have measured the heritability of bill dimensions and of other characters of birds under natural conditions and did find them to be heritable. Also, Arnold (1981) demonstrated that phenotypes with different prey preferences were heritable in garter snakes.

7. Does coevolution lead to patterns in more than two competing species? Does the influence of coevolution extend to guilds or even to entire communities? Reports of community patterns that might be caused by coevolution, perhaps together with other processes, are found in Cody (1968, 1974), Williams (1972), Pianka (1973), Diamond (1973, 1975), Findley (1976), and Cody and Mooney (1978). Challenges to the reality of these patterns and to the use of circumstantial evidence for inferences about coevolution appear in Connor and Simberloff (1979) and Strong et al. (1979) (see Chapter 18 by Simberloff). These challenges are rebutted by Grant and Abbott (1980), Hendrickson (1981), and Diamond and Gilpin (1982).

More generally, it has become clear that a study about the coevolution of competitors must present a *very* extensive body of information to be convincing. The study should include evidence that competition actually occurs, that competition is related to overlap in resource use, that the competition is symmetric or asymmetric, that the competing species actually did coevolve with one another, that the initial condition for the coevolution was either simultaneous invasion or sequential invasion, and that the other species in the system pose only a perturbation to a picture primarily determined by competition and coevolution of the species of interest. It may also become increasingly important to know the dynamics of the renewability of the resources. To date no study is thoroughly satisfactory, although some examples seem very strong.

We shall consider two case studies of the coevolution of competitors. Then we shall review some other systems where the evidence is less complete but where coevolution of competitors does seem to have happened.

THE *HYDROBIA* MUD SNAILS OF LIMFJORD, DENMARK

Most of Denmark's area lies on a peninsula, called Jutland, that extends north from the European mainland (Figure 1). The channel, called Limfjord, that runs east-west across the top of Jutland is the site of possibly the best example of the evolution of character displacement according to the classical picture. The brief account here is drawn from Fenchel (1975), Fenchel and Kofoed (1976), and Fenchel and Christiansen (1977).

The western entrance to the Limfjord from the North Sea has been closed since the last glacial period until historical times. Before the beginning of the last century, the fjord was open only at its eastern end, and the western and central parts consisted of large, interconnected, freshwater lakes. In 1825, following a storm, the western entrance broke open, and has since been kept open artificially for navigation. After the opening of the western entrance, the western parts of the fjord became marine.

With the opening of the western entrance, mud snails of the genus *Hydrobia* entered. Presumably the two species of interest, *H. ulvae* and *H. ventrosa*, entered at approximately the same time. These snails are among the most important deposit-feeding invertebrates of northern European estuaries. They ingest their substrate and assimilate the microorganisms attached to mineral and detrital particles. *H. ulvae* has a preference for higher salinity than *H. ventrosa*, although they show a widely overlapping tolerance range for this and for other environmental factors.

Today, in coastal habitats and at sites in the Limfjord where only one of the two species is present, these species have the same shell size, approximately 3 mm in length. But at sites where they co-occur

388

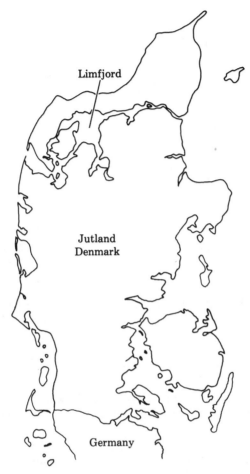

FIGURE 1. Map of Denmark. Notice the Limfjord across northern Jutland.

within the Limfjord, *H. ulvae* has a larger body size (approximately 4 mm) and *H. ventrosa* has a smaller body size (approximately 2.5 mm). Figure 2 offers a summary of these data. Furthermore, the size of the particles that are ingested is correlated with shell length. As Figure 2 also shows, character displacement has led to a partitioning of particles with respect to particle size.

Fenchel and Kofoed (1976) have produced convincing experimental evidence for interspecific competition under laboratory conditions that, to some extent, simulate natural conditions. They have shown that there is inter- and intraspecific density dependence in growth

389

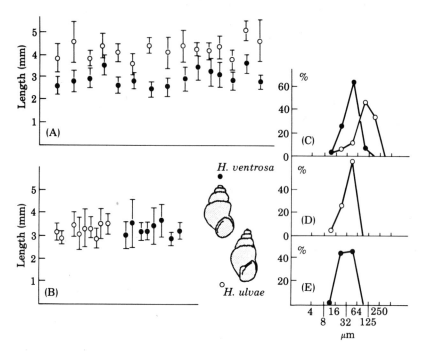

FIGURE 2. Shell lengths of two species of mud snails (with standard deviations) in 15 localities with coexistence (A) and 17 localities with allopatric occurrence (B). The average size distribution (volume %) of ingested food particles of coexisting (C) and allopatric populations of the two *Hydrobia* species (D,E). (From Fenchel and Christiansen, 1977.)

rates of individual snails, and they have detected density dependence in the survivorship of some species. They also have studied the kinetics of resource renewal.

There seem to be two unsolved issues that remain with this example. First, the relationship between the measurements that detect competition in the laboratory and the population dynamics of competition in the field needs to be fleshed out, as does the possible effect of other species. Second, the character displacement that occurs in Limfjord is apparently absent in coastal marine habitats, and the cause of this seeming failure of the evolution of character displacement there, where the population structure is surely different, needs to be worked out.

THE ANOLIS LIZARDS OF THE EASTERN CARIBBEAN

There is a sequence of many small oceanic islands in the eastern Caribbean, as shown in Figure 3. There are populations of *Anolis* lizards on every one of these islands. This particular set of islands has three

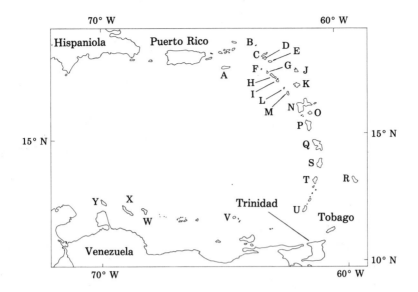

Island	Area (sq. km)	Elev. (m)	Breeding Land Bird Species	Anole Species
A St. Croix	218	1087	20	1
B Sombrero	1	12	NR	1
C Anguilla	91	65	18	1
D St. Maarten	88	424	20	2
E St. Barthelemy	21	302	18	1
F Saba	13	857	20	1
G St. Eustatius	31	598	21	2
H St. Kitts	176	1315	25	2
I Nevis	93	1093	23	2
J Barbuda	161	34	23	2
K Antigua	280	401	23	2
L Redonda	1	304	18	1
M Montserrat	85	913	26	1
N Guadeloupe	1513	1464	34	1
O Marie-Galante	155	203	21	1
P Dominica	790	1443	40	1
Q Martinique	1116	1346	41	1
R Barbados	430	339	18	1
S St. Lucia	616	956	44	1
T St. Vincent	344	1231	40	2
U Grenada	311	838	41	2
V La Blanquilla	47	60	NR	1
W Bonaire	246	243	31	1
X Curacao	448	372	45	1
Y Aruba	179	188	29	1

FIGURE 3. Map of the Lesser Antilles. Notice the absence of a relationship between the number of *Anolis* species on an island and the island area, island elevation, distance from Puerto Rico, distance from Venezuela, or avifaunal diversity. Notice also the two clusters of islands that contain two *Anolis* species. (From Roughgarden et al., 1982.)

391

special characteristics. First, each population of *Anolis* lizards on an island is taxonomically distinct and occurs only on that island (or on the bank containing that island). Second, every island in this system has one or two species of anoles; none has three or more. Third, there is a regular pattern of body sizes among the lizards on these islands, as discussed in more detail below.

The *Anolis* lizard populations on these islands provide an example of the coevolution of competition. They have coevolved with one another, and they do compete. However, the coevolution of body size has not been marked by the divergence of simultaneously colonizing species from one another, as with the character displacement of mud snails in the previous example. Instead, there has been a sequential invasion and extinction of species in an existing resident fauna, and the coevolution of body sizes has typically involved the parallel evolution of species, rather than a divergent evolution. This section is mostly drawn from Roughgarden et al. (1982).

Ecological background

Anolis females have a clutch of only one egg that is typically 0.5 to 1 cm in diameter, with a flexible shell. Once an egg is laid in the ground, there is no subsequent parental care. A hatchling lizard is approximately 1 cm long, not including the tail. It reaches reproductive size in approximately 1 year and approaches the maximum size in 2 to 3 years.

Anolis lizards live in trees and bushes as well as in the crevices of rock piles. They are active throughout the day, beginning 1 to 2 hours after sunrise and ending at dusk. Both sexes are territorial throughout the year. Territorial interactions are usually between members of the same sex. The female territory is mainly a feeding territory. It contains perches from which the females sight prey items, as well as places for sleeping and thermoregulation. The male territory is both a breeding and feeding territory, and it contains the territories of females.

The abundance of *Anolis* on the Caribbean islands is very high when compared with lizard abundance in the semidesert and desert habitats that are usually associated with lizards. A density of 25 lizards per 100 m² is typical, with higher densities (by a factor of two to three) being frequently measured. Moreover these animals occur in all habitats on an island.

The abundance of *Anolis* prompts the question of how the habitat can support so many lizards. Lizards are energetically inexpensive. Many lizards, perhaps as many as 75, are energetically equivalent to one insectivorous bird of the same weight. A lizard draws the energy to heat its body directly from the sun through basking; it does not fuel its body heat from the breakdown of food. Also, an *Anolis* lizard's activity pattern is efficient; it is a "sit and wait predator" that assumes a perch and scans for food. Furthermore, almost all the insectivorous bird species that characteristically obtain food from the places where lizards feed (primarily the ground) are absent in the eastern Caribbean, including ground-feeding forms like robins, towhees, mockingbirds, and so forth (Terborgh and Faaborg, 1980; Adolph and Roughgarden, 1982). There are both snake and avian predators on anoles. The energetic efficiency of lizards, coupled with the absence of ground-feeding insectivorous birds, probably explain most of the abundance of *Anolis* on the Caribbean islands.

Unlike the *Anolis* of the eastern Caribbean, both birds and insects, insofar as the insects can be gauged by the butterfly components, do not show much endemism among the islands. Evidently, the distances separating the eastern Caribbean islands from one another do not constitute nearly as serious a barrier to dispersal for birds and insects as they do for lizards.

How resource use is described

The main resources for anoles are food and the space used for territories. There are many species of insect prey, and dissection of preserved specimens shows that the anoles consume almost every kind of insect found in their habitats. It is not informative to think of food resources of anoles in terms of the specific or family names of their insect prey. What has proved informative is to consider the size and length of the insect prey. Insect size is important to lizards because it is directly related to the food value of a prey item—the bigger the better up to the point where it cannot be swallowed. Insect size is also related to abundance, with small insects generally being more common than large insects.

There is a relation between the size of a lizard and the average size of its prey. For a given set of circumstances, large lizards take larger prey, on the average, than small lizards (Schoener and Gorman, 1968; Roughgarden, 1974). We do not understand the basis of this relationship. Why is the average prey length of a lizard that is 6 cm long approximately 3 mm? The answer to this question perhaps awaits a more theoretical science of functional morphology.

The use of insect resources is quantified as the distribution of insect sizes eaten by lizards of a given body size. When this information

393

is not available, the body sizes of the lizards themselves are taken as indicators of their average prey sizes.

For the use of space, there are similar considerations. Two characteristics of territorial space are clearly important to anoles in the eastern Caribbean: whether the microclimate in the space is hot or cold, and whether the space is located near the ground or high in the crown of the vegetation. Anoles do not live preferentially on particular species of trees or bushes. They occupy cactus, agave, gum trees, manchineel trees, rock piles, fence posts, deciduous and evergreen vegetation, and so forth. What is important about the space is where it is and what its microclimate is. Also, within a species, and usually among closely related species, a lizard's body temperature is an indicator of the thermal microclimate that the lizard inhabits (Roughgarden et al., 1981).

On islands in the southeastern Caribbean that contain two species of anoles, territorial space is partitioned by microclimate and not by height above the ground. In the northeastern Caribbean, space is partitioned by height and not by microclimate (Roughgarden et al., 1982). This type of observation has been termed "niche axis complementarity" by Schoener (1974a). A possible explanation based on competition theory is developed in Pacala and Roughgarden (1982a).

Anoles compete

A controversial issue in the ecology of vertebrates has been whether interspecific competition is strong enough to be an important force in determining their distribution and abundance. We now have good evidence that strong present-day interspecific competition occurs between the two *Anolis* species on one of the eastern Caribbean islands.

The studies indicating strong present-day competition come from the island of St. Maarten. The two species on this island are very similar in body size, more so than on any other island in the eastern Caribbean. The details have been summarized in Roughgarden et al. (1983).

There are three lines of circumstantial evidence for competition between these species. First, the distribution of the small lizard (*A. wattsi pogus*) on St. Maarten is curiously restricted to low hills (420 meters elevation) in the center of the island. The slightly larger lizard (*A. gingivinus*) occurs everywhere on the island, including the central hills. This distribution is curious because the close relatives of *A. wattsi pogus* on other islands occur down to sea level in all habitats. The absence of *A. wattsi* from sea-level habitats on St. Maarten is not caused by a physiological inability to survive and breed there or by the occurrence of more, or special, predators there. This observation suggests that *A. wattsi* may be excluded from the sea-level region of St. Maar-

394

ten by strong present-day competition with *A. gingivinus.*

Second, where *A. gingivinus* co-occurs with *A. wattsi*, it has a generally lower abundance than where it is the only species present. Finally, *A. gingivinus* shows a shift in perch position where it co-occurs with *A. wattsi.* These observations suggest an effect of the smaller lizard on the larger one.

Furthermore, there is direct experimental evidence for competition. There are offshore cays near St. Maarten that have only *A. gingivinus* on them. We have introduced *A. wattsi* to places on a cay from which the resident *A. gingivinus* were largely removed and also to other places where the resident *A. gingivinus* were left undisturbed. The survival rate of the introduced *A. wattsi* during the 6 months after their introduction is nearly twice as high if the resident *A. gingivinus* are removed than if they are left undisturbed. Moreoever, the introduced *A. wattsi* established territories only in marginal habitat when the *A. gingivinus* were left undisturbed but established territories throughout the study sites when the *A. gingivinus* were largely removed.

The experimental evidence for an effect of *A. wattsi* against *A. gingivinus* comes from observations on the rates of body growth for individual lizards (Pacala and Roughgarden, 1982b). We built in natural habitats enclosures that served to prevent immigration and emigration from a given area. By stocking such enclosures with different densities we determined the effect of density on the body-growth rates of individually marked lizards. The experiments used 60 *A. gingivinus* in each of two enclosures and 60 *A. gingivinus* plus 100 *A. wattsi* in each of two other enclosures. These densities are similar to natural ones. The growth rates of the *A. gingivinus* individuals who were together with the *A. wattsi* were approximately one-half that of the *A. gingivinus* individuals who were by themselves. The terminal body size was the same, and only the growth rate to that size was affected.

The present evidence for competition does not provide a comparison of the strength of interspecific and intraspecific competition. But it does show that strong interspecific competition exists. The mechanism of the interspecific competition is complex. The growth-rate experiments suggest, though not conclusively, that exploitative competition for food is important because growth rates of lizards in the field are affected by food abundance (Roughgarden and Fuentes, 1977; Dunham, 1978; Stamps and Tanaka, 1981). The cay introduction experiments suggest that the acquisition of territories is also an important factor.

St. Maarten is an exceptional island in that the body sizes of the two lizard species there are uniquely close. Growth experiments like those done on St. Maarten have also been conducted on St. Eustatius. The two species of *Anolis* there are more different in body size and in their characteristic perch heights than those on St. Maarten. These experiments have shown that the smaller lizard on St. Eustatius has only a very small effect on the growth rate of the larger lizard. This result is important because it supports the assumption that the strength of competition between two species increases as their overlap in the use of food and space resources increases.

Finally, it is reasonable to assume that the competition between individuals with a given difference in body size is asymmetrical, with the larger lizard exerting a stronger effect on the smaller lizard than vice versa. As yet, there is no experimental evidence for this assumption. Nonetheless, the average total volume of food a lizard eats is an increasing function of its size. Hence, for a given overlap in resource use, a larger lizard takes more food away from a smaller lizard than vice versa. Furthermore, the interference component of the competition, that is, disputes over territories, also favors larger lizards over smaller lizards.

The biogeographical pattern of body size and species number

There are 16 islands in the eastern Caribbean system with one native population of *Anolis*. On 15 of these, the lizards have evolved the same body size, called the "solitary size." This size is approximately 60 mm in adult males and approximately 50 mm in adult females. These 15 species differ in many other respects including color, physiological abilities like tolerance to thermal and water stress, and courtship behavior. Yet they share the same body size. There is one exception to this rule for the single-species islands. The island of Marie Galante has a population of lizards that are large, with adult males frequently over 100 mm in length. There is, however, no single-species island that contains lizards whose maximum size is smaller than the 60-mm standard for single-species islands.

There are eight islands with two native populations in the eastern Caribbean system. On six of these, there is a population of large lizards, with adult male sizes over 100 mm. On these six islands, the other species is either smaller than the solitary size or very close to the solitary size, but none is bigger than the solitary size. On these six islands, the two species have a substantial difference in body size.

One of the two exceptions among the islands with two species is St. Maarten. Here one of the species, *A. wattsi pogus*, is slightly smaller than the solitary size and is found only in the central hills. The other species, *A. gingivinus*, is at the solitary size itself. The closest relatives of *A. gingivinus* are large lizards on the other nearby two-species islands (Lazell, 1972). Both *A. gingivinus* and its relatives occur in all habitats. The closest relatives of *A. wattsi pogus* are the small lizards on the other nearby two-species islands (Lazell, 1972). These relatives of *A. wattsi pogus* occur in all habitats, whereas *A. wattsi pogus* has a restricted distribution. The difference in size between the two species of St. Maarten is very slight and there is strong present-day competition between them.

The other exception is St. Eustatius, which is perhaps best described as an intermediate between St. Maarten and the other nearby two-species islands. One species is slightly smaller than the solitary size, and the other is only somewhat bigger than the solitary size because its males rarely exceed 90 mm in length.

Thus, there is a complex relation between the body sizes of lizards and the number of lizard species on an island. Although there are some "rules" that describe the great majority of islands, there are definite exceptions to these rules.

Posing the problem

The processes of invasion, competition, and coevolution occur among the anoles of the eastern Caribbean. Hence, these may be the processes that have produced the community structure and biogeography of the *Anolis* communities there. Specifically, there has been no *Anolis* speciation within these small islands. The closest relative of each species is a species on a nearby island, not the other species of the same island, and thus it is invasion that brings a species to an island. Competition for food and space occurs between the species on an island. The more similar the species are to one another in their resource use, the stronger the competition. The species are together in the same habitats and have been together long enough that each has acquired taxonomic distinctness during that time. Because the species have been together for a long time and do compete, they have undoubtedly coevolved with one another on at least some of the islands.

These are the processes that may have produced the community structure of the eastern Caribbean. And what is that structure? There is a relation between body size and species number throughout the eastern Caribbean. There is a characteristic size for lizards on islands with one species. And there is a large difference in size between the species on islands with two species. Yet there are isolated, but clear, exceptions to this pattern.

So the problem is this: Can processes known to be occurring in the eastern Caribbean explain the observed pattern of community structure? That is, can invasion, competition, and coevolution explain the relation between lizard size and species number, including the exceptional cases?

Faunal buildup

Mathematical theory for the processes of competition, coevolution, and invasion has led to a picture of how these processes could have produced the ecology of *Anolis* in the eastern Caribbean.

The central theme is a turnover of species on islands caused by an extinction during the coevolution of competitors, followed by a reinvasion of the island after the extinction occurs. The extinction results from coevolution with asymmetrical competition. Figure 4 illustrates the overall idea, and the mathematical basis to the theory is discussed in Chapter 3 by Roughgarden.

There are three parts to the cycle. First, a lizard species that has no other *Anolis* competitor evolves to a characteristic size, as observed on most of the islands with one species. The carrying capacity of an island is assumed to be a bell-shaped function of lizard body size, as illustrated in Figure 4A. The body size to which a solitary species evolves is the size corresponding to the highest carrying capacity. Presumably this "solitary size" is the size that yields the highest net catch of insect food relative to a lizard's energy requirements.

Next, the island is invaded by a larger species (Figure 4B). A species that has the solitary size or smaller cannot invade because of competition from the resident. In the model, the size of the invader is the size that produces the highest rate of increase for a propagule that has landed on an island where a resident is already established.

Because of the assumed asymmetry in the competition, only a larger lizard can successfully invade. This large species presumably came from Puerto Rico during the Pleistocene when the sea level was as much as 100 meters lower than it is today. Even then, the dispersers must have rafted across the sea, because there has never been a land bridge connection between the Puerto Rico Bank and the small islands in the Eastern Caribbean south of the Virgin Islands.

Finally, the two species, now together on an island, coevolve as competitors (Figure 4C). The resident's body size shifts to the left to avoid competition from the invader. The invader's body size also shifts to the left, where there are more resources and where the population size of the resident is less than when the invasion began. St. Eustatius is possibly at this stage.

As the resident continues its evolution of a smaller body size, the invader's body size continues to approach the characteristic size of a solitary species, as observed on St. Maarten. But during this coevolution, the resident is eventually driven to extinction by competitive ex-

398

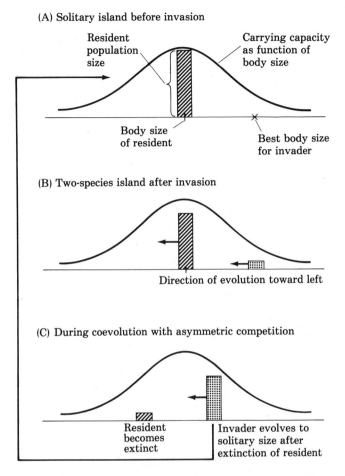

FIGURE 4. Species turnover resulting from invasion followed by an extinction during coevolution. The horizontal axis is a measure of body size, possibly in units of the logarithm of body length. The vertical axis is population size in units of number of individuals.

clusion. When the extinction of the smaller species occurs, the invader has not attained the solitary size; it is still larger than the solitary size, as observed on Marie Galante. According to the model the extinction of the smaller species depends on the asymmetry of the competition. After the extinction of the smaller species, the invader completes its convergence to the solitary size, and the island awaits another invasion.

In this theory it is assumed that changes in average body size occur

399

much more rapidly than changes in the variance of body size within a species. Some predictions may change if this assumption is modified.

This hypothesis of turnover based on coevolutionary extinction of the small species and reinvasion by a large species accords nicely with the biogeographical pattern relating body size to species number. By this hypothesis, all islands with one species should have lizards whose body size is the solitary size or larger, *but not smaller.* An island with one species that is larger than the solitary size is, by this hypothesis, one on which the smaller species has recently become extinct. Marie Galante is possibly just such an island. This is the only island with a single species whose size is different from the solitary size; it is larger than the solitary size. And, in fact, there is no island with a solitary *Anolis* species that is smaller than the solitary size, as predicted. The observation that so many of the islands with one species possess the solitary size suggests that the waiting time of an island for the arrival of a suitably sized invader is very long. Moreover, the islands with one species of solitary size tend to be the more remote islands.

By the coevolution–invasion turnover hypothesis, islands with two species should generally have species with large differences in body size because the buildup to the two-species stage can only be accomplished by invasion with a sufficiently different species to begin with. Any exception to the rule of large differences in body size should be an island just like St. Maarten. Here it is the smaller species, and not the larger species, that has suffered a range contraction. This range contraction is caused by strong present-day competition from a larger species that has itself evolved a smaller body size than its relatives on other nearby islands. As expected by the theory, there is no "anti-St. Maarten" with a large species having a contracted range and a smaller species that occurs throughout all habitats.

The cycle of invasion, coevolution, and extinction also accords with the reported phylogenetic relationships among the anoles of the northeastern Caribbean. According to these reports, the anoles of St. Maarten are ancestral to the other anoles of the region; they are not considered to be recent derivatives of anoles on other nearby islands. Both Lazell (1972), using morphological data, and Gorman and Kim (1976), using electrophoretic data, consider *A. gingivinus* to be ancestral to the large lizards in the region (*A. bimaculatus* spp.) and consider *A. wattsi pogus* to be ancestral to the other subspecies of *A. wattsi.* These ancestral relationships may be attributed to the placement of St. Maarten (and the rest of the Anguilla Bank) in the north, below the Virgin Islands, so that it is a natural stepping stone for dispersal from the Puerto Rico Bank into the islands of the eastern Caribbean.

By the traditional hypothesis of character displacement, St. Maarten would be the stage before character displacement had evolved. If so, the forms on St. Maarten would be derived from, and not ancestral to, the forms on nearby islands. Moreover, the distribution of the species on the island does not suggest a recent invasion. On islands where recent invasions have occurred, including *A. extremus* from Barbados on Bermuda (Wingate, 1965, and personal observation) and *A. aeneus* and *A. trinitatis* from Grenada and St. Vincent, respectively, on Trinidad (Gorman et al., 1971, and personal observation), the newly introduced forms have distributions localized to habitat near the ocean.

On St. Maarten it is reasonable to assume that the most recent species to invade is the species that occupies the habitat near the ocean (*A. gingivinus*) and that the distribution of *A. wattsi* is relictual. If so, it is difficult to envision how *A. gingivinus* could have spread over the entire island if it had arrived with the size it currently has. All known examples of introduced anoles that are the same size as a resident ecological counterpart have failed to spread. Such introductions persist as enclaves, as does *A. extremus* on Bermuda. I have observed other examples of enclaves of introduced anoles in seaside towns in the Dominican Republic. In contrast, introductions do spread if there is no ecological equivalent of the same size, as did *A. bimaculatus leachi* when introduced to Bermuda (Wingate, 1965) and *A. sagrei* on Jamaica (Williams, 1969).

Thus, a model for competition, coevolution, and invasion leads to a hypothesis that there is a turnover where coevolutionarily caused extinction follows an invasion. This hypothesis accords with the rules relating species number to body size as well as all exceptions to those rules and with available phylogenetic and distributional data.

The picture presented is reminiscent of the idea of a "taxon cycle" introduced by E. O. Wilson (1961) to describe turnover over evolutionary time in the ant fauna of South Pacific islands. The idea of a taxon cycle, and the general proposition that cycles of alternating invasions and coevolutionarily caused extinctions, is discussed in more detail in Chapter 3 by Roughgarden.

The unresolved issue

The major unresolved question with the *Anolis* system, like that with the *Hydrobia* system discussed previously, is that the population dynamics of competition under natural conditions have not been determined. Showing that competition and coevolution occur does not, by itself, establish that the models for the coevolution of competitors are good models. The models yield predictions that seem to explain the biogeographical patterns of community structure. But it would be more convincing to know if the population dynamics of competition actually function in the way represented by the models. This issue can be solved in the field though an experimental design using enclosures with populations of lizards started at various initial densities.

ON THE GENERALITY OF THE COEVOLUTION
OF COMPETITORS

The case studies on *Hydrobia* mud snails and eastern Caribbean *Anolis* lizards establish the existence of the coevolution between competitors as a real process in nature and offer hope that the theory for how this process works is on the right track. But how general is the process of the coevolution of competitors? How often are we likely to be correct in interpreting species differences as the result of coevolution between competitors?

Let us revisit the classic example of Darwin's finches on the Galápagos islands. The studies of Abbott et al. (1977) and Grant and Grant (1980) are recent and thorough examinations of the ecology of these finches. They conclude that competition together with the pattern of resource availability among the islands is a likely factor in the evolution of bill size and body size in these birds. The evidence for competition, though extensive, is strictly correlative. Darwin's finches have obviously evolved together since they are found nowhere else and have evolved their present-day phenotypes in the Galápagos. Thus, as an example of coevolved competitors, Darwin's finches do satisfy the criterion of having evolved together, but only satisfy the criterion that competition has been important in this evolution through correlative evidence. Nonetheless, although we might wish for a stronger case, the Darwin's finches remain a likely example of a coevolved group of competitors.

The classic example of the evolution of character displacement— the pair of central Asian nuthatches—is in more difficulty. Grant (1972a, 1975) established that the pattern of geographical variation in bill size in these birds was consistent with other hypotheses that do not involve interspecific competition in those places where the species ranges overlap. The data are also consistent with the original character displacement interpretation, but the case is not strong.

Case (1979) has developed an important example with Mexican whiptail lizards (genus *Cnemidophorus*). He has documented a biogeographical pattern in the body size of several species in this group and has analyzed how coevolutionary competition theory can explain the biogeographical pattern. He has also examined and provisionally rejected three alternative hypotheses, based on divergence in size to enhance mate recognition, sexual selection for body size, and size-selective predation.

402

These additional possible examples suggest that the coevolution of competitors may have occurred in a variety of systems. Still more possible examples can undoubtedly be found. But the existence of species differences is not itself good evidence for the coevolution of competitors, even if interspecific competition does occur. Species differences may result from competition between an invader and members of the resident fauna, such that the most likely invaders are forms sufficiently different from members of the resident fauna. This process will produce a fauna with species differences without any coevolution occurring. The full extent of the generality of coevolution between competitors is far from known at this time, but it is clearly a real process with at least enough generality to be considered as one of the important processes determining community structure.

CHAPTER 18

SIZES OF COEXISTING SPECIES

Daniel Simberloff

INTRODUCTION

Sizes of coexisting species are often thought to reflect coevolved community structure (references in Simberloff and Boecklen, 1981), a notion that can be traced to Brown and Wilson's conception (1956) of competitive character displacement. Having observed several instances where two species' sympatric populations differ more (especially morphologically) than do allopatric populations of the same species, Brown and Wilson suggested that in sympatry each species exerts selective pressure on the other, either by resource competition or by the penalties of infertile interspecific mating, so that coevolutionary divergence results. Focusing on the competitive rather than the reproductive mechanism, Hutchinson (1959) proposed that divergence must proceed until the difference in body (or trophic apparatus) size exceeds some minimum difference compatible with coexistence. On the basis of data on several sets of birds and mammals, Hutchinson tentatively proposed 1.3 as the critical size difference: two related species whose sizes produced a ratio less than 1.3 could not coexist. As noted by Grant (1972a), this proposal and its sequels contributed to a shift in emphasis away from the Gause-Volterra competitive exclusion principle toward questions of limiting similarity (e.g., MacArthur and Levins, 1967).

The claim that coexistence is impossible unless some minimum size ratio obtains has fostered a related contention about community structure: when there are more than two species, natural selection will act on size to produce constant ratios between adjacent size-ranked species, such that the ratios of each species to the next smaller species will be identical (references in Simberloff and Boecklen, 1981). Mac-Arthur (1972), for example, notes that the sizes of the three North American *Accipiter* hawks (Storer, 1966) are such that for each sex the intermediate-sized *A. cooperi* is approximately 2.5 times as large as *A. striatus*, and only 1/2.5 times as large as *A. gentilis*. MacArthur and Levins (1967) had anticipated such a tendency on theoretical grounds. They began with the notion of a limiting similarity L between two coexisting species. This proposal translates into a minimum size ratio and a 1-dimensional food niche indexed by a species' size, and generates the prediction that if two species' sizes differ by more than L and a third species intermediate in size invades, the third species will evolve so that its size is approximately midway between those of the other two species. If, on the other hand, the two original species' sizes are more similar than L, the new intermediate species' size should converge toward whichever original species' size it most resembles. The apparent paradox that competition could lead to convergence is not unique to MacArthur and Levins' model. Schoener (1969a) similarly predicts, but on different grounds, that competition can lead to convergence. His argument is that a large species may be forced, by a smaller generalized competitor, to take more small food items and thus to evolve a smaller size.

Finally, Schoener (1965), Grant (1968), and Keast (1970) believe that minimum size ratios on islands must exceed those on the nearby mainland because competition is more severe on the reduced insular resource base. For example, Grant (1966) observes that in the Tres Marias Islands the only two hummingbird species, *Amazilia rutila* and *Cynanthus latirostris*, have a size ratio of 1.326 for wing length and 1.281 for bill length, whereas the corresponding ratios on the adjacent mainland are 1.116 and 1.095, respectively. The island *Cynanthus* has slightly larger wings but a slightly smaller bill than the mainland population, whereas the island *Amazilia* is greatly increased in both dimensions. A complication is that these species are but two of five hummingbirds on the mainland, and it is conceivable that the changed island sizes may be a response to release from competition with the three absent species rather than intensified competition between the two island species. Grant (1972a) cites examples in which two species originally were sympatric over their entire joint range; one species subsequently expanded its range (say, to an island) and in the new region of allopatry underwent morphological evolution. Grant prefers to label this phenomenon "character release" rather than "character displacement" because it results from release from competition rather than its

405

imposition. For the Tres Marias hummingbirds, the hypothesis of character release rather than increased minimum ratios is perhaps strengthened by the fact that the greatly increased *Amazilia* has one very similar-sized congener on the mainland plus two larger confamilials, whereas the slightly changed *Cynanthus* is much the smallest of the five mainland species.

These three perceived patterns of community structure—minimum size ratios, constant size ratios, and increased island size ratios—will be the focus of this chapter. For all three patterns I will ask how general the pattern is and whether the causes are in fact a coevolutionary response to competition. It should be clear at the outset that all three patterns might obtain yet have nothing to do with evolution. For island faunas, for example, Grant (1969), Strong et al. (1979), Grant and Abbott (1980), and others have noted that a pattern of species' sizes may just reflect the nonrandom survivors of a large number of mainland colonists. For the Tres Marias hummingbirds, it is clear that at least one species has evolved considerably on the islands, but it could still be that the two species are present only because their initial size permitted them to survive together. Even for coexisting species on the mainland, it is conceivable (Grant, 1972a; Slatkin, 1980) that the sizes of species at a particular site have not resulted from coevolutionary adaptation but rather from which species survive of all possible inhabitants. It may well be that both forces— evolution *and* nonrandom colonization—determine the suite of sizes in any community.

MINIMUM SIZE RATIOS

Grant (1972a) proposes that character displacement and character release both require that morphology of a species change under the selective influence of one or more ecologically and/or reproductively similar species. But he adds that simply observing greater difference in sympatry than between allopatric populations need not implicate character shift, because independent clinal variation in two or more species could produce the same pattern. For example, species W in the west and E in the east (Figure 1) are more different in the region of sympatry and so conform to the pattern demanded by character displacement. But the difference in sympatry could merely be a clinal response to gradients in the physical environment rather than a response to the other species. Grant suggests testing for this possibility by dividing data points for each species into two groups, that is, from areas of sympatry and allopatry, respectively. If slopes and intercepts of the two groups do not differ statistically for either species, then Grant argues that the parsimonious interpretation is independent clines based on the physical environment rather than displacement or

406

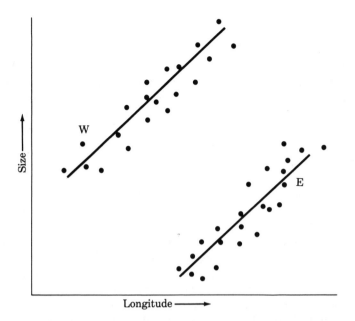

FIGURE 1. Size clines of partially sympatric species, W and E, that fulfill criterion (Brown and Wilson, 1956) for character displacement but for which independent variation is a plausible hypothesis (after Grant, 1972a).

release, because the morphology of each species in allopatry adequately predicts its morphology in sympatry.

I agree that independent clines selected by physical gradients are the parsimonious null hypothesis and that Grant's split-data regressions are the appropriate starting point when sufficient data are available, if only because of the abundant evidence for physically selected clinal variation (e.g., James, 1970). On the other hand, one could argue that even with uniform clines of the sort depicted in Figure 1, the selective agent is interspecific competition in sympatry, and the clines in allopatry are maintained by gene flow from the sympatric region. So an abrupt change in clinal variation in sympatry seems to be presumptive evidence for character shift, but absence of the abrupt change does not prove that character shift has not occurred. Endler (1977) and Dodson and Hallam (1977) review several theoretical models whereby an abrupt change in gene frequencies and the resulting morphology can be produced in the absence of either another species or an abrupt environmental change, simply by the mathematics of the interaction of gene flow and selection; so an abrupt change by itself

407

need not implicate species interaction. If the change coincides closely with the region of sympatry, however, it would seem remarkably coincidental for the change and the sympatry to be independent.

P. R. Grant's review (1972a, 1975) of several well-publicized examples of ecological displacement concludes that in many instances there is no evidence whatever for it, and in no instance do the data satisfactorily accord with the hypothesis. Among more recently published examples, several seem to provide stronger support for the model, but all are beclouded by complicating factors. Fenchel (1975) has found for mud snails (Hydrobiidae) in Scandinavia that when *Hydrobia ventrosa* and *H. ulvae* are alone they are of similar size, whereas when they are sympatric they differ in length by an average ratio of 1.53; Fenchel (1975) also has shown that food particle size is correlated with snail size. From the observations he infers competitive character displacement. However, the data from a number of sites in both allopatry and sympatry are not presented, nor is there evidence that food is limiting (John Gray, personal communication). Huey and Pianka (1974) examined two skink species (*Typhlosaurus*) in the Kalahari Desert. The range of the smaller species (*T. gariepensis*) is contained within the range of the larger (*T. lineatus*), and the larger species is consistently larger in sympatry than in allopatry. Furthermore, morphological change is abrupt rather than clinal, prey size is correlated with body size, and there are differences between allopatric and sympatric *T. lineatus* in proportions of different prey species. However, food is not known to be limiting, and there is a sudden habitat shift between the zones of allopatry (flat desert) and sympatry (dunes). Once again alternative noninteractive hypotheses are possible. Two Chilean foxes (*Dusicyon culpaeus* and *D. griseus*) diverge in mean size with increasing latitude. However, no information is available on whether allopatric size variation predicts that in sympatry, and the species tend to occupy more similar habitats the more different they are in size—a trend that Fuentes and Jaksic (1979) interpret as competitive character displacement. Little is known of diets or whether food is limiting, but Fuentes and Jaksic suggest that the size differences are necessary to allow syntopic coexistence.

Dunham et al. (1979) applied a version of Grant's technique to suckers (*Catostomus*) in the western United States and used multiple regression to see the extent to which the presence or absence of congeneric species and variation in physical environmental variables, respectively, predicted variation in number of gill rakers in various species. Number of gill rakers is curvilinearly related to body length within species. They concluded that there are several likely cases of divergent character displacement. As Dunham et al. note, the functional significance of different numbers of gill rakers must be determined to consolidate this example as a *bona fide* case of competitive

408

character displacement. A similar, but less statistically robust, relationship between number of vertebrae and presence of congeneric species also indicated divergent character displacement, but again the link to competition is obscure. Furthermore, what if anything these fishes compete for is not currently known.

Whether there actually is a minimum size ratio compatible with coexistence and, if so, what that ratio is are far from clear for most communities. That two distinct forces—evolution and nonrandom pattern of initial colonization—may separately or jointly determine a suite of coexisting species' sizes confounds the search for minima. If one were to ignore colonization patterns and assume that evolution could conceivably produce very major size changes, one might ask of a set of coexisting species whether the minimum size ratio is larger than one would have expected if species' sizes evolved independently of one another. This expectation, of course, rests on the null distribution of species' sizes. Simberloff and Boecklen (1981) distributed all species' sizes independently and uniformly on a log-scaled line between the smallest and largest species in a set. With this model, the null probability that the minimum ratio g_1 would be as large as that observed (a) or larger is shown by Irwin (1955) to be

$$\begin{aligned} \Pr(g_1 > a) &= [1 - (n - 1)a]^{n-2} \quad && \text{for } a \le n - 1 \\ &= 0 \quad && \text{for } a > 1/(n - 1) \end{aligned}$$

where n is the number of species in the set. Eighteen published studies held that such a data set demonstrated some particular minimum size ratio required for coexistence and contained sufficient data to test the claim; but Simberloff and Boecklen found that for only one of these would the null hypothesis of random, independent sizes have been rejected at the $p = 0.05$ level in favor of the alternative of an extraordinarily large minimum. For two studies the null hypothesis would have been rejected at $p = 0.05$ in exactly the opposite direction: observed minima too small. In six other studies a large fraction (but not necessarily a majority) of the cases have a very large minimum. If the rejection criterion is raised to 0.30, data from 13 studies generally indicate larger minima than the null model would have predicted, whereas 3 studies lead to exactly the opposite conclusion.

At this point, the above conclusions should be qualified in four ways. First, whether or not the observed minima are surprisingly large, they *are* different; and though most authors state that their results generally confirm Hutchinson's hunch, few find the same minimum (1.3) that he suggested. Instead, critical published ratios include 1.05, 1.10, 1.14, 1.2, 1.4, and 2.0. Furthermore, 3 of the 18

409

studies, plus several others not treated in this context by Simberloff and Boecklen, found ratios of 1.0—identical sizes for at least two species. This was never taken as evidence that there is no minimum ratio compatible with coexistence, but rather was either ignored or the identically sized pairs explained by recourse to *ad hoc* subsidiary hypotheses discussed below—generally that such pairs partition available resources along niche axes other than size.

Second, many papers contending that minimum size ratios structure communities do not present enough data to allow assessment of the contention. Diamond (1973), for example, lists 31 sympatric congeneric pairs of birds "sorting by size" in New Guinea, all of whose weight ratios exceed 1.33. Of course, there are very many more sympatric congeneric pairs of birds in New Guinea than these, including many whose weight ratios are not as large as 1.33. So it seems that the null hypothesis being tested is that all pairs of congeneric species differ in some way, and one way in which they may differ is size. But confirmation of this hypothesis could hardly fail to be forthcoming and cannot by itself provide much support for the contention that competition mandates the size differences.

A related point, which Simberloff and Boecklen (1981) note, is that the reported minima, whether or not unusual, are likely a highly selected set, culled from many data sets whose assemblers sought size patterns and saw nothing they thought worth publishing. Only with recent criticism of the size ratio hypothesis have negative results (e.g., Huey, 1979; Wiens and Rotenberry, 1980) become noteworthy.

Third, as Roth (1981) observes, even when different studies report either similar ratios or just a similar pattern (e.g., apparently large ratios), the similarities are frequently superficial. Some authors cite 1.3 as a modal ratio, others as a minimum, and still others as an optimum. Sometimes the species generating the ratios are congeneric, other times members of the same guild, and other times just taxonomically close. Some authors examine species that are sympatric but allotopic, whereas others demand identical microhabitat. Each study can stand or fall on its own merits, of course, but because they treat such different hypotheses they cannot really indicate a general pattern or mechanism.

Finally, Simberloff and Boecklen's null hypothesis of independent random draws from a uniform distribution on a log-scaled line is particularly unlikely to be rejected on grounds of an extraordinarily large minimum. Two values randomly drawn from a uniform distribution, for example, clearly have an expected difference greater than two values drawn from a modal distribution with the same range. The test is, therefore, a conservative one. On the other hand, no data have yet come to my attention that convincingly indicate that the distribution of logs of sizes in small, coexisting groups of species in one trophic

410

level would be anything but uniform in the absence of interactions. This null distribution would surely be very difficult to deduce from extant data. One would require several groups of coexisting species for which sufficient field research had indicated at most a very minor role for competition. Even then, one could not rule out the past effects of competition. D. Sinclair (personal communication) has found several local communities with sizes not differing from the log-uniform distribution, but the forces acting on these sizes are unknown.

In addition to large sets of species, one might ask of a set of sites each containing two species whether the observed minimum size ratio is larger than chance would have predicted. Again assuming that evolution can achieve any size within specified bounds, Simberloff and Boecklen (1981) modify a test of Pielou and Arnason (1966) that shows that, for each site, the null probability of a larger minimum ratio than that observed is

$$\Pr(x > a) = 1 - 2ab + a^2b^2$$

where a is the observed minimum and b is the factor that scales the logarithmic range of possible sizes to unity. Schoener (1970) and Williams (1972) for anoles of the Lesser Antilles and Barbour (1973) for lake fishes suggest that species pairs in a number of sites have remarkably large size ratios, and for both taxa estimates of largest and smallest possible sizes were provided, allowing the test. Given a null probability P_i for a given site i and the contention that particular values of observed ratios are large, the appropriate test for the entire set of sites is Fisher's combination of independent probabilities, $-2\Sigma \ln P_i$, which should be distributed as χ^2 with $2n$ degrees of freedom and with n equal to the number of sites (Sokal and Rohlf, 1969). If the claim before examination of the data is that all sites' ratios will exceed some specified minimum a, the subsequent observation that all sites' ratios do exceed a has associated null probability $(1 - 2ab + a^2b^2)^n$, with b and n defined as above. For Barbour's fish the contention is clearly the former, and the ratios of pairs of fishes in four lakes have an associated null probability of 0.199. The anoles are discussed below.

CONSTANT SIZE RATIOS

Poole and Rathcke (1979), Roth (1981), and Simberloff and Boecklen (1981) all provide tests for determining whether a sequence of ratios is extraordinarily constant, relative to a null hypothesis of independent sizes. Against ratios of sizes generated by nonrandom algorithms (that is, species' sizes dependent on one another), I have found the latter test, [based on Barton and David's test (1956)] to be by far the most powerful. If there are $n + 1$ species, producing n ratios, g_i is the ith smallest size ratio in a sequence, and $G_{rs} = g_r/g_s$. Barton and David provide the probability distribution for any G_{rs}.

A problem similar to that that bedevils any assessment of the literature on minimum ratios confounds the constant ratio literature: it is highly likely that the reported examples are a selected set, drawn from a number of data sets whose ratios did not appear to the various authors to approximate constancy. As Selvin and Stuart (1966) observe, when a survey of data is "dredged" for examples manifesting a pattern, one cannot assess the probability of the examples as if the dredging had not occurred. Diamond (1973), for example, presents eight species in two genera of fruit pigeons in New Guinea. *Contra* Roth (1981), the seven ratios thus generated *are* remarkably constant, at Pr < 0.001 for a Barton-David test of G_{17}. However, because there are 513 breeding birds in New Guinea (Diamond, 1973) and consequently an enormous number of octets composed of but one or two genera, the observation of one octet that is improbable at the 0.001 level cannot falsify an hypothesis that species' sizes are independent. Even if they were, one might expect to see one octet with nearly constant ratios.

But even if one takes at face value all published claims of ratio constancy for which at least all sizes within the relevant taxon are available, the remarkable result (Simberloff and Boecklen, 1981) is that most data do not reject a random, independent null hypothesis in favor of an alternative of ratio constancy. Of 21 studies, only 4 claims of constancy are sustained for all data at the 0.05 level, whereas another three show a large fraction (but not most) of the data manifesting ratio constancy. Brown's (1975) heteromyid rodents of the Great Basin and Sonoran Deserts serve as a good example of ratio constancy, with the species forming unusually regular sequences at approximately the 0.10 and 0.05 probability levels, respectively. On the other hand, Smith's (1978) reef fish, also thought to be competitively displaced, are highly nonrandom in exactly the wrong direction: too many very small ratios to sustain the null hypothesis. If one raises the rejection level to 0.30, 11 claims of notably constant ratios are generally sustained, though for several there are still a number of specific cases not manifesting the claimed tendency.

T. Schoener (personal communication, 1981) has objected that this entire approach to assessing ratio constancy is illogical because, for example, a trio of species generating two small, but nearly identical, size ratios—say, 1.03 and 1.04—would be found to be highly nonrandom, whereas two large, but highly disparate, ratios—say, 2.0 and 3.0—would not reject a null hypothesis of size independence. Critical

412

ratios have not been deduced but have been chosen empirically. Schoener (1965) suggested 1.14 as indicating significant food size differences (Schoener, 1970), whereas Hutchinson (1959) chose 1.3 as approximately the mean difference that he observed in a few sets of species. It is often stated that the minimum size ratio observed in a given data set is the minimum that competition would allow, but since any set of numbers (if sufficiently precise) has a minimum, the existence of a minimum does not indicate a reason for that particular minimum. So, large ratios seem as worthy of consideration as small ones. In any event, Simberloff and Boecklen (1981) tested just the published assertions of constant ratios, not a new hypothesis that ratios will tend toward constancy only if they are small. However, if one is to follow Schoener's advice and focus on small ratios, it is only fair to disqualify as evidence for competitive displacement examples, such as the hawks cited earlier (MacArthur, 1972), with large ratios that depict remarkable constancy (Pr < 0.04 for the hawks).

SIZE RATIOS ON ISLANDS

Schoener (1965) and Grant (1968, 1969) observe that bill length differences between congeneric species of birds on islands generally exceed those on the mainland. If one considers only the same species on mainland and island, it is far from certain that the island ratios are generally the larger. There are few published comparisons. For the Tres Marias Islands, there are two congeneric pairs (Grant, 1966, 1968): *Myiarchus* and *Vireo*. The island and mainland bill length ratios are the same for both genera. For the West Indies, Grant (1969) observed that of 15 congeneric double invasions of islands, 5 convergences and 5 divergences have occurred, and 5 pairs have the same ratios on islands and mainland. For all of North America and Mexico, Grant (1966) found just three congeneric pairs (two on Newfoundland and one on Cozumel) in which each member had evolved subspecific differences from the mainland form; in two of these the island ratio exceeded the mainland ratio, and in the third there was no difference. For the much greater number of pairs where one or both species do not have subspecific island forms, there is no evidence that island ratios are greater. And the point raised earlier for a heterogeneric pair is equally valid for a congeneric one: one could as well attribute an increased insular ratio, should one be observed, to character release as to character displacement.

If one seeks evidence for competitive character displacement among all congeners found on an island (not just species found on both

413

the island and mainland), the situation is complicated by the fact that one would expect larger ratios on the island even if the island species were but a random proper subset of the mainland pool, simply because there are fewer species (Strong et al., 1979). So one must specify a null expectation for ratios, which in turn requires a model for how an island is colonized.

For islands, the two possible reasons for large size ratios—evolution and unusually dissimilar colonists—may be examined separately, at least if the source area for an island can be denoted. Grant (1966, 1969, 1970) and Grant and Abbott (1980) contend that particular subsets of species found on particular islands are improbably dissimilar, relative to all possible subsets that might have been assembled from the same species pools. Grant (1969) examined five bird genera (*Buteo, Geotrygon, Corvus, Vireo*, and *Dendroica*) in the West Indies and adjacent North and Central America. For *Corvus* and *Dendroica*, the mainland representatives of the island pairs were approximately in the middle of the size ratio distribution for all possible pairs, whereas for the other three genera the mainland forms of the colonizing pair were extraordinarily different from one another. Grant (1970) also examined island colonization by rodents and concluded (as for birds) that competitive interactions have limited the cooccurrence of similar genera; however, here "similarity" meant not size but food niche and other ecological features. In any event, Grant's conclusion that similar genera generally do not colonize together is not statistically supported (Simberloff and Connor, 1981). Finally, Grant (1966) and Grant and Abbott (1980) contend that it is likely that many of the 16 Mexican mainland congeners of Tres Marias Islands birds are precluded from colonizing the islands by being too similar to the insular species. Only two congeneric pairs (noted earlier) have colonized, and these tend to be in the tail of the distribution of bill length differences for the sixteen mainland pairs, but they do not appear to be in the extreme tail. At present, it is not possible to treat this contention precisely because the mainland measurements, by Ridgway and Friedmann (1901–1950), are not all of the same bill dimension nor are they all of the same bill dimension that Grant (1965a) measured.

When one allows for postcolonization evolution, questions of character displacement are not so easily framed because the answer must rest on a more complicated null model. For birds of the Tres Marias, Strong et al. (1979) compared the culmen length ratios within families to the ratios in random subsets (of equal size) from the mainland pool of the same families and found that the island ratios did not differ significantly from random mainland subsets—a result confirmed by Hendrickson (1981). Grant and Abbott (1980) contend that within-family size ratios are not informative because confamilials on average are not very similar ecologically. This may be true, but in the Tres Marias

414

there are only two congeneric pairs, and, as noted earlier, their ratios on island and mainland are the same. For families (Columbidae and Tyrannidae) that have more than two species in the Tres Marias, Simberloff (1982) used Barton-David and Irwin tests to see if bill-size ratios on the islands were more constant and if minima were larger than those on the mainland. Because the null probability of a particular ratio changes with number of species, he compared the probabilities in the two locations of ratios as large or as constant as observed. There did not appear to be significant differences between mainland and island results, though generally the former seemed more extreme—a result exactly opposite to the prediction of competitive character displacement.

Grant (1968) examined thirteen pairs (and one trio) of congeners on various islands; each pair was thought to consist of two descendants of the same mainland species. Bill-length ratios range from 1.16 to 1.50, and Grant interpreted this as evidence that lower ratios would be forbidden by food competition. However, as Grant noted, the usefulness of these data is limited by absence of comparable mainland data. Because every set of numbers has a minimum, the existence of a minimum ratio (1.16) does not by itself tell us why the minimum has the value it does. Lack of comparable mainland data beclouds the significance of similar observations [e.g., Amadon (1953) for the Gulf of Guinea Islands; Grant (1966) for the Comoro Islands] that certain insular birds have large bill-length differences. Grant (1968) also sees evidence for insular character displacement in Schoener's West Indian bird data (1965), but Strong et al. (1979) find 67 island congeneric ratios greater than mainland ones and 54 mainland quotients greater than island ones, which show no difference between island and mainland forms (one-tailed binomial test, Pr = 0.19). This method is far from decisive, however, because it examines all congeneric ratios, not just those between nearest neighbors in a size ranking.

Keast (1970) has explained much of the size variation in Tasmanian versus Victorian birds (Table 1) as response to variation in interspecific competitive pressure for food, but there seems to be an explanation in these terms for every conceivable pattern, so it is difficult to see how this hypothesis could have been falsified. For robins, the single island species is intermediate in bill size between the two mainland species; Keast interprets this as a response to the island bird's occupation of much of the ecological zone used by both mainland species. For warblers in the genus *Sericornis*, by contrast, the bill of the one mainland species is intermediate in size between those of the two island species, and this is cited as reflecting the need for island

415

TABLE 1. Sizes of selected groups of Tasmanian and Victorian birds.[a]

Group	Mainland	Size	Island	Size (mm)
Robins	*Eupsaltria australis*	10.7[b]	*P. rodinogaster*	8.6
	Petroica rosea	8.0		
Warblers	*Sericornis frontalis*	10.5[b]	*S. magnus*	9.8
			S. humilis	11.7
Warblers	*Acanthiza lineata*	7.2[b]		
	A. nana	7.3		
	A. pusilla	7.4	*A. pusilla*	9.3
	A. reguloides	8.3		
			A. ewingi	8.0
Cuckoo shrikes	*Coracina novae-hollandiae*	18.9[b]	*C. novae-hollandiae*	17.4
	C. tenuirostris	17.8		
	C. robusta	16.0		
Owls	*Ninox novae-seelandiae*	231–238[c]	*N. novae-seelandiae*	198–222
	N. connivans	303–325		
	Tyto tenebricosa	295–320		
	T. novae-hollandiae	298–333	*T. novae-hollandiae*	♂319–343
				♀360–387
	N. strenua	398–420		

[a]Data from Keast (1970).
[b]Bill length.
[c]Wing length.

birds' bills to be especially different because of competition for food. One could have said the same thing, with locations reversed, for the robins. For *Acanthiza* warblers, the bill of the one species in common between mainland and island is much larger on the island, whereas the bill of the second island species is larger than all but one of the four mainland species. These size increases Keast ascribes to the greater ecological diversity (range of foods) likely eaten by the island birds (Grant, 1965b). Grant (1965b) attributes this increased food diversity to ecological release rather than to increased competition. For the cuckoo shrikes, the bill of the single island species is much smaller than that of its mainland conspecific—near the middle of the size range of the three mainland species. The decrease in size of the bill of the island species Keast ascribes to generalization of the island species' role to incorporate much of the roles of all three mainland species— the same explanation as for the robins. For owls, the two island species are both found on the mainland, embedded in a five-species set. On the island, the two species have diverged—the larger increasing, the smaller decreasing—and Keast feels the divergence is due to increasing competition. So one sees that every contingency is explained by

416

either increased competition for food or fewer competitors on islands: size increase on islands, size decrease on islands, divergence on islands, lack of divergence on islands. Keast (1968) develops a similar scenario for honeyeaters *(Melithreptus)*. Size ratios on Tasmania are greater, reflecting increased competition for a poorer resource base, and the lone species on Kangaroo Island has increased bill size because of lower competition because a larger species is absent.

Finally, Higuchi (1980) has combined observations on the number of combinations of woodpecker species seen on Japanese islands and the sizes of component species of missing combinations to conclude that similar sizes preclude species from coexisting. Eleven species produce only five two-species combinations (on 13 islands) of the 15 possible. The independent-colonization model of Simberloff and Connor (1981) shows that this result alone is not surprising: 20% of the time one would have expected five or fewer combinations. But the absent pairs constitute morphologically similar species, whereas the five observed pairs all have larger size ratios. Simberloff (1982) suggests this may be a statistical artifact rather than a biological mechanism. All five observed pairs contain *Dendrocopus kizuki*, much the smallest species and also much the most frequent. Because it is most frequent, combinations including it are far more likely even under an independent-colonization model; and because it is smallest, these are the very combinations that generate the largest size ratios. It may be that its small size releases it from competition with the other species and thus accounts for its common occurrence, but one would require independent evidence of the phenomenon and not just the sizes and geographic patterns.

WEST INDIES ANOLES

The *Anolis* lizards of the West Indies are one of two archipelagic faunas (the other is the Galápagos finches) for which character displacement has most often been invoked as an explanation for sizes of coexisting species. Schoener (1969b, 1970), Williams (1969, 1972), and Roughgarden et al. (1981) all propose scenarios of this sort (see also Chapter 17 by Roughgarden). Over 70 species inhabit at least 75 islands, with numbers ranging from just one species, on at least 40 islands, to 28 on Cuba. Because the Bahamas and other small islands north of the Greater Antilles have faunas that were established rather recently by Pleistocene sea level changes (Williams, 1969; Roughgarden, 1979), attention has focused on the Greater Antilles and especially the Lesser Antilles.

The Lesser Antilles all contain either one or two species; species on the one-species islands are mostly very similar in size (henceforth denoted "solitary size"), whereas species on the two-species islands are usually of very different sizes (Schoener, 1969b, 1970; Williams, 1972; Roughgarden et al., 1981). The sizes of solitary anoles given by Schoener (1970) are depicted in Table 2. For the Lesser Antilles, all sizes are between 16.8 mm and 22.3 mm, except for *A. marmoratus* on Marie-Galante (28.4 mm). Tobago, whose lone species (*A. richardi*) is found in the Lesser Antilles, would be a second exception, because its size is 30.8 mm. If one were to add the Cayman Islands, also believed to have an old *Anolis* fauna, Grand Cayman's species is in the range of the solitary size, whereas Cayman Brac's *A. sagrei* is too small (14.9 mm). It is unlikely that such uniformity of size would arise by chance colonization from a larger pool of anoles. Schoener (1969b) notes that only 6 of 40 one-species islands have lizards outside a size range that encompasses only 10 of the 42 species on Cuba and Hispaniola (Pr < 0.001). Furthermore, the same solitary size seems to obtain in

TABLE 2. Solitary anoles (*Anolis*) in the West Indies.[a]

Island	Species	Size[b]
St. Croix	*acutus*[c]	18.3
Anguilla	*gingivinus*[c]	18.8
St. Bartholomew	*gingivinus*[c]	19.7
Saba	*sabanus*[c]	18.3
Redonda	*nubilis*[c]	21.1
Montserrat	*lividis*[c]	18.7
La Desirade	*marmoratus*[c]	19.9
Guadeloupe	*marmoratus*[c]	18.9
Terre de Bas	*marmoratus*[c]	21.3
Terre de Haut	*marmoratus*[c]	21.5
Marie-Galante	*marmoratus*[c]	28.4
Dominica	*oculatus*[c]	22.3
Martinique	*roquet*[d]	19.2
Barbados	*extremus*[d]	18.4
St. Lucia	*lucieae*[d]	21.4
Grenadines	*aeneus*[d]	16.8
Tobago	*richardi*[d]	30.8
Grand Cayman	*conspersus*	19.0
Cayman Brac	*sagrei*	14.9

[a]Data from Schoener (1970).
[b]Size is mean head length (in mm) of largest third of adult males.
[c]*bimaculatus* subgroup.
[d]*roquet* subgroup.

418

islands colonized by at least two different stocks (*roquet* and *bimaculatus*). So either selective colonization or subsequent evolution apparently produces the uniformity; and Schoener (1969b) and later authors believe evolution is the primary force. Exactly what it is that makes species smaller than 16.8 mm or greater than 22.3 mm unsuitable solitary island colonists has never been satisfactorily explained, though the islands are ecologically very different from one another. Schoener (1969b) explains the convergence of solitary size, as well as sexual dimorphism in size of solitary species, as a function of intraspecific resource competition, primarily for food.

Undersized exceptions to the solitary size—*sagrei* on Cayman Brac and in the Bahamas—might reflect thermodynamic considerations or resource poverty or seasonality (Schoener 1969b). Similarly, Roughgarden and Fuentes (1977) attributed interisland size variation in the smallest of the "normal" solitary anoles—*aeneus*—to amount of available food. If food supply really does determine size variation, then the uniformity of solitary sizes implies a uniformity of food supply. Several possible explanations are offered by Schoener (1969b) and Williams (1972) for the two very large solitary lizards *A. marmoratus* on Marie-Galante and *A. richardi* on Tobago. *Anolis richardi* may be too recent on Tobago to have evolved a solitary size; a smaller species may have gone extinct on Marie-Galante; there may be two distinct stable size optima, one more likely to be achieved than the other.

Table 3 lists anoles on two-species islands in the Lesser Antilles plus Little Cayman. Schoener (1970), Williams (1972), and Roughgarden et al. (1981) all emphasize the differences between members of a sympatric pair, but suggest different reasons—all involving character displacement. Before discussing them, I should note that it is not certain that the species are remarkably different relative to the independent, random-size null model described earlier. Simberloff and Boecklen (1981) used Williams' data (1972, Table 4) and determined that if one assumes an anole can evolve to any size between the maximum and minimum "feasible" anoles specified by Williams, ratios on the islands occupied by two species of the *bimaculatus* group have a combined null probability of 0.038 if a specific minimum is being tested or 0.536 if no prior value is specified. For the three islands with pairs of species in the *roquet* group, the combined null probability is 0.176 if a specific minimum is being tested or 0.631 if no particular minimum is specified in advance.

Williams' model is simplest, resting on divergent competitive character displacement. He finds the smallest ratio—on St. Maarten (1.45; Schoener, 1970)—to be associated with a habitat and geo-

TABLE 3. Anoles on two-species islands in the West Indies.[a]

Island	Species	Size[b]
St. Maarten	*gingivinus*[c]	19.6
	wattsi[c]	13.5
St. Eustatius	*bimaculatus*[c]	23.6
	wattsi[c]	14.3
St. Kitts	*bimaculatus*[c]	27.9
	wattsi[c]	14.3
Nevis	*bimaculatus*[c]	27.1
	wattsi[c]	14.2
Barbuda	*leachi*[c]	28.6
	wattsi[c]	14.5
Antigua	*leachi*[c]	28.8
	wattsi[c]	14.9
St. Vincent	*richardi*[d]	31.1
	trinitatus[d]	17.1
Grenada	*richardi*[d]	30.8
	aeneus[d]	18.1
Carriacou	*richardi*[d]	29.5
	aeneus[d]	17.2
Bequia	*richardi*[d]	28.4
	aeneus[d]	17.0
Little Cayman	*maynardi*	23.0
	sagrei	14.9

[a]Data from Schoener (1970).
[b]Size as in Table 2.
[c]*bimaculatus* subgroup.
[d]*roquet* subgroup.

graphical restriction for one species (*A. wattsi*) that he feels permits the small ratio, and he records other habitat shifts for at least one member of all sympatric pairs. As noted earlier, a claim that a minimum size difference is not met but that extraordinary separation in another dimension allows the size similarity is problematic. As for the divergence, *A. aeneus* is inconsistent with it. One may observe (Tables 2 and 3) that *A. aeneus* is alone only on various islands in the Grenadines but that it has increased in size on the three islands (Grenada, Carriacou, and Bequia) that it shares with the much larger *A. richardi*. Schoener does not actually state why the observed size ratios are all in the observed range, but he is quite explicit (1969b) that there are circumstances for anoles in which both species might decrease in size once a second species establishes itself on a previously one-species island. Though for other parts of the West Indies Schoener

420

(1969b) finds suitable data to test this notion, in the Lesser Antilles few appropriate data exist. None are strikingly supportive, and the *aeneus* and *richardi* data of Tables 2 and 3 seem if anything to imply a slight increase in *aeneus* on two-species islands. Roughgarden et al. (1981) erected a more complicated scenario for Lesser Antillean anoles whereby a two-species island can arise only when a larger species invades an island inhabited by a smaller one, after which both species evolve to be smaller and the original species ultimately becomes extinct unless habitat segregation arises (see Chapter 17). Roughgarden et al. (1981) found that on the two islands with anoles of the most similar sizes (St. Maarten and St. Eustatius) there is microhabitat separation of the species; Roughgarden et al. propose that similar sizes would generally preclude syntopy. There is no direct evidence for the cyclical evolutionary scheme, but one observation seems mildly supportive. In Table 3, three islands have the *bimaculatus-wattsi* pair, two have the *leachi-wattsi* pair, and three have the *richardi-aeneus* pair. Many of the size differences are very small, but it is interesting that for the latter two pairs the two species get smaller (or larger) together from island to island, whereas for the first pair only *wattsi* on St. Eustatius is slightly too large for this pattern to obtain. One could interpret pairs varying together in terms of the Roughgarden et al. hypothesis: the islands where the species are larger are nearer the beginning of the cycle. On the other hand, there is no direct evidence for the model.

For sizes in larger anole faunas of the Greater Antilles, two explanations have been offered. Williams (1972) applies the same model of divergent character displacement that he erected for the Lesser Antilles to the ten Puerto Rican species and combines it with independent estimates of the times of invasions of the different species. Williams depicts the first divergence to occur as that between the giant *cuvieri* and dwarf *occultus* in the crown of shaded forest. Next he posits the evolution of the intermediate-sized *evermanni*, also in the crown; possibly *evermanni* already existed in its present size, but its size permitted it to invade the crown from another microhabitat. Given these three species, Williams feels the crown is saturated and other anoles will have to invade other microhabitats. The sizes ratios of this trio (1.8, 1.95) are quite constant by the Barton-David test (Pr = 0.085), and it is also true that any other species would generate at least one ratio lower than the smallest seen on two-species Lesser Antilles islands. The next invader—*gundlachi*—is thus forced to live below the crown, that is, on tree trunks or shrubs. Next is *krugi*, whose habitat is similar to that of *gundlachi* but whose size is smaller by a factor of only 1:1.3. This size similarity Williams views as permissible by virtue of microhabitat displacement. Now the shaded forest is full, and further evolution consists of climatically different versions of the five forest species. *Anolis stratulus* is similar to *evermanni* but lives in drier regions. It is also smaller than *evermanni* by a factor of 1.5, which Schoener and Schoener (1971) and Williams (1972) attribute to divergent character displacement after secondary contact between the two species. *Anolis cristatellus* and *cooki* live in successively drier habitats but occupy the same structural microhabitat as *gundlachi*, whereas *pulchellus* and *poncensis* also live in successively drier habitats but occupy the same microhabitat as *krugi*. Throughout, Williams emphasizes the primacy of size differences in determining who will be able to invade and/or what evolution must occur.

Schoener (1970) looked at all pairs of anoles in the Greater Antilles that were quite similar to one another in structural habitat but for which at least one species is somewhere allopatric from the other. He sought evidence on whether character displacement actually occurred and whether it is divergent [as Brown and Wilson (1956) suggested] or usually has both species decrease in size [as Schoener's energetic model (1969a) predicts]. In general he finds that predictions of size decreases are much more often borne out than are predictions of increase, and he sees this trend as supporting his model. A complication is that the interspecific differences may not all be significant. However, those pairs where a larger species decreases in size in sympatry with a smaller species are primarily those where the larger species was very large, as Schoener's model predicts. Similarly, Schoener's model predicts that for trios of species the larger pair should have the larger ratio, and this pattern is confirmed for anoles of three-species islands at the 0.012 level (Simberloff and Boecklen, 1981). [For the bird trios examined by Schoener (1965), this pattern is not manifest, however; $Pr = 0.18$.] Finally, Schoener and Gorman (1968) tie anole interspecific sexual dimorphism to Schoener's energetic model by pointing to the greater size ratio between sexes for *richardi* than for *aeneus* on Grenada. Simberloff and Boecklen (1981) show that even if the sizes of species, and of sexes within species, were independent of one another, the ratio difference observed on Grenada would have obtained at least 22% of the time.

GALÁPAGOS FINCHES

The Galápagos finches, especially *Geospiza*, are the other archipelagic fauna whose sizes are frequently explained as a function of character displacement. Of the 13 finches, 6 belong to *Geospiza*, and these inhabit from 5 (*G. conirostris*) to 18 islands (*G. fuliginosa*) (Abbott et al., 1977). No island has all six species; at least three have five species;

and at least two small islands have just one species. Not only are the six species different from one another in bill size and shape and in other morphological features, but within a species the populations are morphologically different from island to island. Lack (1947) first proposed that the interspecific variation in bill morphology is primarily a result of competitive character displacement: on each island that a species occupies it generally faces a different set of competitors, hence it evolves a bill of different size and shape. Lack modified his views (1969) to include a secondary selective role for differences among islands in vegetative and physical environment, but he continued to emphasize competitive character displacement as the main determinant. Grant (1981, 1982; Grant and Schluter, 1982; Abbott et al., 1977) has brought many more data to bear on the finches and has arrived at a position similar to Lack's.

From Lack's morphological data (1947, Table XXIII), it is clear that there is a fair amount of variation even if the species set is held constant. Lack finds that one five-species set occupies three islands, one four-species set occupies three islands, and one three-species set occupies two islands. Beak depth [the size variable that Grant (1981) focuses on] varies considerably even when potential competitors are the same (Figure 2), though perhaps the variation for a given species tends to be narrower when cooccupants are held constant than for all of its populations together (Table 4). Bowman (1961) contended that most interspecific size variation was generated by food supply rather than by coexisting species; and Abbott et al. (1977) demonstrated that the islands are indeed distinctive in the spectrum of seeds and fruits each presents to the finches. Strong et al. (1979) and Strong and Simberloff (1981) show that all the geospizines (not just *Geospiza*) on a given island tend weakly to vary in bill length and bill shape (beak depth/bill length) in the same direction: an island may tend to have short-billed finches, or thick-billed finches, and so forth. This concordant variation is also consistent with Bowman's hypothesis. It does not preclude interspecific effects, however.

Because the Galápagos finches are not subsets of a mainland pool (as are, say, the birds of the Tres Marias), there is no way to ask if they are nonrandom subsets of all available colonists. One might assume that the inter- as well as intraspecific variation is an evolved trait, an assumption that is consistent with the evidence that the finches arose from one or at most two colonizations (Grant, 1981). In order to see if sizes of the island populations are compatible with an hypothesis of divergent character displacement, Strong et al. (1979) randomly chose for each island the number of species actually observed on the island.

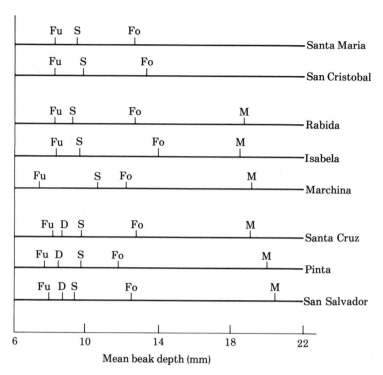

FIGURE 2. Mean beak depths (mm) of species in *Geospiza* combinations that occur on more than one island (data from Lack, 1947). Fu, *fuliginosa*; S, *scandens*; Fo, *fortis*; M, *magnirostris*; D, *difficilis*.

Then, for each drawn species chosen to colonize an island, they randomly drew one of the available races. They then calculated size ratios of contiguous size-ranked species in the random set and compared these to the observed values. When median ratios for replicate simulations are compared to actual ratios (Hendrickson, 1981), one finds for tree finch bill length and depth and ground finch depth no differences between observation and expectation, whereas *Geospiza* bill-length ratios are significantly larger than expected—an observation that is consistent with the divergent displacement model. Grant and Abbott (1980) and Grant and Schluter (1982) object that this classic randomization technique is not likely to be useful for detecting displacement between pairs of species (e.g., *G. scandens* and *G. fortis*) that have almost all of their populations sympatric with one another, because even randomly chosen races would have evolved in sympatry. This fact doubtless lessens the power of the test.

Abbott et al. (1977) also saw divergent competitive character displacement in the fact that *Geospiza* minimum beak depth ratios are

424

TABLE 4. Mean beak depths (in mm) of *Geospiza* in combinations found on more than one island,[a] plus standard deviations within species among island means.

Species	Island	Island	Island	STANDARD DEVIATION for test	STANDARD DEVIATION over all islands
COMBINATION 1	Charles	Chatham			
fortis	12.5	13.2		0.35	0.65
fuliginosa	8.1	8.1		0	0.25
scandens	9.4	9.7		0.15	0.49
COMBINATION 2	Bindloe	S. Albemarle	Jervis		
magnirostris	19.1	18.4	18.6	0.29	0.94
fortis	12.1	13.9	12.6	0.76	0.65
fuliginosa	7.4	8.3	8.2	0.40	0.25
scandens	10.6	9.6	9.2	0.59	0.49
COMBINATION 3	Indefatigable	Abingdon	James		
magnirostris	19.1	20.0	20.5	0.58	0.94
fortis	12.8	11.8	12.5	0.42	0.65
fuliginosa	8.2	7.7	8.0	0.21	0.25
difficilis	8.7	8.5	9.4	0.39	0.48
scandens	9.8	9.7	8.8	0.45	0.49

[a]Data from Lack (1947, Table XXIII).

larger, on average, the fewer the species present. This trend implied to them that small, resource-poor, and species-poor islands could only be shared by very different species. Strong et al. (1979) pointed out that even random species subsets would have the observed negative correlation between number of species and minimum ratio; Hendrickson (1981) found, however, that 12 of 15 minima are above medians for random sets (Pr < 0.04), so it may well be that minimum ratios tend to exceed the expected values. There is no indication, however, that this excess is larger on species-poor islands, so the data are inconsistent with the hypothesis of intensified competition on small islands (Simberloff, 1982). A similar pattern obtains when bill width is considered along with length and depth (Grant and Abbott, 1980; Grant, 1981): size neighbors tend to be more different (using Mahalanobis distance as a measure of difference) than they would be in random sets (this observa-

425

tion is consistent with divergent character displacement); but the excesses are not related to numbers of species present, so there is no apparent intensification of competition on small islands. Also using Mahalanobis distance as a measure of difference, Grant (1982) found that 12 of 14 observed species pairs have sympatric races on average more different than allopatric races. However, the two species pairs most similar morphologically are the two that are more similar for sympatric races than for allopatric races. Grant feels that these two anomalies need not invalidate an hypothesis that competition has heavily influenced bill morphology, and he suggests that habitat differences between these pairs may allow the similar morphology.

Simberloff (1982) performed Barton-David and Irwin tests separately on *Geospiza* and *Camarhynchus*; he was seeking evidence of unusually large minimum ratios or constant ratios and of whether any such trends are more exaggerated on smaller islands. The only significant result was for minimum beak depth ratios in *Geospiza*; these are remarkably large, as a competition hypothesis would have predicted. They are not especially larger on smaller islands, however. Finally, the two *Geospiza* species pairs that are never found together in nature are not particularly similar to one another in either bill length or depth (Simberloff, 1982).

The sum of these results (Simberloff, 1982) is that most tests do not show divergent character displacement, but a substantial minority show *Geospiza* bill morphology to be displaced as competition would predict. A difficulty is that several related hypotheses with correlated tests of statistical significance have been framed, and only certain data have been examined of all that are available. Selvin and Stuart (1966) show that such procedures render significance tests difficult, and it is very likely that the probabilities of reported extreme results are inflated.

CONCLUSIONS

One sees repeatedly that it is very difficult to falsify a hypothesis that size ratios of coexisting species are unusually large because of competition. The problem is that when species do not differ in size as much as the competition hypothesis predicts investigators commonly argue that some other difference that *is* large is the difference that allows size similarity in the face of competition (Simberloff and Boecklen,1981). Rarely is the core contention of competition directly addressed. For example, Grant (1968) suggested that sympatric congeneric birds on islands must differ in bill length by 15% or more. When Mees (1970) noted that two Norfolk Island *Zosterops* actually differ in bill length by approximately 4.5% and not the originally measured 16.7%, Grant (1972b) amended his rule to read, "Sympatric, double-invasion, species pairs of birds on islands differ in at least one bill dimension by 15% or more."

More frequently the critical difference allowing sympatry of similar-sized species is sought in some behavioral or ecological dimension. Examples abound in this chapter, and Simberloff and Boecklen (1981) cite several others. Short (1978) studied sympatry in lowland Malayan woodpeckers and concluded that specialized species with similar foraging modes must differ greatly in size or must shift some aspect of their foraging modes and that all 13 species fulfill this criterion. Enders (1974) believes that a 1.28 size ratio is required by pairs of *Araneus* spiders to avoid competitive exclusion, but there are species pairs more similar than 1.28 in size that coexist by occupying somewhat different habitats.

Because some differences can always be found between any two species, this sort of explanation of exceptions to some size ratio rule comes perilously close to rendering competitively caused character displacement an unfalsifiable hypothesis (Simberloff and Boecklen, 1981). So long as exact predictions are stated in advance and are in principle not trivially true and exceptions are not treated in an egregiously ad hoc fashion, it might be possible to assess some claim of competitive displacement in morphology and/or some other trait(s). Davidson (1977a), Davidson et al. (1980), and Begon and Mortimer (1981) contend that harvester ants that are the same size cannot coexist unless they differ in foraging mode (group or individual). The hypothesis in this instance is very clear (Davidson, 1977a): "Figure 4 [of Davidson, 1977a] provides some suggestion that niche separation based on worker body sizes and colony foraging strategies of coexisting species is greater than would be predicted if combinations of coexisting species occurred at random without respect to these characteristics." Davidson describes 14 common harvester ant species (seven group and seven individual foragers) arranged over ten sites, with species sets and rankings of mean worker body length as in Table 5. To test Davidson's assertion, I randomly drew 2, 3, 4, 5, 6, and 7 species 100 times each from the available pool of 14, each time asking how many nearest-size neighbors of identical foraging type were found. Then I placed the observed number of nearest-size neighbors of identical foraging type in the simulated distributions (Table 6). Fisher's combined probability test shows $\chi^2 \cong -2 \Sigma \ln P_i = 11.667$ (20 degrees of freedom, $0.90 < \Pr < 0.95$), and clearly there is nothing surprising about seeing so few nearest-size neighbors of identical foraging type on ten sites with the observed diversities. If anything, there are more such cases than one would have expected if species were really random and independent of one another in these two traits.

This test was sensible only because all data were provided, the null

TABLE 5. Size rankings and foraging modes of common harvester ants at ten United States Southwest sites.[a]

Species[b]	SITE									
	1	2	3	4	5	6	7	8	9	10
1				G	G	G		G	G	
2							G			
3							G	G	G	
4				G		G	G	G	G	G
5								S	S	S
6						S				
7	G	G		G	G	G	G			
8	S									
9		S	S	S			S			
10								S	S	S
11					S					
12									G	
13						G		G		G
14			S			S		S	S	S
TOTAL	2	2	2	4	3	6	5	7	7	5

[a]Data from Davidson (1977a).
[b]G, Group forager; S, solitary forager.

TABLE 6. Observed numbers of harvester ant size-neighbors of identical foraging type at ten sites, and null probabilities of this many or fewer.

Site	Number of species	Number of size-neighbors of identical foraging types	Null probability of such an extreme result
1	2	0	0.87
2	2	0	0.87
3	2	0	0.87
4	4	0	0.53
5	3	0	0.69
6	6	0	0.17
7	5	4[a]	0.97
8	7	2	0.48
9	7	2	0.48
10	5	0	0.32

[a]Three size-neighbors in sequence; all group foragers.

428

hypothesis was clearly stated, and the trait examined in addition to body size (foraging mode) was well rationalized and exhaustively studied (Davidson, 1977a,b). Furthermore, Brown and Davidson (1977) and Munger and Brown (1981) have experimentally shown that seeds are a limiting resource for these animals in at least some sites. Rarely in the character displacement literature does one find a study with even one of these characteristics, much less with all of them. Consequently, it is difficult even to frame most contentions as falsifiable hypotheses, much less to evaluate them. This problem is compounded as more traits are examined, because even if all traits are specified in advance, the null probability that random species sets would show every species pair differing in at least one trait is very large. That a hypothesis is not operationally falsifiable does not, of course, mean it is incorrect, but it greatly vitiates the importance one can accord a published result. Strong (1980) has emphasized the importance of null hypotheses and falsifiability to evolutionary ecology generally; Huey (1979) has reasoned similarly with respect to character displacement and adding further niche dimensions when size does not sufficiently separate species: "In the absence of attempts to falsify alternative explanations, observations of parapatry or of niche dimension complementarity do not demonstrate conclusively the impact of competition as a force structuring communities." There seems to me little reason to believe that the philosophical approach that has served the physical sciences so well (Popper, 1963) will not be equally enlightening in evolutionary ecology.

Finally, it is possible that studies of character displacement in size are treating an epiphenomenon and that natural selection on a set of species has acted primarily to produce a suite of body shapes, with body size just a correlated character of shape (Simberloff and Boecklen, 1981). Mosimann and James (1979) show that size and shape generally covary and that at most one size variable in a population can be statistically independent of shape. To my knowledge, no study of character displacement has attempted to isolate such a size variable. The usual assumption seems to be that divergence results from selection on size, with shape a passive correlated trait; indeed, this is the *modus operandi* for most work on size (Lande, 1979). Lande (1979) suggests a way to determine which of two correlated traits has been the main target of natural selection, but this approach has not been adopted yet for character displacement. There seems no a priori reason to assume size is more important than shape. Indeed, though for anoles (e.g., Schoener, 1967; Schoener and Gorman, 1968) and harvester ants (Davidson, 1977a) there is good evidence of correlation of food size with body size, Wilson (1975a) suggests that for many animals body size is not likely to be a good index of the food niche, so that size displacement based on competition for food would not be expected. Slatkin (1980) has similarly suggested on genetic grounds that size displacement is less likely than is widely believed. Whether the same caveats obtain for shape is unstudied.

In summary, a variety of patterns have been sought in the size ratios of coexisting species of several taxa and have been ascribed to the coevolutionary pressures of resource competition. Even if the size arrays are not epiphenomenal, the patterns have often not been clearly stated as falsifiable hypotheses, much less tested. When statistically tested, some instances of character displacement are found, but many of the perceived patterns do not emerge and no general trend is manifest. This observation argues that sizes of coexisting species in communities in general are not primarily a coevolutionary product of interspecific competition, though the forces that mold any particular community may include such interactions.

CHAPTER 19

CONVERGENT EVOLUTION AT THE COMMUNITY LEVEL

Gordon H. Orians and Robert T. Paine

Since the time of global biological exploration by plant geographers and ecologists of the past century, it has been suggested that areas in different parts of the world with similar climates have vegetation with similar structures. This finding was the basis for the classification of the Earth's vegetation into types corresponding with, and often being named for, the climatic region in which they were found, for example, Mediterranean Vegetation Type (Schimper, 1903; Grisebach, 1872; Raunkiaer, 1934; Cain, 1944; Good, 1974). Because each regional flora was composed of taxonomically distinct groups of plants, this similarity of overall appearance prompted the fundamental question whether two very similar physical environments would produce structurally and functionally similar ecosystems even if those forces acted on phylogenetically dissimilar organisms. The extent to which community ecology can become a predictive science depends, in part, on the degree to which the answer to that question is affirmative.

Convergent evolution can be said to occur when features of an organism come to resemble features of another organism more closely than was the case among their ancestors. Community-level convergence occurs when features of assemblages in different communities become more similar than they were previously; it depends in part on adaptive trends within the component species, but such features as

431

species richness can converge in the absence of convergence among the component species. Also, species convergence does not guarantee that features that depend on the aggregate properties of species will also converge. Coevolution, on the other hand, is strictly a within-community phenomenon (although convergent similarities may be identified between communities) and can be recognized as an evolved reciprocity of response in the interrelationships of two or more species.

Convergent evolution can be driven by two types of interwoven, but distinct, factors. Organisms can converge because they are evolving in response to similar *physical features* of the environment. These features usually are little influenced by the activities of the organisms themselves, although terrestrial soils and marine substrates, and their respective microclimates, for example, depend strongly upon the organisms living at a site. Organisms can also converge in response to *biological interactions* leading to similar patterns of coevolution in different geographical regions. Thus, it is possible to imagine that coevolution could result in convergence in both the adaptations of unrelated species and in community-level properties of aggregates of species.

One reason for expecting convergence is that available energy is limited and must be allocated to competing functions such as feeding, locomotion, defense, growth, reproduction, and maintenance. If environments in two localities are similar, the best allocation patterns of resources among these competing functions may also be similar. Among ectotherms, the cost of body maintenance and the rates of most biological functions increase with temperature. Thus, their overall energy budgets are temperature sensitive and inversely related to latitude. In terrestrial environments, water is often in short supply, and overall energy budgets, particularly those of plants, may be restricted as much by dryness as by cold.

Community convergence is not an all-or-nothing phenomenon. Various degrees of convergence are possible, and a community can converge strikingly in some features and not at all in others. In its extreme form, the hypothesis of convergence can be stated as (Orians and Solbrig, 1977, vii), "given two regions with identical climate, geology, and topography, ecosystems that are identical in structure and function will result, despite differences in the initial floras and faunas, provided that there has been sufficient time for natural selection to have produced convergence." Even this extreme statement, which is demonstrably false, cannot be tested in an entirely satisfactory way. No two regions on Earth can be matched precisely, nor can we determine just what constitutes "sufficient time." Moreover, ecological communities contain many potentially convergent features, and the importance attached to any one of them is a matter of informed judgment.

432

DIFFICULTIES IN THE STUDY OF CONVERGENCE

Study of convergence requires clarity about the level at which convergence *is* being sought, knowing at what level convergence *should* be sought, and deciding whether a given pattern really represents convergence or is simply the result of some of the basic properties of living systems, such as shared photosynthetic mechanisms, or changes that inevitably accompany differences in size (Maynard Smith, 1968; Western, 1979). Clarity of thinking about these problems does not, however, resolve a number of difficulties faced by any field ecologist who wishes to put either intuitive feelings or data sets derived from comparative work in different parts of the world to a more rigorous test. These difficulties stem from problems that we will discuss in the following sections.

The long time spans of relevant processes

The slowness of evolution means that initial differences among colonists influence their characteristics for a very long time. Therefore, initially different communities that are at two remote sites and that are in the process of converging are likely to differ in many traits. Unfortunately, we don't know the starting dates, how long is long enough, and how that time varies for different traits of communities for many comparisons of potential interest.

Difficulty and limited value of experimentation

Rarely can experiments be continued long enough to reveal evolutionary adjustments among species. We cannot introduce a lizard to an island and observe the evolution of size differences in response to competitive interactions. We can observe immediate responses, such as agonistic interactions, habitat segregation, and short-term changes in population densities, that provide insights into the effects of species arrivals, but longer-term responses are not readily predictable from short-term events.

Some long-term observations are available because many species introductions, most of them the result of human activity, have taken place over the past few millennia, and for many we know their approximate starting dates. For even longer time spans we must make use of "natural experiments," that is, introductions nature has provided for us. Such "experiments" have already played an important role in the study of evolutionary ecology. Their major weakness is the absence of a strict control.

433

How similar is similar enough?

Because we do not expect convergence to identity, either among species or among communities, traits that reflect ancestry always remain. Determining how much convergence might have occurred despite these persisting differences requires comparing and contrasting a large number of units. One technique is to seek similarities among close genetic relatives that occupy nearby, but different, habitats on the same continent and to compare these with nonrelatives that occupy the same habitats on different continents. However, there are problems associated with choice of faunal units and patterns to be compared (Crowder, 1980; Fuentes, 1980; Grant and Abbott, 1980; Hendrickson, 1981; Strong and Simberloff, 1981), and no resolution exists on how to eliminate, or minimize, the important subjective element in such comparisons.

Appropriate units for assessing community convergence

Convergence has been sought in three rather different traits of ecological communities: flow of energy and materials, physical structure, and species richness. These traits do not necessarily covary, and each reveals something different about the evolution of communities.

Overall community productivity in terrestrial systems is similar in communities with similar climates, but this reflects parallel rather than convergent evolution. Productivity in marine systems is much more complex because many assemblages do not depend primarily, or even partially, on *in situ* photosynthesis. Nonetheless, it is useful to look for convergence in patterns of allocation of energy to different processes.

In terrestrial communities physical structure is overwhelmingly determined by plants, and useful comparisons can be made of their growth forms and phenologies. Associated with these patterns are life history and temporal patterns among animals. In contrast, in marine communities, much of the habitat heterogeneity and of its associated consequences is generated by animals.

Most attention has been given to species richness comparisons. In terrestrial communities the extreme richness of animal species and the different skills needed to study different taxa have generated a strong taxonomic orientation. This approach, driven by practical considerations, has influenced the way in which community structure has been investigated. For example, the concept of the guild (Root, 1967) refers to the members of a taxonomically defined specialty group utilizing a common resource by similar techniques, for example, foliage-gleaning insectivorous birds, seed-eating mammals, and flower-visiting bees. The concept also applies to marine environments, except membership

434

is usually polyphyletic. In both domains members of guilds should exert much stronger influences on one another's distribution, abundance, and evolution than on organisms placed in other guilds.

The geographical scale of convergence

Community traits can be measured over several geographical scales. Ecologists recognize four levels, each with its appropriate richness (number of species) or diversity (some weighted measure of species number). The number of species found in a single point-sample is referred to as the *point richness*. The number of species found in a relatively small, uniform plot is referred to as α-richness (α-diversity). The rate at which species change with habitat structure on a local scale is referred to as β-richness (β-diversity). Finally, species replacement rates on a broad geographical scale are referred to as γ-richness (γ-diversity). Division of the space scale into these categories is arbitrary, but different processes are thereby highlighted. β-richness and γ-richness patterns are likely to show the effects of history longer than α-richness patterns. In practice, conclusions drawn from independently sampled communities must be sensitive to these distinctions.

Distinctness of biotas

For convergence to occur, initial community composition must be reasonably distinct, but not too much so (one does not expect a tortoise to become a cow no matter how long it evolves in the absence of competition with ungulates). Ever since the breakup of Pangaea, the continents have experienced some degree of isolation. The southern continents, in particular, have been sufficiently isolated for long enough times to have produced some of the classic cases of convergence, for example, radiation of marsupials and placentals.

Marine environments, on the other hand, are much less isolated. Transatlantic genetic exchanges occur through long-distance dispersal of larvae (Scheltema, 1971). The shores of southern continents and intervening islands share many common species (Knox, 1960; Newman, 1978). In fact, Williams et al. (1981) have concluded that oceanic diatom distribution is what could be expected from a single population inhabiting a relatively homogeneous environment. Ironically, this uniformity underlies the major themes of marine community comparison. Thorson (1957) identified from shallow subtidal sands or muds a number of "parallel level bottom communities." Such assemblages are often similar in species composition, local density,

435

and even life history patterns and can occur from tropical to polar environments. Stephenson and Stephenson (1972) promoted a "universal" scheme for subdividing the communities of rocky shores, with each major zone being identified with one group of organisms. Both approaches reveal numerous genera, and even species, held in common. The lack of biotic independence, probably due to excellent long-range dispersal capabilities, seriously compromises attempts to recognize convergence.

Community saturation

Implicit in community convergence in species richness patterns is the notion that assemblages eventually reach some saturation level, for if richness continues to increase indefinitely over evolutionary time, convergence at any of the scales identified above will be difficult to recognize. Evidence for the existence of long-term species saturation as opposed to short-term *equilibrium* species richness is difficult to come by. In terrestrial communities many introductions have been successful (Elton, 1958), although the most conspicuous examples involve disturbed habitats. The fact that the richest terrestrial plant communities are geologically the oldest (Whittaker, 1977) strongly suggests that plant species accumulate over very long evolutionary time spans. If so, the animals dependent on plants, many of which are specialists, can be expected to change as well.

Similarly, the literature on near-shore marine communities indicates that most are some distance from saturation and that, especially in temperate zones, we are witnessing an increased tempo of successful invasion that involves all trophic levels (Norton, 1976; Crisp, 1958; Quayle, 1964). Even species-rich assemblages are not immune: the eastern Pacific shore community has been augmented by a host of successful invaders with no apparent loss of native flora or fauna (Ricketts et al., 1968). Subtropical communities as well are susceptible; Knight-Jones and Knight-Jones (1980) record two spirorbid worms successfully invading the Canary Islands (28°N) despite competition from nine resident species.

Currently, no techniques exist to evaluate the relative levels of species saturation in comparable communities. Differences in species saturation are not apt to seriously bias productivity estimates, but they are certain to introduce problems of unknown magnitude to species richness and guild comparisons.

Morphological and behavioral flexibility

Members of most plant and animal phyla are indeterminate growers, with body size or mass reflecting local conditions for growth. Because

436

individual size has implications for the relative susceptibility to disturbance and relative ability to coexist, to escape predation, or to consume prey, growth plasticity is of obvious ecological significance.
Most between-community terrestrial comparisons, although sensitive
to within-group differences, are focused on taxa of roughly similar
mean mass whether they be insects, lizards, or birds.

Such within-taxon comparability is often not found in marine
groups. The asteroids of coastal Washington State are species rich;
some are exceptionally large-bodied (Mauzey et al., 1968). Vermeij
(1978) has identified persistent differences in body size and shell morphologies in Atlantic versus Pacific molluscan taxa. In the vicinity of
Cape Town, South Africa, the limpet assemblage (Patellidae and Acmaeidae only) includes 13 species with an average shell length of 52
mm (based on Day, 1969). Approximately the same number of species
occurs in central California; their mean shell length is 26 mm (various
sources). These lengths translate into an order of magnitude difference
in mass, or the same body weight differences separating kangaroo rats
from rabbits or sparrows from quail. Whether such size differences are
of the same ecological significance as size differences among terrestrial
organisms is unknown.

Multiple morphological and behavioral solutions to adaptive problems are common, and they complicate efforts to assess the extent of
convergence. For instance, a variety of innovations enable the vertebrate eye to function adequately underwater (Walls, 1942). Succulence
in desert plants is achieved in various ways (Peet, 1978). Branch (1981)
has documented that most limpets are subject to extensive predation.
Thus, although "flight" responses are common to certain members of
all limpet (Patellidae and Acmaeidae) assemblages, mantle rolling
responses, chemical shell crypsis, and the development of territories
with home scars are not. Without a detailed understanding of the organism's functional morphology and clear identification of its adaptive environment, convergence could be missed or judged to be present
where it is not.

Also, we can identify species for which no taxonomic uncertainty
exists, but that are radically different in different places. The asteroid
Crossaster paposus is primarily a predator of sea stars in the North
Atlantic, whereas in the eastern Pacific basin it consumes holothurians and nudibranchs (Mauzey et al., 1968). Detailed studies of the
carnivorous gastropod *Conus* (Kohn, 1959; Kohn and Nybakken, 1975)
illustrate similar shifts in diet, suggesting that when there is a variety
of adaptive options available to a species pool, there can be no guarantee of ecological constancy in geographically separated communities.

437

BASIC DIFFERENCES BETWEEN MARINE AND TERRESTRIAL ECOSYSTEMS

Patterns of convergence or nonconvergence in marine and terrestrial ecosystems reflect, in part, basic differences in these systems caused primarily by differences in the physical environments.

Competition for sunlight influences plant morphology more strongly than it does that of the major sessile occupants of space in the marine environment, animals that compete both with one another and with benthic algae. Whereas there is root competition among terrestrial plants, benthic algae obtain their nutrients from the water and compete only for access to the water column (and for light only to a lesser extent). Water provides much greater physical support than does air, but the movement of water generates much greater forces. Thus, terrestrial environments with moderate to high precipitation are dominated by woody plants that devote substantial energy to structural carbohydrates and lignins that provide physical support and are resistant to degradation by animals, whereas marine plants allocate resources more to photosynthetic or reproductive tissues than to structural ones. This basic difference between terrestrial and marine vegetation has enormous consequences for the structure of the associated decomposer and herbivore communities. Moreover, because of the large heat capacity of water, and the limited depth to which light can penetrate, the wide variety of regional community types that terrestrial plants exhibit (e.g., Holdridge, 1947) cannot exist in the marine environment. The limited depth of penetration of light into water also means that most marine assemblages depend on a rain of allochthonously fixed carbon, whereas most terrestrial ecosystems are energetically self-sufficient. Our surveys, then, focus on events in the photic zone and deal with consumer rather than decomposer food chains.

Another difference involves two universal aspects of marine communities, especially benthic ones. The most fundamental observation is that while competition for physical space is usually intense (Connell, 1961; Dayton, 1971; Woodin, 1974), monopolization of a resource by competitive dominants tends to be restricted by predation (Paine, 1966) or disturbance (Levin and Paine, 1974). For instance, Grice et al. (1980) have shown experimentally that mesoscale, enclosed planktonic assemblages are powerfully influenced by predators, and Dayton and Hessler (1972), although not without challenge, have made a similar suggestion for the deep sea. Second, most marine algae tend to be vulnerable to herbivores at all life history stages, and, thus, such community attributes as species composition, morphology, and local primary productivity are under strict grazer control (Lubchenco and Gaines, 1981). Grazers exert less control in terrestrial systems today but they might have exerted more control in the recent past before the extermination of the rich Pleistocene mammal communities of all continents except Africa. The presence of critical species characterized by disproportionate influences on community structure and organization and linked by strong interactions to many members of the associated community appears to be a general, though not universal, phenomenon. It seems especially characteristic of aquatic communities (Paine, 1980).

CONVERGENCE AMONG TERRESTRIAL GUILDS

Because of the taxonomic bias of most terrestrial work, we begin with a discussion of some of the best-studied taxonomic guilds and then turn to some cases that have been studied from the point of view of the resource.

Convergence in bird community structure

Ornithologists have long known that different habitat types support characteristically different species and numbers of species of birds. Local experts also know what species to expect in an area simply by looking at the vegetation. Insights into the bases of those intuitive judgments were stimulated by the work of Robert MacArthur and his associates (MacArthur, 1971; MacArthur and MacArthur, 1961; MacArthur et al., 1962), who showed that bird species diversity was strongly correlated with foliage height diversity of the site. Similar results have been obtained by others (Orians, 1969; Recher, 1969, 1971; Schoener, 1971a; Tramer, 1974). α-Diversities are greater for a given foliage profile in tropical regions than in temperate ones, and island communities are impoverished in bird species compared to mainland ones (MacArthur et al., 1966). The larger number of species in tropical communities is due to both a larger array of resources in those communities, as judged by the morphological variability of the birds themselves (Karr and James, 1975), and to the finer division of resources.

The reasons for the correlations between foliage height profiles and bird species richness are still not clear. The number of distinct ways in which prey may be sought, rather than the number of prey species per se, is probably the key factor. Some insights can be derived from communities that have more or fewer species than expected. For example, hot desert communities have more species than expected (Tomoff, 1974), apparently because the life forms of desert shrubs are so varied that different types may provide highly distinct foraging substrates and because some growth forms, especially spinescent shrubs, provide unusually safe nest sites. The addition of just a few individuals of those plant species increases the number of bird species without having any important influence on foliage height profiles. In more humid environments, nest sites are probably more generally available and different plant species do not offer such distinct sites with respect to safety from predators. Fewer bird species than expected are found in forests in which the distribution of leaves is highly uniform vertically. An example is forests dominated by strongly multilayered trees (Horn, 1971), where foraging conditions for birds change very little from the top to the bottom of the canopy (Balda, 1969).

Convergence in bird community structure has been investigated in two major vegetation types of the world: hot deserts and Mediterranean scrub. For hot deserts the best information is available for Sonoran Desert in Arizona and the similar Monte of Argentina (Orians and Solbrig, 1977). These two areas support similar numbers of species of birds, and in both areas bird species richness bears the same relationship to plant cover diversity, a measure of the proportion of total plant cover found in each of the following groups of perennial plants: herbs; small, nonspinescent shrubs; spinescent shrubs, cacti; and small trees. Both diversity of foraging substrates and nest sites are strongly correlated with this measure, so it is difficult to distinguish which is contributing most to the similarities. That genuine convergence and not parallel evolution is involved is shown by the large number of analogous species pairs drawn from distinct avian families.

Studies of the structure of Mediterranean-type bird communities are more complete because there are comparisons for more continents (North and South America and Africa) and because data on α- and β-diversities are also available (Cody, 1975). In California, Chile, and South Africa, α-diversities are more similar than β- and γ-diversities. There are more species of birds at the grassland extreme in Chile, more species at intermediate sites in Africa, but more species in the taller forests in California. Also, species turnovers (β-diversity) occur more rapidly in Africa and most slowly in Chile. Africa has large expanses of savannah and scrub habitats, a great richness of bird species in those habitats, and high species turnover with slight changes in vegetation structure. Chile and California have much less scrub and far fewer species adapted to such habitats. California, on the other hand, has extensive woodlands and forests, vegetation that is scarce in both Chile and southern Africa.

These observations suggest that long-term opportunities for speciation, correlated with the geographical extent of different vegetation types, may exert important influences on species richness patterns, especially at the β-level but also at the α-level as well.

Convergence in lizard community structure

Studies of lizard community structure differ from those on birds in that there are fewer species of lizards in any region than there are of birds, making it more difficult to obtain reliable measurements of species turnovers along habitat gradients. On the other hand, there is much more precise information on the foods of lizards because of the ease with which they can be captured and the stomach contents analyzed. For both groups the best available comparisons are for hot deserts and Mediterranean vegetation.

Pianka and his associates (Pianka, 1966, 1967, 1969, 1971, 1973, 1975; Pianka and Huey, 1971) have carried out extensive studies of the structure of lizard communities in the Sonoran Desert of North America, the Kalahari Desert of southern Africa, and the deserts of central Australia. In all three areas the number of lizard species is positively correlated with the standard deviation in mean annual precipitation; this finding suggests that variability in productivity is favorable to lizards or unfavorable to their competitors and predators. In other respects the lizard communities do not seem to be convergent in their major features. The number of lizard species per site varies from a mean of 6.9 for the North American sites (range, 4–10), 14.7 for the Kalahari sites (range, 11–18), to 28.3 for the Australian sites (range, 18–40). The relative poverty of North American communities is due to the absence of subterranean and arboreal nocturnal species and to many fewer nocturnal terrestrial and diurnal terrestrial species. Lizard diets are also quite different in the three deserts, termites being especially important in Africa and vertebrates in Australia.

The reasons for these striking differences are not clear, but Pianka (1975) believes that Australian lizards in part take over ecological roles filled by other taxa in African and North American deserts. For example, the pygopodid and varanid lizards in Australia replace certain snakes and carnivorous mammals, taxa that are relatively impoverished in Australia. There are, however, more "typical" lizards in Australia than in the other two continents, and Pianka suggests, from indirect evidence, that competition among lizards and birds may be involved. Whatever the reasons, it is evident that very little convergence has occurred in these desert lizard communities.

CONVERGENCE AMONG TROPHIC GUILDS IN TERRESTRIAL COMMUNITIES

The resources of any ecological community are seldom exploited exclusively or even primarily by a single taxonomic group. Plants produce a variety of structures, leaves, flowers, pollen, nectar, sap, fruits, seeds, and wood that are sufficiently different in their physical and chemical properties that they are usually exploited by rather distinct groups of animals. Some of these structures, such as nectar and fleshy

441

fruits, have evolved to attract animals, whereas others are heavily defended. The distinctiveness of these plant-produced resources and the animals exploiting them has made a guild approach, based on type of plant tissue, a profitable one. These studies have provided the best evidence of competition among organisms belonging to very different taxa in terrestrial communities.

Photosynthetic plants—the dominant trophic guild in terrestrial communities

The amount of solar energy reaching the Earth's surface varies in a predictable way with latitude, allowing an excellent match for this key variable among geographically remote sites. Moreover, the amount of solar energy actually captured by photosynthesis is limited by several key processes, including the absorption spectra of chlorophyll. With the uniformity of photosynthetic mechanisms among terrestrial plants, combined with the tight linkage between transpirational loss and CO_2 entrance, it is not surprising that yearly total gross photosynthesis can be reasonably predicted from a knowledge of potential evapotranspiration (Rosenzweig, 1968). Ryahchikov (1968) demonstrated a tight correlation between yearly plant growth (tons per hectare per year) and a measure termed the hydrothermal productivity potential, which is derived from estimates of precipitation, growing season, and annual radiation balance. The success of these regressions points to the strong dependence of total photosynthesis upon rather simple climatic relationships. In fact, Holdridge (1947) based his global classification of plant communities on similar criteria.

However, we consider this relationship a parallel and not a convergent one. The "convergence" is the result of the possession of a common mechanism for capture of solar energy shared by all vascular plants regardless of their taxonomic affinities. Nonetheless, the relationship between yearly gross photosynthesis and climate is valuable in the study of true convergence because it guarantees that if climates are well matched the energetic resources for ecological communities should also be well matched.

Other features of plant community structure, however, are not similarly constrained by common mechanisms. There is no necessary reason for photosynthesis to be accomplished by plants of the same growth form, height, or density. Similarity in these measures, the ones that originally attracted the early plant geographers, are more informative of evolutionary processes than is total photosynthesis.

Plant growth form. A profitable way of viewing the evolution of plant growth forms is in terms of the carbon budgets of individual plants (Mooney and Dunn, 1970). Fitness is, of course, not measured in energetic units, but a plant capable of increasing its annual carbon

442

budget has more energy to devote to competition, defense, and reproduction, activities that do contribute to total fitness. If regional climate exerts a dominant influence on carbon-gaining strategies, then convergence in those features of plant form that affect carbon gain should follow.

The key role of the constraints imposed by Mediterranean climates with their mild, wet winters and hot, dry summers has been treated extensively by Mooney and Dunn (1970) and Mooney et al. (1975), who give compelling arguments for the observed dominance of evergreen-sclerophyllous shrubs in these climates. Within regions with Mediterranean climates, there are important local differences that allow more precise testing of the theories. For example, in both Chile and California there is a zone with mild winter temperatures but low annual rainfall and a long, severe drought during the warmer half of the year adjacent to the ocean. Further inland there is greater seasonality of temperatures, somewhat higher precipitation, but, nonetheless, long summer droughts. At higher elevations, winter temperatures are substantially lower, precipitation is greater, and the summer drought less severe. In both areas the coastal zone is dominated by drought-deciduous shrubs. Further inland broadleaf evergreens predominate, whereas at higher elevations winter deciduous trees and shrubs become common (Mooney, 1977). Carbon-budget models successfully predict where changes between these vegetation types occur (Miller, 1979). Similarities are also found in leaf size, leaf-area index, leaf duration, specific leaf weight, and photosynthetic pathway among plants occupying comparable positions on the climatic gradients on the two continents (Mooney, 1977). Similar community-level patterns include seasonal pattern of leaf, twig, and litter production and the presence of a lush herb community in winter and early spring. An important difference is the much longer period of shrub growth in Chile than in California.

Latitudinal and altitudinal gradients in plant community structure. Information about changes in plant communities along latitudinal and altitudinal gradients provides a useful supplement to data gathered by comparing specific sites matched climatically. For example, Jordan (1971) found that the ratio of energy bound in long-lived tissues to that bound in short-lived tissues increases with latitude, probably in response to the greater stresses on high-latitude trees. Also the caloric concentrations of tissues of trees increase with latitude. The adaptive significance of the underlying, carbon-allocation patterns that lead to these community patterns is not well understood. Leigh (1975) showed that in tropical forests leaf size decreases with

443

altitude at a rate of approximately 4.5% for every degree centigrade drop in the average temperature of the coldest month. Leigh interprets this change in terms of the reduction in transpiration rates with altitude in such humid regions. The existence of the same trend in all tropical regions, despite major differences in the taxonomic affinities of the dominant trees, suggests that true convergence is probably involved.

Plant species richness. In contrast to patterns in growth form, repeatable patterns in the number of plant species in comparable regions have not been found (see Whittaker, 1977, for a general review). A particularly good data base is available for Mediterranean vegetation on five different continents. African and Australian plant communities are far richer in species than their California and Chilean counterparts (Werger et al., 1972a,b; Whittaker, 1977; Werger, 1973; Whittaker and Niering, 1965, 1968; Parsons and Cameron, 1974; Westman, 1975; Mooney and Dunn, 1970; Parson and Moldenke, 1975).

Comparable measures of β-richness are not available, but the literature strongly indicates that in the South African fynbos both α-richness and β-richness are remarkably high. There has been a great proliferation of species in some genera, culminating in over 600 species in the genus *Erica* alone (Walter, 1973). Part of the reason for these differences may be the much greater age of the African and Australian Mediterranean climates compared to those of Chile and California, which may date only from the Miocene (Axelrod, 1973).

Nonconvergence in species richness is also found in temperate deciduous forests, those of eastern North America and eastern Asia being much richer than those of Europe. Again, a historical explanation seems likely (Whittaker, 1977). Thus, the only community property that might plausibly be affected by coevolution, that is, species richness, does not show convergence; the convergence observed in terrestrial plant communities is of individual species characteristics such as growth form and leaf size, which may simply reflect the adaptation of individual species to physical factors.

The seed-eating guild of hot deserts

The guild of animals for which the best global comparisons are available is the seed-eating guild of hot deserts. Three groups of animals, each with rather different characteristics, dominate this guild. Granivorous birds are highly mobile and concentrate in areas where seeds are locally abundant (Reitt and Pimm, 1976; Pulliam and Parker, 1979). Ants and rodents both maintain permanent populations in local habitats and store seeds for times of scarcity. Rodents are large and have high individual energy requirements during periods of food shortages, whereas ants are small. Ant colonies can survive even if large numbers of their members starve during periods of low food availability. Field experiments have demonstrated the importance of competition among rodents and ants in Arizona (Brown et al., 1979; Davidson et al., 1980). Birds, as exploiters of temporary patches of high seed densities, appear to influence the other two much less. Competition among ants and rodents has also been shown to influence the relative abundances of desert annual plants (Inouye et al., 1980); this process provides feedback between the exploiters and the composition of their prey.

For both ants and rodents, the numbers of species increase along a gradient of increasing resource productivity, but the number of species present at any site is always less than the number whose geographical distribution make them potential colonists of those sites. For both groups, communities consist of species of different sizes. In ants, body size influences exploitation of seeds of different sizes (Davidson, 1977a,b), but in rodents it influences the microhabitats where the animals can safely forage. Larger rodent species are more adept at escaping predators and can forage at greater distances from cover (Brown, 1975; Brown and Lieberman, 1973; Hutto, 1978; Reichman and Obsertein, 1977).

Despite the existence of strong competition for food, convergence in desert granivore community structure on different continents is limited for several reasons. First, the evolution of seed size is probably more strongly influenced by competition among plants, particularly at seedling establishment, than it is by size-selective seed predation (Baker, 1972; Salisbury, 1942; C. C. Smith, 1975). Second, even though rodents and ants compete with one another, there is as yet no evidence that this competition influences the *numbers* of species in each group that live together or their specific properties. Competition experiments are too short to clarify this point, but it will be useful to gather evidence of movements of new species into areas from which one or the other of the two groups has been excluded.

Rodents able to live on metabolic water derived from seeds are found in most hot deserts of the world, but not in Argentina (Mares, 1976). There are fossils of South American bipedal marsupials (Argyrolagidae) that were probably granivores, but they became extinct at the time of the great faunal interchange between North and South America, possibly as a result of the arrival of northern placental carnivores (Simpson, 1970), which also caused the extinction of the indigenous marsupial carnivores. However, neither species richness nor overall abundances of ants are higher in Argentine deserts than in Arizona. This is quite unexpected, given the immediate and dramatic response of ants to the removal of rodents from experimental plots in

445

Arizona. No convincing theory exists to explain this anomaly, and, unfortunately, a thorough comparative study of the characteristics of seeds of desert plants in the two localities has not yet been undertaken.

The folivore guild

Major patterns among communities of folivorous insects have been summarized by Lawton and Strong (1981). These patterns are interpreted very differently by different people. However, competition by means of resource depression has long been considered to be rare in communities of folivorous insects (Andrewartha and Birch, 1954). Some cases of competition among folivores have been discovered, but their frequency is so low that it is difficult to invoke it as a major process molding current structures of folivorous insect communities.

The situation is less clear with respect to resource quality. Nitrogen levels in plant tissues are highly variable (McNeill and Southwood, 1978); levels are often too low to support adequate growth of insects (Friend, 1958), and many insects shift host species in response to changing nitrogen levels (Gibson, 1980). In addition, plants devote considerable energy to the production of defensive compounds that exert important effects on the feeding, growth, fecundity, and digestion efficiency of folivorous insects (D. A. Levin, 1976; Rhoades and Cates, 1976; Rhoades, 1979). Plants are, in turn, involved with complex trade-offs in their allocation of energy reserves (Cates and Orians, 1975; Gigon, 1979), and they change their allocations in response to physical stress (T. C. R. White, 1974, 1978) and grazing pressure itself (Haukioja and Niemela, 1979; Rhoades, 1983).

The consistent finding that plants with more complex structures support more species of herbivorous insects than plants with simpler structures (Lawton, 1978; Lawton and Strong, 1981; Southwood et al., 1979; Schultz, 1978) suggests that the number of distinct hiding places on a plant may be a major determinant of the number of species of small herbivores that can coexist on it. Whether parasitoids, which are so common in most insect communities (Askew, 1980; Holmes and Price, 1980; Price, 1973), are similarly affected by habitat complexity is unknown.

There are only a few attempts to determine the degree of convergence in the structure of folivore communities. One approach is to examine the community on a single plant species or group of closely related species in different parts of the world. This has been done for insects on bracken fern by Lawton (1982), who found little evidence of convergence. Work on insect communities on desert shrubs in Arizona and Argentina revealed similar patterns of tissue preferences and similar correlations between the number of species of herbivores and plant structure (Orians and Solbrig, 1977), even though the morphologies and behavior of the component species were quite different.

446

The flower-visitor guild

Although mutualistic interactions between flowers and their visitors have attracted considerable attention, much less attention has been given to multispecies interactions and how competition among flowers for visitors and among visitors for floral resources act together to mold community structure. There are strong reasons for believing that the interactions are complex and asymmetrical (Kodric-Brown and Brown, 1979). There is little evidence that pollinator availability limits seed set, and total overlap in use of pollinators does not prevent plants from coexisting. In fact, sympatric species of plants sometimes converge to use the same pollinators (Grant and Grant, 1965, 1968).

From the perspective of the flower visitors, however, floral resources are as depletable as any other food resource, and interference and depletion are regularly observed (Kodric-Brown and Brown, 1979; Chapter 13 by Feinsinger). Depending upon suitability of the pollinators as movers of pollen, changes in flowers that favor one type of visitor over another, such as shape, color, and time of nectar production, may be favored. Larger flower visitors can dominate smaller ones, but birds cannot effectively defend resources against bees. Community-level convergence in the flower and flower-visitor community was examined in Arizona and Argentina (Orians and Solbrig, 1977), but little evidence of convergence was uncovered. In all measures, such as percentage of species that are self-compatible, importance of different categories of flower visitors, floral morphology, total production of floral rewards, and species richness of flower visitors, the two communities were strikingly different.

Summary of terrestrial community convergence

The preceding discussion indicates that evidence exists for a limited amount of convergence in plant community structure and in species richness patterns for a few taxa. In most cases, however, either little convergence has occurred or the nature and extent of available information does not allow such convergence to be detected. These results are summarized briefly in Table 1. Much of the convergence that is evident appears to consist of convergent adaptations or similar physiological responses of individual species, rather than similar patterns of diversity of resource utilization that might be attributed to coevolution.

A MARINE PERSPECTIVE ON CONVERGENCE

Although the subject of convergence has substantial appeal for students of terrestrial plant and animal assemblages, the subject has

TABLE 1. Summary of evidence of terrestrial community convergence.

Community or community component	Temporal pattern of productivity	Plant growth forms	Phenology	α richness	β richness	γ richness
			CONVERGENCE OF			
Plant communities						
Mediterranean	Yes	Yes	Yes	No	No	No
Hot desert	—	Yes	—	—	—	—
Bird communities				Yes	Perhaps	Little
Lizard communities						
Mediterranean				Yes	Yes	—
Hot desert				No	No	—
Seed-eating guild of hot deserts				No	No	—
Folivores				No	No	No
Flower visitors in hot deserts				No	No	No

generally been ignored in temperate zone marine communities. Why this is so reveals subtle dimensions of the topic.

Ecological equivalence and functional groups

Many near-shore communities share similar organizing processes, which provide an environment in which the important biological processes are comparable. Marine ecologists have identified a limited number of resource utilization patterns (Woodin and Jackson, 1979), in which the principal axes are feeding mode, relative mobility, and extent of coloniality. Each major group contains large numbers of species that therefore become (roughly) members of the same guild (Root, 1967) in that they exploit the environment in similar ways. Woodin and Jackson (1979) stress the varied phyletic composition of these guilds. For instance, the mobile grazers may be composed of echinoderms, crustaceans, and mollusks; the sessile, colonial suspension feeders are usually drawn from at least six phyla. Although subdivision within guilds by body morphology is possible, such refinements do not reduce substantially the overall taxonomic complexity. Thus, in both Washington State and Chile, a diverse group of sponges, colonial tunicates, crustose coralline algae, bryozoa, and coelenterates compete for space on rock surfaces under an algal canopy.

448

An extreme case, illustrating the difficulty of comparing guilds between different regions, has been examined by Paine and Suchanek (1983). In Chile, the chordate *Pyura praeputialis* (a large solitary tunicate) occupies a niche comparable in many ways to that of the bivalve mollusk *Mytilus californianus* along the shores of western North America. Both are large-bodied, competitively superior, long-lived species capable of monopolizing space in the midintertidal zone. Moreover, they overlap extensively in many of the subtle details of relation to conspecifics and predators, larval settlement patterns, and possible dependence on an associated biota. Other *Pyura* species are present in both communities, as are other mussels. However, the ecological attributes of *P. praeputialis* are much more similar to those of *M. californianus* than either is to its congener in the other continent. In communities where functional roles can potentially be filled by members of a different phylum, taxonomic similarity will provide inadequate guidelines for guild analysis.

Convergence within the kelp community

One prevalent assemblage found along most cold temperate shores where sufficient nutrients and rocky substrates are available is the kelp community, with primary membership drawn from the Laminariales and Durvilleacea. *Laminaria, Macrocystis, Ecklonia,* and *Durvillea* are the principals, with individual plants attaining lengths in excess of 20 m. Where true kelps are missing, as in the Antarctic, another brown algal order predominates, with individual plants characterized by blades 10 m long and 1 m wide (Moe and Silva, 1977). We would hesitate to identify these situations as convergent unless large body size itself is a sufficient indicator.

There may be some evidence for convergence in productivity in these communities. Where cool nutrient-rich waters occur (Nova Scotia, western Europe, California, western South Africa, southern New Zealand), the primary productivity is high, approaching 2000 g C $m^{-2} yr^{-1}$ (C. Hay, 1977; Mann, 1973; Velimirov et al., 1977). Although the figures are impressively similar, it is disquieting to note that *Gelidium*, being farmed by the limpet *Patella cochlear*, is as productive as a kelp (Branch and Branch, 1981) and that the epiflora of commercially grown mussels may even exceed these figures (Lapointe et al., 1981). Moreover, Duggins (1980) has calculated that in coastal Alaska total kelp bed productivity varies from 0 g C $m^{-2} yr^{-1}$ in the presence of normal urchin densities to 1774–4806 g C in their absence. These maximal rates tend to approach the productivity limit to photosynthesis set by natural light conditions in aqueous systems of 15 g C m^2 d^{-1} (Ryther, 1955). Productivity seems to depend on a uniform capacity for photosynthesis and immediate environmental factors, rather than on taxon, morphology, or geography.

Assemblages of large brown algae show substantial geographical variation in taxonomic composition. Southern hemisphere assemblages contain roughly the same number of genera and species, but these tend to be the same (*Ecklonia, Macrocystis, Durvillea*) in all communities, presumably because of the opportunity for exchange provided by the westwind drift (Knox, 1960). North Atlantic and possibly northwestern Pacific assemblages are not substantially different, although the dominant genera are *Laminaria* and *Alaria*. However, in the northeastern Pacific, two to four times as many species exist locally as in the southern hemisphere, and such monotypic genera as *Postelsia, Nereocystis, Lessoniopsis, Hedophyllum*, and *Pterygophora* are prominent, in addition to *Macrocystis* and *Laminaria*. Although productivities are comparable in all these macroalgal assemblages, their species richnesses are not (Table 2).

TABLE 2. The number of species and genera of large brown, intertidal or shallow, subtidal algae characteristic of specific sites on moderately exposed rocky shorelines. The biogeographical provinces are based on Vermeij (1978). Algal lists come in part from Kuhnemann (1971), Nagai (1940), and Womersley (1967).

Province and site	Latitude	Genera	Species
Cold Temperate North Atlantic			
Nova Scotia	44°N	5	6
Ireland	52°N	3	5
Cold Temperate North Pacific			
Southern Kuriles	44°N	5	7
Outer coast, Washington State	48°N	12	17
Central California	36°N	10	13
Cold Temperate South America			
Deseado, Argentina	48°S	2	3
Montemar, Chile	33°S	3	4
Cold Temperate South Africa			
West side, Cape Peninsula	34°S	3	3
Cold Temperate Australia			
Tasmania	42°S	4	5
Cold Temperate New Zealand			
Kaikoura	42°S	4	5

450

Kelp forests, like terrestrial forests, provide escape space, structural heterogeneity, sites for reproduction, and a varied set of food and nutrient resources. One might surmise, then, that the relationship between α-diversity and habitat structure shown by MacArthur and MacArthur (1961) might apply to fish associated with kelp beds, but this is not the case (Quast, 1971). Fronds of these brown algae floating near the surface also provide a suitable environment for an epibiota. *Saccorhiza* in Ireland are encrusted predominantly by the bryozoan *Membranipora membranacea*, with up to 60% cover (Ebling et al., 1948). The same bryozoan species may cover 90% of the photosynthetic surface of *Macrocystis* in California (Wing and Clendenning, 1971). However, in South Africa the kelp canopy is dominated by epiphytic algae, and *M. membranacea* is a minor component (Allen and Griffiths, 1981).

Limpets sometimes excavate pits in the kelp tissue. Such scarring is known from only 3 of the 10 communities listed in Table 2 and involved *Collisella pelta* and *Acmaea insessa* from the northern Pacific (Black, 1976) and *Scurria scurra* from Chile (Vermeij, 1978). It is possible that the European limpet *Patina pellucida* also forms an association with a laminarian. Choat and Black (1979) discuss the geographical nonuniformity of such limpet–laminarian associations and conclude that there are no consistent patterns of niche partitioning. Certain limpets are also morphologically specialized to utilize kelps by cruising up and down the stripes, feeding on diatoms and other microalgae. Despite the worldwide presence of stipitate kelp and limpets, this ecological role is only seen in the South African *Patella compressa* (Branch, 1971) and the eastern Pacific *Acmaea instabilis*. Comparable opportunities have not elicited similar responses elsewhere. We conclude that convergence within the limpet–kelp relationship is far from complete.

Algal turf assemblages

Convergent patterns are seen in algal adaptive responses to herbivore presence. An abundant literature documents that plants with $CaCO_3$ skeletons are less susceptible to grazing than those without such depositions. In temperate zones, where invertebrate grazers predominate, coralline algae with 80–85% $CaCO_3$ usually are the major algal understory group (Paine and Vadas, 1969). In tropical regions, with the addition of fish as major consumers of plant biomass (Bakus, 1969), heavily calcified brown and green algae also occur. The pattern of carbonate deposition is universal, is expressed both within and be-

451

tween major algal groups, and is directly related to herbivore activity. It provides probably the cleanest example of morphological convergence driven by a known selective influence for marine communities. By itself, however, it does not constitute community-level convergence. Hay (1981), however, has provided a possible example of the latter. In many tropical communities members of all three algal phyla grow as "turfs," a physiognomy found predominately in shallow reef habitats characterized by dense herbivore populations. His measurements indicate that grazers essentially defend the turf species from overgrowth by competitively more vigorous, more productive, and yet more palatable algae, and that the convergent turf morphology, though reducing individual productivity, provides an effective means of retaining space.

Herbivore guilds

If grazers have provided a major influence on algal evolutionary trends, it is reasonable to anticipate that some reciprocity exists. Furthermore, because the resource base in different temperate regions is comparable because of the taxonomic and overt morphological similarities of the floras, one might expect that similar grazer guilds would develop.

Sea urchins constitute one of the more adequately studied groups of temperate zone grazers. They appear ubiquitous on rocky shores and often are abundant in the intertidal and shallow subtidal zones. Their removal, whether through commercial exploitation, by natural predators, or in controlled manipulations inevitably produces a dramatic response in the algal community regulated by them (Duggins, 1980; Fricke, 1979; Estes and Palmisano, 1974; Jones and Kain, 1967; Kitching and Ebling, 1961; Paine and Vadas, 1969). Does this represent convergence or simply common heritage and the possession of a superior scraping organ?

In most temperate communities grazer guild membership is drawn from a broad array of higher taxa: annelids, gastropods, chitons, echinoderms, crustacea, and an occasional fish. Table 3 provides a comparison of the macroscopic invertebrates in three such guilds, based on what a competent observer might be expected to encounter during the course of fieldwork of moderate duration and intensity; more extensive lists could be compiled by a specialist focusing on a particular taxon. Table 3 shows little agreement in within-taxon species richness: *Fissurella* is species-rich in Chile, chitons in the northeastern Pacific. If body mass were considered, northeastern Pacific urchins and chitons, Chilean fissurellids, and African limpets would all be exceptional.

Some of these differences might be resolved by deleting species in-

TABLE 3. Membership in the macroscopic invertebrate grazer and anemone guilds on three west-facing, continental, rocky, intertidal shores.

Guild	W. Washington State (48°N)	Chile (33°S)	Cape Town (34°S)
Invertebrate herbivore			
Herbivorous starfish	1	1	1
Patelliform limpets	8	11	16
Fissurellid limpets	1	9	1
Chitons	10	9	3
Herbivorous snails	6	8	7
Hermit crabs	3	2	1
Herbivorous crabs	5	8	1
Sea urchins	3	2	1
Total	37	50	31
Sea anemones	11	14	11

advertently assigned to the wrong trophic group. For instance, some of the crabs, chitons, fissurellids, and even one urchin might really be "closet" carnivores. The difference in total membership might diminish if amphipods and herbivorous fish were included. However, such fine tuning would not alter the domination of the individual assemblages by certain higher taxa of extraordinary species richness or mass. These lists do not constitute strong evidence for community convergence, and it seems as likely that chance or historical factors are responsible for both the similarities and the conspicuous differences.

Anemone guilds

Sea anemones constitute a conspicuous and unique group of marine consumers: though some harbor and derive energy from symbiotic algae, the majority are essentially sessile, solitary carnivores. The species richness of these is comparable (Table 3) in several cold temperate, rocky, intertidal comparisons (South African list based on Day, 1969; Chile on Sebens and Paine, 1978). Furthermore, the largest-bodied species in each assemblage subsist primarily on macroscopic animal prey dislodged by wave action: *Anthopleura* in western North America (Dayton, 1973), *Isoulactis* and *Phymactis* in Chile (Sebens and Paine, 1978), and *Pseudactinia* and *Bunodactis* in South Africa

453

(R. T. Paine, unpublished). The guilds are similar in α-diversity, size distribution, and feeding behavior, but not in reproductive mode or the relative development of aggressive behavior.

MUTUALISM IN TEMPERATE ZONE MARINE COMMUNITIES

We have been comparing groups of species, assuming that examination of the group itself should prove sufficient to reveal convergent properties. This would not be the case if mutualistic relationships, defined as either facultative or obligate interdependence in which neither participant suffers a net loss of fitness, were at all common. We have chosen to discuss temperate zone examples in which both participants are free-living, approximately equal in mass, and in different trophic levels or phyla. They thus differ in some important aspects from the obligate symbioses characteristic of coral–zooxanthellae relationships in tropical waters.

Grazer-dependent algae

In a previous section we developed an overview of marine community organization based on two primary features: the susceptibility of benthic algae to grazers and the presence of strong, controlling interactions. These phenomena have generated conditions favorable for coevolution between benthic algae and their grazers. These grazer-dependent species seem to be general inhabitants of many rocky shore communities. For example, both Lubchenco (1978) and Slocum (1980) have shown that fleshy red algae benefit from gastropod grazing as does the persistence and recovery of coralline algae (Paine, 1980). Crustose coralline algae, moreover, apparently signal their presence to generalized consumers by producing a chemical larval recruitment cue (Morse et al., 1980). Although the signal may also be recognized by highly detrimental specific grazers as well, herbivores are a necessary component of the environment of these dependent species. In addition, in certain coralline algae, morphological adaptations to grazing protect both the meristematic and reproductive tissues (Steneck, 1982a) while simultaneously exposing expendible superficial cells.

Protection from enemies and animal–animal mutualisms

Most temperate zone animal–animal mutualisms share the common feature of being expressly related to protection from predators and in being facultative. In fact, the absence of obligate involvements in these near-shore situations suggests the existence of unknown evolutionary constraints. Bloom (1975) and Forester (1979) have examined sponge–scallop interdependencies in which each of the symbionts is

454

threatened by predators, nudibranchs and sea stars, respectively. Sponges interfere with attachment of starfish tubefeet; this interference benefits scallops; scallops reciprocate by increasing the supply of suspended nutrient for the sponge or by removing them from contact with nudibranchs. Vance (1978) has identified a clam–hydroid mutualism in which starfish again fill the role of predator. Osman and Haugsness (1981) document a hydroid–bryozoan partnership driven by predation by nudibranchs and flatworms. The reciprocity inherent in all the above mutualisms implies that some evolutionarily meaningful units of community structure involve more than one trophic level.

One model for the evolution of mutualisms is that of Roughgarden (1975), based on the tropical anemone–anemonefish observations of Verwey (1930). Although the ecological forces promoting these interspecific relationships are generally unknown, the anemone clearly provides protection and the fish is known to feed its host. None of the possible pairings between five species each of damselfish and anemones is obligate. In this case a developmental limit may be imposed by an anemone mimic, the corallimorphan *Amplexidiscus*, which is capable of attracting and consuming anemonefish (Hamner and Dunn, 1980).

Summary of convergence in marine benthic systems

Convergence can be sought at a variety of levels. Thus, abundant examples exist of common adaptive responses to single, overwhelming selective pressures such as predation (Vermeij, 1978). But such examples pertain to particular morphological structures, not to properties of entire assemblages. The search for community convergence has been unsatisfying; convergence may not even be identifiable in rocky shore communities for a complex of reasons. All the higher taxa are ancient, with pedigrees extending at least to the Cambrian. Many of the species have excellent larval or propagule dispersal abilities. Even slowly dispersing species may be transported great distances by rafting. Thus, the possibility of biotic distinctiveness (i.e., the assumption of community independence) is substantially reduced by the equivalent of a species level panmixis. Many of the species remain poorly known or even undescribed, and therefore the quality of comparison is diminished by the absence of specific natural historical detail and knowledge of functional morphology. Furthermore, many marine organisms are characterized by an enormous latitude of expression in the ecologically significant areas of morphology, body mass, and behavior. This flexibility increases the acuity of local adaptation; it makes between-community comparisons more difficult. In fact, if near-shore

455

assemblages are ecologically unsaturated, this flexibility should produce a varied and unpredictable array of responses. Finally, marine communities are organized around highly interactive, polyphyletic, functional groupings that tend not to share common morphological traits. Hence, community comparisons must be of the "apples and oranges" variety, with their implied imperfections. Coevolved mutualisms exist, but because they tend to be peripheral rather than central to organization, they remain obscure and seemingly unimportant.

We conclude this section with two thoughts. It may be too late to detect community convergence in temperate marine assemblages because of their antiquity and relative taxonomic uniformity, which suggests that after some time interval increasing similarity masks the adaptive process. Or, if community-level convergence has occurred and is detectable, it will be found in universal patterns of interaction that lack the conventional taxonomic and morphological common denominators.

CONCLUSIONS

On land there has been sufficient isolation of continental land masses to have provided many cases where terrestial biotas have evolved more or less independently for long enough to produce convergence. Those biotas, though distinct, are nonetheless similar enough that convergence should not be prevented because gross differences in morphology and physiology of original colonizers imposed impenetrable "adaptive valleys" between existing states and potentially convergent ones.

The best-known examples of striking convergence in terrestrial communities are exactly the ones first noted by plant geographers, namely, the physiognomic structure of plant communities. The reasons are that the physical environment exerts a powerful and direct influence upon plant morphology. It is when strong biological interactions become dominant that convergent patterns are more difficult to detect. Plant species richness, in contrast to plant growth forms, is not convergent. There is as yet no comprehensive theory or data base that allows us to assess the relative roles of direct competition, habitat heterogeneity, competition for pollinators and propagule dispersers, and selective grazing by herbivores in determining plant species richness. There is also strong evidence that plant species richness may continue to increase over long periods of evolutionary time.

If such an important component of community structure as plant species richness fails to converge on different continents, communities of animals dependent on those plants should also fail to converge in those characteristics that depend heavily on plant species richness. The cases where significant convergence can be demonstrated are ones in which, as in the case of birds, the animals respond most strongly to

456

the *structure* of the plant community rather than its species composition. In contrast, the species-specificity of many herbivorous insects (Gilbert and Smiley, 1978) and of insect parasites reduces the chances of finding convergence in many features of these communities, particularly where it has most often been sought—species richness.

It is evident that we suffer from a poverty of both theory and data. Competition theory, which has formed the basis for so many efforts to detect patterns in natural communities, is heavily grounded in assumptions that make it of limited use in dealing with organisms like plants (Schaffer and Leigh, 1976). The situation with respect to predator–prey theory is similar. Once we leave single prey–single predator models, our insights rapidly diminish. We understand little about predation in highly structured environments and how this might act to influence prey species that are similar to one another. Predation may determine limits to similarities among coexisting species, just as competition can, but appropriate theories and tests are yet to be made. The results may also depend strongly on how the prey compete among themselves. All of the cases known in which predators act to increase species richness occur among space-competing prey (Connell, 1971; Harper, 1969; Janzen, 1970; Paine, 1966, 1969). There is little evidence that predation on other groups increases richness even though it is clear that it can have major impacts on abundances and may reduce species richness. This perception may reflect ignorance rather than reality, but it deserves further attention.

For both competition and predator–prey models, attention has been directed primarily toward the short-term demographic consequences of the interactions and not upon the long-term, evolutionary ones that are more important for community convergence. The short-term effect of competition might be habitat segregation whereas the long-term effect might be evolution of size differences and coexistence. Similarly, the short-term effect of predation might be to eliminate a prey species from certain habitats whereas the longer-term influence might be evolution of defensive measures that allow the prey to regain some of the lost area. Both new theoretical developments and detailed field studies, especially those making use of the opportunities provided by similar climates on different continents, are badly needed.

A variety of direct and indirect lines of evidence strongly suggests that predation as a force molding marine communities is inversely proportional to latitude (see Vermeij, 1978, for an extensive review). This has apparently been an important factor in the rates of evolution of tropical marine assemblages, their species richness, and their susceptibility to extinction. Evidence of increased importance of predation at

457

low latitudes in terrestrial communities has also been assembled but it has proved more difficult to demonstrate its influence upon community structure. Nonetheless, such broad patterns of the relative importance of different selective agents, if substantiated, provide the basis for both predictions and tests of possible convergent patterns in community structure and functioning.

Our evaluations have suggested that convergent properties can be molded by both pervasive physical and biological influences. Species obviously must respond to both, although the emphasis can vary substantially. Currently mankind is altering environments in drastic ways: commercially valuable, foreign species are being added; equally we are now faced with the highest rate of species extinction in the history of life. The very notion of convergent evolution at the community level assumes that the adaptive environments are complete and natural. This is clearly no longer true, and it casts a long shadow over what has been an important approach to understanding evolutionary processes.

THE STUDY OF
COEVOLUTION

Douglas J. Futuyma and
Montgomery Slatkin

The chapters in this volume describe evolutionary aspects of several kinds of interactions among species: competition; exploitation of one species by another (predator–prey systems, parasite–host systems, and some cases of mimicry); and exploitation of two or more species by each other (other cases of mimicry, and mutualisms). Each of these interactions raises questions specific to that interaction. More general questions, however, have recurred throughout the book: Have specific associations between evolving lineages persisted through long periods of geologic time? Have interacting species evolved mutual adaptations, and does such evolution depend on their long-term association? Are species specially adapted to one or a few other species, or to larger sets of interacting species? Does coevolution proceed by continual, simultaneous genetic refinement of mutual adaptations over long periods of time? Has coevolution been important in shaping community structure?

Few general answers have emerged for these very general questions. The evidence of systematics (Chapter 4) and of paleontology (Chapters 14–16) indicates that some lineages have interacted for very long periods of time, and others have formed associations only recently. Adaptation of a species to other species is abundantly evident, but evidence of reciprocal adaptation of particular species, one to another, is sparse except in some mutualistic associations (Chapters 6, 13, and 14) and a few parasite–host associations (Chapters 7–10). Unilateral adaptation is more easily demonstrated than coevolution, strictly defined.

459

The most prevalent pattern, in fact, seems to be adaptation of a species to a suite of often phylogenetically diverse species, to which it holds a similar ecological relationship (Chapters 10, 11, and 13), i.e., diffuse coevolution. Studies of geographical variation (Chapters 8, 10, and 12) and of genetic changes in populations that have recently encountered one another (Chapters 5, 7, and 9) indicate that populations may adapt rapidly to biotic interactions, but little evidence for prolonged coevolutionary change can be gleaned from the fossil record, in which species appear to persist with little change and to be replaced by unrelated forms that may or may not be better adapted to the interaction (Chapters 15 and 16). Coevolution among small groups of strongly interacting species seems in a few instances to affect community structure (Chapters 17 and 18), but the evidence that coevolution has played an important role in resource partitioning among large numbers of species or in the development of parallel structure in independently formed communities is equivocal or weak (Chapters 18 and 19). On the whole, the evidence suggests that if major patterns of evolution and community structure have been affected at all by coevolution, diffuse coevolution rather than refined mutual adaptation of pairs of species has played the dominant role.

These conclusions cast light on four major questions about coevolution that Ehrlich and Raven posed in 1964.

1. "Without recourse to long-term experimentation on single systems, what can be learned about the coevolutionary responses of ecologically intimate organisms?" We conclude, as Ehrlich and Raven did from the relationships between butterfly feeding patterns and plant chemistry, that studies of systematics, morphology, behavior, and biochemistry can say a great deal about the length of time an association has persisted, the extent to which interacting lineages radiate in parallel, and the extent to which their association promotes the evolution of new adaptations. Studies of systematics, functional morphology, and ecology can in concert indicate whether adaptations arise by pairwise coevolution or diffuse coevolution. However, few, if any, studies have rigorously combined these kinds of information; ecological studies of adaptation are usually divorced from the phylogenetic studies that might indicate when the adaptations developed.

2. "Are predictive generalities about community evolution attainable?" At least at present it is hard to be optimistic on this count. An elaborate theory of the coevolution of pairs of species exists (Chapter 3) and may have predictive power in some instances (Chapter 17), but the pace of coevolution seems so uneven (Chapters 15 and 16) and the interactions among species so diverse and so variable from place to place and time to time (Chapters 11 and 13)

460

that consistent community patterns may not be easily predictable and in any case are hard to discern (Chapters 18 and 19).

3. "In the absence of a fossil record can the patterns discovered aid in separating the rate and time components of evolutionary change in either or both groups?" Ehrlich and Raven were pessimistic about the prospects of determining whether, say, the association of a group of insects with a group of plants was the product of rapid recent evolution or of slow, progressive modification over a longer term. Although none of the authors in this volume has addressed this question in detail, we feel there is reason for guarded optimism. Modern techniques of phylogenetic analysis show promise of determining the order in which different groups have radiated; the "molecular clock" (Kimura, 1968; Wilson et al., 1977), although still a subject of controversy, may yet prove useful in determining absolute times of divergence; and it should thereby be possible to determine when different associations of species came into being, especially if geologic data such as continental drift can be brought to bear (Chapter 4). Determining the rapidity of coevolutionary change should be possible in at least some instances.

4. "Do studies of coevolution provide a reasonable starting point for the understanding of community evolution in general?" We are rather less optimistic on this point than Ehrlich and Raven were. "Understanding community evolution" embraces a multitude of questions. Does the diversification of a group of species promote the diversification of predators or parasites? Almost certainly. Does coevolution increase or reduce species diversity? This is probably too broad a question to be meaningful. Diversification of one group of species may provide resources for another group, thereby enhancing diversity, but coevolution between predators and their prey or between competitors may in theory either promote or retard extinction (Chapters 3 and 9). There is far less empirical information on extinction than on diversification. Moreover, the species composition of a community depends not only on *in situ* evolution, but also on extinction and immigration; the pace of coevolution may be sporadic and its direction dependent on temporary changes in species composition; and coevolutionary interactions are often so diffuse that the direction of evolution may be hard to predict. Thus, the factors affecting community evolution are so enormously complex that it is hard for us to envisage a predictive evolutionary theory of community structure, and the existing theory appears so far not to have been very successful (Chapters 18 and 19). We can envisage such a theory being successful for strongly interacting

461

subsets of the community network, but the relationship between coevolutionary theory and a more global theory of community structure seems quite tenuous.

We should like, in closing, to identify some other questions that seem not to have been fully answered. These apply both to specific ecological interactions and to coevolution in general.

Character displacement among competing species is perhaps the most fully developed theory of coevolution and has received extensive empirical attention; yet surprisingly few unequivocal cases of character displacement have been described. Why? Is competition a less potent force than has been supposed? Or do species compete with so many other species that coevolutionary change is sluggish (Chapter 3) and, hence, unimportant? The theory of competing species indicates that coevolved competitors should jointly use resources more efficiently and have higher productivity in aggregate, than groups of species that have not coevolved. Do they? None of the authors in this volume has addressed this point.

How do mutualisms arise? Do they generally evolve from parasitic systems, as seems sometimes to be true (Chapters 7 and 13)? Or do they also arise from commensal relationships? Under what conditions is a host, from which a guest derives benefit, likely to derive benefit from its guest? What conditions favor the evolution of specialized versus diffuse mutualistic interactions?

Is extinction or a coevolutionary steady state the more common outcome in those parasite/host systems in which selection favors greater virulence in the parasite? Can the frequency of extinction in such cases be assessed? What are the genetic dynamics of such systems in natural populations? If coevolved parasite–host associations are more stable than novel ones, is this chiefly because the host has evolved to be resistant or because the parasite has evolved to be benign? This very fundamental question seems not to have been adequately answered. What are the relative roles of individual selection and group selection and of multiple versus single infection, population subdivision, and the mode of transmission between hosts in the evolution of virulence or avirulence in parasites? These questions, as May and Anderson indicate (Chapter 9), are only beginning to be answered.

For predator–prey systems, similarly, there seems to be no clear answer to the question whether coevolution affects coexistence. Are long-term associations stable only because of the extinction of unstable associations, or has coevolution lent added stability? Do multiple interactions among species, which cause coevolution to be diffuse if it occurs at all, prevent predators from becoming so proficient as to extinguish their prey? As for parasite–host systems, we need methods for determining how frequently coevolution causes extinction

462

of predators or prey. Finally, there exist contrasting theories of predator–prey evolution that need to be assessed. Schaffer and Rosenzweig (1978) have proposed that evolutionary change, though occurring continually at a low rate, may be so balanced in predators and their prey as not to destabilize the interaction. Maynard Smith (1976; see also Stenseth, Maynard Smith and Haigh, 1976), in contrast, has proposed that systems of interacting species may either undergo continual evolution, with extinction a frequent consequence (the "Red Queen" hypothesis of Van Valen, 1973), or else achieve both ecological and genetic stability, so that coevolution comes to a halt. Are predator–prey systems stable over geologic time, then, because they coevolve or because they do not? Should the paleontological examples described in Chapters 15 and 16 be interpreted in this light?

These questions may be subsumed under several more general questions about coevolution that emerge from the discussions in this book.

Is coevolution predictable? It may be premature to hope for a predictive theory, given that even the documentation of coevolution is a problem with which many of the authors in this volume have struggled. Nevertheless, there now exists a substantial body of coevolutionary theory (Chapters 3 and 9) which, as most of the chapters in this book illustrate, has hardly been applied or tested in practice. It is very possible, as Roughgarden (Chapter 3) suggests, that models of coevolution may not be highly predictive unless they describe particular forms of interaction more precisely. It is also likely that, like population genetics theory, the models will explain individual evolutionary events or be capable of short-term prediction, without predicting the pattern of such events in geologic time. Whether or not a globally predictive theory of coevolution can be developed is uncertain.

Do systems of interacting species engage in continual coevolution, if only at a low rate, or do they arrive at genetic stasis? Theoretical models (e.g., Schaffer and Rosenzweig, 1978) seem to indicate that either is possible, and the data advanced by various authors in this book similarly favor both points of view. Although the paleontological chapters in this volume are both by advocates of the punctuated equilibrium model of evolution, it does seem that the fossil record offers little evidence of sustained directional coevolution; yet the neontological studies show that populations often adapt rapidly to the species with which they interact. If the punctuated pattern in the fossil record can generally be explained by short episodes of directional selection followed by long episodes of stabilizing selection (Charlesworth et al., 1982), it is necessary to explain why the fossil record of predator–prey systems should show stasis, when the arms race, or

Red Queen, model (Van Valen, 1973) supposes continual directional selection.

This paradox may be more apparent than real. We need to explain the apparent stability of species interactions only if they are actually stable. It may be, however, that coevolution causes extinction more frequently than we suppose. Imagine, for example, that the rate of coevolution between predators and prey depends on mutation rates and is quite slow. The rate of extinction of species that fall behind in the race might in fact be high enough to be explained by occasional improvements in their enemies that are too slight to be discerned in the fossil record. Slight genetic changes in predators, for example, might alter predator–prey dynamics enough to cause the occasional extinction of a prey species, while the predator switches to alternative prey and remains little changed. It would be useful to have a theory of coevolution-induced extinction rates and more detailed data on rates of extinction.

Thus, the apparent conflict between neontological and paleontological data raises the question: Can a precise theory of diffuse coevolution be devised? Almost all coevolutionary theory, including the arms race models of predator–prey interactions, applies to pairs of species. Exclusive pairwise interactions, however, are unusual in nature, except for a scattering of mutualisms and parasite–host combinations (Chapters 6, 7, 8, 10, 11, and 13). It is in one-on-one (or few-on-few) interactions that the evidence of coevolution is most pronounced, and in which, for all we know, the rate of coevolution-driven extinction may be high. [Perhaps the missing parasites in otherwise congruent phylogenies of parasites and hosts (Chapter 4) are evidence of extinction.] But, in general, species do not contend with one competitor, predator, prey, or mutualist; they contend with many. Their adaptations to any one other species must be compromised by their direct or indirect interactions with others, and this multiplicity of interactions surely imposes stabilizing selection (Slobodkin, 1974). Although local populations may respond quickly to differences in the composition of interacting species, changes in community composition may occur too frequently to allow for prolonged evolution in any one direction. Almost all the authors in this volume have emphasized that diffuse coevolution is likely to be far more common than pairwise coevolution, but models of diffuse coevolution have hardly been developed. The models will no doubt be complex; but it may be the very complexity of the interactions among species that accounts for the variable pace of coevolution.

ACKNOWLEDGMENTS

Chapter 5/BRUCE R. LEVIN AND RICHARD E. LENSKI
"We wish to thank Susan Hattingh and Margaret Riley for useful comments, and Karen Morse for processing the words. We profited greatly from our discussion about some of this material with Frank Stewart and Allan Campbell and from conversations [Bruce R. Levin] had with Raymond Devoret and Michael Yarmolinsky." The authors' research was supported by a grant from the National Institutes of Health (GM 19848).

Chapter 6/LEE EHRMAN
"I wish to acknowledge Professor John Preer, Indiana University, who read and reread the section on protozoa. Mrs. Joan Probber, Research Associate at the State University of New York at Purchase, and John Boltri competently helped with all aspects of the compilation of the chapter." The author's research on *Drosophila paulistorum* was supported by a grant from the National Science Foundation (PCM 79-102062).

Chapter 8/JOHN C. HOLMES
"The preparation of this manuscript and the work leading to the development of my ideas has been supported by the National Sciences and Engineering Research Council of Canada. The editors and the parasitology group at the University of Alberta, especially J. Aho and T. M. Stock, provided useful comments on earlier drafts."

Chapter 11/DANIEL H. JANZEN
"The development of the manuscript profited from discussions with H. Howe, C. Herrera, N. Wheelwright, G. Orians, W. Hallwachs, F. Gill, D. McKey, P. Parker, G. Stevens, M. Gonzalez-Espinosa, P. Martin, R. Wells, D. Futuyma, W. Freeland, W. Haber and C. Jordano." The author's research was supported by a grant from the National Science Foundation (DEB 80-11558).

Chapter 12/LAWRENCE E. GILBERT
"I am very grateful to Professor E. B. Ford, F. R. S., for making space available to me in his laboratory during 1966 and 1967, and to the Oxford University Hope Collection Librarian, Mrs. Audrey Smith, who made available the early literature on mimicry that stimulated my con-

tinued interest in the subject. Over the years my colleagues W. W. Benson, K. S. Brown, and J. R. G. Turner have greatly influenced my thinking on mimicry and *Heliconius* through correspondence, discussions, and publications. The Honorable Miriam Rothschild deserves special thanks for her continuing encouragement and for the example she sets for all naturalists curious about mimicry. Finally, I thank Sharon Bramblett, who typed the manuscript, and P. J. DeVries and M. C. Singer, who read and commented on the first draft."

Chapter 14/Geerat J. Vermeij

"I am extremely grateful to D. E. Gill and to the editors for unabashedly criticizing and seeking improvements in this manuscript. R. Fritz, D. Inouye, R. K. Kropp, E. C. Dudley, and E. Zipser also helped in ironing out difficulties and inconsistencies in the paper. L. G. Eldredge at the University of Guam Marine Laboratory kindly placed his large library of reprints at my disposal. I am also grateful to the National Science Foundation, which enabled me to carry out studies on the agents of selection that have been important in the evolution of intimate associations."

Chapter 19/Gordon H. Orians and Robert T. Paine

"This chapter has benefited from the criticism and advice of T. R. E. Southwood, John Lawton, Ronald Ydenberg, Kate Lessels, and George Branch. Our research in a variety of communities has been funded by the U.S. National Science Foundation. We both recognize with gratitude the important freedoms provided by fellowships from the John Simon Guggenheim Memorial Foundation."

LITERATURE CITED

(Numbers in parentheses at the end of each reference indicate the chapter or chapters in which the work is cited.)

Abbott, H. G. and T. F. Quink. 1970. Ecology of eastern white pine seed caches made by small forest mammals. Ecology 51: 271–278. (11)

Abbott, I., L. K. Abbott and P. R. Grant. 1977. Comparative ecology of Galapagos ground finches (*Geospiza* Gould): Evaluation of the importance of floristic diversity and interspecific competition. Ecol. Monogr. 47: 151–184. (17, 18)

Abdel-Rahman, M. 1977. Patterns of hormones, respiration and ripening enzymes during development, maturation and ripening of cherry tomato fruits. Physiol. Plant. 39: 115–118. (11)

Abele, L. G. 1976. Comparative species richness in fluctuating and constant environments: coral associated decapod crustaceans. Science 192: 461–463. (14)

Abrams, P. 1975. Limiting similarity and the form of the competition coefficient. Theor. Pop. Biol. 8: 356–375. (17)

Abrams, P. 1980. Some comments on measuring niche overlap. Ecology 61: 44–49. (17)

Ackerman, J. D. 1981. The phenological relationships of male euglossine bees (*Hymenoptera*: Apidae) and their orchid fragrance hosts. Ph.D. dissertation, Florida State University. (13)

Adams, J. 1927. The germination of the seeds of some plants with fleshy fruits. Amer. J. Bot. 14: 415–428. (11)

Adams, J., T. Kinney, S. Thompson, L. Rubin and R. B. Helling. 1979. Frequency-dependent selection for plasmid-containing cells of *Escherichia coli*. Genetics 91: 627–637. (5)

Adams, M. H. 1959. *Bacteriophages*. Interscience, New York. (5)

Addicott, J. F. 1979. A multispecies aphid-ant association: density dependence and species-specific effects. Can. J. Zool. 57: 558–569. (3)

Addicott, J. F., 1981. Stability properties of 2-species models of mutualism: simulation studies. Oecologia 49: 42–49. (3)

Adey, W. H. 1965. The genus *Clathromorphum* in the Gulf of Maine. Hydrobiologia 26: 559–573. (15)

Adey, W. H. 1973. Temperature control of reproduction and productivity in subarctic coralline algae. Phycologia 12: 111–118. (15)

Adolph, S. C. and J. Roughgarden. 1982. Foraging by Passerine birds and *Anolis* lizards on St. Eustatius, Lesser Antilles, and implications for interclass competition. Oecologia. (In press) (17)

Agosin, M. and A. S. Perry. 1974. Microsomal mixed-function oxidases. In *The Physiology of Insecta*, Vol. V, Chapter 10, M. Rockstein (ed.), pp. 538–596. Academic Press, New York. (10)

Ahmadjian, V. 1966. Lichens. In *Symbiosis*, Vol. I, S. M. Henry (ed.), pp. 35–97. Academic Press, New York. (7)

Ahmadjian, V. 1967. *The Lichen Symbiosis*. p. 152. Blaisdell Publishing Co., Waltham, Massachusetts. (7)

Ahmadjian, V. 1970. The lichen symbiosis: its origin and evolution. In *Evolutionary Biology*, Vol. 4, T. Dobzhansky, M. K. Hecht and W. C. Steere (eds.), pp. 163–184. Appleton-Century-Crofts, New York. (7)

Ahmadjian, V. 1980. Separation and artificial synthesis of lichens. In *Cellular Interactions in Symbiosis and Parasitism*, C. B. Cook, P. W. Pappas and E. D. Rudolph (eds.), pp. 3–25. Ohio State University Biosciences Colloquia No. 5, Ohio State University Press, Columbus, Ohio. (7)

Ahmadjian, V. and H. Heikkila. 1970. The culture and synthesis of *Endocarpon pusillum* and *Staurothele clopina*. Lichenologist 4: 259–267. (7)

Ahmadjian, V. and J. B. Jacobs. 1981. Relationship between fungus and alga in the lichen *Cladonia cristatella* Tuck. Nature 289: 169–172. (7)

467

Ahmadjian, V., J. B. Jacobs and L. A. Russel. 1978. Scanning electron microscope study of early lichen synthesis. Science 200: 1062-1064. (7)

Ainsworth, G. C. 1971. *Ainsworth and Bisby's Dictionary of the Fungi.* Commonwealth Mycological Institute, Kew, Surrey. (7)

Aker, C. L. and D. Udovic. 1981. Oviposition and pollination behavior of the yucca moth, *Tegeticula maculata* (Lepidoptera: Prodoxidae), and its relation to the reproductive biology of *Yucca whipplei* (Agavaceae). Oecologia 49: 96-101. (13)

Alberch, P. 1980. Ontogenesis and morphological diversification. Amer. Zool. 20: 653-667. (2)

Alexander, M. 1971. *Microbial Ecology.* John Wiley & Sons, New York. (7)

Alexandre, D. Y. 1978. Le rôle disséminateur des éléphants en forêt de Tai, Côte d'Ivoire. La Terre et la Vie 32: 47-72. (11)

Ali-Kahn, Z. 1978. Pathological changes in the lymphoreticular tissues of Swiss mice infected with *Echinococcus granulosus* cysts. Z. Parasitenkd. 58: 47-54. (8)

Allee, W. C., A. E. Emerson, O. Park, T. Park and K. P. Schmidt. 1949. *Principles of Animal Ecology.* W. B. Saunders, Philadelphia. (2)

Allen, D. F. 1979. *Wolves of Minong.* Houghton Mifflin, Boston. (16)

Allen, J. C. and C. L. Griffiths. 1981. The fauna and flora of a kelp bed canopy. S. Afr. J. Zool. 16: 80-84. (19)

Allen, L. G. 1938. Phytobezoar. Amer. J. Roentgenol. 39: 67-74. (11)

Allison, A. C. 1982. Co-evolution between hosts and infectious disease agents, and its effects on virulence. In *Population Biology of Infectious Disease Agents (Dahlem Konferenzen)*, R. M. Anderson and R. M. May (eds.), pp. 245-267. Springer-Verlag, Berlin. (8, 9)

Alston, F. H. and J. B. Briggs. 1977. Resistance genes in apple and biotypes of *Dysaphis devecta.* Ann. Appl. Biol. 87: 75-81. (10)

Altmann, S. A. and J. Altmann. 1970. *Baboon Ecology.* University of Chicago Press, Chicago. (8)

Amadon, D. 1953. Avian systematics and evolution in the Gulf of Guinea. Bull. Amer. Mus. Nat. Hist. 100: 397-431. (18)

Amin, O. M. and J. M. Burrows. 1977. Host and seasonal associations of *Echinorhynchus salmonis* (Acanthocephala: Echinorhynchidae) in Lake Michigan fishes. J. Fish. Res. Bd. Canada 34: 325-331. (8)

Anderson, E. S. 1968. The ecology of transferable drug resistance in the Enterobacteria. Annu. Rev. Microbiol. 22: 131-180. (5)

Anderson, R. C. 1972. The ecological relationships of meningeal worm and native cervids in North America. J. Wildl. Dis. 8: 304-310. (8)

Anderson, R. M. 1978. The regulation of host population growth by parasitic species. Parasitology 76: 119-157. (8)

Anderson, R. M., H. Jackson, R. M. May and T. Smith. 1981. The population dynamics of fox rabies in Europe. Nature 289: 765-771. (9)

Anderson, R. M. and R. M. May. 1978. Regulation and stability of host-parasite population interactions: I. Regulatory processes. J. Anim. Ecol. 47: 219-247. (3, 9)

Anderson, R. M. and R. M. May, 1979. Population biology of infectious diseases: Part I. Nature 280: 361-367. (8, 9)

Anderson, R. M. and R. M. May. 1981. The population dynamics of microparasites and their invertebrate hosts. Phil. Trans. Roy. Soc. B. 291: 451-524. (9)

Anderson, R. M. and R. M. May. 1982. Coevolution of hosts and parasites. Parasitology. (In press) (9)

Anderson, W. W. 1971. Genetic equilibrium and population growth under density-regulated selection. Amer. Natur. 105: 489-498. (2, 9)

Andrewartha, H. G. and L. C. Birch. 1954. *The Distribution and Abundance of Animals.* University of Chicago Press, Chicago. (19)

Andrews, E. A. 1935. Shell repair in the snail, *Neritina.* J. Exp. Zool. 70: 75-107. (14)

Annen, J. M., P. Kohler and J. Eckert. 1981. Cytotoxicity of *Echinococcus granulosus* cyst fluid in vitro. Z. Parasitenkd. 65: 79-88. (8)

Anonymous. 1960. List of plants eaten by various types of mammals. Koedoe 3: 1-205. (11)

Anonymous. 1972. *Genetic Vulnerability of Major Crops.* National Academy of Sciences, Washington, D.C. (7)

468

Applegate, R. D., L. L. Rogers, D. A. Casteel and J. M. Novak. 1979. Germination of cow parsnip seeds from grizzly bear feces. Mammalogy 60: 655. (11)

Arber, W. and S. Linn. 1969. DNA modification and restriction. Ann. Rev. Biochem. 38: 467–500. (5)

Argenzio, R. A., J. E. Lowe, D. W. Pickard and C. E. Stevens. 1974. Digesta passage and water exchange in the equine large intestine. Amer. J. Physiol. 226: 1035–1042. (11)

Arneodo, A., P. Coullet, J. Preyraud and C. Tresser. 1982. Strange attractors in Volterra equations for species in competition. J. Math. Biol. 14: 153–158. (3)

Arnold, S. J. 1981. Behavioral variation in natural populations II. The inheritance of feeding response in crosses between geographic races of the garter snake, *Thamnophis elegans*. Evolution 35: 510–515. (17)

Ashlock, P. D. 1974. The uses of cladistics. Ann. Rev. Ecol. Syst. 5: 81–99. (4)

Askew, R. R. 1980. The diversity of insect communities in leaf-mines and plant galls. J. Anim. Ecol. 49: 817–829. (19)

Atchley, W. R. and J. J. Rutledge. 1980. Genetic components of size and shape. I. Dynamics of components of phenotypic variability and covariability during ontogeny in the laboratory rat. Evolution 34: 1161–1173. (2)

Atsatt, P. R. and D. J. O'Dowd. 1976. Plant defense guilds. Science 193: 24–29. (13)

Atwood, K. C., L. K. Schneider and F. J. Ryan. 1951. Periodic selection in *Escherichia coli*. Proc. Natl. Acad. Sci. USA 37: 146–155. (5)

Au, A.C.S. and R. C. Ko. 1979. Cross-resistance between *Trichinella spiralis* and *Angiostrongylus cantonensis* in laboratory rats. Z. Parasitenkd. 59: 161–168. (8)

Auclair, J. L. 1978. Biotypes of the pea aphid, *Acyrthosiphon pisum*, in relation to plants and chemically defined diets. Ent. exp. appl. 24: 12–16. (10)

Auda, H., H. Al-Wandawi and L. Al-Adhami. 1976. Protein and amino acid composition of three varieties of Iraqi dates at different stages of development. J. Agric. Food Chem. 24: 365–366. (11)

Auerbach, M. J. and S. D. Hendrix. 1980. Insect-fern interactions: macrolepidopteran utilization and species-area association. Ecol. Entomol. 5: 99–104. (10)

Auerbach, M. J. and D. R. Strong. 1981. Nutritional ecology of *Heliconia* herbivores: experiments with plant fertilization and alternative hosts. Ecol. Monogr. 51: 63–83. (10)

August, P.V. 1981. Fig fruit consumption and seed dispersal by *Artibeus jamaicensis* in the llanos of Venezuela. Biotropica (supplement) 13: 70–76. (11)

Auriault, C., M. Joseph, J. P. Dessaint and A. Capron. 1980. Inactivation of rat macrophages by peptides resulting from cleavage of IgG by *Schistosoma* larvae proteases. Immunology Letters 2: 135–139. (8)

Austin, S., M. Ziese and N. Sternberg. 1981. A novel role for site-specific recombination in maintenance of bacterial replicons. Cell 25: 729–736. (18)

Axelrod, D. I. 1973. History of the mediterranean ecosystem in California. In *Mediterranean Type Ecosystems: Origin and Structure*. Ecological Studies 7: 225–277, R. di Castri and H. A. Mooney (eds.). Springer-Verlag, New York. (19)

Ayala, F. J. (ed). 1976. *Molecular Evolution*. Sinauer Associates, Sunderland, Massachusetts. (7)

Bailey, J. A. 1968. Rate of food passage by caged cottontails. J. Mammal. 49: 340–342. (11)

Bailey, N. J. T. 1975. *The Mathematical Theory of Infectious Diseases*, 2nd Ed. Macmillan, New York. (9)

Baird, J. W. 1980. The selection and use of fruits by birds in an eastern forest. Wilson Bull. 92: 63–73. (11)

Baker, E. C. S. 1913. *Indian Pigeons and Doves*. Witherby, London. (11)

Baker, H. G. 1961. The adaptations of flowering plants to nocturnal and crepuscular pollinators. Quart. Rev. Biol. 36: 64–73. (13)

Baker, H. G. 1963. Evolutionary mechanisms in pollination biology. *Science* 139: 877–883. (15)

Baker, H. G. 1965. Characteristics and modes of origins in weeds. In *The Genetics of Colonizing Species*, H. G. Baker and G. L. Stebbins (eds.), pp. 147–168. Academic Press, New York. (13)

Baker, H. G. 1972. Seed weight in relation to environmental conditions in California. Ecology 53: 947–1010. (19)

Baker, H. G., R. W. Cruden and I. Baker. 1971. Minor parasitism in pollination biology and its community function: the case of *Ceiba acuminata*. BioScience 21: 1127–1129. (13)

Baker, H. G. and P. D. Hurd. 1968. Intrafloral ecology. Ann. Rev. of Entomol. 13: 385–415. (13)

Bakker, R. T. 1977. Cycles of diversity and extinction. In *Patterns of Evolution*, A. Hallam (ed.), pp. 431–478. Elsevier, Amsterdam. (16)

Bakker, R. T. and J. Bickert. 1983. Predation pressure, survivorship, and habitat change among early Cenozoic ungulates. (In preparation) (16)

Bakker, R. T. and D. Schankler. 1983. Immigrant floods and punctuated evolution in Eocene primates and predators. (In preparation) (16)

Bakski, B. K., M. A. R. Reddy, S. Singh, P. C. Pandey and S. N. Mukherjee. 1968. Khair seedling mortality in plantations. Ind. For. 94: 659–661. (11)

Bakus, G. J. 1969. Energetics and feeding in shallow marine waters. Intern. Rev. Gen. Exp. Zool. 4: 275–369. (19)

Bakus, G. J. 1974. Toxicity in holothurians: a geographical pattern. Biotropica 6: 229–236. (14)

Bakus, G. J. 1981. Chemical defense mechanisms on the Great Barrier Reef, Australia. Science 211: 497–499. (14)

Bakus, G. J. and G. Green. 1974. Toxicity in sponges and holothurians. Science 185: 951–953. (14)

Balda, R. P. 1969. Foliage use by birds of the oak-juniper woodland and ponderosa pine forest in southeastern Arizona. Condor 71: 399–412. (19)

Balda, R. P. 1980. Are seed caching systems co-evolved? Proc. XVII Int. Cong. Ornithol. 1185–1191. (11)

Balgooyen, T. G. and L. M. Moe. 1972. Dispersal of grass fruits—an example of endornithochory. Amer. Midl. Nat. 90: 454–455. (11)

Barbehenn, K. R. 1969. Host-parasite relationships and species diversity in mammals: an hypothesis. Biotropica 1: 29–35. (8)

Barbour, C. D. 1973. A biogeographical history of *Chirostoma* (Pisces: Atherinidae): a species flock from the Mexican plateau. Copeia 3: 533–556. (18)

Barile, M. F. and Razin, S. (eds.). 1978. *The Mycoplasmas.* Academic Press, New York. (6)

Barnes, H. and H. T. Powell. 1953. The growth of *Balanus balanoides* (L.) and *B. crenatus* Brug. under varying conditions of submersion. J. Mar. Biol. Assoc. (U.K.) 32: 107–128. (16)

Barnett, R. J. 1977. The effect of burials by squirrels on germination and survival of oak and hickory nuts. Amer. Midl. Nat. 98: 319–329. (11)

Baron, G. and P. Jolicoeur. 1980. Brain structure in Chiroptera: some multivariate trends. Evolution 34: 386–393. (2)

Barrett, J. A. 1978. A model of epidemic development in variety mixtures. In *Plant Disease Epidemiology*, P. R. Scott and A. Bainbridge (eds.), pp. 129–137. Blackwell Scientific, Oxford. (7)

Barrett, J. A. 1980. Pathogen evolution in multilines and variety mixtures. Z. Pflkrankh. Pflschutz. 87: 383–396. (7)

Barrett, J. A. 1981. The evolutionary consequences of monoculture. In *Genetic Consequences of Man-Made Change*, J. A. Bishop and L. M. Cook (eds.), pp. 209–248. Academic Press, London. (7)

Barriga, O. O. 1980. Responses of B-cells to mitogens and antigen in mice receiving isogenic splenocytes from animals treated with *Trichinella* extract. J. Parasitol. 66: 730–734. (8)

470

Bartel, M. H. 1965. The life cycle of *Rallietina (R). loeweni* Bartel and Hensen, 1964 (Cestoda) from the black-tailed jackrabbit, *Lepus californicus melanotis* Mearns. J. Parasitol. 51: 800–806. (8)

Barton, D. E., and F. N. Davidson. 1956. Some notes on ordered random intervals. J. Roy. Stat. Soc., B, 18: 79–94. (18)

Barton, L. V. 1965. Viability of seeds of *Theobroma cacao* L. Contrib. Boyce Thompson Inst. 23: 109–122. (11)

Basch, P. 1975. An interpretation of snail-trematode infection rates: specificity based on concordance of compatible phenotypes. Int. J. Parasitol. 5: 449–452. (8)

Bates, H. W. 1862. Contributions of an insect fauna of the Amazon Valley. Trans. Linn. Soc. London. 23: 495–566. (1, 12)

Baudoin, M. 1975. Host castration as a parasite strategy. Evolution 29: 335–352. (8)

Bawa, K. S. 1980. Evolution of dioecy in flowering plants. Ann. Rev. Ecol. Syst. 11: 15–39. (13)

Bawa, K. S. and J. H. Beach. 1981. Evolution of sexual systems in flowering plants. Ann. Miss. Bot. Gard. 68: 254–274. (13)

Bayssade-Dufour, C. 1979. Variations du système sensoriel de la cercaire de *Schistosoma mansoni*. Intéret éventuel en épidémiologie. Ann. Parasitol. 54: 593–614. (8)

Beale, G. H. 1980. The genetics of drug resistance in malaria parasites. Bull. W.H.O. 58: 799–804. (8)

Beattie, A. J. and D. C. Culver. 1981. The guild of myrmecochores in the herbaceous flora of West Virginia forests. Ecology 62: 107–115. (11)

Beck, S. D. 1965. Resistance of plants to insects. Ann. Rev. Entomol. 10: 207–232. (10)

Beck, S. D. and J. C. Reese. 1976. Insect-plant interactions: nutrition and metabolism. Rec. Adv. Phytochem. 10: 41–92. (10)

Begon, M. and M. Mortimer. 1981. *Population Ecology: A Unified Study of Animals and Plants*. Sinauer Associates, Sunderland, Massachusetts. (18)

Behnke, J. M., P. W. Bland and D. Wakelin. 1977. Effect of the expulsion phase of *Trichinella spiralis* on *Hymenolepsis diminuta* infection in mice. Parasitology 75: 79–88. (8)

Behnke, J. M., D. Wakelin and M. M. Wilson. 1978. *Trichinella spiralis*: delayed rejection in mice concurrently infected with *Nematospiroides dubius*. Exp. Parasitol. 46: 121–130. (8)

Benacerraf, B. 1981. Role of MHC gene products in immune regulation. Science 212: 1229–1238. (8)

Bennett, F. J., I. G. Kagan, N. A. Barnicot and J. C. Woodburn. 1970. Helminth and protozoal parasites of the Hadza of Tanzania. Trans. R. Soc. Trop. Med. Hyg. 64: 857–880. (8)

Bennett, P. M. and M. H. Richmond. 1978. Plasmids and their possible influence on bacterial evolution. *The Bacteria* Vol. VI. Academic Press, New York. (5)

Benson, W. W. 1972. Natural selection for Müllerian mimicry in *Heliconius erato* in Costa Rica. Science 176: 936–939. (4, 10, 12)

Benson, W. W., K. S. Brown and L. E. Gilbert. 1975. Coevolution of plants and herbivores: Passion flower butterflies. Evolution 29: 659–680. (4, 10, 12)

Bentley, B. L. 1976. Plants bearing extrafloral nectaries and the associated ant community: interhabitat differences in the reduction of herbivore damage. Ecology 57: 815–820. (4)

Bentley, B. L. 1977. Extrafloral nectaries and protection by pugnacious bodyguards. Ann. Rev. Ecol. Syst. 8: 407–427. (3, 10)

Berenbaum, M. 1980. Adaptive significance of midgut pH in larval Lepidoptera. Amer. Natur. 115: 138–146. (10)

Berenbaum, M. R. 1981. Patterns of furanocoumarin production and insect herbivory in a population of wild parsnip (*Pastinaca sativa* L.). Oecologia 49: 236–244. (10)

Berenbaum, M. R. and P. Feeny. 1981. Toxicity of angular furanocoumarins to swallowtail butterflies: escalation in a coevolutionary arms race? Science 212: 927–929. (10)

Bernays, E. A. 1978. Tannins: an alternative viewpoint. Ent. exp. appl. 24: 44–53. (10)

Bernays, E. A., D. Chamberlain, and P. McCarthy. 1980. The differential effects of ingested tannic acid on different species of Acridoidea. Ent. exp. appl. 28: 158–166. (10)

Bernays, E. A. and R. F. Chapman. 1978. Plant chemistry and acridoid feeding behavior. In *Biochemical Aspects of Plant and Animal Coevolution*, J. B. Harborne (ed.), pp. 99–141. Academic Press, New York. (10)

Bernstein, R. 1979. Evolution of niche breadth in populations of ants. Amer. Natur. 114: 533–544. (17)

Bertness, M. D. 1981a. Interference, exploitation, and sexual components of competition in a tropical hermit crab assemblage. J. Exp. Mar. Biol. Ecol. 49: 189–202. (14)

Bertness, M. D. 1981b. The influence of shell-type on hermit crab growth rate and clutch size (Decapoda, Anomura). Crustaceana 40: 197–205. (14)

Bertness, M. D. 1981c. Pattern and plasticity in tropical hermit crab growth and reproduction. Amer. Nat. 117: 754–773. (14)

Bertness, M. D. 1981d. Conflicting advantages in resource utilization: the hermit crab housing dilemma. Amer. Nat. 118: 432–437. (14)

Bertness, M. D., S. D. Garrity and S. C. Levings. 1981. Predation pressure and gastropod foraging: a tropical-temperate comparison. Evolution 35: 995–1007. (14)

Bertram, B. C. R. 1978. Living in groups: predators and prey. In *Behavioural Ecology: An Evolutionary Approach*, J. R. Krebs and N. B. Davies (eds.). Blackwell Scientific, Oxford. (15)

Bertram, B. C. R. 1979. Serengeti predators and their social systems. In *Serengeti, Dynamics of an Ecosystem*, A.R.E. Sinclair and M. Norton-Griffiths (eds.). University of Chicago Press, Chicago. (16)

Bethel, W. M. and J. C. Holmes. 1973. Altered evasive behavior and responses to light in amphipods harboring acanthocephalan cystacanths. J. Parasitol. 59: 945–956. (8)

Bethel, W. M. and J. C. Holmes. 1977. Increased vulnerability of amphipods to predation owing to altered behavior induced by larval acanthocephalans. Can. J. Zool. 55: 110–115. (8)

Bews, J. W. 1917. The plant succession in the thorn veldt. S. Afric. J. Sci. 14: 153–172. (11)

Bierzychudek, P. 1981. *Asclepias, Lantana*, and *Epidendrum*: a floral mimicry complex? Biotropica 13 (supplement): 54–58. (13)

Biffen, R. H. 1905. Mendel's laws of inheritance and wheat breeding. J. Agric. Soc. 1: 4–48. (7)

Biffen, R. H. 1907. Studies in the inheritance of disease resistance. J. Agric. Soc. 2: 109–128. (7)

Bisset, N. G. and A. K. Choudhury. 1974. Alkaloids and iridoids from *Strychnos nux-vomica* fruit. Phytochemistry 13: 265–269. (11)

Bjornhag, G. 19?. Separation and delay of contents in the rabbit colon. Swedish J. Agric. Res. 2: 125–136. (11)

Bjornhag, G. and I. Sperber. 1977. Transport of various food components through the digestive tract of turkeys, geese and guinea fowl. Swedish J. Agric. Res. 7: 57–66. (11)

Black, R. 1976. The effects of grazing by the limpet, *Acmaea insessa*, on the kelp, *Egregia laevigata*, in the intertidal zone. Ecology 57: 265–277. (14, 19)

Blau, P. A., P. Feeny, L. Contardo and D. S. Robson. 1978. Allylglucosinolate and herbivorous caterpillars: a contrast in toxicity and tolerance. Science 200: 1296–1298. (10)

Bloom, B. R. 1979. Games parasites play: how parasites evade immune surveillance. Nature 279: 21–26. (8)

Bloom, S. A. 1975. The motile escape responses of a sessile prey: a sponge-scallop mutualism. J. Exp. Mar. Biol. Ecol. 17: 311–321. (3, 19)

Boag, P. and P. Grant. 1978. Heritability of external morphology in Darwin's finches. Nature 774: 793–794. (17)

Bonaccorso, F. J. 1979. Foraging and reproductive ecology in a Panamanian bat community. Bull. Florida State Mus. Biol. Sci. 24: 359–408. (11)

Bonaccorso, F. J., W. Glanz and C. M. Sanford. 1980. Feeding assemblages of mammals at fruiting *Dipteryx panamensis* (Papilionaceae) trees: seed predation, dispersal and parasitism. Rev. Biol. Trop. 28: 61–72. (11)

472

Bonaccorso, F. J. and C. M. Sanford. 1979. Seed dispersal of the almendro tree, *Dipteryx panamensis* (Papilionaceae), in Panama. (Unpublished manuscript) (11)

Boss, K. J. 1971. Critical estimate of the number of Recent Mollusca. Harvard Univ. Mus. Comp. Zool. Occas. Pap. 3: 81-135. (15)

Bossema, I. 1979. Jays and oaks: an eco-ethological study of a symbiosis. Behaviour 70: 1-11. (11)

Boucher, D. H. and V. L. Sork. 1979. Early drops of nuts in response to insect infestation. Oikos 33: 440-443. (11)

Bowman, R. I. 1961. Morphological variation and adaptation in the Galapagos finches. Univ. California Publ. Zool. 59: 1-302. (18)

Bown, T. M. and M. J. Kraus. 1981. Vertebrate fossil-bearing paleosol units (Willwood Formation, Lower Eocene, Northwest Wyoming, U.S.A.). Palaeog. Palaeoclim. Palaeoec. 34: 31-56. (16)

Bowsher, A. L. 1955. Origin and adaptation of platyceratid gastropods. Kansas Univ. Paleont. Contrib. 17 (Mollusca, article 5): 1-11. (14)

Bradbury, J. W. 1977. Lek mating behavior in the hammer-headed bat. Z. Tierpsychol. 45: 225-255. (11)

Branch, G. M. 1971. The ecology of *Patella* Linnaeus from the Cape Peninsula, South Africa. Zool. Afr. 6: 1-38. (19)

Branch, G. M. 1981. The biology of limpets: physical factors, energy flow, and ecological interactions. Oceanogr. Mar. Bird. Ann. Rev. 19: 235-380. (19)

Branch, G. M. and M. Branch. 1981. *The Living Shores of Southern Africa*. C. Strunk Publishers, Cape Town. (19)

Brattsten, L. B. 1979. Biochemical defense mechanisms in herbivores against plant allelochemicals. In *Herbivores: Their Interaction with Secondary Plant Metabolites*, G. A. Rosenthal and D. H. Janzen (eds.), pp. 200-270. Academic Press, New York. (10)

Brattsten, L. B., C. F. Wilkinson and T. Eisner. 1977. Herbivore-plant interactions: mixed-function oxidases and secondary plant substances. Science 196: 1349-1352. (10)

Brecht, P. E., L. Keng, C. A. Bisogni and H. M. Munger. 1976. Effect of fruit portion, stage of ripeness and growth habit on chemical composition of fresh tomatoes. J. Food Sci. 41: 945-948. (11)

Bremermann, H. J. 1980. Sex and polymorphism as strategies in host-pathogen interactions. J. Theoret. Biol. 87: 671-702. (8, 9)

Bremermann, H. J. 1982. On optimal virulence levels in host-pathogen interactions. (Unpublished manuscript) (9)

Briggs, J. B. 1965. The distribution, abundance, and genetic relationships of four strains of the rubus aphid (*Amphorophora rubi* (Kalt.)) in relation to raspberry breeding. J. Hort. Sci. 40: 109-117. (10)

Brink, D. E. 1980. Reproduction and variation in *Aconitum columbianum* (Ranunculaceae), with emphasis on California populations. Amer. J. Botany 67: 263-273. (13)

Brink, D. and J. M. J. DeWet. 1980. Interpopulation variation in nectar production in *Aconitum columbianum* (Ranunculaceae). Oecologia 47: 160-163. (13)

Broda, P. 1979. *Plasmids*. W. H. Freeman, San Francisco. (5)

Bromley, R. G. 1975. Comparative analysis of fossil and recent echinoid bioerosion. Palaeontology 18: 725-739. (14, 15)

Brooks, D. R. 1977. Evolutionary history of some plagiorchioid trematodes of anurans. Syst. Zool. 26: 277-289. (4, 8)

Brooks, D. R. 1978a. Systematic status of proteocephalid cestodes from reptiles and amphibians in North America with descriptions of four new species. Proc. Helm. Soc. Wash. 45: 1-28. (8)

Brooks, D. R. 1978b. Evolutionary history of the cestode order Proteocephalidea. Syst. Zool. 27: 312-323. (4)

Brooks, D. R. 1979a. Testing hypotheses of evolutionary relationships among parasites: the digeneans of crocodiles. Amer. Zool. 19: 1225-1238. (4)

Brooks, D. R. 1979b. Testing the context and extent of host-parasite coevolution. Syst. Zool. 28: 299–307. (4, 8)

Brooks, D. R. 1980a. Allopatric speciation and non-interactive parasite community structure. Syst. Zool. 29: 192–203. (10)

Brooks, D. R. 1980b. Revision of the Acanthostominae (Digenea: Cryptogonimidae). Zool. Proc. Linn. Soc. 70: 1–70. (4)

Brooks, D. R. 1981. Hennig's parasitological method: a proposed solution. Syst. Zool. 30: 229–250. (4)

Brooks, D. R. and D. R. Glen. 1982. Pinworms and primates: a case study in coevolution. Proc. Helm. Soc. Wash. 49: 76–85. (4)

Brooks, D. R. and C. Mitter. 1983. Phylogenetics and coevolution. In *Fungus/Insect Relationships: Perspectives in Ecology and Evolution*, Q. Wheeler and M. Blackwell (eds.). Columbia University Press, New York. (In press) (4)

Brooks, D. R., T. B. Thorson and M. A. Mayes. 1981. Freshwater stingrays (Potamotrygonidae) and their helminth parasites: testing hypotheses of evolution and coevolution. In *Advances in Cladistics: Proceedings of the First Meeting of the Willi Hennig Society*, V. A. Funk and D. R. Brooks (eds.), pp. 147–176. New York Botanical Garden, New York. (4)

Brower, J. V. Z. 1958. Experimental studies of mimicry in some North American butterflies. Part II. *Battus philenor* and *Papilio troilus, P. polyxenes* and *P. glaucus.* Evolution 12: 123–136. (12)

Brower, L. P. 1963. Mimicry Symposium. *Proceedings of the XVI International Congress of Zoology*, Washington, D.C., pp. 145–186. (12)

Brower, L. P., J. V. Z. Brower and C. T. Collins. 1963. Experimental studies of mimicry: 7. Relative palatability and Müllerian mimicry among Neotropical butterflies of the subfamily Heliconiinae. *Zoologica* (N.Y.) 48: 65–84. (12)

Brower, L. P., M. E. Edmunds and C. M. Moffitt. 1975. Cardenolide content and palatability of a population of *Danaus chrysippus* butterflies from West Africa. J. Entomology 49: 183–196. (12)

Brower, L. P. and S. C. Glazier. 1975. Localization of heart poisons in the monarch butterfly. Science 188: 19–25. (12)

Brower, L. P., W. N. Ryerson, L. L. Coppinger and S. C. Glazier. 1968. Ecological chemistry and the palatability spectrum. Science 161: 1349–1351. (10)

Brown, A. R., R. B. Crandall and C. A. Crandall. 1976. Increased IgG catabolism as a possible factor in the immunosuppression produced in mice infected with *Heligmosomoides polygyrus.* J. Parasitol. 62: 169–171. (8)

Brown, J. H. 1975. Geographical ecology of desert rodents. In *Ecology and Evolution of Communities*, M. L. Cody and J. M. Diamond (eds.), pp. 315–341. Harvard University Press, Cambridge, Massachusetts. (18, 19)

Brown, J. H. and D. W. Davidson. 1977. Competition between seed-eating rodents and ants in desert ecosystems. Science 196: 880–882. (18)

Brown, J. H., D. W. Davidson and O. J. Reichman. 1979. An experimental study of competition between seed-eating desert rodents and ants. Amer. Zool. 19: 1129–1143. (11)

Brown, J. H. and A. Kodric-Brown. 1979. Convergence, competition, and mimicry in a temperate community of hummingbird-pollinated flowers. Ecology 60: 1022–1035. (3, 13)

Brown, J. H. and G. A. Lieberman. 1973. Resource utilization and coexistence of seed-eating rodents in sand dune habitats. Ecology 54: 788–797. (19)

Brown, J. H., O. J. Reichman and D. W. Davidson. 1979. Granivory in desert ecosystems. Ann. Rev. Ecol. Syst. 10: 201–227. (11, 19)

Brown, K. S. 1972. The heliconians of Brazil, Part III. Ecology and biology of *Heliconius nattereri*, a key primitive species near extinction, and comments on the evolutionary development of *Heliconius* and *Eueides*. Zoologica (N.Y.) 57: 41–69. (4)

Brown, K. S. 1981. The biology of *Heliconius* and related genera. Ann. Rev. Entomol. 26: 427–456. (4, 12)

Brown, K. S. and W. W. Benson. 1974. Adaptive polymorphism associated with multiple Müllerian mimicry in *Heliconius numata*. Biotropica 6: 205–228. (12)

Brown, K. S. and W. W. Benson. 1977. Evolution in modern Amazonian nonforest islands: *Heliconius hermathena*. Biotropica 9: 95–117. (12)

474

Brown, W. L., Jr. and E. O. Wilson. 1956. Character displacement. Syst. Zool. 7: 49-64. (1, 8, 17, 18)

Browning, J. A. 1972. Corn, wheat, rice, man: endangered species. J. Environ. Quality 1: 209-211. (7)

Browning, J. A. 1980. Genetic protective mechanisms of plant-pathogen populations: their co-evolution and use in breeding for resistance. In *Biology and Breeding for Resistance to Arthropods and Pathogens of Cultivated Crops*, M. K. Harris (ed.), pp. 52-75. Texas A & M University Press, College Station, Texas. (7)

Bruce, A. J. 1976. Coral reef Caridae and "commensalism." Micronesica 12: 83-98. (14)

Bruce, A. J. 1977. The hosts of the coral-associated Indo-West-Pacific pontoniine shrimps. Atoll. Res. Bull. 205: 1-19. (14)

Bruce, R. G. and D. Wakelin. 1977. Immunological interactions between *Trichinella spiralis* and *Trichuris muris* in the intestine of the mouse. Parasitology 74: 163-173. (8)

Brues, C. T. 1920. The selection of food-plants by insects with special reference to lepidopterous larvae. Amer. Natur. 54: 313-332. (4)

Brues, C. T. 1924. The specificity of food-plants in the evolution of phytophagous insects. Amer. Natur. 58: 127-144. (10)

Brunner, H., R. V. Harris and R. L. Amor. 1976. A note on the dispersal of seeds of blackberry (*Rubus procerus* P. J. Muell.) by foxes and emus. Weed Research 16: 171-173. (11)

Bullock, S. H. 1976. Consequences of limited seed dispersal within simulated annual populations. Oecologia 24: 247-256. (11)

Bullock, S. H. 1978. Plant abundance and distribution in relation to types of seed dispersal in chaparral. Madroño 25: 104-105. (11)

Bulmer, M. G. 1974. Density-dependent selection and character displacement. Amer. Natur. 108: 45-58. (3, 10, 17)

Burdon, J. J., D. R. Marshall and N. H. Luig. 1981. Isozyme analysis indicates that a virulent cereal rust pathogen is a somatic hybrid. Nature 293: 565-566. (7)

Burnet, M. and D. O. White. 1972. *Natural History of Infectious Disease*, 4th Ed. Cambridge University Press, Cambridge, England. (7, 9)

Burtt, B. D. 1929. A record of fruits and seeds dispersed by mammals and birds from the Singida district of Tanganyika territory. J. Ecol. 17: 351-355. (11)

Bush, A. O. 1980. Faunal similarity and infracommunity structure in the helminths of the lesser scaup. Ph.D. Thesis, Department of Zoology, University of Alberta. (8)

Bush, G. L. 1966. The taxonomy, cytology, and evolution of the genus *Rhagoletis* in North America (Diptera, Tephritidae). Bull. Museum Comp. Zool. Harvard Coll. 134: 431-562. (10)

Bush, G. L. 1974. The mechanism of sympatric host race formation in the true fruit flies (Tephritidae). In *Genetic Mechanisms of Speciation in Insects*, M. J. D. White (ed.), pp. 3-23. Australia and New Zealand Book Co., Sydney. (10)

Bush, G. L. 1975. Sympatric speciation in phytophagous parasitic insects. In *Evolutionary Strategies of Parasitic Insects and Mites*, P. W. Price (ed.), pp. 187-206. Plenum Press, New York. (4)

Butterworth, E. W. 1982. A study of the structure and organization of intestinal helminth communities in 10 species of waterfowl (Anatinae). Ph.D. Thesis, Department of Zoology, University of Alberta. (8)

Bychowsky, B. E. 1957. *Monogenetic Trematodes, their Systematics and Phylogeny*. Engl. Transl. by A.I.B.S., Washington, D.C. (8)

Cahalane, V. H. 1942. Caching and recovery of food by the western fox squirrel. J. Wildlife Management 6: 338-352. (11)

Cain, A. J. 1964. The perfection of animals. In *Viewpoints in Biology*, J. C. Carthy and C. L. Duddington (eds.), pp. 36-63. Butterworths, London. (16)

Cain, A. J. and P. M. Sheppard. 1954. Natural selection in Cepaea. Genetics 39: 89-116. (3)

Cain, R. F. 1972. Evolution of the fungi. Mycologia 44: 1–14. (7)

Cain, S. A. 1944. *Foundations of Plant Geography*. Harper & Row, New York. (19)

Callenbach, J. A. 1951. Rescue wheat and its resistance to wheat stem saw-fly attack. J. Econ. Ent. 44: 999–1001. (10)

Calos, M. and J. Miller. 1980. Transposable genetic elements. Cell 20: 579–595. (5)

Cameron, T. W. M. 1964. Host specificity and evolution of helminthic parasites. Adv. in Parasitol. 2: 1–34. (4)

Campbell, A. 1961. Conditions for the existence of bacteriophage. Evolution 15: 153–165. (5)

Campbell, A. 1981. Evolutionary significance of accessory DNA elements in bacteria. Ann. Rev. Microbiol. 35: 55–83. (5)

Campbell, B. C. and J. S. Duffy. 1979. Tomatine and parasitic wasps: potential incompatibility of plant antibiosis with biological control. Science 205: 700–702. (10)

Cant, J. G. H. 1979. Dispersal of *Stemmadenia donnell-smithii* by birds and monkeys. Biotropica 11: 122. (11)

Carpenter, F. L. 1979. Competition between hummingbirds and insects for nectar. American Zoologist 19: 1105–1114. (13)

Carpenter, G. D. H. 1942. Observations and experiments in Africa by the late C. F. M. Swynnerton on wild birds eating butterflies and the preferences shown. Proc. Linn. Soc. 154: 10–46. (12)

Carpenter, G. D. H. and E. B. Ford. 1933. *Mimicry*. Methuen, London. (12)

Carroll, C. R. and C. A. Hoffman. 1980. Chemical feeding deterrent mobilized in response to herbivory, and counteradaptation by *Epilachna tredecimnotata*. Science 209: 414–416. (10)

Cartier, J. J., A. Isaak, R. H. Painter and E. L. Sorenson. 1965. Biotypes of pea aphids, *Acyrthosiphon pisum* (Harris), in alfalfa clones. Can. Ent. 97: 754–760. (10)

Cartier, J. J. and R. H. Painter. 1956. Differential reactions of two biotypes of the corn leaf aphid to resistant and susceptible varieties, hybrids, and selections of sorghums. J. Econ. Ent. 49: 498–508. (10)

Carvajal, J. and M. D. Dailey. 1975. Three new species of *Echeneibothrium* (Cestoda: Tetraphyllidea) from the skate, *Raja chilensis* Guichenot, 1848, with comments on mode of attachment and host specificity. J. Parasitol. 61: 89–94. (8)

Case, T. 1979. Character displacement and coevolution in some *Cnemidophorus* lizards. Fortschr. Zool. 25: 235–282. (17)

Case, T. 1982. Coevolution in resource-limited competition communities. Theor. Pop. Biol. 21: 69–91. (3, 17)

Castro, G. A., C. Malone and S. Smith. 1980. Systemic anti-inflammatory effect associated with enteric trichinellosis in the rat. J. Parasitol. 66: 407–412. (8)

Castro, P. 1971. Nutritional aspects of the symbiosis between *Echinoecus pentagonus* and its host in Hawaii, *Echinothrix calamaris*. In *Aspects of the Biology of Symbiosis*, T. C. Cheng (ed.), pp. 229–247. University Park Press, Baltimore. (14)

Castro, P. 1976. Brachyuran crabs symbiotic with scleractinian corals: a review of their biology. Micronesica 12: 99–110. (14)

Cates, R. G. 1980. Feeding patterns of monophagous, oligophagous, and polyphagous insect herbivores: the effect of resource abundance and plant chemistry. Oecologia 46: 22–31. (10)

Cates, R. G. and G. H. Orians. 1975. Successional status and the palatability of plants to generalized herbivores. Ecology 56: 410–418. (19)

Caugant, D. A., B. R. Levin and R. K. Selander. 1981. Genetic diversity and temporal variation in the *E. coli* population of a human host. Genetics 98: 476–490. (5)

Cavalcante, P. B. 1976. *Frutas Comestiveis da Amazonia*. INPA, Belem, Brasil. (11)

Cavanaugh, C. M., S. L. Gardiner, M. L. Jones, H. W. Jannasch and J. B. Waterbury. 1981. Prokaryotic cells in the hydrothermal vent tube worm *Riftia pachyptila* Jones: possible chemotrophic symbionts. Science 213: 340–342. (14)

Cavener, D. 1979. Preference for ethanol in *Drosophila melanogaster* associated with the alcohol dehydrogenase polymorphism. Behavioral Genetics 9: 359–365. (10)

Chabaud, A. 1959. Remarques sur l'évolution et la taxonomie chez les nématodes parasites de vertébrés. Proc. Int. Congr. Zool. (15th), London, pp. 679–680. (8)

476

Chakrabarty, A. M. 1976. Plasmids in *Pseudomonas*. Ann. Rev. Gen. 10: 7-30. (5)

Chao, L. 1979. The population biology of colicinogenic bacteria; a model for the evolution of allelopathy. Ph.D. dissertation, University of Massachusetts. (5)

Chao, L. and B. R. Levin. 1981. Structured habitats and the evolution of anticompetitor toxins in bacteria. Proc. Natl. Acad. Sci. USA 78: 6324-6328. (5)

Chao, L., B. R. Levin and F. M. Stewart. 1977. A complex community in a simple habitat: an experimental study with bacteria and phage. Ecology 58: 369-379. (1, 5)

Chapman, F. M. 1935. Jose. Two months from the life of a Barro Colorado coati. Natural History Magazine 35: 299-308. (11)

Charlesworth, B. 1971. Selection in density-regulated populations. Ecology 52: 469-474. (2)

Charlesworth, B. and D. Charlesworth. 1973. A study of linkage disequilibrium in *Drosophila melanogaster*. Genetics 73: 351-359. (2, 10)

Charlesworth, B., R. Lande and M. Slatkin. 1982. A neo-Darwinian commentary on macroevolution. Evolution 36: 474-498. (2, Ep)

Charlesworth, D. and B. Charlesworth. 1975. Theoretical genetics of Batesian mimicry I. Single locus models. J. Theoretical Biology 55: 283-303. (3, 12)

Charnov, E. L. 1979. Simultaneous hermaphroditism and sexual selection. Proc. Natl. Acad. Sci. USA 76: 2480-2484. (13)

Chase, V. and P. H. Raven. 1975. Evolutionary and ecological relationships between *Aquilegia formosa* and *A. pubescens* (Ranunculaceae), two perennial plants. Evolution 29: 474-486. (13)

Chasen, F. N. 1939. *The Birds of the Malay Peninsula*. Witherby, London. (11)

Chatfield, C. and L. I. McLaughlin 1928. Proximate composition of fresh fruits. U.S.D.A. Circular No. 50. (11)

Cheke, A. S., W. Nanakorn and C. Yankoses. 1979. Dormancy and dispersal of seeds of secondary forest species under the canopy of a primary tropical rain forest in northern Thailand. Biotropica 11: 88-95. (11)

Chemsak, J. A. 1963. Taxonomy and bionomics of the genus *Tetraopes*. Univ. Calif. Publ. Ent. 30: 1-90. (4)

Chew, F. S. 1980. Foodplant preferences of *Pieris* caterpillars (Lepidoptera). Oecologia 46: 347-353. (10)

Chew, F. S. and R. K. Robbins. 1982. Egg laying in butterflies. In *Biology of Butterflies*. Sympos. Roy. Ent. Soc. Lond. XI. (In press) (4)

Chew, F. S. and J. E. Rodman 1979. Plant resources for chemical defense. In *Herbivores, Their Interaction with Secondary Plant Metabolites*, G. A. Rosenthal and D. H. Janzen (eds.), pp. 271-307. Academic Press, New York. (4)

Chin, K. M. 1979. Aspects of the epidemiology and genetics of the foliar pathogens, *Erysiphe graminis* f. sp. *hordei*, in relation to infection of homogeneous and heterogeneous populations of the barley host (*Hordeum vulgare*). Ph.D. Thesis, University of Cambridge. (7)

Choat, J. H. and R. Black. 1979. Life histories of limpet and the limpet-laminarian relationship. J. Exp. Mar. Biol. Ecol. 41: 25-50. (19)

Christiansen, F. B. and T. M. Fenchel. 1977. *Theories of Populations in Biological Communities*. Springer-Verlag, Berlin. (3)

Christie, P. R., D. Wakelin and M. M. Wilson. 1979. The effect of the expulsion phase of *Trichinella spiralis* on *Hymenolepis diminuta* infection in rats. Parasitology 78: 323-330. (8)

Chrome, F. H. J. 1975. The ecology of fruit pigeons in tropical Northern Queensland. Aust. Wildl. Res. 2: 155-185. (11)

Claridge, M. F. and J. den Hollander. 1980. The "biotypes" of the rice brown planthopper, *Nilaparvata lugens*. Ent. exp. appl. 27: 23-30. (10)

Claridge, M. F. and M. R. Wilson. 1978a. British insects and trees: a study in island biogeography or insect/plant coevolution? Amer. Natur. 112: 451-456. (10)

Claridge, M. F. and M. R. Wilson. 1978b. Oviposition behaviour as an ecological factor in woodland canopy leafhoppers. Ent. exp. appl. 24: 101-109. (10)

Clark, J., J. R. Beerbower and K. K. Kietzke. 1967. Oligocene sedimentation, stratigraphy, paleoecology and paleoclimatology. Fieldiana: Geology Mem. 5: 1–158. (15)

Clark, W. H. and P. L. Comanor. 1973. The use of western harvester ant *Pogonomyrmex occidentalis* (Cresson) seed stores by heteromyid rodents. Occasional Paper Biol. Soc. Nevada 34: 1–6. (11)

Clark, W. H. and P. L. Comanor. 1976. The northern desert horned lizard, *Phrynosoma platyrhinos platyrhinos*, as a predator of the western harvester ant, *Pogonomyrmex occidentalis*, and a dispersal agent for *Eriogonum baileyi*. Idaho Acad. Sci. 12: 9–12. (11)

Clarke, B. 1969. The evidence for apostatic selection. Heredity 24: 347–352. (3)

Clarke, B. 1972. Density-dependent selection. Amer. Natur. 106: 1–13. (13)

Clarke, B. 1976. The ecological genetics of host-parasite relationships. In *Genetic Aspects of Host-Parasite Relationships*, A. E. R. Taylor and R. Muller (eds.), Symp. Brit. Soc. Parasitol. 14: 87–103. (3, 7, 8, 9)

Clarke, B. C. 1975. The causes of biological diversity. Sci. Amer. 233: 50–60. (9)

Clarke, B. C. 1979. The evolution of genetic diversity. Proc. Roy. Soc. Lond. B. 205: 453–474. (7)

Clemens, E. T., C. E. Stevens and M. Southworth. 1975. Sites of organic acid production and pattern of digesta movement in the gastrointestinal tract of swine. J. Nutrition 105: 759–768. (11)

Cody, M. L. 1968. On the methods of resource division in grassland communities. Amer. Natur. 102: 107–148. (17)

Cody, M. L. 1971. Ecological aspects of reproduction. In *Avian Biology*, Vol. I, D. S. Farner and J. R. King (eds.), pp. 461–512. Academic Press, New York. (3)

Cody, M. L. 1974. *Competition and the Structure of Bird Communities*. Princeton University Press, Princeton, New Jersey. (1, 17)

Cody, M. L. 1975. Towards a theory of continental species diversities: bird distributions over Mediterranean habitat gradients. In *Ecology and Evolution of Communities*, M. L. Cody and J. M. Diamond (eds.), pp. 214–257. Harvard University Press, Cambridge, Massachusetts. (19)

Cody, M. L. and H. A. Mooney. 1978. Convergence versus nonconvergence in Mediterranean-climate ecosystems. Ann. Rev. Ecol. Syst. 9: 265–321. (3, 17)

Cole, B. J. 1981. Overlap, regularity, and flowering phenologies. Amer. Natur. 117: 993–997. (13)

Colwell, R. K. 1973. Competition and coexistence in a simple tropical community. Amer. Natur. 107: 737–760. (13)

Conn, E. E. 1979. Cyanide and cyanogenic glycosides. In *Herbivores: Their Interaction with Secondary Plant Metabolites*, G. A. Rosenthal and D. H. Janzen (eds.), pp. 387–412. Academic Press, New York. (10)

Connell, J. H. 1961. The influence of interspecific competition and other factors on the distribution of the barnacle *Chthamalus stellatus*. Ecology 42: 710–723. (19)

Connell, J. H. 1971. On the role of natural enemies in preventing competitive exclusion in some marine animals and in rain forest trees. In *Dynamics of Populations*, P. J. den Boer and G. Gradwell (eds.). Pudoc, Wageningen. (19)

Connell, J. 1980. Diversity and the coevolution of competitors, or the ghost of competition past. Oikos 35: 131–138. (17)

Connor, E. F., S. H. Faeth, D. Simberloff and P. A. Opler. 1980. Taxonomic isolation and the accumulation of herbivorous insects: a comparison of introduced and native trees. Ecol. Entomol. 5: 205–211. (10, 13)

Connor, E. F. and D. Simberloff. 1979. The assembly of species communities: chance or competition? Ecology 60: 1132–1140. (13, 17)

Conover, M. R. 1979. Effect of gastropod shell characteristics and hermit crabs on shell epifauna. J. Exp. Mar. Biol. Ecol. 40: 81–94. (14)

Constant-Desportes, M., J. P. Couland and M. Payet. 1976. Considérations sur le portage de l'antigène Australie en Martinique. Bull. Soc. Path. Exot. 69: 382–388. (8)

Cook, R. E. 1982. Attractions of the flesh. Nat. Hist. Mag. 91: 21–24. (11)

Cooke, R. C. 1977a. *The Biology of Symbiotic Fungi*. John Wiley & Sons Ltd., London. (7)

478

Cooke, R. C. 1977b. *Fungi, Man and His Environment.* Longmans, London. (7)

Corbet, A. S. and H. M. Pendlebury. 1978. *The Butterflies of the Malay Peninsula.* 3rd Ed., revised by J. N. Eliot. Oliver and Boyd, Edinburgh. (4)

Cornell, H. V. and J. O. Washburn. 1979. Evolution of the richness-area correlation for cynipid gall wasps on oak trees: a comparison of two geographic areas. Evolution 33: 257–274. (4)

Counce, S. J. and D. F. Poulson. 1962. Developmental effects of the sex-ratio agent in embryos of *Drosophila willistoni.* J. Exp. Zool. 151: 17–31. (6)

Counce, S. J. and D. F. Poulson. 1966. The expression of maternally-transmitted sex ratio condition (SR) in the strains of *Drosophila melanogaster.* Genetica 37: 364–390. (6)

Cowen, R. 1970. Analogies between the recent bivalve *Tridacna* and the fossil brachiopods *Lyttoniacea* and *Richthofeniacea.* Paleogeogr., Paleoclimatol., Paleoecol. 8: 329–344. (14)

Cowles, R. B. 1936. The relation of birds to seed dispersal of the desert mistletoe. Madroño 3: 352–356. (11)

Crepet, W. J. 1979. Insect pollination: a paleontological perspective. BioScience 29: 102–108. (13)

Crile, M. D. and D. P. Quiring. 1940. A record of the body weight and certain organ and gland weights of 3690 animals. Ohio J. Sci. 40: 219–259. (16)

Crisp, D. J. 1958. The spread of *Elminius modestus* in Northwest Europe. Jour. Mar. Biol. Assoc. UK 37: 483–520. (19)

Crofton, H. D. 1971. A quantitative approach to parasitism. *Parasitology* 63: 179–193. (9)

Crowder, L. B. 1980. Ecological convergence of community structure: a neutral model analysis. Ecology 61: 194–198. (19)

Cruden, R. W. 1972a. Pollination biology of *Nemophila menziesii* (Hydrophyllaceae) with comments on the evolution of oligolectic bees. Evolution 26: 373–389. (13)

Cruden, R. W. 1972b. Pollinators in high-elevation ecosystems: relative effectiveness of birds and bees. Science 176: 1439–1440. (13)

Cruden, R. W., S. Kinsman, R. E. Stockhouse II and Y. B. Linhart. 1976. Pollination, fecundity, and the distribution of moth-flowered plants. Biotropica 8: 204–210. (13)

Cruz, A., 1974. Feeding assemblages of Jamaican birds. Condor 76: 103–107. (11)

Cruz, A. 1981. Bird activity and seed dispersal of a montane forest tree (*Dunalia arborescens*) in Jamaica. Biotropica (supplement) 13: 34–44. (11)

Culver, D. C. and A. J. Beattie. 1978. Myrmecochory in *Viola*: dynamics of seed-ant interactions in some West Virginia species. J. Ecol. 66: 53–72. (11)

Cuttress, C., D. M. Ross and L. Sutton. 1970. The association of *Calliactis tricolor* with its pagurid, calappid and majid partners in the Caribbean. Canad. J. Zool. 48: 371–376. (14)

Czaplinski, B. 1975. Hymenolepididae parasitizing wild mute swans *Cygnus olor* (Gm.) of different age in Poland. Acta Parasitol. Pol. 23: 305–327. (8)

Czyhrinciw, N. 1969. Typical fruit technology. Adv. Food Res. 16: 153–214. (11)

DaCosta, C. P. and C. M. Jones. 1971. Cucumber beetle resistance and mite susceptibility controlled by the bitter gene in *Cucumis sativa* L. Science 172: 1145–1146. (10)

Dafni, A. and Y. Ivri. 1981a. Floral mimicry between *Orchis israelitica* Baumann and Dafni (Orchidaceae) and *Bellevalia flexuosa* Boiss. (Liliaceae). Oecologia 49: 229–232. (13)

Dafni, A. and Y. Ivri. 1981b. The flower biology of *Cephalanthera longifolia* (Orchidaceae)—pollen imitation and facultative floral mimicry. Plant System. Evol. 137: 229–240. (13)

Damian, R. T. 1979. Molecular mimicry in biological adaptation. In *Host-Parasite Interfaces*, B. B. Nickol (ed.), pp. 103–126. Academic Press, New York. (5)

Daniel, M. J. 1976. Feeding by the short-tailed bat (*Mystacina tuberculata*) on fruit and possibly nectar. New Zealand J. Zool. 3: 391–398. (11)

Darley-Hill, S. and W. C. Johnson 1981. Acorn dispersal by the blue jay (*Cyanocitta cristata*). Oecologia 50: 231–232. (11)

Darwin, C. 1859. *On the Origin of Species*. John Murray, London. (1, 2)

Darwin, C. 1896. *On the Origin of Species*, 6th Ed. D. Appleton, New York. (12)

Dausset, J. 1981. The major histocompatibility complex in man: past, present and future concepts. Science 213: 1469–1474. (8)

Dauterman, W. C. and E. Hodgson. 1978. Detoxication mechanisms in insects. In *Biochemistry of Insects*, M. Rockstein (ed.), pp. 541–577. Academic Press, New York. (10)

Davey, R. P. and D. C. Reanney. 1980. Extrachromosomal genetic elements and the adaptive evolution of bacteria. In *Evolutionary Biology*, Vol. 13, M. K. Hecht, W. C. Steere, and B. Wallace (eds.), pp. 113–147. Plenum Press, New York. (5)

Davidge, C. 1977. Baboons as dispersal agents for *Acacia cyclops*. Zoologica Africana 12: 249–250. (11)

Davidson, D. W. 1977a. Species diversity and community organization in desert seed-eating ants. Ecology 58: 711–724. (18, 19)

Davidson, D. W. 1977b. Foraging ecology and community organization in desert seed-eating ants. Ecology 58: 725–737. (18, 19)

Davidson, D. W., J. H. Brown and R. S. Inouye. 1980. Competition and the structure of granivore communities. BioScience 30: 233–238. (18, 19)

Davidson, D. W. and S. R. Morton. 1981. Competition for dispersal in ant-dispersed plants. Science 213: 1259–1261. (11)

Davies, J. S., J. G. Hall, G. A. T. Targett and M. Murray. 1980. The biological significance of the immune response with special reference to parasites and cancer. J. Parasitol. 66: 705–721. (8)

Davies, S. J. J. F. 1978. The food of emus. Aust. J. Ecol. 3: 411–422. (11)

Dawkins, M. 1971. Perceptual changes in chicks: another look at the "search image" concept. Anim. Behav. 19: 566–574. (10)

Dawkins, R. and J. R. Krebs. 1979. Arms races between and within species. Proc. Roy. Soc. London B. 205: 489–511. (1, 14)

Day, J. H. 1969. *A Guide to Marine Life on South African Shores*. A. A. Balkema, Cape Town. (19)

Day, P. R. 1974. *Genetics of Host-Parasite Interactions*. W. H. Freeman, San Francisco. (1, 3, 7, 8, 9)

Dayton, P. K. 1971. Competition, disturbance, and community organization: the provision and subsequent utilization of space in a rocky intertidal community. Ecol. Monog. 41: 351–389. (19)

Dayton, P. K. 1973. Two cases of resource partitioning in an intertidal community: making the right prediction for the wrong reason. Amer. Natur. 107: 662–670. (19)

Dayton, P. K. and R. R. Hessler. 1972. Role of biological disturbance in maintaining diversity in the deep sea. Deep Sea Res. 19: 199–208. (19)

De Angelis, D. L., E. W. Stiles, W. C. Johnson, D. M. Sharpe and R. K. Schreiber. 1977. A model for the dispersal of seeds by animals. Environmental Sci. Div. Pub. No. 1053, Oak Ridge National Lab., Oak Ridge, Tennessee. (11)

Dean, D. A., M. A. Bukowski and A. W. Cheever. 1981. Relationship between acquired resistance, portal hypertension, and lung granulomas in ten strains of mice infected with *Schistosoma mansoni*. Am. J. Trop. Med. Hyg. 30: 806–814. (8)

Dean, R. E. 1980. Passage rate of alfalfa through the digestive tract of elk. J. Wild. Manag. 44: 272–273. (11)

DeLeon, J. R. and B. O. L. Duke. 1966. Experimental studies on the transmission of Guatemalan and West African strains of *Onchocerca volvulus* by *Simulium ochraceum, S. metallicum*, and *S. callidum*. Trans. R. Soc. Trop. Med. Hyg. 60: 735–752. (8)

Della Bruna, C. and B. Xenia. 1976. *Nippostrongylus brasiliensis* in mice: reduction of worm burden and prolonged infection induced by presence of *Nematospiroides dubius*. J. Parasitol. 62: 490–491. (8)

480

Delyamure, S. L. 1955. *The Helminthofauna of Marine Animals in the Light of their Ecology and Phylogeny.* Izd. Akad. Nauk SSSR, Moscow. (Translation TT67-51202, U.S. Department of Commerce, Springfield, Virginia.) (9)

Demerec, M. and U. Fano. 1945. Bacteriophage-resistant mutants in *Escherichia coli.* Genetics 30: 119-136. (5)

Dennis, W. 1930. Rejection of wormy nuts by squirrels. J. Mammalogy 11: 195-201. (11)

den Hollander, J. and P. K. Pathak. 1981. The genetics of the "biotypes" of the rice brown planthopper, *Nilaparvata lugens.* Ent. exp. appl. 29: 76-86. (10)

Depéret, C. 1905. L'évolution des Mammiferes tertiaires; l'importance des migrations (Eocène). C. R. Acad. Sci. Paris 141: 702-741. (16)

Dethier, V. G. 1947. *Chemical Insect Attractants and Repellents.* Blakiston, Philadelphia. (10)

Dethier, V. G. 1954. Evolution of feeding preferences in phytophagous insects. Evolution 8: 33-54. (10)

Dethier, V. G. 1970. Chemical interactions between plants and insects. In *Chemical Ecology,* E. Sondheimer and J. B. Simeone (eds.), pp. 83-102. Academic Press, New York. (10)

Dethier, V. G. 1976. The importance of stimulus patterns for host-plant recognition and acceptance. In *The Host-Plant in Relation to Insect Behavior and Reproduction,* T. Jermy (ed.), pp. 67-70. Plenum Press, New York. (10)

Diamond, J. M. 1973. Distributional ecology of New Guinea birds. Science 179: 759-769. (17, 18)

Diamond, J. M. 1975. Assembly of species communities. In *Ecology and Evolution of Communities,* M. L. Cody and J. M. Diamond (eds.), pp. 342-444. Harvard University Press, Cambridge, Massachusetts. (17)

Diamond, J. and M. Gilpin. 1982. Examination of the "null" model of Conner and Simberloff for species co-occurrence on islands. Oecologia 52: 64-74. (17, 18)

Dice, L. R. and P. M. Blossom. 1937. Studies of the mammalian ecology in southwestern North America, with special attention to the colors of desert mammals. Carnegie Inst. Wash. Publ. 485: 1-129. (2)

Dickerson, G. E. 1955. Genetic slippage in response to selection for multiple objectives. Cold Spring Harbor Symp. Quant. Biol. 20: 213-224. (2)

Dinoor, A. 1974. Role of wild and cultivated plants in the epidemiology of plant diseases in Israel. Ann. Rev. Phytopath. 12: 413-436. (7)

Dinoor, A. 1977. Oat crown rust resistance in Israel. In *The Genetic Basis of Epidemics in Agriculture,* P. R. Day (ed.), Annals of N.Y. Acad. Sci. 287: 357-366. (7)

Dixey, F. A. 1897. Mimetic attraction. Trans. Ent. Soc. Lond. 1897: 317-332. (12)

Dixey, F. A. 1909. On Müllerian mimicry and diaposematism. Trans. Ent. Soc. Lond. 1909: 559-583. (12)

Dobson, A. 1983. Parasite virulence: a survey of the trends in some 300 parasite-host associations. (In preparation) (9)

Dobzhansky, T. 1970. *Genetics of the Evolutionary Process.* Columbia Univ. Press, New York. (2)

Dobzhansky, T., O. Pavlovsky, and J. R. Powell. 1976. Partially successful attempt to enhance reproductive isolation between semispecies of *Drosophila paulistorum.* Evolution 30: 201-212. (6)

Dobzhansky, T. and B. Spassky. 1967. An experiment on migration and simultaneous selection for several traits in *Drosophila pseudoobscura.* Genetics 59: 411-425. (2)

Docters van Leeuwen, W. M. 1954. On the biology of some Javanese Loranthaceae and the role birds play in their life-history. Beaufortia 4: 105-207. (11)

Dodson, C. H., R. L. Dressler, H. G. Hills, R. M. Adams and N. H. Williams. 1969. Biologically active compounds in orchid fragrances. Science 164: 1243-1249. (13)

Dodson, M. M. and A. Hallam. 1977. Allopatric speciation and the fold catastrophe. Amer. Natur. 111: 415-433. (18)

Dogiel, V. A. 1964. *General Parasitology.* Oliver and Boyd, Edinburgh and London. (4)

Dolinger, P. M., P. R. Ehrlich, W. L. Fitch and D. E. Breedlove. 1973. Alkaloid and predation patterns in Colorado lupine populations. Oecologia 13: 191–204. (10)

Doolittle, W. F. and C. Sapienza. 1980. Selfish genes, the phenotype paradigm and genome evolution. Nature 284: 601–603. (5)

Dorst, J. 1947. Le role disséminateur des oiseaux dans la vie des plantes. La Terre et la Vie 94: 106–119. (11)

Dove, W. F. 1971. Biological inferences. In *The Bacteriophage Lambda*, A. D. Hershey (ed.), pp. 297–312. Cold Spring Harbor Laboratory, Cold Spring Harbor, New York. (5)

Dressler, R. L. 1968. Pollination by euglossine bees. Evolution 22: 202–210. (13)

Dressler, R. L. 1981. *The Orchids: Natural History and Classification.* Harvard University Press, Cambridge, Massachusetts. (13)

Dudley, E. C. and G. J. Vermeij. 1978. Predation in time and space: drilling in the gastropod *Turritella.* Paleobiology 4: 436–441. (14)

Duggins, D. O. 1980. Kelp beds and sea otters: an experimental approach. Ecology 61: 447–453. (19)

Duke, B. O. L., P. J. Moore and J. R. DeLeon. 1967. *Onchocerca-Simulium* complexes. V. The intake and subsequent fate of microfilariae of a Guatemalan strain of *Onchocerca volvulus* in forest and Sudan-savannah forms of a West African *Simulium damnosum.* Ann. Trop. Med. Parasitol. 61: 332–337. (8)

Dunham, A. E. 1978. Food availability as a proximate factor influencing growth rates in the iguanid lizard, *Sceloporus merriami.* Ecology 59: 770–778. (17)

Dunham, A. E., G. R. Smith and J. N. Taylor. 1979. Evidence for ecological character displacement in western American catastomid fishes. Evolution 33: 877–896. (18)

Dunn, D. F. 1981. The clownfish sea anemones: Stichodactylidae (Coelenterata: Actinaria) and other sea anemones symbiotic with pomacentrid fishes. Trans. Amer. Phil. Soc. 71: 1–115. (14)

Dunn, D. F., D. M. Devaney and B. Roth. 1980. *Stylobates*: a shell-forming sea anemone (Coelenterata, Anthozoa, Actiniidae). Pac. Sci. 34: 379–388. (14)

Eastop, V. F. 1973a. Deductions from the present day host plants of aphids and related insects. In *Insect/plant Relationships*, H. F. van Emden (ed.), pp. 157–178. Symp. Roy. Ent. Soc. London 6, London. (10)

Eastop, V. F. 1973b. Biotypes of aphids. In *Perspectives in Aphid Biology*, A. D. Lowe (ed.), pp. 40–51. Ent. Soc. N. Z. Bull. No. 2. (10)

Ebenman, B. and S. Nilsson. 1982. Components of niche width in a territorial bird species: habitat utilization in males and females of the chaffinch (*Fringilla coelebs*) on islands and mainland. Amer. Natur. 119: 331–344. (13)

Ebling, F. J., J. A. Kitching, R. D. Purchon and R. Bassindale. 1948. The ecology of Lough Ine rapids with special reference to water currents. 2. The fauna of the *Saccorhiza* canopies. J. Anim. Ecol. 17: 203–244. (19)

Echols, H. 1972. Developmental pathways for the temperate phage: lysis vs. lysogeny. Ann. Rev. Gen. 6: 157–190. (5)

Edgar, J. A., C. C. J. Culvenor and T. E. Pliske. 1974. Coevolution of danaid butterflies with their host plants. Nature 250: 646–648. (4)

Edlin, G. L. Lin and R. Bitner. 1977. Reproductive fitness of P1, P2, and Mu lysogens of *Escherichia coli.* J. Virol. 21: 560–564. (5)

Edmunds, G. F., Jr. and D. N. Alstad. 1978. Coevolution in insect herbivores and conifers. Science 199: 941–945. (10)

Edson, K. M., S. B. Vinson, D. B. Stoltz and M. D. Summers. 1981. Virus in a parasitoid wasp: suppression of the cellular immune response in the parasitoid's host. Science 211: 582–583. (8)

Edson, W. L. G. 1918. Fruits for birds. Bird-lore 17: 448–449. (11)

Edwards, P. J. and S. D. Wratten. 1980. *Ecology of Insect-Plant Interactions.* Edward Arnold, London. (10)

Ehrlich, P. R. 1958. The comparative morphology, phylogeny and higher classification of the butterflies (Lepidoptera: Papilionoidea). Univ. Kansas Sci. Bull. 39: 305–370. (4)

482

Ehrlich, P. R. and P. H. Raven. 1964. Butterflies and plants: a study in coevolution. Evolution 18: 586–608. (1, 4, 10, 15, 19, Ep)

Ehrlich, P. R. and P. H. Raven. 1969. Differentiation of populations. Science 165: 1228–1232. (2)

Ehrlich, P. R., R. White, M. C. Singer, W. W. McKechnie and L. E. Gilbert. 1975. Checkerspot butterflies: a historical perspective. Science 188: 221–228. (2)

Ehrman, L. 1961. The genetics of sexual isolation in *Drosophila paulistorum*. Evolution 16: 1025–1038. (6)

Ehrman, L. 1965. Direct observation of sexual isolation between allopatric and between sympatric strains of the different *Drosophila paulistorum* races. Evolution 19: 459–464. (6)

Ehrman, L. and S. Daniels. 1975. Pole cells of *Drosophila paulistorum*: embryologic differentiation with symbionts. Aust. J. Biol. Sci. 28: 133–144. (6)

Ehrman, L. and R. P. Kernaghan. 1971. Microorganismal basis of infectious hybrid male sterility in *D. paulistorum*. Journal of Heredity 62: 67–71. (6)

Ehrman, L. and P. A. Parsons. 1981. *Behavior Genetics and Evolution*. McGraw-Hill, New York. (6)

Ehrman, L. and J. R. Powell. 1981. The *Drosophila willistoni* species group. In *Genetics and Biology of Drosophila*, Vol. 3b, M. Ashburner, H. L. Carson and J. N. Thompson (eds.). Academic Press, New York. (In press) (6)

Eichler, D. A. 1971. Studies on *Onchocerca gutterosa* (Neumann, 1910) and its development in *Simulium ornatum* (Meigen, 1818). II. Behaviour of *S. ornatum* in relation to the transmission of *O. gutturosa*. J. Helminth. 45: 259–270. (8)

Eichler, W. 1941a. Wirtsspezifität und stammesgeschichtliche Gleichläufigkeit (Fahrenholz Regel) bei Parasiten in allgemeinen und bei Mallophagen im besonderen. Zool. Anzeig. 132: 254–292. (4)

Eichler, W. 1941b. Korrelation in der Stammesentwicklung von Wirten und Parasiten. Zeitschr. Parasiten. 19: 24. (4)

Eichler, W. 1948. Some rules in ectoparasitism. Ann. Mag. Nat. Hist. 12: 588–598. (4)

Eickwort, G. C. and H. S. Ginsberg. 1980. Foraging and mating behavior in Apoidea. Ann. Rev. Entomol. 25: 421–446. (13)

Eisenberg, J. F. 1981. *The Mammalian Radiations*. University of Chicago Press, Chicago. (15)

Eisenberg, J. F. and M. Lockhart. 1972. An ecological reconnaisance of Wilpattu National Park, Ceylon. *Smithson. Contrib. Zool.* 101: 1–118. (16)

Eisenberg, J. F. and D. E. Wilson. 1978. Relative brain size and feeding strategies in the Chiroptera. Evolution 32: 740–751. (11)

Eisner, T. 1964. Catnip: its *raison d'être*. Science 146: 1318–1320. (10)

Eldredge, N. and S. J. Gould. 1972. Punctuated equilibria: an alternative to phyletic gradualism. In *Models in Paleobiology*, T. J. M. Schopf (ed.), pp. 82–115. W. H. Freeman, San Francisco. (2)

Ellingboe, A. H. 1978. A genetic analysis of host-parasite interactions. In *The Powdery Mildews*, D. M. Spencer (ed.), pp. 159–181. Academic Press, London. (7, 8)

Elliot, G. F. 1964. Tertiary solenoporacean algae and the reproductive structures of the Solenoporaceae. Palaeontology 7: 695–702. (15)

Elliot, J. N. 1973. The higher classification of the Lycaenidae (Lepidoptera): a tentative arrangement. Bull. Br. Mus. Nat. Hist. (Ent.) 28: 373–505. (4)

Elner, R. W. and D. G. Raffaelli. 1980. Interactions between two marine snails, *Littorina rudis* Maton and *Littorina nigrolineata* Gray, a predator, *Carcinus maenus* (L.), and a parasite, *Microphallus similis* Jägerskiold. J. Exp. Mar. Biol. Ecol. 43: 151–160. (14)

Elton, C. S. 1958. *The Ecology of Invasions by Animals and Plants*. Methuen, London. (19)

Eltringham, H. 1916. On specific and mimetic relationships in the genus *Heliconius*. Trans. Ent. Soc. Lond. 1916: 101–155. (12)

Emmons, L. 1975. Ecology and behavior of African rainforest squirrels. Ph.D. Thesis, Cornell University, Ithaca, New York. (11)

483

Emmons, L. H. 1980. Ecology and resource partitioning among nine species of rain forest squirrels. Ecol. Monogr. 50: 31-54. (11)

Emsley, M. G. 1965. Speciation in *Heliconius*: morphology and geographic distribution. Zoologica (NY) 50: 191-254. (4)

Enders, F. 1974. Vertical stratification in orb-web spiders and a consideration of other methods of coexistence. Ecology 55: 317-328. (18)

Endler, J. A. 1977. *Geographic Variation, Speciation and Clines*. Princeton University Press, Princeton, New Jersey. (2, 18)

Endler, J. A. 1978. A predator's view of animal color patterns. Evol. Biol. 11: 317-364. (3)

Erickson, J. M. and P. Feeny. 1974. Sinigrin: a chemical barrier to the black swallowtail butterfly, *Papilio polyxenes*. Ecology 55: 103-111. (10)

Eshed, N. and A. Dinoor. 1981. Genetics of pathogenicity in *Puccinia coronata*: the host range among grasses. Phytopathology 71: 156-163. (7)

Eshel, I. 1972. On the neighbor effect and the evolution of altruistic traits. Theor. Popul. Biol. 3: 257-277. (3)

Estes, J. A. and J. F. Palmisano. 1974. Sea otters: their role in structuring nearshore communities. Science 185: 1058-1060. (19)

Estes, J. R. and R. W. Thorp. 1975. Pollination ecology of *Pyrrhopappus carolinianus* (Compositae). Amer. J. Botany 62: 148-159. (13)

Euzet, L. 1957. Cestodes de Selacians. In *Premier Symposium sur la Spécificité Parasitaire des Parasites de Vertébrés*, J. G. Baer (ed.), pp. 259-269. Int. Union Biol. Sci., Ser. B, No. 32, Paris. (8)

Ewens, W. J. 1979. *Mathematical Population Genetics*. Springer-Verlag, New York. (2)

Ewens, W. J. and M. W. Feldman. 1976. The theoretical assessment of selective neutrality. In *Population Genetics and Ecology*, S. Karlin and E. Nevo (eds.), pp. 305-307. Academic Press, New York. (7)

Faegri, K. and L. van der Pijl. 1979. *The Principles of Pollination Ecology*, 3rd Ed. Pergamon Press, New York. (13)

Fahrenholz, H. 1913. Ectoparasiten und Abstammungslehre. Zoologischer Anzeiger, Leipzig 41: 371-374. (4)

Falconer, D. S. 1960. *Introduction to Quantitative Genetics*. Ronald Press, New York. (2, 3)

Falkow, S. 1975. *Infectious Multiple Drug Resistance*. Pion, London. (5)

Farris, J. S. 1970. Methods for computing Wagner Trees. Syst. Zool. 19: 83-92. (4)

Farris, J. S. 1979. The information content of the phylogenetic system. Syst. Zool. 28: 483-520. (4)

Farris, J. S. 1981. (Comment on the paper by Simberloff et al.) In *Vicariance Biogeography: A Critique*. G. Nelson and D. E. Rosen (eds.), pp. 73-84. Columbia University Press, New York. (4)

Feare, C. J. 1971. Predation of limpets and dog-whelks by oystercatchers. Bird Study 18: 121-129. (14)

Feder, H. M. 1966. Cleaning symbiosis in the marine environment. In *Symbiosis*, Vol. I., M. Henry (ed.)., pp. 327-380. Academic Press, New York. (14)

Feeny, P. P. 1970. Seasonal changes in oak leaf tannins as a cause of spring feeding by winter moth caterpillars. Ecology 51: 656-681. (10)

Feeny, P. P. 1976. Plant apparency and chemical defense. In *Biochemical Interactions Between Plants and Insects*, J. Wallace and R. Mansell (eds.), Rec. Adv. Phytochem. 10: 1-40. (10)

Feinsinger, P. 1976. Organization of a tropical guild of nectarivorous birds. Ecol. Monogr. 46: 257-291. (13, 17)

Feinsinger, P. 1978. Ecological interactions between plants and hummingbirds in a successional tropical community. Ecol. Monog. 48: 269-287. (13)

Feinsinger, P. 1980. Asynchronous migration patterns and the coexistence of tropical hummingbirds. In *Migrant Birds in the Neotropics: Ecology, Behavior, Distribution, and Conservation*, A. Keast and E. S. Morton (eds.), pp. 411-419. Smithsonian Institution Press, Washington, D.C. (13)

484

Feinsinger, P. and S. B. Chaplin. 1975. On the relationship between wing disc loading and foraging strategy in hummingbirds. Amer. Natur. 109: 217–224. (13)

Feinsinger, P. and R. K. Colwell. 1978. Community organization among neotropical nectar-feeding birds. Amer. Zoologist 18: 779–795. (13)

Feinsinger, P., R. K. Colwell, J. Terborgh and S. B. Chaplin. 1979. Elevation and the morphology, flight energetics, and foraging ecology of tropical hummingbirds. Amer. Natur. 113: 481–497. (13)

Feinsinger, P. and L. A. Swarm. 1982. "Ecological release," seasonal variation in food supply, and the hummingbird *Amazilia tobaci* on Trinidad and Tobago. Ecology 63: 1574–1587. (13)

Feinsinger, P., J. A. Wolfe and L. A. Swarm. 1982. Island ecology: reduced hummingbird diversity and the pollination biology of plants, Trinidad and Tobago, West Indies. Ecology 63: 494–506. (13)

Felbeck, H., J. J. Childress and G. N. Somero. 1981. Calvin-Benson cycle and sulphide oxidation enzymes in animals from sulphide-rich habitats. Nature 293: 291–293. (14)

Felsenstein, J. 1976. The theoretical population genetics of variable selection and migration. Ann. Rev. Gen. 10: 253–280. (10)

Felsenstein, J. 1981. Scepticism towards Santa Rosalia, or why are there so few kinds of animals? Evolution 35: 124–138. (10)

Fenchel, T. 1975. Character displacement and coexistence in mud snails (Hydrobiidae). Oecologia 20: 19–32. (17, 18)

Fenchel, T. and F. Christiansen. 1977. Selection and interspecific competition. In *Measuring Selection in Natural Populations*, Vol. 19 in Lecture Notes in Biomathematics, F. Christiansen and T. Fenchel (eds.), pp. 477–498. Springer-Verlag, New York. (3, 17)

Fenchel, T. and L. Kofoed. 1976. Evidence for exploitative interspecific competition in mud snails (Hydrobiidae). Oikos 27: 367–376. (17)

Fenner, F. 1965. Myxoma virus and *Oryctolagus cuniculus*: two colonising species. In *The Genetics of Colonizing Species*, H. G. Baker and G. L. Stebbins (eds.), pp. 485–499. Academic Press, New York. (7)

Fenner, F. 1971. Evolution in action: Myxomatosis in the Australian wild rabbit. In *Topics in the Study of Life, The Bio Source Book*, A. Kramer (ed.), pp. 463–471. Harper & Row, New York. (3)

Fenner, F. and K. Myers. 1978. Myxoma virus and myxomatosis in retrospect: the first quarter century of a new disease. In *Viruses and Environment*, E. Kurstak and K. Maramorosch (eds.), pp. 539–570. Academic Press, New York. (9)

Fenner, F. and F. N. Ratcliffe. 1965. *Myxomatosis*. Cambridge University Press, Cambridge. (1, 8, 9)

Fenton, M. and T. H. Fleming. 1976. Ecological interactions between bats and nocturnal birds. Biotropica 8: 104–110. (11)

Ferraroni, J. J., C. A. Speer, J. Hayes and M. Suzaki. 1981. Prevalence of chloroquine-resistant *falciparum* malaria in the Brazilian Amazon. Am. J. Trop. Med. Hyg. 30: 526–530. (8)

Fife, P. C. and L. A. Peletier. 1981. Clines induced by variable selection and migration. Proc. R. Soc. Lond. B. 214: 99–123. (3)

Findley, J. 1976. The structure of bat communities. Amer. Natur. 110: 129–139. (17)

Fine, P. E. M. 1975. Vectors and vertical transmission: an epidemiologic perspective. Ann. N.Y. Acad. Sci. 266: 173–194. (6)

Fine, P. E. M. 1981. Vertical transmission of pathogens of invertebrates. Comparative Pathobiology. (In press) (6)

Fischbeck, G., E. Schwarzback, Z. Sobel and I. Wahl. 1976. Types of protection against barley powdery mildew in Germany and Israel selected from *Hordeum spontaneum*. In *Barley Genetics*, Vol. III, pp. 351–360. Proc. 3rd Int. Barley Genet. Symp., Garching, 1975. (7)

Fishelson, L. 1974. Ecology of the northern Red Sea crinoids and their epi- and endozoic fauna. Mar. Biol. 26: 183–192. (14)

Fisher, A. G. and M. A. Arthur. 1979. Secular variations in the pelagic realm. Tulsa Okla. Soc. Econ. Paleont. Mineral. Spec. Pub. 25: 19–50. (16)

Fisher, R. A. 1927. On some objections to mimicry theory; statistical and genetic. Trans. Ent. Soc. Lond. 1927: 269–278. (12)

Fisher, R. A. 1928a. The possible modification of the response of the wild type to recurrent mutations. Amer. Nat. 62: 115–126. (7)

Fisher, R. A. 1928b. Two further notes on the origin of dominance. Amer. Natur. 62: 571–574. (7)

Fisher, R. A. 1930. *The Genetical Theory of Natural Selection.* Clarendon Press, Oxford. (2, 3, 7, 12)

Fitch, W. M. 1971. Toward defining the course of evolution: minimum change for a specific tree topology. Syst. Zool. 20: 406–416. (4)

Fleming, R. 1980. Selection pressures and plant pathogens: robustness of the model. Phytopathology 70: 175–178. (9)

Fleming, R. 1982. Stability properties of simple gene-for-gene relationships. (Unpublished manuscript) (9)

Fleming, T. H. 1971. *Artibeus jamaicensis.* Delayed embryonic development in a neotropical bat. Science 171: 402–404. (11)

Fleming, T. H. 1979. Do tropical frugivores compete for food? Amer. Zool. 19: 1157–1172. (11)

Fleming, T. H. 1981. Fecundity, fruiting pattern, and seed dispersal in *Piper amalago* (Piperaceae), a bat-dispersed tropical shrub. Oecologia 51: 42–46. (11)

Fleming, T. H. and E. R. Heithaus 1981. Frugivorous bats, seed shadows, and the structure of tropical forests. Biotropica (supplement) 13: 45–53. (11)

Fleming, T. H., E. R. Heithaus and W. B. Sawyer. 1977. An experimental analysis of the food location behavior of frugivorous bats. Ecology 58: 619–627. (11)

Fletcher, M., P. T. LoVerde and D. S. Woodruff. 1981. Genetic variation in *Schistosoma mansoni*: enzyme polymorphisms in populations from Africa, southwest Asia, South America, and the West Indies. Am. J. Trop. Med. Hyg. 30: 406–421. (8)

Flor, H. H. 1942. Inheritance of pathogenicity in *Melampsora lini.* Phytopathology 32: 653–669. (7)

Flor, H. H. 1955. Host-parasite interaction in flax rust—its genetic and other implications. Phytopathology 45: 680–685. (3, 7, 9)

Flor, H. H. 1956. The complementary genic systems in flax and flax rust. Adv. Genet. 8: 29–54. (7, 9, 10)

Flügel, E. 1977. Environmental models for upper Paleozoic benthic calcareous algal communities. In *Fossil Algae*, E. Flügel (ed.), pp. 314–343. Springer-Verlag, New York. (15)

Ford, E. B. 1971. *Ecological Genetics*, 3rd Ed. Chapman and Hall, London. (12)

Forester, A. J. 1979. The association between the sponge *Halichondria panicea* (Pellas) and scallop *Chlamys varia* (L.): a commensal-protective mutualism. J. Exp. Mar. Biol. Ecol. 36: 1–10. (19)

Foster, M. S. 1976. Nesting biology of the long-tailed manakin. Wilson Bulletin 88: 400–420. (11)

Foster, M. S. 1977a. Odd couples in manakins: a study of social organization and cooperative breeding in *Chiroxiphia linearis.* Amer. Natur. 11: 845–853. (11)

Foster, M. S. 1977b. Ecological and nutritional effects of food scarcity on a tropical frugivorous bird and its fruit source. Ecology 58: 73–85. (11)

Foster, M. S. 1978. Total frugivory in tropical passerines: a reappraisal. Tropical Ecology 19: 131–154. (11)

Foster, R. 1973. Seasonality of fruit production and seed fall in a tropical forest ecosystem. Ph.D. Thesis, Duke University, Durham, North Carolina. (11)

Fox, J. F. 1974. Coevolution of white oak and its seed predators. Ph.D. Dissertation, University of Chicago, Chicago. (11)

Fox, L. R. 1981. Defense and dynamics in plant-herbivore systems. Amer. Zool. 21: 853–864. (1, 10)

Fox, L. R. and B. J. Macauley. 1977. Insect grazing on *Eucalyptus* in response to variation in leaf tannins and nitrogen. Oecologia 29: 145–162. (10)

486

Fox, L. R. and P. A. Morrow. 1981. Specialization: species property or local phenomenon? Science 211: 887–893. (10, 13)

Fraenkel, G. 1959. The raison d'être of secondary plant substances. Science 121: 1466–1470. (10)

Fraenkel, G. 1969. Evaluation of our thoughts on secondary plant substances. Ent. exp. appl. 12: 473–486. (10)

Frazer, B. D. 1972. Population dynamics and recognition of biotypes in the pea aphid (Homoptera: Aphididae). Can. Entomol. 104: 1729–1733. (10)

Freeland, W. J. 1976. Pathogens and the evolution of primate sociality. Biotropica 8: 12–24. (8)

Freeland, W. J. 1979a. Mangabey (Cercocebus albigena) social organization and population density in relation to its food use and availability. Fol. Primat. 32: 43–56. (11)

Freeland, W. J. 1979b. Primate social groups as biological islands. Ecology 60: 719–728. (8, 9)

Freeland, W. J. 1980. Mangabey (Cercocebus albigena) movement patterns in relation to food availability and fecal contamination. Ecology 61: 1297–1303. (8)

Freeman, R. and K. Wright 1960. Factors concerned with the epizootiology of Capillaria hepatica (Bancroft, 1893) (Nematoda) in a population of Peromyscus maniculatus in Algonquin Park, Canada. J. Parasitol. 46: 373–382. (8)

Freese, C. H. 1977. Food habits of white-faced capuchins Cebus capucinus L. (Primates: Cebidae) in Santa Rosa National Park, Costa Rica. Brenesia 10/11: 43–56. (11)

Freese, C. H. 1978. The behavior of white faced capuchins (Cebus capucinus) at a dry-season waterhole. Primates 19: 275–286. (11)

Fricke, A. H. 1979. Kelp grazing by the common sea urchin Parechinus angulosis Leske in False Bay, Cape. S. Afr. J. Zool. 14: 143–148. (19)

Friend, W. G. 1958. Nutritional requirements of phytophagous insects. Ann. Rev. Entomol. 3: 57–74. (19)

Frith, H. J., F. H. J. Crome and T. O. Wolfe 1976. Food of fruit-pigeons in New Guinea. Emu 76: 49–58. (11)

Frost, P. G. H. 1980. Fruit-frugivore interactions in a South Africa coastal dune forest. Proceed. XVIIth Inter. Ornith. Congr., Berlin, pp. 1179–1184. (11)

Frost, T. M. and C. E. Williamson. 1980. In situ determination of the effect of symbiotic algae on the growth of the freshwater sponge Spongilla lacustris. Ecology 61: 1361–1370. (14)

Fuentes, E. R. 1980. Convergence of community structure: neutral model vs. field data. Ecology 61: 198–200. (19)

Fuentes, E. R. and F. M. Jaksic. 1979. Latitudinal size variation of Chilean foxes: tests of alternative hypotheses. Ecology 60: 43–47. (18)

Furuya, S., K. Sakamoto, T. Asano, S. Takahashi and K. Kameoka. 1978. Effects of added dietary sodium polyacrylate on passage rate of markers and apparent digestibility by growing swine. J. Anim. Sci. 47: 159–165. (11)

Futuyma, D. J. 1970. Variation in genetic response to interspecific competition in laboratory populations of Drosophila. Amer. Natur. 104: 239–252. (1)

Futuyma, D. J. 1976. Food plant specialization and environmental predictability in Lepidoptera. Amer. Natur. 110: 285–292. (10)

Futuyma, D. J. 1979. Evolutionary Biology. Sinauer Associates, Sunderland, Massachusetts. (4)

Futuyma, D. J. 1983. Selective factors in the evolution of host choice by phytophagous insects. In Herbivorous Insects: Host-seeking Behavior and Mechanisms, S. Ahmad (ed.). Academic Press, New York. (In press) (10)

Futuyma, D. J. and F. Gould. 1979. Associations of plants and insects in a deciduous forest. Ecol. Monogr. 49: 33–50. (10)

Futuyma, D. J., S. L. Leipertz and C. Mitter. 1981. Selective factors affecting clonal variation in the fall cankerworm Alsophila pometaria (Lepidoptera: Geometridae). Heredity 47: 161–172. (10)

Futuyma, D. J. and G. C. Mayer. 1980. Non-allopatric speciation in animals. Syst. Zool. 29: 254–271. (10)

Futuyma, D. J. and S. S. Wasserman. 1981. Food plant specialization and feeding efficiency in the tent caterpillars *Malacosoma disstria* Hübner and *M. americanum*
(Fabricius). Ent. exp. appl. 30: 106–110. (10)

Gadgil, M. and O. Solbrig. 1972. The concept of r- and K- selection: evidence from wild
flowers and some theoretical considerations. Amer. Natur. 106: 14–31. (3)
Gadgil, R. L. and P. D. Gadgil. 1971. Mycorrhyza and litter decomposition. Nature 233:
133. (7)
Gair, R., J. E. E. Jenkins and E. Lester. 1976. *Cereal Pests and Diseases*. Farming Press
Ltd., Ipswich, England. (7)
Galdikas, B. M. F. 1982. Orangutans as seed dispersers at Tanjung Puting, Central
Kalimantan: implications for conservation. (In preparation) (11)
Gallun, R. L. 1977. The genetic basis of Hessian fly epidemics. Ann. N.Y. Acad. Sci. 287:
223–229. (1, 10)
Gallun, R. L., K. J. Starks and W. D. Guthrie. 1975. Plant resistance to insects attacking
cereals. Ann. Rev. Ent. 20: 337–357. (10)
Gardner, A. L. 1977. Feeding habits. In *Biology of Bats of the New World Family Phyl-
lostomatidae*, Part II, R. J. Baker, J. K. Jones and D. C. Carter (eds.), pp. 293–350.
Special Publ. Mus., Texas Tech. Univ., Lubbock, Texas 13: 1–364. (11)
Garth, J. S. 1974. On the occurrence in the eastern tropical Pacific of Indo-West Pacific
decapod crustaceans commensal with reef-building corals. Proc. 2nd Intern. Coral
Reef Symp. 1: 397–404. (14)
Gautier-Hion, A., L. H. Emmons and G. Dorst. 1980. A comparison of the diets of three
major groups of primary consumers of Gabon (primates, squirrels and ruminants).
Oecologia 45: 182–189. (11)
Gelfand, L. J., and J. F. McDonald. 1980. Relationship between ADH activity and be-
havioral response to environmental alcohol in *Drosophila*. Behav. Genet. 10:
237–249. (10)
Gerechter-Amitai, Z. K. 1973. Stemrust, *Puccinia graminis*, on cultivated and wild
grasses in Israel. Ph.D. Thesis, Hebrew University, Jerusalem. (Quoted by Brown-
ing, 1980) (7)
Gerrits-Heybroek, E. M., W. M. Herrebout, S. A. Ulenberg and J. T. Wiebes. 1978.
Host plant preferences of five species of small ermine moths (Lepidoptera,
Yponomeutidae). Ent. exp. appl. 24: 360–368. (10)
Gibbs, R. D. 1974. *Chemotaxonomy of Flowering Plants*. McGill-Queens University
Press, Montreal. (10)
Gibson, C. W. D. 1980. Niche use patterns among Stenodemini (Heteroptera: Miridae) of
limestone grassland, and an investigation of the possibility of interspecific competi-
tion between *Notostira elongata* Geoffroy and *Megaloceraea recticornia* Geoffroy.
Oecologia 47: 352–364. (19)
Gigon, A. 1979. CO_2 gas exchange, water relations, and convergence of Mediterranean
shrub-types from California and Chile. Oecologia Plantarum 14: 129–150. (19)
Gilbert, L. E. 1971. Butterfly-plant coevolution: Has *Passiflora adenopoda* won the
selectional race with heliconiine butterflies? Science 172: 585–586. (1, 10, 12)
Gilbert, L. E. 1975. Ecological consequences of a coevolved mutualism between but-
terflies and plants. In *Coevolution of Animals and Plants*, L. E. Gilbert and P. H.
Raven (eds.), pp. 210–240. University of Texas Press, Austin, Texas. (10, 12)
Gilbert, L. E. 1976. Postmating female odor in *Heliconius* butterflies: A male contrib-
uted antiaphrodisiac? Science 193: 419–420. (12)
Gilbert, L. E. 1977. The role of insect-plant coevolution in the organization of ecosys-
tems. In *Comportement des Insectes et Milieu Trophique*, V. Labeyrie (ed.), pp.
399–413. Coll. Int. C.N.R.S. 265. (10)
Gilbert, L. E. 1979. Development of theory in the analysis of insect-plant interactions.
In *Analysis of Ecological Systems*, D. Horn, R. Mitchell and G. Stairs (eds.), pp.
117–154. Ohio State University Press, Columbus. (10)
Gilbert, L. E. 1980. Food web organization and the conservation of neotropical diversity.
In *Conservation Biology*, M. E. Soulé and B. A. Wilcox (eds.), pp. 11–33. Sinauer
Associates, Sunderland, Massachusetts. (12)

488

Gilbert, L. E. 1983. The biology of butterfly communities. In *The Biology of Butterflies*, XI Symposium of the Royal Entomological Society of London, R. Vane-Wright and P. Ackery (eds.). (In press) (12)

Gilbert, L. E. and P. R. Ehrlich. 1970. The affinities of the Ithomiinae and Satyrinae. J. Lep. Soc. 24: 297-300. (4)

Gilbert, L. E. and P. H. Raven (eds.). 1975. *Coevolution of Animals and Plants.* University of Texas Press, Austin. (1, 8)

Gilbert, L. E. and M. C. Singer. 1975. Butterfly ecology. Ann. Rev. Ecol. Syst. 6: 365-397. (10)

Gilbert, L. E. and J. T. Smiley. 1978. Determinants of local diversity in phytophagous insects: host specialists in tropical environments. In *Diversity of Insect Faunas*, L. A. Mound and N. Waloff (eds.), pp. 89-104. Blackwell Scientific, London. (4, 10, 12, 19)

Gill, D. E. 1974. Intrinsic rates of increase, saturation density and competitive ability. II. The evolution of competitive ability. Amer. Natur. 118: 103-116. (5, 6)

Gill, F. B. 1971. Ecology and evolution of the sympatric mascarene white-eyes, *Zosterops borbonica* and *Zosterops olivacea.* Auk 88: 35-60. (11)

Gillespie, J. H. 1975. Natural selection for resistance to epidemics. Ecology 56: 493-495. (3, 9)

Ginzburg, L. 1977. The equilibrium and stability for n alleles under density-dependent selection. J. Theor. Biol. 68: 545-550. (3)

Gipson, P. S. 1974. Food habits of coyotes in Arkansas. J. Wildl. Manag. 38: 848-853. (11)

Gipson, P. S. and J. A. Sealander. 1976. Changing food habits of wild *Canis* in Arkansas with emphasis on coyote hybrids and feral dogs. Amer. Midl. Nat. 95: 249-253. (11)

Givnish, T. J. 1980. Ecological constraints on the evolution of breeding systems in seed plants: dioecy and dispersal in gymnosperms. Evolution 34: 959-972. (11, 13)

Glynn, P. W. 1976. Some physical and biological determinants of coral community structure in the Eastern Pacific. Ecol. Monogr. 46: 431-456. (14)

Glynn, P. W. 1977. Interactions between *Acanthaster* and *Hymenocera* in the field and laboratory. Proc. Third Intern. Coral Reef Symp. 1: 209-215. (14)

Glynn, P. W. 1980. Defense by symbiotic Crustacea of host corals elicited by chemical cues from predator. Oecologia 47: 287-290. (14)

Goh, B. S. 1979. Stability in models of mutualism. Amer. Natur. 113: 261-275. (3)

Goldschmidt, V. 1928. Vererbungsversuche mit den biologischen Arten des Antherenbrandes (*Ustilago violacea*). Ein Beitrag zur Frage der parasitären Spezialisierung. Z. Bot. 21: 1-90. (Quoted by Day, 1974). (7)

Goldstein, J. L. and T. Swain. 1963. Changes in tannins in ripening fruits. Phytochemistry 2: 371-383. (11)

Gondwe, A. T. D. 1976. Cyanogenesis in passion fruit. 1. Detection and quantifications of cyanide in passion fruit (*Passiflora edulis* Sims.) at different stages of fruit development. E. Afr. For. J. 42: 117-120. (11)

Good, R. 1974. *Geography of Flowering Plants.* Longman, London. (19)

Gorman, G. C. and Y. S. Kim. 1976. *Anolis* lizards of the Eastern Caribbean: a case study in evolution II: Genetic relationship and genetic variation of the *bimaculatus* group. Syst. Zool. 25: 62-77. (17)

Gorman, G. C., P. Licht, H. C. Dessauer and J. O. Boos. 1971. Reproductive failure among the hybridizing *Anolis* lizards of Trinidad. Syst. Zool. 20: 1-18. (17)

Gottlieb, F. J., G. M. Simmons, L. Ehrman, B. Inocencio, J. Kocka and N. Somerson. 1981. Characteristics of the *Drosophila paulistorum* male sterility agent in a secondary host, *Ephestia kuehniella.* Appl. Environ. Microbiol. 42: 838-842. (6)

Gottsberger, G. 1978. Seed dispersal by fish in the inundated regions of Humaita, Amazonia. Biotropica 10: 170-183. (11)

Gotz, P., A. Boman and H. G. Boman. 1981. Interactions between insect immunity and an insect-pathogenic nematode with symbiotic bacteria. Proc. R. Soc. London B. 212: 333-350. (8)

489

Gould, F., C. R. Carroll and D. J. Futuyma. 1982. Cross-resistance to pesticides and plant defenses: a study of the two-spotted spider mite. Ent. exp. appl. 31: 175-180. (10)

Gould, S. J. 1966. Allometry and size in ontogeny and phylogeny. Biol. Rev. 41: 587-640. (2)

Gould, S. J. 1977. *Ontogeny and Phylogeny.* Harvard University Press, Cambridge. (2)

Gould, S. J. 1980. Is a new and general theory of evolution emerging? Paleobiology 6: 119-130. (2)

Gould, S. J. 1981. Palaeontology plus ecology as palaeobiology. In *Theoretical Ecology: Principles and Applications,* 2nd Ed., R. M. May (ed.), pp. 295-317. Sinauer Associates, Sunderland, Massachusetts. (9)

Gould, S. J. and N. Eldredge. 1977. Punctuated equilibria: the tempo and mode of evolution reconsidered. Paleobiology 3: 115-151. (2, 16)

Gould, S. J. and R. C. Lewontin. 1979. The spandrels of San Marco and the Panglossian paradigm: a critique of the adaptationist programme. Proc. R. Soc. Lond. B 205: 581-598. (2)

Goulding, M. 1980. *The Fishes and the Forest.* University of California Press, Berkeley. (11)

Grant, B. R. and P. R. Grant. 1981. Exploitation of *Opuntia* cactus by birds on the Galapagos. Oecologia 49: 179-187. (11)

Grant, K. A. and V. Grant. 1968. *Hummingbirds and Their Flowers.* Columbia University Press, New York. (13, 19)

Grant, P. R. 1965a. A systematic study of the terrestrial birds of the Tres Marias Islands, Mexico. Postilla 90: 1-106. (18)

Grant, P. R. 1965b. The adaptive significance of some island size trends in birds. Evolution 19: 355-367. (18)

Grant, P. R. 1966. Ecological incompatibility of bird species on islands. Amer. Natur. 100: 451-462. (18)

Grant, P. R. 1968. Bill size, body size and the ecological adaptations of bird species to competitive situations on islands. Syst. Zool. 17: 319-333. (18)

Grant, P. R. 1969. Colonization of islands by ecologically dissimilar species of birds. Can. J. Zool. 47: 41-43. (18)

Grant, P. R. 1970. Colonization of islands by ecologically dissimilar species of mammals. Can. J. Zool. 48: 545-553. (18)

Grant, P. R. 1971. Variation of tarsus length in island and mainland situations. Evolution 25: 599-614. (17)

Grant, P. R. 1972a. Convergent and divergent character displacement. Biol. Journ. Linn. Soc. 4: 39-68. (17, 18)

Grant, P. R. 1972b. Bill dimensions of the three species of *Zosterops* on Norfolk Island. Syst. Zool. 21: 289-291. (18)

Grant, P. R. 1975. The classical case of character displacement. Evol. Biol. 8: 237-337. (17, 18)

Grant, P. R. 1979. Ecological and morphological variation of Canary Island blue tits, *Parus caeruleus* (Aves: Paridae) Biol. J. Linn. Soc. 11: 103-129. (17)

Grant, P. R. 1981. Speciation and the adaptive radiation of Darwin's finches. Amer. Sci. 69: 653-663. (18)

Grant, P. R. 1982. The role of interspecific competition in the adaptive radiation of Darwin's finches. In *Patterns of Evolution in Galapagos Organisms,* A. Levinton and R. I. Bowman (eds.). Spec. Publ. 1, Amer. Assoc. Adv. Sci., Wash., D.C. (In press) (18)

Grant, P. R. and I. Abbott. 1980. Interspecific competition, island biogeography and null hypotheses. Evolution 34: 332-341. (17, 18, 19)

Grant, P. R. and B. R. Grant. 1980. The breeding and feeding characteristics of Darwin's finches on Isla Genovesa, Galápagos. Ecol. Monogr. 50: 381-410. (17)

Grant, P. R. and D. Schluter. 1982. Interspecific competition inferred from patterns of guild structure. In *Ecological Communities: Conceptual Issues and the Evidence,* D. R. Strong, D. Simberloff and L. G. Abele (eds.). Princeton University Press, Princeton, New Jersey. (In press) (18)

Grant, V. 1975. *Genetics of Flowering Plants.* Columbia University Press, New York. (13)

490

Grant, V. and K. A. Grant. 1965. *Flower Pollination in the Phlox Family*. Columbia University Press, New York. (13, 19)

Grassle, J. F. 1977. Slow recolonisation of deep-sea sediments. Nature 265: 618–619. (14)

Greenberg, R. 1981. Frugivory in some migrant tropical forest wood warblers. Biotropica 13: 215–223. (11)

Greenhall, A. M. 1956. The food of some Trinidad fruit bats (*Artibeus* and *Carollia*). J. Agr. Soc. Trin. Tob. 56: 3–26. (11)

Gregory, D. P. 1963–64. Hawkmoth pollination in the genus *Oenothera*. Aliso 5: 357–384, 385–419. (13)

Grice, G. D., R. P. Harris, M. R. Reeve, J. F. Heinbokel and C. O. Davis. 1980. Large-scale enclosed water column ecosystems. An overview of foodweb I. Jour. Mar. Biol. Assoc., U.K. 60: 401–414. (19)

Grisebach, A. H. R. 1872. *Die Vegetation der Erde nach ihrer klimatischen Anordnung*. U. Engelmann, Leipzig. (19)

Groth, J. V. and C. Person. 1977. Genetic interdependence of host and parasite in epidemics. Ann. N.Y. Acad. Sci. 287: 97–106. (7)

Grun, P. 1976. *Cytoplasmic Genetics and Evolution*. Columbia University Press, New York. (16)

Grundlach, H. 1968. Brutvorsorge, Brutpflege, Verhaltensontogenese und Tagesperiodik beim europäischen Wildschwein (*Sus scrofa* L). Z. *Tierpsychol* 25: 955–995. (16)

Guevara, S. and A. Gomez-Pompa. 1972. Seeds from surface soils in a tropical region of Veracruz, Mexico. J. Arnold Arboretum 53: 312–335. (11)

Gupta, P. D. and A. J. Thorsteinson. 1960. Food plant relationships of the diamond-back moth *Plutella maculipennis* (Curt.) II. Sensory regulation of oviposition by the adult female. Ent. exp. appl. 3: 305–314. (10)

Gupta, R. and R. Banerji. 1967. Studies in taxonomy and ecology of *Bursera delpechiana* Poiss. ex Engl. in India. J. Bombay Nat. Hist. Soc. 64: 49–54. (11)

Guttman, S. I., T. K. Wood and A. A. Karlin. 1981. Genetic differentiation along host plant lines in the sympatric *Enchenopa binotata* Say complex (Homoptera: Membracidae). Evolution 35: 205–217. (10)

Gwynne, M. D. 1969. Notes on the nutritive values of *Acacia* pods in relation to *Acacia* seed distribution by ungulates. East African Wildlife Journal 7: 176–178. (11)

Habbu, M. K. 1960. The present position of plague in India. Ind. J. Path. Bet. 3: 123. (9)

Hairston, N. G. 1962. Population ecology and epidemiological problems. In *Bilharziasis*, G. E. W. Wolstenholme and M. O'Connor (eds.), pp. 36–62. CIBA Foundation Symposium, Churchill, London. (8)

Hairston, N. G. 1980. Evolution under interspecific competition: field experiments on terrestrial salamanders. Evolution 34: 409–420. (1)

Haldane, J. B. S. 1928. *Possible Worlds*. Harper & Bros., New York. (2)

Haldane, J. B. S. 1932. *The Causes of Evolution*. Longmans, Green & Co., London. (2)

Haldane, J. B. S. 1949. Disease and evolution. La Ricerca Sci. Suppl. 19: 68–76. (9)

Haldane, J. B. S. 1954. The statics of evolution. In *Evolution as a Process*, J. Huxley, A. C. Hardy and E. B. Ford (eds.), pp. 109–121. Allen and Unwin, London. (7)

Hale, M. E. 1967. *The Biology of Lichens*. Contemporary Biology Series (E. J. W. Barrington and A. J. Willis, eds.). Edward Arnold Ltd., London. (7)

Hallam, T. G. 1980. Effects of cooperation on competitive systems. J. Theoret. Biol. 82: 415–423. (3)

Hamilton, W. D. 1964a. The genetical evolution of social behavior. I. J. Theoret. Biol. 7: 1–16 (2)

Hamilton, W. D. 1964b. The genetical evolution of social behavior II. J. Theoret. Biol. 7: 17–52. (2)

Hamilton, W. D. 1966. The moulding of senescence by natural selection. J. Theoret. Biol. 12: 12–45. (2)

Hamilton, W. D. 1980. Sex versus non-sex versus parasite. Oikos 35: 282–290. (8, 9)

Hamilton, W. D. 1982. Pathogens as causes of genetic diversity in their host populations. In *Population Biology of Infectious Disease Agents* (Dahlem Konferenzen), R. M. Anderson and R. M. May (eds.), pp. 269–296. Springer-Verlag, Berlin and New York. (In press) (8, 9)

Hamilton, W. D. and R. M. May. 1977. Dispersal in stable habitats. Nature 279: 578–581. (11)

Hamilton, W. J., R. Buskirk and W. H. Buskirk. 1977. Intersexual dominance and differential mortality of gemsbok *Oryx gazella* at Namib desert waterholes. Madqua 10: 5–19. (11)

Hammerberg, B., C. Dangler and J. F. Williams. 1980. *Taenia taeniaeformis*: chemical composition of parasite factors affecting coagulation and complement cascades. J. Parasitol. 66: 569–576. (8)

Hamner, W. M. and D. F. Dunn. 1980. Tropical corallimorpharia (Coelenterata: Anthozoa): feeding by envelopment. Micronesica 16: 37–42. (19)

Hand, C. 1975. Behavior of some New Zealand sea anemones and their molluscan and crustacean hosts. N. Z. J. Mar. Fresh-W. Res. 9: 509–527. (14)

Hanlon, R. T. and L. Kaufman. 1976. Associations of seven West Indian reef fishes with sea anemones. Bull. Mar. Sci. 26: 225–232. (14)

Hanna, R. E. B. 1980. *Fasciola hepatica*: an immunofluorescent study of antigenic changes in the tegument during development in the rat and the sheep. Exp. Parasitol. 50: 155–170. (8)

Harada, E. 1969. On the interspecific association of a snapping shrimp and gobioid fishes. Publ. Seto Mar. Biol. Lab. 16: 315–334. (14)

Harborne, J. B. (ed.) 1978. *Biochemical Aspects of Plant and Animal Coevolution*. Academic Press, New York. (10)

Hare, J. D. and D. J. Futuyma. 1978. Different effects of variation in *Xanthium strumarium* L. (Compositae) on two insect seed predators. Oecologia 39: 109–120. (10)

Harper, J. L. 1969. The role of predation in vegetational diversity. Brookhaven Symp. Biol. 22: 48–62. (19)

Harper, J. L. 1977. *Population Biology of Plants*. Academic Press, New York. (3)

Harper, J. L., P. H. Lovell and K. G. Moore. 1970. The shapes and sizes of seeds. Ann. Rev. Ecol. Syst. 1: 327–356. (11)

Harris, A. W., D. W. A. Mount, C. R. Fuerst and L. Siminovitch. 1967. Mutations in bacteriophage Lambda affecting host cell lysis. Virology 32: 553–569. (5)

Harris, L. G. 1973. Nudibranch associations. Current Topics in Comparative Pathobiology 2: 213–315. (14)

Harvey, P. H., T. H. Clutton-Brock and G. M. Mace. 1980. Brain size and ecology in small mammals and primates. Proc. Natl. Acad. Sci. USA 77: 4387–4389. (2)

Hassell, M. P. 1978. *The Dynamics of Arthropod Predator-prey Systems*. Princeton University Press, Princeton, New Jersey. (3)

Hassell, M. P. and C. Huffaker. 1969. Regulatory processes and population cyclicity in laboratory populations of *Anagasta kühniella* (Zeller) (Lepidoptera: Phycitidae). III. The development of population models. Res. Pop. Ecol. 11: 186–210. (1)

Hastings, E. F. 1966. Yield and chemical analyses of fruit produced by selected deer-browse plants in a loblolly-shortleaf pine-hardwood forest. Ph.D. Thesis, Louisiana State University, Baton Rouge. (11)

Hatchett, J. H. and R. L. Gallun. 1970. Genetics of the ability of the Hessian fly, *Mayetiola destructor*, to survive on wheats having different genes for resistance. Ann. Ent. Soc. Amer. 63: 1400–1407. (10)

Hathway, D. E. 1959. Myrobalans: an important tanning material. Trop. Sci. 1: 85–106. (11)

Hatton, H. 1938. Essais de bionomie explicative sur quelques espèces intercôtidales d'algues et d'animaux. Ann. Inst. Oceanogr. Monaco. 17: 241–348. (15)

Haukioja, E. and P. Niemelä. 1977. Retarded growth of a geometrid larva after mechanical damage to leaves of its host tree. Ann. Zool. Fenn. 14: 48–52. (10)

Haukioja, E. and P. Niemelä. 1979. Birch leaves as a resource for herbivores: seasonal occurrence of increased resistance in foliage after mechanical damage of adjacent leaves. Oecologia 39: 151–159. (10, 19)

492

Haukioja, E., P. Niemelä, L. Iso-Irrari, H. Ojala and E. Aro. 1978. Birch leaves as a resource for herbivores. I. Variation in the suitability of leaves. Rep. Kevo Subarctic Res. Sta. 14: 5-12. (10)

Hawking, F. 1975. Circadian and other rhythms of parasites. Adv. Parasitol. 13: 123-182. (8)

Hawksworth, D. L. 1973. Ecological factors and species delimitation in the lichens. In *Taxonomy and Ecology*, V. H. Heywood (ed.), pp. 31-69. Academic Press, London. (7)

Hay, C. H. 1977. A biological study of *Durvillaea antarctica* (Chamisse) Harriot and *D. willana* Lindauer in New Zealand. Ph.D. Thesis, University of Canterbury, Christchurch. (19)

Hay, M. E. 1981. The functional morphology of turf-farming seaweeds: persistence in stressful marine habitats. Ecology 62: 739-750. (19)

Hay, M. E. and P. J. Fuller 1981. Seed escape from heteromyid rodents: the importance of microhabitat and seed preference. Ecology 62: 1395-1399. (11)

Hayes, W. 1968. *The Genetics of Bacteria and their Viruses.* John Wiley & Sons, New York. (5)

Heaney, L. R. and R. W. Thorington. 1978. Ecology of neotropical red-tailed squirrels, *Sciurus granatensis*, in the Panama Canal Zone. J. Mammalogy 59: 846-851. (11)

Hedgpeth, J. W. (ed.). 1957. *Treatise on Marine Ecology and Paleoecology.* Geological Society of America, Memoir 67, New York City. (3)

Hedin, P. A., F. G. Maxwell and J. N. Jenkins. 1974. Insect plant attractants, feeding stimulants, repellents, deterrents, and other related substances affecting insect behavior. In Proc. Summer Inst. Biol. Control Mount. Insects Dis., F. G. Maxwell and F. A. Harris (eds.), pp. 494-527. University Press of Mississippi, Jackson. (10)

Hedrick, P. W., M. E. Ginevan and E. P. Ewing. 1976. Genetic polymorphism in heterogeneous environments. Ann. Rev. Ecol. Syst. 7:1-32. (10)

Heim de Balsac, H. and N. Mayaud 1930. Compléments à l'étude de la propagation du Gui (*Vicium album* L.) par les oiseaux. Alauda 2: 474-493. (11)

Heinrich, B. 1975a. Bee flowers: a hypothesis on flower variety and blooming times. Evolution 29: 325-334. (13)

Heinrich, B. 1975b. Energetics of pollination. Ann. Rev. Ecol. Syst. 6: 139-170. (13)

Heinrich, B. 1975c. The role of energetics in bumblebee flower interrelationships. In *Coevolution of Animals and Plants*, L. E. Gilbert and P. H. Raven (eds.), pp. 141-158. University of Texas Press, Austin. (13)

Heinrich, B. 1976a. The foraging specializations of individual bumblebees. Ecol. Monogr. 46: 105-128. (13)

Heinrich, B. 1976b. Resource partitioning among some eusocial insects: bumblebees. Ecology 57: 874-899. (13)

Heinrich, B. 1979a. *Bumblebee Economics.* Harvard University Press, Cambridge, Massachusetts. (13)

Heinrich, B. 1979b. "Majoring" and "minoring" by foraging bumblebees, *Bombus vagans*: an experimental analysis. Ecology 60: 245-255. (13)

Heinrich, B. 1979c. Resource heterogeneity and patterns of movement in foraging bumblebees. Oecologia 55: 39-45. (13)

Heinrich, B. and P. H. Raven. 1972. Energetics and pollination ecology. Science 176: 597-602. (13)

Heiser, C. B. 1973. Variation in the bottle gourd. In *Tropical Forest Ecosystems in Africa and South America: A Comparative Review*, B. J. Meggers, E. S. Ayensu and W. D. Duckworth (eds.), pp. 121-128. Smithsonian Inst. Press, Washington, D.C. (11)

Heithaus, E. R. 1979. Flower-feeding specialization in wild bee and wasp communities in seasonal neotropical habitats. Oecologia 42: 179-194. (13)

Heithaus, E. R. 1981. Seed predation by rodents on three ant-dispersed plants. Ecology 62: 136-145. (11)

Heithaus, E. R. and T. H. Fleming. 1978. Foraging movements of a frugivorous bat, *Carollia perspicillata* (Phyllostomatidae). Ecol. Monogr. 48: 127–143. (11)

Heithaus, E. R., T. H. Fleming and P. A. Opler. 1975. Foraging patterns and resource utilization in seven species of bats in a seasonal neotropical forest. Ecology 56: 841–854. (11, 13)

Heithaus, E. R., D. C. Culver and A. J. Beattie. 1980. Models of some ant-plant mutualisms. Amer. Natur. 116: 347–361. (3)

Hendrickson, J. A., Jr. 1981. Community-wide character displacement reexamined. Evolution 35: 794–809. (17, 18, 19)

Hendrikse, A. 1979. Activity patterns and sex pheromone specificity as isolating mechanisms in eight species of *Yponomeuta* (Lepidoptera: Yponomeutidae). Ent. exp. appl. 25: 172–180. (10)

Hennig, W. 1966. *Phylogenetic Systematics.* University of Illinois Press, Urbana. (4)

Hering, E. M. 1954. Die Nahrungswahl phytophager Insekten. Verh. deutsch. Ges. angew Ent. 13: 29–38. (4)

Herrebout, W. M., P. J. Kuijten and J. T. Wiebes. 1976. Small ermine moths of the genus *Yponomeuta* and their host relationships (Lepidoptera, Yponomeutidae). In *The Host-Plant in Relation to Insect Behavior and Reproduction*, T. Jermy (ed.), pp. 91–94. Plenum Press, New York. (10)

Herrera, C. M. 1981a. Fruit variation and competition for dispersers in natural populations of *Smilax aspera.* Oikos 36: 51–58. (11)

Herrera, C. M. 1981b. Fruit food of robins wintering in southern Mediterranean scrubland. Bird Study 28: 115–122. (11)

Herrera, C. M. 1981c. Are tropical fruits more profitable to dispersers than temperate ones? Amer. Natur. 118: 897–907. (11)

Herrera, C. M. 1982a. Defense of ripe fruits from pests: its significance in relation to plant-disperser interactions. Amer. Natur. 120: 218–241. (11)

Herrera, C. M. 1982b. Seasonal variation in the quality of fruits and diffuse coevolution between plants and avian dispersers. Ecology 63: 773–785. (11)

Herrera, C. M. and P. Jordano. 1981. *Prunus mahaleb* and birds: the high-efficiency seed dispersal system of a temperature fruiting tree. Ecol. Monogr. 51: 203–218. (11)

Hershey, A. D. 1946. Spontaneous mutations in bacterial viruses. Cold Spring Harbor Symposia Quant. Biol. 11: 67–77. (5)

Hespenheide, H. 1973a. Ecological inferences from morphological data. Ann. Rev. Ecol. Syst. 4: 213–224. (17)

Hespenheide, H. A., 1973b. A novel mimicry complex: beetles and flies. J. Ent. 48: 49–56. (12)

Hespenheide, H. 1975. Prey characteristics and predator niche width. In *Ecology and Evolution of Communities*, M. L. Cody and J. M. Diamond (eds.), pp. 158–180. Harvard University Press, Cambridge, Massachusetts. (3, 17)

Heumann, G. A. 1926. Mistletoe-birds as plant distributors. Emu 26: 110–111. (11)

Hibler, C. P. and J. L. Adcock. 1971. Elaeophorosis. In *Parasitic Diseases of Wild Mammals*, J. W. Davis and R. C. Anderson (eds.), pp. 263–278. Iowa State University Press, Ames. (8)

Hickey, L. J. and J. A. Wolfe. 1975. The bases of angiosperm phylogeny: vegetative morphology. Ann. Miss. Bot. Gard. 62: 583–589. (4)

Higginbotham, J. D. and C. A. M. Hough. 1977. Useful taste properties of amino acids and proteins. In *Sensory Properties of Food*, G. G. Birch, J. G. Brennan and K. J. Parker (eds.), pp. 129–149. Applied Science Publishers, London.

Highsmith, R. C. 1980a. Geographic patterns of coral bioerosion: a productivity hypothesis. J. Exp. Mar. Biol. Ecol. 46: 177–196. (14)

Highsmith, R. C. 1980b. Burrowing by the bivalve *Lithophaga curta* in the living reef coral *Montipora berryi* and a hypothesis of reciprocal larval recruitment. Mar. Biol. 56: 155–162. (14)

Higuchi, H. 1980. Colonization and coexistence of woodpeckers in the Japanese Islands. Misc. Repts. Yamashina Inst. for Ornithology 12: 139–156. (18)

Hildreth, K. C. and V. Ahmadjian. 1981. A study of *Trebouxia* and *Pseudotrebouxia* isolates from different lichens. Lichenologist 13: 65–86. (7)

Hills, H. G., N. H. Williams and C. H. Dodson. 1972. Floral fragrances and isolating mechanisms in the genus *Catasetum* (Orchidaceae). Biotropica 4: 61–67. (13)

Hitchins, P. M. 1968. Records of plants eaten by mammals in the Hluhluwe Game Reserve, Zululand. Lammergeyer 8: 31-39. (11)

Hiura, U. 1978. Genetic basis of formae speciales in *Erysiphe graminis* D.C. In *The Powdery Mildews*, D. M. Spencer (ed.), pp. 101-128. Academic Press, London. (7)

Hladik, A. and C. M. Hladik 1969. Rapports trophiques entre végétation et primates dans la forêt de Barro Colorado (Panama). La Terre et la Vie 1: 25-117. (11)

Hladik, C. M. 1973. Alimentation et activité d'un groupe de chimpanzés réintroduits en forêt Gabonaise. La Terre et la Vie 27: 343-413. (11)

Hladik, C. M. and A. Hladik 1967. Observations sur le rôle des primates dans la dissémination des végétaux de la forêt Gabonaise. Biologia Gabonico 3: 43-58. (11)

Hladik, C. M. and A. Hladik 1972. Disponibilités alimentaires et domaines vitaux des primates a Ceylan. La Terre et la Vie 26: 149-215. (11)

Hladik, C. M., A. Hladik, J. Bousset, P. Valdebouze, G. Viroben and J. Delort-laval. 1971. Le régime alimentaire des primates de l'ile de Barro Colorado (Panama). Folia Primat. 16: 85-122. (11)

Hochachka, P. W. and G. N. Somero. 1973. *Strategies of Biochemical Adaptation.* W. B. Saunders, Philadelphia. (14)

Hochreutiner, G. 1899. Dissémination des graines par les poissons. Bull. de l'Herbier Boissier 3: 459-466. (11)

Hoelzel, F. 1930. The rate of passage of inert materials through the digestive tract. Amer. J. Physiol. 92: 466-497. (11)

Hoffman, R. S. 1976. Intertidal distribution and movement in *Collisella strigatella.* West. Soc. Malacol. 9: 18. (15)

Holdridge, L. R. 1947. Determination of world plant formations from simple climatic data. Science 105: 367-368. (19)

Holloway, B. W. 1979. Plasmids that mobilize bacterial chromosome. Plasmid 2: 1-19. (5)

Holloway, J. D. and P. D. N. Hebert. 1979. Ecological and taxonomic trends in macrolepidopteran host plant selection. Biol. J. Linn. Soc. 11: 229-251. (10)

Holmes, J. C. 1973. Site selection by parasitic helminths: interspecific interactions, site segregation, and their importance to development of helminth communities. Can. J. Zool. 51: 333-347. (8)

Holmes, J. C. 1976. Host selection and its consequences. In *Ecological Aspects of Parasitology*, C. R. Kennedy (ed.), pp. 21-39. North-Holland, Amsterdam. (8)

Holmes, J. C. 1979. Parasite populations and host community structure. In *Host-Parasite Interfaces*, B. B. Nickol (ed.), pp. 27-46. Academic Press, New York. (8)

Holmes, J. C. 1982. Impact of infectious disease agents on the population growth and geographical distribution of animals. In *Population Biology of Infectious Diseases*, R. M. Anderson and R. M. May (eds.), pp. 37-51. Springer-Verlag, Berlin and New York. (8, 9)

Holmes, J. C. and W. M. Bethel. 1972. Modification of intermediate host behavior by parasites. In *Behavioral Aspects of Parasite Transmission*, E. U. Canning and C. A. Wright (eds.), pp. 123-149. Zool. J. Linn. Soc. Suppl. 1. (8, 14)

Holmes, J. C., R. P. Hobbs and T. S. Leong. 1977. Populations in perspective: community organization and regulation of parasite populations. In *Regulation of Parasite Populations*, G. W. Esch (ed.), pp. 209-245. Academic Press, New York. (8)

Holmes, J. C. and P. W. Price. 1980. Parasite communities: the roles of physiology and ecology. Syst. Zool. 29: 203-213. (19)

Home, E. 1812. On the different structures and situations of the solvent glands in the digestive organs of birds, according to the nature of their food and particular modes of life. Phil. Trans. Royal Soc. London 102: 394-404. (11)

Honer, M. R. 1961. Some observations on the ecology of *Hydrobia stagnorum* (Gmelin) and *Hydrobia ulvae* (Pennant), and the relationship ecology-parasitofauna. Basteria 25: 7-29. (14)

Honigberg, B. M. 1970. Protozoa associated with termites and their role in digestion. In *Biology of Termites*, Vol. II., K. Krishna and F. M. Weesner (eds.), pp. 1–36. Academic Press, New York. (4)

Hopwood, D. A. and M. J. Merrick. 1977. Genetics of antibiotic production. Bact. Rev. 41: 595–635. (5)

Horn, H. S. 1971. *The Adaptive Geometry of Trees*. Princeton University Press, Princeton, New Jersey. (19)

Horne, M. T. 1970. Coevolution of *Escherichia coli* and bacteriophage in chemostat culture. Science 168: 992–993. (5)

Horsfall, J. G. and E. B. Cowling. 1978. Some epidemics man has known. In *Plant Disease*, Vol. II., J. G. Horsfall and E. B. Cowling (eds.), pp. 17–32. Academic Press, London. (7)

Horvitz, C. C. and A. J. Beattie. 1980. Ant dispersal of *Calathea* (Marantaceae) seeds by carnivorous ponerines (Formicidae) in a tropical rain forest. Amer. J. Bot. 67: 321–326. (11)

Hough, C. A. M. 1978. Antibodies to thaumatin as a model of the sweet taste receptor. Nature 271: 381–383. (11)

House, W. L., J. R. Sanburn and A. M. Rhodes. 1976. Western corn rootworm adult and spotted cucumber beetle associations with *Cucurbita* and cucurbitacins. Env. Entomol. 5: 1043–1048. (10)

Houston, D. C. 1979. The adaptations of scavengers. In *Serengeti, Dynamics of an Ecosystem*, A. R. E. Sinclair and M. Norton Griffiths (eds.). University of Chicago Press, Chicago. (16)

Howe, H. F. 1977. Bird activity and seed dispersal of a tropical wet forest tree. Ecology 58: 539–550. (11)

Howe, H. F. 1979. Fear and frugivory. Amer. Natur. 114: 925–931. (11)

Howe, H. F. 1980. Monkey dispersal and waste of a neotropical fruit. Ecology 61: 944–959. (11)

Howe, H. F. 1981a. Dispersal of a neotropical nutmeg (*Virola sebifera*) by birds. Auk 98: 88–98. (11)

Howe, H. F. 1981b. Fruit production and animal activity in two tropical trees. In *The Ecology of a Tropical Forest: Seasonal Rhythms and Long-term Changes*, E. Leigh, A. S. Rand and D. M. Windsor (eds.), p. 11. Smithsonian Institution Press, Washington, D. C. (11)

Howe, H. F. and D. DeSteven 1979. Fruit production, migrant bird visitation, and seed dispersal of *Guara glabra* in Panama. Oecologia 39: 185–196. (11)

Howe, H. F. and G. F. Estabrook, 1977. On intraspecific competition for avian dispersers in tropical trees. Amer. Natur. 111: 817–832. (11)

Howe, H. F. and R. B. Primack. 1975. Differential seed dispersal by birds of the tree *Casearia nitida* (Flacourtiaceae). Biotropica 7: 278–283. (11)

Howe, H. F. and G. A. Vande Kerckhove. 1979. Fecundity and seed dispersal of a tropical tree. Ecology 60: 180–189. (11)

Howe, H. F. and G. A. Vande Kerckhove. 1981a. Nutmeg dispersal by tropical birds. Science 210: 925–927. (11)

Howe, H. F. and G. A. Vande Kerckhove. 1981b. Removal of wild nutmeg (*Virola surinamensis*) crops by birds. Ecology 62: 1093–1106. (11)

Hsiao, T. H. 1969. Chemical basis of host selection and plant resistance in oligophagous insects. Ent. exp. appl. 12: 777–788. (10)

Hsiao, T. H. 1978. Host plant adaptation among geographic populations of the Colorado potato beetle. Ent. exp. app. 24: 237–247. (10)

Hubbell, S. P. 1979. Tree dispersion, abundance and diversity in a tropical deciduous forest. Science 203: 1299–1309. (11)

Huettel, M. D. and G. L. Bush. 1972. The genetics of host selection and its bearing on sympatric speciation in *Procecidochares* (Diptera: Tephritidae). Ent. exp. appl. 15: 465–480. (10)

Huey, R. B. 1979. Parapatry and niche complementarity of Peruvian desert geckos (*Phyllodactylus*): The ambiguous role of competition. Oecologia 38: 249–259. (18)

Huey, R. B. and E. R. Pianka. 1974. Ecological character displacement in a lizard. Amer. Zool. 14: 1127–1136. (18)

496

Hughes, R. N. and R. W. Elner. 1979. Tactics of a predator, *Carcinus maenas*, and morphological responses of the prey, *Nucella lapillus*. J. Anim. Ecol. 48: 65–78. (14)

Huheey, J. 1976. Studies on warning coloration and mimicry VII. Evolutionary consequences of a Batesian-Muellerian spectrum: a model for Muellerian mimicry. Evolution 30: 86–93. (3)

Hulme, A. C. (ed.). 1970. *The Biochemistry of Fruits and Their Products*, Vol. 1, Academic Press, New York. (10, 11)

Hulme, A. C. (ed.). 1971. *The Biochemistry of Fruits and Their Products*, Vol. 2. Academic Press, New York. (10, 11)

Humes, A. G. 1979. Coral-inhabiting copepods from the Moluccas, with a synopsis of cyclopods associated with scleractinian corals. Can. Biol. Mar. 20: 87–107. (14)

Hungate, R. E. 1975. The rumen microbial ecosystem. Ann. Rev. Ecol. Syst. 6: 39–66. (8)

Hunt, T. J. 1956. Death of *Testudo elegans* from intestinal obstruction. British J. Herpetology 2: 35. (11)

Hutchinson, G. E. 1959. Homage to Santa Rosalia, or why are there so many kinds of animals? Amer. Natur. 93: 145–159. (1, 18)

Hutto, R. L. 1978. A mechanism for resource allocation among sympatric heteromyid rodent species. Oecologia 33: 115–126. (19)

Huxley, J. S. 1932. *Problems of Relative Growth*. Methuen & Co., London. (2)

Huxley, J. S. 1942. *Evolution, the Modern Synthesis*. Allen and Unwin, London. (2)

Huxley, J. S. 1966. Introduction (to a discussion on ritualization of behaviour in man and animals). Phil. Trans. R. Soc. 251B: 249–271. (2)

Hylander, W. L. 1975. Incisor size and diet in anthropoids with special reference to Cercopithecidae. Science 189: 1095–1098. (11)

Ikeda, H. and J. Tomizawa. 1968. Prophage P1, an extrachromosomal replication unit. Cold Spring Harbor Symp. Quant. Biol. 33: 791–798. (5)

Inglis, W. G. 1971. Speciation in parasitic nematodes. Advances in Parasitology 9: 185–223. (4)

Inouye, D. W. 1977. Species structure of bumblebee communities in North America and Europe. In *The Role of Arthropods in Forest Ecosystems*, W. J. Mattson (ed.), pp. 35–40. Springer-Verlag, New York. (13)

Inouye, D. W. 1978. Resource partitioning in bumblebees: experimental studies of foraging behavior. Ecology 59: 672–678. (13)

Inouye, D. W. 1980. The effect of proboscis and corolla tube length on patterns and rates of flower visitation by bumblebees. Oecologia 45: 197–201. (13)

Inouye, R. S., G. S. Byers and J. H. Brown. 1980. Effects of predation and competition on survivorship and fecundity of desert annuals. Ecology 61: 1344–1351. (19)

Irwin, J. O. 1955. A unified derivation of some well-known frequency distributions of interest in biometry and statistics. J. Roy. Stat. Soc. (A) 118: 389–404. (18)

Ivlev, V. S. 1961. *Experimental Feeding Ecology of Fishes*, Yale University Press, New Haven, Connecticut. (17)

Iwagaka, I. and K. Kudo. 1977. Fruit quality relating to the location of fruit in a canopy of satsuma mandarin tree. Bull. Shikoku Agric. Exper. Sta. No. 30: 17–23. (11)

Jackson, J. B. C. 1977. Competition on marine hard substrata: the adaptive significance of solitary and colonial strategies. Amer. Natur. 111: 743–767. (14)

Jackson, J. F. 1981. Seed size as a correlate of temporal and spatial patterns of seed fall in a neotropical forest. Biotropica 13: 121–130. (11)

Jaenike, J. 1978a. On optimal oviposition behavior in phytophagous insects. Theor. Pop. Biol. 14: 350–356. (10)

497

Jaenike, J. 1978b. An hypothesis to account for the maintenance of sex within populations. Evol. Theory 3: 191–194. (9)

Jaenike, J. and R. K. Selander. 1979. Ecological generalism in *Drosophila falleni*: genetic evidence. Evolution 33: 741–748. (10)

Jaenike, J. and R. K. Selander. 1980. On the question of host races in the fall webworm, *Hyphantria cunea*. Ent. exp. appl. 27: 31–37. (10)

Jain, S. K. 1976. The evolution of inbreeding in plants. Ann. Rev. Ecol. Syst. 7: 469–495. (13)

James, F. C. 1970. Geographic size variation in birds and its relationship to climate. Ecology 51: 365–390. (18)

James, P. W. and A. Henssen. 1976. The morphology and taxonomic significance of cephalodia. In *Lichenology: Progress and Problems*, D. H. Brown, D. L. Hawkworth and R. H. Bailey (eds.). Academic Press, London. (7)

Janis, C. 1976. The evolutionary strategy of the Equidae and the origins of rumen and cecal digestion. Evolution 30: 757–774. (11)

Jansen, E. F. 1965. Ethylene and polyacetylenes. In *Plant Biochemistry*, J. Bonner and J. E. Varner (eds.), pp. 641–664. Academic Press, New York. (11)

Janzen, D. H. 1966. Coevolution of mutualism between ants and acacias in Central America. Evolution 20: 249–275. (1, 10, 12, 15)

Janzen, D. H. 1969. Seed-eaters versus seed size, number, toxicity and dispersal. Evolution 23: 1–27. (11)

Janzen, D. H. 1970. Herbivores and the number of tree species in tropical forests. Amer. Natur. 104: 501–528. (10, 11, 19)

Janzen, D. H. 1971a. Seed predation by animals. Ann. Rev. Ecol. Syst. 2: 465–492. (11)

Janzen, D. H. 1971b. The fate of *Scheelea rostrata* fruits beneath the parent tree: predispersal attack by bruchids. Principes 15: 89–101. (11)

Janzen, D. H. 1971c. Euglossine bees as long-distance pollinators of tropical plants. Science 171: 203–205. (13)

Janzen, D. H. 1972. Escape in space by *Sterculia apetala* seeds from the bug *Dysdercus fasciatus* in a Costa Rican deciduous forest. Ecology 53: 350–361. (11)

Janzen, D. H. 1974. Tropical blackwater rivers, animals, and mast fruiting by the Dipterocarpaceae. Biotropica 6: 69–103. (11)

Janzen, D. H. 1975. Behavior of *Hymenaea courbaril* when its predispersal seed predator is absent. Science 189: 145–147. (11)

Janzen, D. H. 1976. Why bamboos wait so long to flower. Ann. Rev. Ecol. Syst. 7: 347–391. (11)

Janzen, D. H. 1977a. Why fruits rot, seeds mold, and meat spoils. Amer. Natur. 111: 691–713. (11)

Janzen, D. H. 1977b. Promising directions of study in tropical animal-plant interactions. Ann. Missouri Bot. Gard. 64: 706–736. (11)

Janzen, D. H. 1977c. A note on optimal mate selection by plants. Amer. Natur. 111: 365–371. (13)

Janzen, D. H. 1978a. The size of a local peak in a seed shadow. Biotropica 10: 78. (11)

Janzen, D. H. 1978b. Seeding patterns of tropical trees. In *Tropical Trees as Living Systems*, P. B. Tomlinson and M. H. Zimmerman (eds.), pp. 83–128. Cambridge University Press, New York. (11)

Janzen, D. H. 1978c. Complications in interpreting the chemical defenses of trees against tropical arboreal plant-eating vertebrates. In *The Ecology of Arboreal Folivores*, G. G. Montgomery (ed.), pp. 73–84. Smithsonian Institution Press, Washington, D. C.. (11)

Janzen, D. H. 1979a. How to be a fig. Ann. Rev. Ecol. Syst. 10: 13–51. (11, 13)

Janzen, D. H. 1979b. Why food rots. Nat. Hist. Mag. 88: 60–66. (11)

Janzen, D. H. 1979c. New horizons in the biology of plant defenses. In *Herbivores: Their Interaction with Secondary Plant Metabolites*, G. A. Rosenthal and D. H. Janzen (eds.), pp. 331–350. Academic Press, New York. (10)

Janzen, D. H. 1980a. When is it coevolution? Evolution 34: 611–612. (1, 10, 11)

Janzen, D. H. 1980b. Heterogeneity of potential food abundance for tropical small land birds. In *Migrant Birds in the Neotropics*, A. Keast and E. S. Morton (eds.), pp. 545–552. Smithsonian Institution Press, Washington, D.C. (11)

498

Janzen, D. H. 1981a. Guanacaste tree seed-swallowing by Costa Rican range horses. Ecology 62: 587–592. (11)

Janzen, D. H. 1981b. *Enterolobium cyclocarpum* seed passage rate and survival in horses, Costa Rican Pleistocene seed dispersal agents. Ecology 62: 593–601. (11)

Janzen, D. H. 1981c. Digestive seed predation by a Costa Rican Baird's tapir. Biotropica (Supplement) 13: 59–63. (11)

Janzen, D. H. 1981d. *Ficus ovalis* seed predation by an orange-chinned parakeet (*Brotogeris jugularis*) in Costa Rica. Auk 98: 841–844. (11)

Janzen, D. H. 1981e. Evolutionary physiology of personal defense. In *Physiological Ecology: An Evolutionary Approach to Resource Use*, C. R. Townsend and P. Calow (eds.), pp. 145–164. Blackwell Scientific, Oxford, England. (11)

Janzen, D. H. 1982a. Differential seed survival and passage rates in cows and horses, surrogate Pleistocene dispersal agents. Oikos 38: 150–156. (11)

Janzen, D. H. 1982b. Ecological distribution of chlorophyllous developing embryos among perennial plants in a tropical deciduous forest. Biotropica. (In press) (11)

Janzen, D. H. 1982c. Attraction of *Liomys* mice to horse dung and the extinction of this response. Animal Behaviour 30:483–489. (11)

Janzen, D. H. 1982d. Horse response to *Enterolobium cyclocarpum* (Leguminosae) fruit crop size in a Costa Rican deciduous forest pasture. Brenesia. (In press) (11)

Janzen, D. H. 1982e. Variation in average seed size and fruit seediness in a fruit crop of a guanacaste tree (Leguminosae: *Enterolobium cyclocarpum*). Amer. J. Botany 69: 1169–1178 (11)

Janzen, D. H. 1982f. Cenizero tree (Leguminosae: *Pithecellobium saman*) delayed fruit development in Costa Rican deciduous forests. Amer. J. Bot. 69:1269–1276. (11)

Janzen, D. H. 1982g. Natural history of guacimo fruits (Sterculiaceae: *Guazuma ulmifolia*) with respect to consumption by large mammals. Amer. J. Bot. 69: 1240–1250. (11)

Janzen, D. H. 1982h. Removal of seeds from dung by tropical rodents: influence of habitat and amount of dung. Ecology 63: 1887–1900. (11)

Janzen, D. H. 1982i. Large herbivore dispersal of small seeds: the plant is the fruit. Amer. Natur. (Submitted) (11)

Janzen, D. H. and M. L. Higgins 1979. How hard are *Enterolobium cyclocarpum* seeds? Brenesia 16: 61–67. (11)

Janzen, D. H. and P. S. Martin. 1982. Neotropical anachronisms: the fruits the gomphotheres left behind. Science 215: 19–27. (11)

Janzen, D. H., G. A. Miller, J. Hackforth-Jones, C. M. Pond, K. Hooper and D. P. Janos. 1976. Two Costa Rican bat generated seed shadows of *Andira inermis* (Leguminosae). Ecology 56: 1068–1075. (11)

Jarman, P. J. 1976. Damage to *Acacia tortilis* seeds eaten by impala. E. Afr. Wildl. J. 14: 223–225. (11)

Jarrett, E. and H. Bazin. 1974. Elevation of total serum IgE in rats following helminth parasite infection. Nature 251: 613–614. (8)

Jayakar, S. D. 1970. A mathematical model for interaction of gene frequencies in a parasite and its host. Theor. Pop. Biol. 1: 140–164. (3)

Jeffords, M. R., J. R. Sternberg and G. P. Waldbauer. 1979. Batesian mimicry: field demonstration of the survival value of pipevine swallowtail and monarch color patterns. Evolution 33: 275–286. (12)

Jenkins, D. C. 1975. The influence of *Nematospiroides dubius* on subsequent *Nippostrongylus brasiliensis* infections in mice. Parasitology 71: 349–355. (8)

Jenkins, S. N. and J. M. Behnke. 1977. Impairment of primary expulsion of *Trichurus muris* in mice concurrently infected with *Nematospiroides dubius*. Parasitology 75: 71–78. (8)

Jerison, H. F. 1973. *Evolution of the Brain and Intelligence*. Academic Press, New York. (15)

Jermy, T. 1966. Feeding inhibitors and food preference in chewing phytophagous insects. Ent. exp. appl. 9: 1-2 (10)

Jermy, T. (ed.) 1976a. *The Host-Plant in Relation to Insect Behavior and Reproduction.* Plenum Press, New York. (10)

Jermy, T. 1976b. Insect-host-plant relationship—co-evolution or sequential evolution? Symp. Biol. Hung. 16: 109–113. (4, 10)

Jermy, T. and A. Szentesi. 1978. The role of inhibitory stimuli in the choice of oviposition site by phytophagous insects. Ent. exp. appl. 24: 458–471. (10)

Jimbo, S. and H. O. Schwassman. 1967. Feeding behavior and daily emergence pattern of *Artibeus jamaicensis*. Atlas Simp. Biota Amazon. 5: 239–253. (11)

Joern, A. and L. R. Lawlor. 1980. Food and microhabitat utilization by grasshoppers from arid grasslands: comparisons with neutral models. Ecology 61: 591–599. (10)

Johnson, J. H. 1960. Paleozoic Solenoporaceae and related red algae. Quart. Colorado School of Mines 55: 1–77. (15)

Johnson, T. 1953. Variation in the rusts of cereals. Biol. Rev. 28: 105–157. (7)

Jones, C. 1970. Stomach contents and gastro-intestinal relationships of monkeys collected in Rio Muni, West Africa. Mammalia 34: 107–117. (11)

Jones, C. 1972a. Comparative ecology of three pteropid bats in Rio Muni, West Africa. J. Zool. Lond. 167: 353–370. (11)

Jones, C. 1972b. Natural diets of wild primates, Part I. In *Pathology of Simian Primates*, R. N. T-W-Fiennes (ed.), pp. 58–77. Karger, Basel. (11)

Jones, D. A. 1973. Coevolution and cyanogenesis. In *Taxonomy and Ecology*, V. H. Heywood (ed.), pp. 213–242. Academic Press, London. (10)

Jones, D. A., R. J. Keymer and W. M. Ellis. 1978. Cyanogenesis in plant and animal feeding. In *Biochemical Aspects of Plant and Animal Coevolution*, J. B. Harborne (ed.), pp. 21–24. Academic Press, London. (10)

Jones, N. S. and J. M. Kain. 1967. Subtidal algal colonization following the removal of *Echinus*. Helgolander wiss. Meeresunter. 15: 460–466. (19)

Jordan, C. F. 1971. A world pattern in plant energetics. Amer. Sci. 59: 425–433. (19)

Jordan, P. A., P. C. Shelton, and D. L. Allen. 1967. Numbers, turnover, and social structure of the Isle Royale wolf population. Amer. Zool. 7: 233–252. (8)

Jordan, P., J. D. Christie and G. O. Unrau. 1980. Schistosomiasis transmission with particular reference to possible ecological and biological methods of control. A review. Acta Tropica 37: 95–135. (8)

Jordano, P. 1982. Migrant birds and the seed-dispersal system of southern Spanish blackberries. Oikos 38:183–193. (11)

Joslyn, A. M. and J. L. Goldstein. 1964. Astringency of fruits and fruit products in relation to phenolic content. Adv. Food. Res. 13: 179–217. (11)

Jouventin, P. 1975. Observations sur la socio-ecologie du mandrilli. La Terre et la Vie 29: 493–532. (11)

Jouventin, P., G. Pasteur and J. P. Cambefort. 1977. Observational learning of baboons and avoidance of mimics: exploratory tests. Evolution 31: 214–218. (11)

Jurand, A., J. R. Preer, and B. M. Rudman. 1978. Further investigations on the prelethal effects of the killing action of kappa containing killer stocks of *Paramecium aurelia*. J. Exp. Zoology 201: 25–47. (6)

Kaikini, N. S. 1968. Cultivation and management of *Bursera delpichiana* in Mysore. State. Ind. For. 94: 32–36. (11)

Kakati, B. N. and C. K. Rajkonwar. 1972. Some observations on the reproductive behavior of *Rhinoceros unicornis*. Ind. For. 98: 357–358. (11)

Kaneda, C. and R. Kisimoto. 1979. Studies of varietal resistance to brown planthopper in Japan. In *Brown Planthopper: Threat to Rice Production in Asia*, International Rice Research Institute, pp. 209–218. Los Baños, Laguna, Philippines. (10)

Kantak, G. E. 1979. Observations on some fruit-eating birds in Mexico. Auk 96: 183–186. (11)

Karlin, S. 1979. Models of multifactorial inheritance I. Multivariable formulations and basic convergence results. Theor. Popul. Biol. 15: 308–355. (3)

Karr, J. R. 1976. Seasonality, resource availability and community diversity in tropical bird communities. Amer. Natur. 110: 973–994. (11)

500

Karr, J. R. and F. C. James. 1975. Eco-morphological configurations and convergent evolution in species and communities. In *Ecology and Evolution of Communities*, M. L. Cody and J. M. Diamond (eds.), pp. 258-291. Belknap Press, Harvard University Press, Cambridge, Massachusetts. (19)

Kay, Q. O. N. 1978. The role of preferential and assortative pollination in the maintenance of flower colour polymorphisms. In *The Pollination of Flowers by Insects*, A. J. Richards (ed.), pp. 175-190. Academic Press, London. (13)

Kazacos, K. R. 1975. Increased resistance in the rat to *Nippostrongylus brasiliensis* following immunization against *Trichinella spiralis*. Vet. Parasitol. l: 165-174. (8)

Kearn, G. C. 1971. The physiology and behavior of the monogenean skin parasite *Entobdella soleae* in relation to its host (*Solea solea*). In *Ecology and Physiology of Parasites*, A. M. Fallis (ed.), pp. 161-187. University of Toronto Press, Toronto. (8)

Kearn, G. C. 1976. Body surface of fishes. In *Ecological Aspects of Parasitology*, C. R. Kennedy (ed.), pp. 185-208. North-Holland Publ. Co., Amsterdam. (8)

Keast, A. 1958. The influence of ecology on variation in the mistletoe-bird (*Dicaeum hirundinaceum*). Emu 58: 195-206. (11)

Keast, A. 1968. Competitive interactions and the evolution of ecological niches as illustrated by the Australian honeyeater genus *Melithreptus* (Meliphagidae). Evolution 22: 762-784. (18)

Keast, A. 1970. Adaptive evolution and shifts in niche occupation in island birds. Biotropica 2: 61-75. (18)

Keast, A. 1977. Historical biogeography of the marsupials. In *The Biology of Marsupials*, B. Stonehouse, and D. Gilmore (eds.), pp. 69-95. Macmillan, London. (16)

Kellogg, D. E. 1980. Character displacement and phyletic change in the evolution of the radiolarian subfamily Artiscinae. Micropaleontology 26: 196-210. (1)

Kellogg, V. L. 1896. New Mallophaga, I. Proc. Calif. Acad. Sci., vol. 6. (4)

Kemp, A. C. 1975. The duration of natural treehole nests. Ostrich 46: 118. (11)

Kemper, J. T. 1982. The evolutionary effect of endemic infectious disease: continuous models for an invariant pathogen. J. Math. Biol. 15: 65-77. (9)

Kendall, D. G. 1956. Deterministic and stochastic epidemics in closed populations. *Proc. Third Berkeley Symp. Math. Statist. and Prob.* 4: 149-165. (9)

Kennedy, J. S. 1965. Mechanisms of host plant selection. Ann. Appl. Biol. 56: 317-322. (10)

Kennedy, M. J. 1980. Geographical variation in some representatives of *Haematoloechus* Looss, 1899 (Trematoda, Haematoloechidae) from Canada and the United States. Can. J. Zool. 58: 1151-1167. (8)

Kennedy, M. W. 1980. Immunologically mediated, non-specific interactions between the intestinal phases of *Trichinella spiralis* and *Nippostrongylus brasiliensis* in the mouse. Parasitology 80: 6172. (8)

Kermack, W. O. and A. G. McKendrick. 1927. A contribution to the mathematical theory of epidemics. Proc. Roy. Soc. A. 115: 700-721. (9)

Kernaghan, R. P. 1971. The ultrastructure of the organism associated with hybrid sterility in *D. paulistorum*. Drosophila Inf. Serv. 47: 69-70. (6)

Kethley, J. B. and D. E. Johnston. 1975. Resource tracking patterns in bird and mammal ectoparasites. Misc. Publ. Ent. Soc. Amer. 9: 231-236. (4)

Kier, P. M. 1974. Evolutionary trends and their functional significance in the post-Paleozoic echinoids. J. Paleont., Paleont. Soc. Mem. 5, 48, Part II of II: 1-95. (14, 15)

Killip, E. P. 1938. The American species of Passifloraceae. Publ. Field Mus. Nat. Hist. (Bot.) 19: 1-613. (4)

Kiltie, R. A. 1981. Distribution of palm fruits on a rain forest floor: why white-lipped peccaries forage near objects. Biotropica 13: 141-145. (11)

Kimsey, L. S. 1980. The behaviour of male orchid bees (Apidae, Hymenoptera, Insecta) and the question of leks. Anim. Behav. 28: 996-1004. (13)

Kimura, M. 1968. Evolutionary rate at the molecular level. Nature 217: 624-626. (2, Ep)

King, J. L. and T. Jukes. 1969. Non-Darwinian evolution. Science 164: 788-798. (2)

Kinsey, A. C. 1923. The gall wasp genus *Neuroterus*. Indiana Univ. Stud. 10: 1-50. (4)

Kinsey, A. C. 1930. The gall wasp genus *Cynips*: a study in the origin of species. Indiana Univ. Stud. 16: 1-577. (4)

Kirby, H. 1937. Host-parasite relations in the distribution of Protozoa in termites. Univ. Calif. Publ. Zool. 41: 189-212. (4)

Kircher, H. W., W. B. Heed, J. S. Russell and J. Grove. 1967. Senita cactus alkaloids; their significance to Sonoran Desert *Drosophila* ecology. J. Ins. Physiol. 13: 1864-1874. (10)

Kistner, D. H. 1969. The biology of termitophiles. In *Biology of Termites*, Vol. 1, K. Krishna and F. M. Weesner (eds.), pp. 525-557. Academic Press, New York. (4)

Kitching, J. A. and F. J. Ebling. 1961. The ecology of Lough Ine. XI. The control of algae by *Paracentrotus lividus* (Echinoidea). J. Anim. Ecol. 30: 373-383. (19)

Klein, L. L. and D. J. Klein. 1973. Observations on two types of neotropical primate intertaxa associations. Amer. J. Phys. Anthro. 38: 649-653. (11)

Kleinman, D. G. and J. F. Eisenberg. 1973. Comparisons of canid and felid social systems from an evolutionary perspective. Anim. Behav. 21: 637-659. (15)

Klite, P. D. 1965. Intestinal bacterial flora and transit time of three Neotropical bat species. J. Bacter. 90: 375-379. (11)

Klots, A. B. 1931. A generic revision of the Pieridae. Entomol. Amer., n.s. 12: 139-242. (4)

Kluge, A. G. and J. S. Farris. 1969. Quantitative phyletics and the evolution of anurans. Syst. Zool. 18: 1-32. (4)

Knerer, G. and C. E. Atwood. 1973. Diprionid sawflies; polymorphism and speciation. Science 179: 1090-1099. (10)

Knight, R. L., E. Keep and J. B. Briggs. 1960. Genetics of resistance to *Amphorophora rubi* (Kalt.) in the raspberry. II. The genes A_2-A_7 from the American variety. Genet. Res. Cam. 1: 319-331. (10)

Knight-Jones, E. W. and P. Knight-Jones. 1980. Pacific spirorbids in the east Atlantic. Jour. Mar. Biol. Assoc. U.K. 60: 461-464. (19)

Knox, G. A. 1960. Littoral ecology and biogeography of the southern oceans. Proc. Roy. Soc. London (B) 152: 567-624. (19)

Kodric-Brown, A. and J. H. Brown. 1979. Competition between distantly related taxa in the coevolution of plants and pollinators. Amer. Zool. 19: 1115-1127. (19)

Koehn, R. K. and W. F. Eanes. 1978. Molecular structure and protein variation within and among populations. Evol. Biol. 11: 39-100. (7)

Kogan, M. 1977. The role of chemical factors in insect/plant relationships. Proc. Internat. Cong. Entomol. 15: 211-227. (10)

Kohn, A. J. 1959. The ecology of *Conus* in Hawaii. Ecol. Monogr. 29: 47-90. (10)

Kohn, A. J. and J. W. Nybakken. 1975. Ecology of *Conus* on eastern Indian Ocean fringing reefs. Mar. Biol. 29: 211-234. (19)

Kotecki, N. R. 1970. Circulation of the cestode fauna of Anseriformes in the Municipal Zoological Garden in Warszawa. Acta Parasitol. Pol. 17: 329-355. (8)

Kott, P. 1980. Algal-bearing didemnid ascidians in the Indo-West-Pacific. Mem. Qd. Mus. 20: 1-47. (14)

Krebs, J. R. 1973. Behavioral aspects of predation. In *Perspectives in Ethology*, P. P. G. Bateson and P. H. Klopfer (eds.), pp. 73-111. Plenum Press, New York. (10)

Krefting, L. W. and E. L. Roe. 1949. The role of some birds and mammals in seed germination. Ecol. Monogr. 19: 269-286. (11)

Krieger, R. I., P. P. Feeny and C. F. Wilkinson. 1971. Detoxification enzymes in the guts of caterpillars: an evolutionary answer to plant defenses? Science 172: 579-581. (10)

Krishna, K., 1970. Taxonomy, phylogeny, and distribution of termites. In *Biology of Termites*, Vol. II, K. Krishna and F. M. Weesner (eds.), pp. 127-153. Academic Press, New York and London. (4)

Krishnamurthy, G. V., H. L. Jain and B. S. Bhatia. 1960. Changes in the physico-chemical composition of mangoes during ripening after picking. Food Science (Mysore) 9: 277-279. (11)

Kristensen, N. P. 1975. Remarks on the family-level phylogeny of butterflies (Insecta, Lepidoptera, Rhopalocera). Z. zool. Syst. Evolutionsforsch. 14: 25-33. (4)

502

Kristensen, N. P. and E. S. Nielsen. 1980. The ventral diaphragm of primitive (non-ditrysian) Lepidoptera. A morphological and phylogenetic study. Z. zool. Syst. Evolutionsforsch. 18: 123-146. (4)

Kropp, R. K. and C. Birkeland. 1982. Comparison of crustacean associates of *Pocillopora verrucosa* from a high island and an atoll. Proc. 4th Intern. Coral Reef Symp., Manila, 1981, Vol. 2, pp. 627-632. (14)

Krull, W. H. and C. R. Mapes. 1952. Studies on the biology of *Dicrocoelium dendriticum* (Rudolphi, 1819) Looss, 1899 (Trematoda:Dicrocoeliidae), including its relation to the intermediate host, *Cionella lubrica* (Muller). VII. The second intermediate host of *Dicrocoelium dendriticum*. Cornell Vet. 42: 603-604. (8)

Kruuk, H. 1972. *The Spotted Hyaena*. University of Chicago Press, Chicago. (16)

Kubo, I. and K. Nakanishi. 1977. Insect antifeedants and repellants from African plants. In *Host Plant Resistance to Plants*, P. A. Hedin (ed.), pp. 165-178. Amer. Chemical Society Symposium Series No. 62, Washington, D. C. (11)

Kulman, H. M. 1971. Effects of insect defoliation on growth and mortality of trees. Ann. Rev. Entomol. 16: 289-324. (10)

Kung, C. 1970. The election transport system of kappa particles from *Paramecium aurelia* stock 51. J. Gen. Microbiol. 61: 371-378. (6)

Kung, C. 1971. Aerobic respiration of kappa particles from *Paramecium aurelia*. J. Protozool. 18: 328-332. (6)

Kurtén, B. 1957. Mammal migrations, Cenozoic stratigraphy, and the age of Peking Man. J. Paleont. 31: 215-227. (16)

Kurtén, B. 1963. Notes on some Pleistocene mammal migrations from the Palaearctic to the Nearctic. Eiszeit. u. Gegen. 14: 96-103. (16)

Kwak, M. M. 1979. Effects of bumblebee visits on the seed sets of *Pedicularis, Rhinanthus* and *Melampyrum* (Scrophulariaceae) in the Netherlands. Acta Botanica Neerlandica 28: 177-195. (13)

Labeyrie, V. 1978. Reproduction of insects and coevolution of insects and plants. Ent. exp. appl. 24: 296-304. (10)

Lack, D. 1947. *Darwin's Finches*. Cambridge University Press, Cambridge. (1, 17, 18)

Lack, D. 1969. Subspecies and sympatry in Darwin's finches. Evolution 23: 252-263. (1)

Lackie, A. M. 1980. Invertebrate immunity. Parasitology 80: 393-412. (8)

Lamprey, H. F. 1967. Notes on the dispersal and germination of some tree seeds through the agency of mammals and birds. East African Wildlife Journal 5: 179-180. (11)

Lamprey, H. F., G. Halevy and S. Makacha. 1974. Interactions between *Acacia*, bruchid seed beetles and large herbivores. East African Wildlife Journal 12: 81-85. (11)

Land, H. C. 1963. A tropical feeding tree. Wilson Bull. 75: 199-200. (11)

Lande, R. 1976. Natural selection and random genetic drift in phenotypic evolution. Evolution 30: 314-334. (16)

Lande, R. 1979. Quantitative genetic analyses of multivariate evolution, applied to brain:body size allometry. Evolution 33: 402-416. (16, 18)

Langguth, A. 1975. Ecology and evolution in the South American canids. In *The Wild Canids*, M. W. Fox (ed.). Van Nostrand Reinhold, New York. (16)

Langley, C. H., D. B. Smith and F. M. Johnson. 1978. Analysis of linkage disequilibria between allozyme loci in natural populations of *Drosophila melanogaster*. Genet. Res. 32: 215-229. (2, 10)

Lanner, R. M. and S. B. Vander Wall. 1980. Dispersal of limber pine seed by Clark's nutcracker. J. Forestry 78: 637-639. (11)

Lanzani, A. and G. Jacini. 1973. Attuali conoscenze sui composti solforati ad ozione gozzigena presenti nel seme di colza. La Revista Italiana delle Sostanze Grasse 50: 277-285. (11)

Lapointe, B. E., F. X. Neill and J. M. Fuente. 1981. Community structure, succession and production of seaweeds associated with mussel rafts in the Ria de Arosa, N.W. Spain. Marine Ecology Program Series 5: 243-253. (19)

Laurence, B. R. and F. R. N. Pester. 1967. Adaptation of a filarial worm, *Brugia patei*, to a new mosquito host, *Aedes togoi*. J. Helminth. 41: 365-392. (8)

Laverty, T. M. 1980. The flower-visiting behavior of bumblebees: floral complexity and learning. Can. J. Zool. 58: 1324-1335. (13)

Law, R. and D. H. Lewis. 1982. Biotic environments and the maintenance of sex—some evidence from mutualistic symbioses. Biol. J. Linn. Soc. (In press) (7)

Lawlor, L. R. 1979. Direct and indirect effects on *n*-species competition. Oecologia 43: 355-364. (3)

Lawlor, L. R. and J. Maynard Smith. 1976. The coevolution and stability of competing species. Amer. Natur. 110: 79-99. (2, 17)

Lawrence, G. J., K. W. Shepherd and G. M. E. Mayo. 1981a. Fine structure of genes controlling pathogenicity in flax rust *Melampsora lini*. Heredity 46: 297-313. (7)

Lawrence, G. J., G. M. E. Mayo and K. W. Shepherd. 1981b. Interactions between genes controlling pathogenicity in the flax rust fungus. Phytopathology 71: 12-19. (7)

Lawton, J. H. 1978. Host-plant influences on insect diversity: the effects of space and time. In *Diversity of Insect Faunas*, L. A. Mound and N. Waloff (eds.). pp. 105-125. Symp. Royal Ent. Soc. Lond. 9: 41-55. (10, 19)

Lawton, J. H. 1982. Non-competitive populations, non-convergent communities and vacant niches: the herbivores on bracken. In *Ecological Communities: Conceptual Issues and the Evidence*, D. R. Strong, D. Simberloff and L. G. Abele (eds.). Princeton University Press. (In press) (19)

Lawton, J. H. and D. Schröder. 1978. Some observations on the structure of phytophagous insect communities: the implications for biological control. Proc. IV Internat. Symp. Biol. Control of Weeds., pp. 57-73. Inst. Food Agr. Sci., Univ. Fla., Gainesville. (10)

Lawton, J. H. and D. R. Strong, Jr. 1981. Community patterns and competition in folivorous insects. Amer. Natur. 118: 317-338. (10, 19)

Lazell, J. 1972. The *Anolis* (Sauria, Iguanidae) of the Lesser Antilles. Bull. Mus. Comp. Zool. Harvard Univ. 143: 1-115. (17)

Lebedev, B. I. 1977. O nekotorix aspiektax ekologii i evolutsii Monogenoidea. [On some aspects of biology and evolution of monogeneans (Monogenea, Platyhelminthes).] Zhurn. Obshchei Biol. 40: 271-281. (In Russian, Eng. summ.) (8)

Leck, C. F. 1969. Observations of birds exploiting a Central American fruit tree. Wilson Bulletin 81: 264-269. (11)

Leck, C. F. 1970a. The seasonal ecology of fruit and nectar eating birds in lower Middle America. Ph.D. Thesis, Cornell University, Ithaca, New York. (11)

Leck, C. F. 1970b. Comments on the seasonality of fruiting in the Neotropics. Amer. Natur. 104: 583-584. (11)

Leck, C. F. 1971a. Overlap in the diet of some neotropical birds. The Living Bird 10: 89-106. (11)

Leck, C. F. 1971b. Measurement of social attractions between tropical passerine birds. Wilson Bulletin 83: 278-283. (11)

Leck, C. F. 1972a. Seasonal changes in feeding pressure of fruit- and nectar-eating birds in Panama. Condor 74: 54-60. (11)

Leck, C. F. 1972b. Observations of birds at *Cecropia* trees in Puerto Rico. Wilson Bulletin 84: 498-500. (11)

Leck, C. F. 1972c. The impact of some North American migrants at fruiting trees in Panama. Auk 89: 842-850. (11)

Leck, C. F. and S. Hilty. 1968. A feeding congregation of local and migratory birds in the mountains of Panama. Bird-Banding 34: 318. (11)

Lee, S. H. 1957. The life cycle of *Skrjabinoptera phrynosoma* (Ortlepp) Schultz, 1927 (Nematoda: Spiruroidea), a gastric nematode of Texas horned toads, *Phrynosoma cornutum*. J. Parasitol. 43: 66-75. (8)

Leete, E. 1973. Esophageal cancer. Science 179: 228. (11)

Leigh, E. G., Jr. 1975. Structure and climate in tropical rain forest. Ann. Rev. Ecol. Syst. 6: 67-86. (19)

Leighton, M. and D. R. Leighton. 1981. The relationship of size of feeding aggregate to size of food patch: howler monkeys (*Alouatta palliata*) feeding in *Trichilia cipo* fruit trees on Barro Colorado Island. Biotropica. (In press) (11)

LeJambre, L. F. 1981. Hybridization of Australian *Haemonchus placei* (Place, 1893), *Haemonchus contortus cayugensis* (Das and Whitlock, 1960) and *Haemonchus contortus* (Rudolphi, 1803) from Louisiana. Int. J. Parasitol. 11: 323–330. (8)

LeJambre, L. F., W. M. Royal and P. J. Martin. 1979. The inheritance of thiabendazole resistance in *Haemonchus contortus*. Parasitology 78: 107–119. (8)

Lekagul, B. and J. A. McNeeley. 1977. *Mammals of Thailand.* Kurusapha Ladprao Press, Bangkok. (16)

León, J. and B. Charlesworth. 1976. Ecological versions of Fisher's fundamental theorem of natural selection. Adv. Appl. Probab. 8: 639–641. (3)

Leonard, K. J. 1969. Genetic equilibria in host-pathogen systems. Phytopathology 59: 1858–1863. (7)

Leonard, K. J. 1977. Selection pressures and plant pathogens. Ann. N.Y. Acad. Sci. 287: 207–222. (7, 9)

Leonard, K. J. and R. J. Czochor. 1980. Theory of genetic interactions among populations of plants and their pathogens. Ann. Rev. Phytopathol. 18: 237–258. (7)

Leong, T. S. and J. C. Holmes. 1981. Communities of metazoan parasites in open water fishes of Cold Lake, Alberta. J. Fish Biol. 18: 693–713. (8)

Leppick, E. E. 1957. Evolutionary relationship between entomophilous plants and anthemophilous insects. Evolution 11: 466–481. (15)

Lerner, I. M. 1958. *The Genetic Basis of Selection.* John Wiley & Sons, New York. (2)

Levene, H. 1953. Genetic equilibrium when more than one ecological niche is available. Amer. Natur. 87: 331–333. (10)

Levin, B. R. 1980. Conditions for the existence of R-plasmids in bacterial populations. In *Antibiotic Resistance: Transposition and Other Mechanisms*, S. Mitsuhashi, L. Rosival and V. Kromery (eds.), pp. 197–202. Springer Verlag, Berlin. (5)

Levin, B. R. 1981. Periodic selection, infectious gene exchange and the genetic structure of *E. coli* populations. Genetics 99: 1–23. (5)

Levin, B. R. et al. 1982. Evolution of parasites and hosts (group report). In *Population Biology of Infectious Diseases*, R. M. Anderson and R. M. May (eds.), pp. 213–243. Springer Verlag, Berlin and New York. (9)

Levin, B. R. and W. L. Kilmer. 1974. Interdemic selection and the evolution of altruism: a computer simulation study. Evolution 28: 527–545. (5)

Levin, B. R., M. L. Petras and D. I. Rassmussen. 1969. The effect of migration on the maintenance of a lethal polymorphism in the house mouse. Amer. Natur. 103: 647–661. (2)

Levin, B. R. and V. A. Rice. 1980. The kinetics of transfer of nonconjugative plasmids by mobilizing conjugative factors. Genet. Res. 35: 241–259. (5)

Levin, B. R. and F. M. Stewart. 1980. The population biology of bacterial plasmids: a priori conditions for the existence of mobilizable nonconjugative factors. Genetics 87: 209–228. (5)

Levin, B. R., F. M. Stewart and L. Chao. 1977. Resource limited growth, competition and predation: a model and some experimental studies with bacteria and bacteriophage. Amer. Natur. 111: 3–24. (5)

Levin, B. R., F. M. Stewart and V. A. Rice. 1979. The kinetics of conjugative plasmid transmission: fit of simple mass action models. Plasmid 2: 247. (5)

Levin, D. A. 1973. The role of trichomes in plant defense. Quart. Rev. Biol. 48: 3–15. (10)

Levin, D. A. 1975. Pest pressure and recombination systems in plants. Amer. Natur. 109: 437–451. (10)

Levin, D. A. 1976. The chemical defenses of plants to pathogens and herbivores. Ann. Rev. Ecol. Syst. 6: 67–86. (10, 13, 19)

Levin, D. A. 1978. The origin of isolating mechanisms in flowering plants. Evol. Biol. 11: 185–317. (13)

505

Levin, D. A. and H. W. Kerster. 1974. Gene flow in seed plants. Evol. Biol. 7: 139–220. (11, 13)

Levin, S. A. 1976. Population dynamic models in heterogeneous environments. Ann. Rev. Ecol. Syst. 7: 287–310. (13)

Levin, S. A. 1978. On the evolution of ecological parameters. In *Ecological Genetics: The Interface*, P. Brussard (ed.), pp. 3–26. Springer Verlag, New York. (3)

Levin, S. A. 1983. Some approaches to the modelling of coevolutionary interactions. In *Coevolution*, M. Nitecki (ed.). University of Chicago Press, Chicago. (In press) (9)

Levin, S. A. and R. T. Paine. 1974. Disturbance, patch formation and community structure. Proc. Natl. Acad. Sci. USA 71:2744–2747. (19)

Levin, S. A. and D. Pimentel. 1981. Selection of intermediate rates of increase in parasite-host systems. Amer. Natur. 117: 308–315 (3, 8, 9)

Levin, S. A. and L. A. Segel. 1982. Models of the influence of predation on aspect diversity in prey populations. J. Math. Biol. 14: 253–284. (3, 9)

Levin, S. A. and J. D. Udovic. 1977. A mathematical model of coevolutionary populations. Amer. Natur. 111: 657–675. (9)

Levins, R. 1968. *Evolution in Changing Environments*. Princeton University Press, Princeton, New Jersey. (1, 2, 13, 14)

Levins, R. 1975. Evolution in communities near equilibrium. In *Ecology and Evolution of Communities*, M. L. Cody and J. M. Diamond (eds.), pp. 16–50. Harvard University Press, Cambridge, Massachusetts. (3)

Levins, R. and R. MacArthur. 1969. An hypothesis to explain the incidence of monophagy. Ecology 50: 910–911. (10)

Leviton, P. J. 1978. Resource partitioning by gastropods of the genus *Conus* of subtidal Indo-Pacific coral reefs: the significance of prey size. Ecology 59: 614–631. (17)

Lewin, R. 1981. Seeds of change in evolutionary development. Science 214: 42–44. (11)

Lewin, R., L. Cheng and L. Lafargue. 1980. Prochlorophytes in the Caribbean. Bull. Mar. Sci. 30: 744–745. (14)

Lewis, D. H. 1973a. Concepts in fungal nutrition and the origin of biotrophy. Biol. Rev. 48: 261–278. (7)

Lewis, D. H. 1973b. The relevance of symbiosis to taxonomy and ecology with particular reference to mutualistic symbioses and the exploitation of marginal habitats. In *Taxonomy and Ecology*, V. H. Heywood (ed.), pp. 151–172. Academic Press, London. (7)

Lewis, D. H. 1974. Micro-organisms and plants: the evolution of parasitism and mutualism. In *Evolution in the Microbial World*, 24th Symposium of Society for General Microbiology. Cambridge University Press, Cambridge. (7)

Lewis, J. W. 1981a. On the coevolution of pathogen and host: I. General theory of discrete time coevolution. J. Theor. Biol. 93: 927–951. (3, 9)

Lewis, J. W. 1981b. On the coevolution of pathogen and host: II. Selfing hosts and haploid pathogens. J. Theor. Biol. 93: 953–985. (9)

Lewontin, R. C. 1962. Interdeme selection controlling a polymorphism in the house mouse. Amer. Natur. 96: 65–78. (2)

Lewontin, R. C. 1965. Selection for colonizing ability. In *The Genetics of Colonizing Species*, H. G. Baker and G. L. Stebbins (eds.), pp. 77–94. Academic Press, New York. (8)

Lewontin, R. C. 1970. The units of selection. Ann. Rev. Ecol. Syst. 1: 1–18. (2)

Lewontin, R. C. 1974. *The Genetic Basis of Evolutionary Change*. Columbia University Press, New York. (2)

Lieberman, M. and D. Lieberman. 1980. The origin of gardening as an extension of infrahuman seed dispersal. Biotropica 12: 316–317. (11)

Lieberman, D., J. B. Hall, M. D. Swaine and M. Lieberman. 1979. Seed dispersal by baboons in the Shai Hills, Ghana. Ecology 60: 65–75. (11)

Lin, L., R. Bitner and G. Edlin. 1977. Increased reproductive fitness of *Escherichia coli* Lambda lysogens. J. Virol. 21: 554–559. (5)

Lincicome, D. R. 1971. The goodness of parasitism: a new hypothesis. In *Aspects of The Biology of Symbiosis*, T. Cheng (ed.), pp. 139–228. University Park Press, Baltimore. (8)

Linhart, Y. B. 1973. Ecological and behavioral determinants of pollen dispersal in hummingbird-pollinated *Heliconia*. Amer. Natur. 107: 511–523. (13)

506

Linhart, Y. B., R. B. Mitton, K. B. Sturgeon and M. L. Davis. 1981. Genetic variation in space and time in a population of ponderosa pine. Heredity 46: 407–426. (13)

Linsley, E. G. 1958. The ecology of solitary bees. Hilgardia 27: 543–599. (13)

Linsley, E. G. 1978. Temporal patterns of flower visitation by solitary bees, with particular reference to the southwestern United States. J. Kansas Entomol. Soc. 51: 531–546. (13)

Linsley, E. G. and P. D. Hurd. 1959. Ethological observations on some bees of southeastern Arizona and New Mexico (Hymenoptera: Apoidea). Entomologists' News 70: 63–68. (13)

Linsley, E. G. and J. W. MacSwain. 1958. The significance of floral constancy among bees of the genus *Diadasia* (Hymenoptera, Anthophoridae). Evolution 12: 219–223. (13)

Linsley, E. G. and J. S. MacSwain. 1959. Ethology of some *Ranunculus* insects with emphasis on competition for pollen. Univ. Calif. Publ. Entomol. 16: 1–46. (13)

Lister, B. 1976. The nature of niche expansion in West Indian *Anolis* lizards II. Evolutionary components. Evolution 30: 677–692. (17)

Livingston, R. B. 1972. Influence of birds, stones and soil on the establishment of pasture juniper, *Juniperus communis*, and red cedar, *J. virginiana*, in New England pastures. Ecology 53: 1141–1147. (11)

Llewellyn, J. 1956. The host-specificity, micro-ecology, adhesive attitudes and comparative morphology of some trematode gill parasites. J. Mar. Biol. Assn. U.K. 35: 113–127. (8)

Llewellyn, J. 1957. Host specificity in monogenetic trematodes. In *Premier Symposium sur la Specificite parasitaire des parasites de Vertébrés*, J. G. Baer (ed.), pp. 199–212. Int. Union Biol. Sci., Ser. B, No. 32, Paris. (8)

Lloyd, D. G. 1979. Parental strategies of angiosperms. N.Z. J. Bot. 17: 595–606. (13)

Lloyd, H. G. 1968. Observations on nut selection by a hand-reared grey squirrel (*Sciurus carolinensis*). J. Zool. 155: 240–244. (11, 13)

Lloyd, J. E. 1965. Aggressive mimicry in *Photuris*: firefly femmes fatales. Science 149: 653–654. (12)

Lockard, R. B. and J. S. Lockard. 1971. Seed preference and buried seed retrieval of *Dipodomys deserti*. J. Mammal. 52: 219–222. (11)

Long, C. A. 1976. Evolution of mammalian cheek pouches and a possible discontinuous origin of a higher taxon (Geomyoidea). Amer. Natur. 110: 1093–1097. (11)

Long, E., M. Levin, G. Targett and M. Doenhoff. 1981. Factors affecting the acquisition of resistance against *Schistosoma mansoni* in the mouse. VIII. Failure of concurrent infections with *Plasmodium chabaudi* to affect resistance to reinfection with *S. mansoni*. Ann. trop. Med. Parasitol. 75: 79–86. (8)

Losey, G. S., Jr. 1974. Cleaning symbiosis in Puerto Rico with comparison to the tropical Pacific. Copeia 1974: 960–970. (3, 14)

Losey, G. S., Jr. 1978. The symbiotic behavior of fishes. In *The Behavior of Fish and Other Aquatic Animals*, D. I. Mostofsky (ed.), pp. 1–31. Academic Press, New York. (3)

Lu, K. L. and M. R. Mesler. 1981. Ant dispersal of a neotropical forest floor gesneriad. Biotropica 13: 159–160. (11)

Lubchenco, J. 1978. Plant species diversity in a marine intertidal community: importance of herbivore food preference and algal competitive abilities. Amer. Natur. 112: 23–39. (19)

Lubchenco, J. and S. D. Gaines. 1981. A unified approach to marine plant-herbivore interactions. I. Populations and communities. Ann. Rev. Ecol. Syst. 12: 405–437. (19)

Lundberg, J. G. 1972. Wagner networks and ancestors. Syst. Zool. 21: 398–413. (4)

Luria, S. E. 1945. Mutations of bacterial viruses affecting their host range. Genetics 30: 84–99. (5)

Luria, S. E. and M. Delbrück. 1943. Mutations of bacteria from virus sensitivity to virus resistance. Genetics 28: 491–511. (5)

Lwoff, A. 1953. Lysogeny. Bacter. Rev. 17: 269–337. (5)

Lyell, C. 1830. *Principles of Geology*, Vol. 1, John Murray, London. (16)

Lyell, K. 1881. *Life, Letters, and Journals of Sir Charles Lyell*, Bart., Vol. I. John Murray, London. (1)

Lyon, D. G. and C. Chadek. 1971. Exploitation of nectar resources by hummingbirds, bees (*Bombus*), and *Diglossa baritula* and its role in the evolution of *Penstemon kunthii*. Condor 73: 246–248. (13)

Mabry, T. J. and J. E. Gill. 1979. Sesquiterpene lactones and other terpenoids. In *Herbivores: Their Interaction with Secondary Plant Metabolites*, G. A. Rosenthal and D. H. Janzen (eds.), pp. 501–537. Academic Press, New York. (10)

MacArthur, R. H. 1962. Some generalized theorems of natural selection. Proc. Natl. Acad. Sci. USA 48: 1893–1897. (3)

MacArthur, R. H. 1971. Patterns of terrestrial bird communities. In *Avian Biology*, Vol. 1, D. S. Farner and J. King (eds.), pp. 189–221. Academic Press, New York. (19)

MacArthur, R. H. 1972. *Geographical Ecology*. Harper & Row, New York. (1, 18)

MacArthur, R. H. and R. Levins. 1967. The limiting similarity, convergence and divergence of coexisting species. Amer. Natur. 101: 377–385. (18)

MacArthur, R. H. and J. W. MacArthur. 1961. On bird species diversity. Ecology 42: 594–598. (19)

MacArthur, R. H., J. W. MacArthur and J. Preer. 1962. On bird species diversity. II. Predictions of bird censuses from habitat measurements. Amer. Natur. 96: 167–174. (19)

MacArthur, R. H. and E. R. Pianka. 1966. On optimal use of a patchy environment. Amer. Natur. 100: 603–609. (2)

MacArthur, R. H., H. Recher and M. Cody. 1966. On the relation between habitat selection and species diversity. Amer. Natur. 100: 319–332. (19)

Macdonald, W. W. and C. P. Ramachandran. 1965. The influence of the gene f (filarial susceptibility, *Brugia malayi*) on the susceptibility of *Aedes aegypti* to seven strains of *Brugia*, *Wuchereria* and *Dirofilaria*. Ann. Trop. Med. Parasitol. 59: 64–73. (8)

MacInnis, A. J. 1976. How parasites find hosts: some thoughts on the inception of host-parasite integration. In *Ecological Aspects of Parasitology*, C. R. Kennedy (ed.), pp. 3–20. North-Holland Publ. Co., Amsterdam. (8)

MacIntosh, N. W. G. 1975. The Origin of the Dingo. In *The Wild Canids*, M. W. Fox (ed.). Van Nostrand Reinhold, New York. (16)

Macior, L. W. 1968. Pollination adaptation in *Pedicularis groenlandica*. Amer. J. Botany 55: 927–932. (13)

Macior, L. W. 1971. Coevolution of plants and animals—systematic insight from plant-insect interaction. Taxon 20: 17–28. (13)

Macior, L. W. 1978. Pollination ecology of vernal angiosperms. Oikos 30: 452–460. (13)

Madison, M. 1979. Protection of developing seeds in neotropical Araceae. Aroideana 2: 52–61. (11)

Magure, Y. P. and N. F. Haard. 1975. Fluorescent product accumulation in ripening fruit. Nature 258: 599–600. (11)

Malogolowkin, C. 1958. Maternally inherited "sex-ratio" conditions in *Drosophila willistoni* and *Drosophila paulistorum*. Genetics 43: 276–286. (6)

Malogolowkin, C., G. G. Carvalho and M. C. Da Paz. 1960. Interspecific transfer of the "sex-ratio" condition in *Drosophila*. Genetics 45: 1553–1557. (6)

Malogolowkin, C. and D. F. Poulson. 1957. Infective transfer of maternally inherited abnormal sex-ratio in *Drosophila willistoni*. Science 126: 32. (6)

Malogolowkin, C., D. F. Poulson and E. Y. Wright. 1959. Experimental transfer of maternally inherited sex-ratio in *Drosophila willistoni*. Genetics 44: 59–74. (6)

Mann, K. H. 1973. Seaweeds: their productivity and strategy of growth. Science 182: 975–981. (19)

Manter, H. W. 1955. The zoogeography of trematodes of marine fishes. Exp. Parasitol. 4: 62–86. (4, 8)

Manter, H. W. 1966. Parasites of fishes as biological indicators of recent and ancient conditions. In *Host-Parasite Relationships*, J. E. McCauley (ed.), pp. 59–71. Proc. 26th Biology Colloq., Oregon State University Press, Corvallis. (4)

Marchant, J. 1916. *Alfred Russel Wallace: Letters and Reminiscences.* Harper and Bros., New York. (16)

Mares, M. A. 1976. Convergent evolution of desert rodents: multivariate analysis and zoogeographic implications. Paleobiology 2: 39-63. (19)

Margolis, L. and J. R. Arthur. 1979. *Synopsis of the Parasites of Fishes of Canada.* Fish. Res. Bd. Can., Bull. 199. Ottawa. (8)

Margulis, L. 1970. *Origin of Eukaryotic Cells.* Yale University Press, New Haven. (6)

Margulis, L. 1974. Introduction to origin and evolution of the eukaryotic cell. Taxon 23: 225-226. (6)

Margulis, L. 1976. Genetic and evolutionary consequences of symbiosis. Experimental Parasitology 39: 277-349. (6)

Mariscal, R. N. 1971. Experimental studies on the protection of anemone fishes from sea anemones. In *Aspects of the Biology of Symbiosis*, J. C. Cheng (ed.), pp. 283-315. University Park Press, Baltimore. (14, 18)

Marques, E. J. and L. E. de Magalhaes. 1973. The frequency of SR female of *Drosophila nebulosa* in a natural population. Dros. Inf. Serv. 50: 87. (6)

Marsh, J. A. 1970. Primary productivity of reef building calcareous red algae. Ecology 51: 255-263. (15)

Marshall, G. A. K. 1908. On diaposematism, with reference to some limitations of the Müllerian hypothesis of mimicry. Trans. Ent. Soc. Lond. 1908: 93-142. (12)

Martin, D. R. 1969. Lecithodendriid trematodes from the bat, *Peropteryx kappleri* in Columbia, including discussions of allometric growth and significance of ecological isolation. Proc. Helm. Soc. Wash. 36: 250-260. (8)

Marx, D. H. 1972. Ectomycorrhizae as biological deterrents to pathogenic root infections. Ann. Rev. Phytopath. 10: 429-454. (7)

Massieu, G., A. R. Quiles and R. O. Cravioto. 1956. Nuevos datos sobre el contenido de vitamina C de *Malpighia puricifolia* L., procedente de Yucatan. Ciencia (Mexico) 15: 206-207. (11)

Matessi, C. and R. Cori. 1972. Models of population genetics of Batesian mimicry. Theor. Pop. Biol. 3: 41-68. (3)

Matessi, C. and S. Jayakar. 1976. Model of density-frequency-dependent selection for the exploitation of resources. I. Intraspecific competition. In *Population Genetics and Ecology*, S. Karlin and E. Nevo (eds.), pp. 707-721. Academic Press, New York. (17)

Mather, K. 1955. Polymorphism as an outcome of disruptive selection. Evolution 9: 52-61. (10)

Matthew, W. D. 1937. Paleocene faunas of the San Juan Basin, New Mexico. Amer. Philos. Soc. 30. (16)

Maugh, T. H. 1981. Why does sickle trait persist? Science 211: 266. (8)

Mauzey, K. P., C. Birkland and P. K. Dayton. 1968. Feeding behavior of asteroids and escape responses of their prey in the Puget Sound region. Ecology 49: 603-619. (19)

Maxwell, F. G. and P. R. Jennings (eds.). 1980. *Breeding Plants Resistant to Insects.* John Wiley & Sons, New York. (10)

May, R. M. 1973. *Stability and Complexity in Model Ecosystems.* Princeton University Press, Princeton, New Jersey. (1, 3)

May, R. M. 1979. Bifurcations and dynamic complexity in ecological systems. Ann. N.Y. Acad. Sci. 316: 517-529. (9)

May, R. M. 1982a. Parasitic infections as regulators of animal populations. Amer. Sci. 71 (in press). (9)

May, R. M. 1982b. Introduction. In *Population Biology of Infectious Diseases*, R. M. Anderson and R. M. May (eds.), pp. 1-12. Springer Verlag, New York. (9)

May, R. M. and R. M. Anderson. 1978. Regulation and stability of host-parasite population interactions: II. Destabilizing processes. J. Anim. Ecol. 47: 249-267. (3, 9)

May, R. M. and R. M. Anderson. 1979. Population biology of infectious diseases: II. Nature 280: 455-461. (9)

May, R. M. and R. M. Anderson. 1983. Frequency and density dependent effects in the coevolution of hosts and parasites. Proc. Roy. Soc. Lond. B (in press). (9)

509

May, R. M. and W. J. Leonard. 1975. Nonlinear aspects of competition between three species. SIAM J. Appl. Math 29: 243–253. (3)

Maynard Smith, J. 1964. Group selection and kin selection. Nature 201: 1145–1147. (2)

Maynard Smith, J. 1966. Sympatric speciation. Amer. Natur. 100: 637–650. (10)

Maynard Smith, J. 1968. *Mathematical Ideas in Biology*. Cambridge University Press, Cambridge. (2, 19)

Maynard Smith, J. 1976. A comment on the Red Queen. Amer. Natur. 110: 325–330. (Ep)

Maynard Smith, J. 1978. *The Evolution of Sex*. Cambridge University Press, Cambridge. (9)

Maynard Smith, J. 1981. Will a sexual population evolve to an ESS? Amer. Natur. 117: 1015–1018. (2)

Maynard Smith, J. and G. R. Price. 1973. The logic of animal conflict. Nature 248: 15–18. (2, 9)

Mayr, E. 1947. Ecological factors in speciation. Evolution 1: 263–288. (10)

Mayr, E. 1963. *Animal Species and Evolution*. Harvard University Press, Cambridge, Massachusetts. (2)

Mazingue, C., D. Camus, J. P. Dessaint, M. Capron and A. Capron. 1980. In vitro and in vivo inhibition of mast cell degranulation by a factor from *Schistosoma mansoni*. Int. Arch. Allergy Appl. Immunol. 63: 178–189. (8)

McAtee, W. L. 1906. Virginia creeper as a winter food for birds. Auk 23: 346–347. (11)

McAtee, W. L. 1947. Distribution of seeds by birds. Amer. Mid. Nat. 38: 214–223. (11)

M'Closkey, R. T. 1980. Spatial patterns in sizes of seeds collected by four species of heteromyid rodents. Ecology 61: 486–489. (11)

McClure, H. E. 1966. Flowering, fruiting and animals in the canopy of a tropical rain forest. Malayan Forester 29: 182–203. (11)

McConnell, D. J., D. G. Searcy and J. G. Sutcliffe. 1978. A restriction enzyme, Thal, from the therophilic mycoplasma, *Thermoplasma acidophilium*. Nucl. Acid. Res. 5: 1729–1739. (5)

McDade, L. A. and S. Kinsman. 1980. The impact of floral parasitism in two neotropical hummingbird-pollinated plant species. Evolution 34: 944–958. (13)

McDiarmid, R. W., R. E. Ricklefs and M. S. Foster. 1977. Dispersal of *Stemmadenia donnell-smithii* (Apocynaceae) by birds. Biotrop. 9: 9–25. (11)

McKey, D. 1975. The ecology of coevolved seed dispersal systems. In *Coevolution of Animals and Plants*, L. E. Gilbert and P. H. Raven (eds.), pp. 159–191. University of Texas Press, Austin. (11)

McKey, D. 1979. The distribution of secondary compounds within plants. In *Herbivores: Their Interaction with Secondary Plant Metabolites*, G. A. Rosenthal and D. H. Janzen (eds.), pp. 55–133. Academic Press, New York. (10)

McLaughlin, J. J. A. and P. A. Zahl. 1966. Endozoic algae. In *Symbiosis*, Vol. I, S. M. Henry (ed.), pp. 257–297. Academic Press, New York. (14)

McNeill, S. and T. R. E. Southwood. 1978. The role of nitrogen in the development of insect/plant relationships. In *Biochemical Aspects of Plant and Animal Coevolution*, J. B. Harborne (ed.), pp. 77–98. Academic Press, New York. (19)

McNeill, W. H. 1976. *Plagues and Peoples*. Doubleday, New York. (9)

McVittie, R. 1979. Changes in the social behaviour of South Western African cheetah. Madoqua 2(3): 179–184. (15)

Mead-Briggs, A. R. 1977. The European rabbit, the European rabbit flea and myxomatosis. In *Applied Biology*, II, T. H. Coaker (ed.), pp. 183–261. Academic Press, London. (7)

Mead-Briggs, A. R. and J. A. Vaughan. 1975. The differential transmissibility of myxoma virus strains of differing virulence grades by the rabbit flea *Spilopsyllus cuniculi* (Dale). J. Hyg. 75: 237–247. (9)

Mech, L. D. 1970. *The Wolf*. Natural History Press, New York. (16)

Medway, L. 1972. Phenology of a tropical rain forest in Malaya. Biol. J. Linn. Soc. 4: 117–146. (11)

Mees, G. F. 1970. Amendment to Grant's paper on species-pairs of birds. Syst. Zool. 19: 196–198. (18)

510

Meeuse, A. D. J. 1973. Co-evolution of plant hosts and their parasites as a taxonomic tool. In *Taxonomy and Ecology*, V. H. Heywood (ed.), pp. 289–316. Academic Press, London. (7)

Meijer, W. 1969. Fruit trees in Sabah (North Borneo). Malayan Forester 32: 252–265. (11)

Mellett, J. S. 1977. Paleobiology of North American *Hyaenodon* (Mammalia, Creodonta). Contrib. Vert. Evol. S. Karger, Basel. (16)

Mellor, P. S. 1974. Studies on *Onchocerca servicalis* Railliet and Henry, 1910: IV. Behaviour of the vector *Culicoides tuberculosus* in relation to the transmission of *Onchocerca cervicalis*. J. Helminth. 48: 283–288. (8)

Menge, B. A. 1978a. Predation intensity in a rocky intertidal community: relation between predator foraging activity and environmental harshness. Oecologia 34: 1–16. (14)

Menge, B. A. 1978b. Predation intensity in a rocky intertidal community: effect of an algal canopy, wave action and dessication on predator feeding rates. Oecologia 34: 17–35. (14)

Mertz, D. B. 1971. Life history phenomena in increasing and decreasing populations. In *Statistical Ecology, Vol. 2, Sampling and Modeling Biological Populations and Population Dynamics*, G. P. Patil, E. C. Pielou and W. E. Waters (eds.), pp. 361–392. Pennsylvania State University Press, University Park. (8)

Metcalf, M. 1929. Parasites and the aid they give in problems of taxonomy, geographical distribution, and paleogeography. Smithson. Misc. Coll. 81: 1–36. (4)

Meyer, F. H. 1966. Mycorrhyza and other plant symbioses. In *Symbiosis*, Vol. I, S. M. Henry (ed.), pp. 171–255. Academic Press, New York. (7)

Meynell, G. G. 1973. *Bacterial Plasmids*. MIT Press, Cambridge, Massachusetts. (5)

Michener, C. D. 1965. A classification of the bees of the Australian and South Pacific regions. Bull. Amer. Mus. Nat. Hist. 130: 1–362. (13)

Mickevich, M. 1978. Comments on the recognition of convergence and parallelism on Wagner trees. Syst. Zool. 27: 239–242. (4)

Mickevich, M. 1981. Quantitative phylogenetic biogeography. In *Advances in Cladistics. Proceedings of the First Meeting of the Willi Hennig Society*. V. A. Funk and D. R. Brooks (eds.), pp. 209–222. New York Botanical Garden, New York. (4)

Miller, P. L. 1979. Quantitative plant ecology. In *Analysis of Ecological Systems*, D. J. Horn, R. D. Mitchell and G. R. Stairs (eds.), pp. 179–231. Ohio State University Press, Columbus. (19)

Miller, R. B. 1978. The pollination ecology of *Aquilegia elegantula* and *A. caerulea* (Ranunculaceae) in Colorado. Amer. J. Botany 65: 406–414. (13)

Miller, R. B. 1981. Hawkmoths and the geographic patterns of floral variation in *Aquilegia caerulea*. Evolution 35: 763–774. (13)

Milton, K. 1980. *The Foraging Strategy of Howler Monkeys*. Columbia University Press, New York. (11)

Milton, K. and F. R. Dintzis. 1981. Nitrogen-to-protein conversion factors for tropical plant samples. Biotropica 13: 177–181. (11)

Misra, K. and T. R. Seshadri. 1968. Chemical composition of the fruits of *Psidium guava*. Phytochemistry 7: 64–65. (11)

Mitchell, G. F. 1979a. Effector cells, molecules and mechanisms in host-protective immunity to parasites. Immunology 38: 209–223. (8)

Mitchell, G. F. 1979b. Responses to infection with metazoan and protozoan parasites in mice. Adv. Immunol. 28: 451–511. (8)

Mitchell, G. F. and E. Handman. 1977. Studies on immune responses to larval cestodes in mice: a simple mechanism of non-specific immunosuppression in *Mesocestoides corti*-infected mice. Austr. J. Exp. Biol. Med. Sci. 55: 615–622. (8)

Mitchell, R. D. 1981. Insect behavior, resource exploitation, and fitness. Ann. Rev. Entomol. 26: 373–396. (10)

511

Mitsuhashi, S. 1971. Epidemiology of bacterial drug resistance. In *Transferable Drug Resistance Factor*, S. Mitsuhashi (ed.), pp. 1–23. University Park Press, Baltimore. (5)

Mitter, C. 1980. The thirteenth annual Numerical Taxonomy Conference. Syst. Zool. 29: 177–190. (4)

Mitter, C. and D. J. Futuyma. 1979. Population genetic consequences of feeding habits in some forest Lepidoptera. Genetics 92: 1005–1021. (10)

Mitter, C. and D. J. Futuyma. 1983. An evolutionary-genetic view of host plant utilization by insects. In *Herbivores and their Variable Hosts in Natural and Managed Systems*, R. F. Denno and M. S. McClure (eds.), Academic Press, New York. (In press) (4, 10)

Mitter, C., D. J. Futuyma, J. C. Schneider and J. D. Hare. 1979. Genetic variation and host plant relations in a parthenogenetic moth. Evolution 33: 777–790. (10)

Miyake, S. 1978. *The Crustacean Anomura of Sagami Bay*. Biological Laboratory, Imperial Household, Tokyo. (14)

Mode, C. J. 1958. A mathematical mode for the co-evolution of obligate parasites and their hosts. Evolution 12: 158–165. (3, 7, 9)

Mode, C. J. 1961. A generalized model of a host-pathogen system. Biometrics 17: 386–404. (7, 9)

Moe, R. L. and P. C. Silva. 1977. Antarctic marine flora: uniquely devoid of kelps. Science 196: 1206–1208. (19)

Mogford, D. J. 1978. Pollination and flower colour polymorphism, with special reference to *Cirsium palustre*. In *The Pollination of Flowers by Insects*, A. J. Richards (ed.), pp. 191–199. Academic Press, London. (13)

Molez, N. 1976. Adaptation alimentaire du galago d'Allen aux milieux forestiers secondaires. La Terre et la Vie 30: 210–228. (11)

Monod, J. 1949. The growth of bacterial cultures. Ann. Rev. Microbiol. 3: 371–394. (5)

Moodie, R. L. 1967. General considerations of the evidences of pathological conditions found among fossil animals. In *Diseases in Antiquity*, D. Brothwell and A. T. Sandison (eds.), pp. 31–46. C. C. Thomas, Springfield, Illinois. (9)

Mooney, H. A. 1977. *Convergent Evolution in Chile and California*. Dowden, Hutchinson & Ross, Stroudsburg, Pennsylvania. (19)

Mooney, H. A. and E. L. Dunn. 1970. Convergent evolution of Mediterranean-climate evergreen sclerophyll shrubs. Evolution 24: 292–303. (19)

Mooney, H. A., A. T. Harrison and P. A. Morrow. 1975. Environmental limitations of photosynthesis on a California evergreen shrub. Oecologia 19: 293–301. (19)

Moore, J. M. 1952. Competition between *Drosophila melanogaster* and *Drosophila simulans*. II. The improvement of competitive ability through selection. Proc. Natl. Acad. Sci. USA 38: 813–817. (1)

Moqbel, R. and D. Wakelin. 1979. *Trichinella spirallis* and *Strongyloides ratti*: immune interaction in adult rats. Exp. Parasitol. 47: 65–72. (8)

Moran, V. C. 1980. Interactions between phytophagous insects and their *Opuntia* hosts. Ecol. Entomol. 5: 153–164. (10)

Morgan-Davies, A. M. 1960. The association between impala and olive baboon. J. East African Natural History Society 23: 297–298. (11)

Morris, D. 1962. The behavior of the green acouchi (*Myoprocta pratti*) with special reference to scatter hoarding. Proc. Zool. Soc. London 139: 701–702. (11)

Morrison, D. 1975. The foraging behavior and feeding ecology of a neotropical fruit bat, *Artibeus jamaicensis*. Ph.D. Thesis, Cornell University, Ithaca, New York. (11)

Morrison, D. W. 1978a. Foraging ecology and energetics of the frugivorous bat *Artibeus jamaicensis*. Ecology 59: 716–723. (11)

Morrison, D. W. 1978b. Lunar phobia in a neotropical fruit bat, *Artibeus jamaicensis* (Chiroptera: Phyllostomidae). Animal Behaviour 26: 852–855. (11)

Morrison, D. W. 1978c. On the optimal searching strategy for refuging predators. Amer. Natur. 112: 925–934. (11)

Morrison, D. W. 1978d. Influence of habitat on the foraging distance of the fruit bat, *Artibeus jamaicensis*. J. Mammal. 59: 622–624. (11)

Morrison, D. W. 1979. Apparent male defense of tree hollows in the fruit bat, *Artibeus jamaicensis*. J. Mammal. 60: 11–15. (11)

512

Morrison, D. W. 1980. Foraging and day-roosting dynamics of canopy fruit bats in Panama. J. Mammal. 61: 20–29. (11)

Morse, D. E., N. Hooker, H. Duncan and L. Jensen. 1980. γ-Aminobutyric acid, a neurotransmitter, induces planktonic abalone larvae to settle and begin metamorphosis. Science 204: 407–410. (19)

Morton, E. S. 1973. On the evolutionary advantages and disadvantages of fruit-eating in tropical birds. Amer. Natur. 107: 8–22. (11)

Mosimann, J. E. and F. C. James. 1979. New statistical methods for allometry with application to Florida red-winged blackbirds. Evolution 33: 444–459.

Motten, A. F., D. R. Campbell, D. E. Alexander and H. L. Miller. 1981. Pollination effectiveness of specialist and generalist visitors to a North Carolina population of *Claytonia virginica*. Ecology 62: 1278–1287. (13)

Motulsky, A. G. 1975. Glucose-6-phosphate dehydrogenase and abnormal haemoglobin polymorphism—evidence regarding malarial selection. In *The Role of Natural Selection in Human Evolution*, F. M. Salzano, (ed.), pp. 271–291. North-Holland Publ. Co., Amsterdam. (8)

Moulton, J. C. 1909. On some of the principal mimetic (Müllerian) combinations of tropical butterflies. Trans. Ent. Soc. Lond. 1909: 585–606. (12)

Mulcahy, D. L. 1979. The rise of the angiosperms: a genecological factor. Science 206: 20–23. (13, 14)

Müller, F. 1879. *Ituna* and *Thyridis*; a remarkable case of mimicry in butterflies. (translated by R. Mendola). Proc. Ent. Soc. London 1879: 20–29. (1, 12)

Müller, F. P. 1971. Isolationsmechanismen zwischen sympatrischen bionomischen Rassen am Beispiel der Erbsenblattlaus *Acyrthosiphon pisum* (Harris) (Homoptera: Aphididae). Zool. J. Syst. 98: 131–152. (10)

Müller, F. P. 1980. Wirtspflanzen, Generationenfolge und reproduktive Isolation infraspezifischer Formen von *Acyrthosiphon pisum*. Ent. exp. appl. 28: 145–157. (10)

Munger, J. C. and J. H. Brown. 1981. Competition in desert rodents: an experiment with semipermeable exclosures. Science 211: 510–512. (11, 18)

Munroe, E. and P. R. Ehrlich. 1960. Harmonization of concepts of higher classification of the Papilionidae. J. Lep. Soc. 14: 169–175. (4)

Munsell, H. E., L. O. Williams, L. P. Guild, C. B. Troescher, G. Nightingale and R. S. Harris. 1949. Composition of food plants of Central America. I. Honduras. Food Research 14: 144–164. (11)

Murgatroyd, J. G. 1970. Yellow-billed hornbill brings an egg to nest. Honeyguide No. 63: 25. (11)

Murie, A. 1940. Ecology of the coyote. In *The Yellowstone*, Fauna Series No. 4, Conservation Bulletin No. 4. U.S. Dept. of Interior National Park Service, Washington, D.C. (15)

Murie, J. O. 1977. Cues used for cache-finding by agoutis (*Dasyprocta punctata*). J. Mammal. 58: 95–96. (11)

Murray, J., A. Murray, M. Murray and C. Murray. 1978. The biological suppression of malaria: an ecological and nutritional interrelationship of a host and two parasites. Am. J. Clin. Nutr. 31: 1363–1366. (8)

Murton, R. K. 1971. The significance of specific search image in the feeding behavior of the wood-pigeon. Behavior 40: 10–42 (10)

Muskin, A. and A. J. Fischgrund. 1981. Seed dispersal of *Stemmadenia* (Apocynaceae) and sexually dimorphic feeding strategies by *Ateles* in Tikal, Guatemala. Biotropica (Supplement) 13: 78–80. (11)

Mutere, J. A. 1973. On the food of the aegyptian fruit bat, *Rousettus aegyptiacus* E. Geoffroy. Period. Biol. 75: 159–162. (11)

Nagy, S. and P. E. Shaw (eds.). 1980. *Tropical and Subtropical Fruits*. AVI Publishing Co., Westport, Connecticut. (11)

Nagylaki, T. 1975. Conditions for the existence of clines. Genetics 80: 595–615. (3)

Nakamura, R. R. 1980. Plant kin selection. Evolutionary Theory 5: 113–118. (13)

Neill, W. T. 1971. *The Last of the Ruling Reptiles.* Columbia University Press, New York. (4)

Nelson, G. 1974. Historical biogeography: an alternative formalization. Syst. Zool. 23: 555–558. (4)

Nelson, G. and N. I. Platnick. 1981. *Systematics and Biogeography: Cladistics and Vicariance.* Columbia University Press, New York. (4)

Nelson, J. E. 1965. Movements of Australian flying foxes (Pteropidae: Megachiroptera). Aust. J. Zool. 13: 53–73. (11)

Nelson, R. R. 1979. Some thoughts on the coevolution of plant pathogenic fungi and their hosts. In *Host-Parasite Interfaces*, B. B. Nickol (ed.), pp. 17–25. Academic Press, New York. (8)

Newcomb, E. H. and T. D. Pugh. 1975. Blue-green algae associated with ascidians of the Great Barrier Reef. Nature 253: 533–534. (14)

Newman, W. A. 1960. The paucity of intertidal barnacles in the tropical Western Pacific. Veliger 2: 89–94. (14)

Newman, W. A. 1979. On the biogeography of balanomorph barnacles of the southern ocean including new balanid taxa, a subfamily, two genera and three species. Proc. Int. Symp. Mar. Biog. Ecol. Vol 1: 279–306. (19)

Newman, W. A. and H. S. Ladd. 1974. Origin of coral-inhabiting balanids (Cirripedia, Thoracica). Verhandl. Naturf. Ges. Basel. 84: 381–396. (14)

Newman, W. A. and A. Ross. 1976. Revision of the balanomorph barnacles; including a catalog of the species. Mem. San Diego Soc. Nat. Hist. 9: 1–108. (14)

Newman, W. A. and S. M. Stanley. 1981. Competition wins out overall. Paleobiol. 7: 561–569. (15)

Newton, M. and A. M. Brown. 1930. A study of the inheritance of spore colour and pathogenicity in crosses between physiologic forms of *Puccinia graminis*. Sci. Agr. 10: 775–798. (7)

Ng, F. S. P. 1975. Germination of fresh seeds of Malaysian trees. II. Malaysian Forester 38: 171–176. (11)

Ng, F. S. P. 1978. Strategies of establishment in Malayan forest trees. In *Tropical Trees as Living Systems*, P. B. Tomlinson and M. H. Zimmerman (eds.), pp. 129–162. Cambridge University Press, Cambridge. (11)

Ng, F. S. P. 1980. Germination ecology of Malaysian woody plants. Malaysian Forester 43: 406–437. (11)

Nicholson, A. J. 1927. A new theory of mimicry in insects. Austral. Zool. 5: 10. (12)

Niederhauser, J. S. 1961. Genetic studies of *Phytophthera infestans* and *Solanum* species in relation to late-blight resistance in the potato. Proceedings of IX Internat. Botanical Congress, Montreal, 1959. Recent Advances Bot. Vol. I, (University of Toronto Press), pp. 491–497. (7)

Nielsen, J. K. 1978. Host plant discrimination within Cruciferae: feeding responses of four leaf beetles (Coleoptera; Chrysomelidae) to glucosinolates, cucurbitacins, and cardenolides. Ent. exp. appl. 24: 41–54. (10)

Nielson, M. W. and H. Don. 1974. Probing behavior of biotypes of the spotted alfalfa aphid on resistant and susceptible clones. Ent. exp. appl. 17: 477–486. (10)

Noack, D. 1968. A regulatory model for steady-state conditions in populations of lysogenic bacteria. J. Theor. Biol. 18: 1–18. (5)

Noble, E. R. and G. A. Noble. 1976. *Parasitology: The Biology of Animal Parasites*, 4th Ed. Lea and Febiger, Philadelphia. (1, 2, 8)

Noble, J. C. 1975a. The effects of emus (*Dromeius novaehollandiae* Latham) on the distribution of the nitre bush (*Nitraria billardieri* DC.) J. Ecol. 63: 979–984. (11)

Noble, J. C. 1975b. Difference in size of emus on two contrasting diets on the riverside plain of New South Wales. Emu 75: 35–37. (11)

Norris, D. M. and M. Kogan. 1980. Biochemical and morphological bases of resistance. In *Breeding Plants Resistant to Insects*, F. G. Maxwell and P. R. Jennings (eds.), pp. 23–61. John Wiley & Sons, New York. (10)

Norton, T. A. 1976. Why is *Sargassum muticum* so invasive? Brit. Phycol. Jour. 11: 197–198. (19)

Novick, R. P. 1969. Extrachromosomal inheritance in bacteria. Bact. Revs. 33: 217–225. (5)

Novick, R. P. 1974. Bacterial plasmids. In *Handbook of Microbiology*, Vol. IV, A. L. Laskin and H. A. Lechevalier (eds.), pp. 537–586. CRC Press, Cleveland. (5)

Nur, U. 1970. Evolutionary rates of models and mimics in Batesian mimicry. Amer. Natur. 104: 477–486. (3)

O'Dowd, D. J. and M. E. Hay. 1980. Mutualism between harvester ants and a desert ephemeral: seed escape from rodents. Ecology 61: 531–540. (11)

Ogilvie, B. M. 1974. Antigenic variation in the nematode *Nippostrongylus brasiliensis*. In *Parasites in the Immunized Host: Mechanisms of Survival*, pp. 81–90. CIBA Foundation Symp. No. 25. (8)

Ognev, B. 1963. *Mammals of the U.S.S.R.* Nauk, Moscow. (16)

Oishi, K. 1971. Spirochete-mediated abnormal sex-ratio (SR) condition in *Drosophila*: a second virus associated with spirochetes and its use in the study of the SR condition. Genet. Res. 18: 45–56. (6)

Oishi, K. and D. F. Poulson. 1970. A virus associated with SR-spirochetes of *Drosophila nebulosa*. Proc. Natl. Acad. Sci. USA 67: 1565–1572. (6)

Olson, S. L. and K. E. Blum. 1968. Avian dispersers of plants in Panama. Ecology 49: 565–566. (11)

Opler, P. A. 1974. Oaks as evolutionary islands for leaf-mining insects. Amer. Sci. 62: 67–73. (10)

Oppenheimer, J. R. 1977. Forest structure and its relation to activity of the capuchin monkey (*Cebus*). In *Use of Non-human Primates in Biomedical Research*, M. R. N. Prasad and T. C. A. Kumar (eds.), pp. 74–84. Indian Nat. Sci. Acad., New Delhi, India. (11)

Orenstein, R. I. 1973. Colourful plumage in tropical birds. Avicultural Magazine 79: 119–122. (11)

Orians, G. H. 1969. The number of bird species in some tropical forests. Ecology 50: 783–801. (19)

Orians, G. H. and O. T. Solbrig (eds.). 1977. *Convergent Evolution in Warm Deserts*. Dowden, Hutchinson & Ross, Stroudsburg, Pennsylvania (1, 19)

Osborn, H. F. 1909. *The Age of Mammals*. Macmillan, New York. (16)

Osman, R. W. and J. A. Haugsness. 1981. Mutualism among sessile invertebrates: a mediator of competition and predation. Science 211: 846–848. (3, 14, 19)

Osmaston, H. A. 1965. Pollen and seed dispersal in *Chlorophora excelsa* and other Moraceae, and in *Parkia filicoidea* (Mimosaceae) with special reference to the role of the fruit bat, *Eidolon helrum*. Commonw. For. Rev. 44: 97–105. (11)

Oster, G. F., A. Ipaktchi and S. Rocklin. 1976. Phenotypic structure and bifurcation behavior of population models. Theor. Pop. Biol. 10: 365–382. (9)

Ostler, W. K. and K. T. Harper. 1978. Floral ecology in relation to plant species diversity in the Wasatch Mountains of Utah and Idaho. Ecology 59: 848–861. (13)

Ottesen, E. A. and R. W. Poindexter. 1980. Modulation of the host response in human schistosomiasis. II. Humoral factors which inhibit lymphocyte proliferative responses to parasite antigens. Am. J. Trop. Med. Hyg. 29: 592–597. (8)

Ouchi, S., C. Hibino, H. Oku, M. Fujiwara and H. Nakbayashi. 1979. The induction of resistance or susceptibility. In *Recognition and Specificity in Plant Host-Parasite Interactions*, J. M. Daly and I. Uritani (eds.), pp. 49–65. Japan Scientific Societies Press, Tokyo and University Park Press, Baltimore. (7)

Owen, D. F. and D. O. Chanter. 1968. Population biology of tropical African butterflies. 2. Sex ratio and polymorphism in *Damaus chrysippus*. Revue Zool. Bot. Afr. 78: 81–97. (12)

515

Pacala, S. and J. Roughgarden. 1982a. The evolution of resource partitioning in a multi-dimensional resource space. Theoret. Pop. Biol. 22: 127–145. (3, 17)

Pacala, S. and J. Roughgarden. 1982b. Resource partitioning and interspecific competition in two two-species insular *Anolis* lizard communities. Science 217: 444–446. (17)

Paige, C. J. and G. B. Craig, Jr. 1975. Variation in filarial susceptibility among East African populations of *Aedes aegypti*. J. Med. Ent. 12: 485–493. (8)

Paine, R. T. 1966. Food web complexity and species diversity. Amer. Natur. 100: 65–76. (3, 19)

Paine, R. T. 1969. A note on trophic complexity and community stability. Amer. Natur. 103: 91–93. (19)

Paine, R. T. 1980. Food webs: linkage, interaction, strength and community infrastructure. J. Anim. Ecol. 49: 667–685. (15, 19)

Paine, R. T. 1981. The forgotten roles of disturbance and predation. Paleobiology 7: 553–560. (15)

Paine, R. T. and T. H. Suchanek. 1983. Convergence of ecological processes between independently evolved competitive dominants: a tunicate-mussel comparison. Evolution. (In press) (19)

Paine, R. T. and R. L. Vadas. 1969. The effects of grazing by sea urchins, *Strongylocentrotus* spp., on benthic algal populations. Limnol. Oceanogr. 14: 710–719. (15, 19)

Painter, R. H. 1951. *Insect Resistance in Crop Plants*. University Press of Kansas, Lawrence. (10)

Painter, R. H. and M. D. Pathak. 1962. The distinguishing features and significance of the four biotypes of the corn leaf aphid *Rhopalosiphum maidis* (Fitch). Proc. XI Int. Congr. Entomol. 11: 110–115. (10)

Palmer, A. R. 1982. Predation and parallel evolution: recurrent parietal plate reduction in balanomorph barnacles. Paleobiology 8: 31–44. (14)

Panda, N. 1979. *Principles of Host-plant Resistance to Insect Pests*. Allenheld, Osmun, and Co., New York. (10)

Papageorgis, C. 1975. Mimicry in neotropical butterflies. Amer. Scient. 63: 522–532. (12)

Paperna, I. 1964. Competitive exclusion of *Dactylogrus extensus* by *Dactylogrus vastator* (Trematoda: Monogenea) on the gills of reared carp. J. Parasitol. 50: 94–98. (8)

Papp, A., H. Zapfe, F. Bachmayer and A. F. Tauber. 1947. Lebensspuren mariner Krebse. K. Akad. Wissensch. Wien, Mathem. Naturwissensch. Klasse, Sitzber. 155: 281–317. (14)

Park, T. and M. Lloyd. 1955. Natural selection and the outcome of competition. Amer. Natur. 89: 235–240. (1)

Parrish, J. A. D. and F. A. Bazzaz. 1979. Difference in pollination niche relationships in early and late successional plant communities. Ecology 60: 597–610. (13)

Parson, D. J. and A. R. Moldenke. 1975. Convergence in vegetation structure along analogous climatic gradients in California and Chile. Ecology 56: 950–957. (19)

Parsons, R. F. and D. G. Cameron. 1974. Maximum plant species diversity in terrestrial communities. Biotropica 6: 202–203. (19)

Pathak, M. D. 1970. Genetics of plants in pest management. In *Concepts of Pest Management*, R. L. Rabb and F. E. Guthrie (eds.), North Carolina State University Press, Raleigh. (10)

Pathak, M. D. 1977. Defense of the rice crop against insect pests. In *The Genetic Basis of Epidemics in Agriculture*, P. R. Day (ed.), pp. 287–295. Ann. N.Y. Acad. Sci. (10)

Pathak, M. D. and R. C. Saxena. 1976. Insect resistance in crop plants. Curr. Adv. Plant Sci. 8: 1233–1252. (10)

Patterson, B. and R. Pasqual. 1972. The fossil mammal fauna of South America. In *Evolution, Mammals and Southern Continents*, A. Keast, F. C. Erk and B. Glass (eds.). State University of New York Press, Albany. (16)

Patton, W. K. 1967. Studies on *Domecia acanthophora*, a commensal crab from Puerto Rico with particular reference to modification of the coral host and feeding habits. Biol. Bull. 132: 56–67. (14)

Payne, R. B. 1977. The ecology of brood parasitism in birds. Ann. Rev. Ecol. Syst. 8: 1–28. (12)

Paynter, M. J. B. and H. R. Bungay III. 1969. Dynamics of coliphage infections. In *Fermentation Advances*, D. Perlman, (ed.), pp. 323–326. Academic Press, New York. (5)

Peet, R. K. 1978. Ecosystem convergence. Amer. Natur. 112: 441–444. (19)

Perry, A. E. and T. H. Fleming. 1980. Ant and rodent predation on small animal-dispersed seeds in a dry tropical forest. Brenesia 17: 11–22. (11)

Person, C. 1959. Gene-for-gene relationships in host: parasite systems. Canad. J. Bot. 37: 1101–1130. (10)

Person, C. 1966. Genetic polymorphism in parasitic systems. Nature 212: 266–267. (3, 9)

Person, C., J. V. Groth and O. M. Mylyk. 1976. Genetic change in host-parasite populations. Ann. Rev. Phytopathol. 14: 177–188. (7)

Person, C. P. 1967. Genetic aspects of parasitism. Can. J. Bot. 45: 1193–1204. (8)

Peterson, C. H. 1979. Predation, competitive exclusion, and diversity in the soft-sediment benthic communities of estuaries and lagoons. In *Ecological Processes in Coastal and Marine Systems*, R. J. Livingston (ed.), pp. 233–264. Plenum Press, New York. (14)

Peterson, W. L. 1974. *Principles of Economics*. Micro Irwin, Homewood, New Jersey. (2)

Phillips, F. J. 1910. The dissemination of juniper seeds by birds. Forestry Quart. 8: 60–73. (11)

Phillips, P. A. and M. M. Barnes. 1975. Host race formation among sympatric apple, walnut, and plum populations of the codling moth, *Laspeyresia pomonella*. Ann. Ent. Soc. Amer. 68: 1053–1060. (10)

Phillipson, J. D. and S. S. Handa. 1976. Hyoscyamine N-oxide in *Atropa belladonna*. Phytochemistry 15: 605–608. (11)

Pianka, E. R. 1966. Convexity, desert lizards, and spatial heterogeneity. Ecology 47: 1055–1059. (19)

Pianka, E. R. 1967. On lizard species diversity: North American flatland deserts. Ecology 48: 333–351. (19)

Pianka, E. R. 1969. Habitat specificity, speciation and species density in Australian desert lizards. Ecology 50: 498–502. (19)

Pianka, E. R. 1971. Lizard species density in the Kalahari Desert. Ecology 52: 1024–1029. (19)

Pianka, E. R. 1973. The structure of lizard communities. Ann. Rev. Ecol. & Syst. 4: 53–74. (17, 19)

Pianka, E. R. 1975. Niche relations of desert lizards. In *Ecology and Evolution of Communities*, M. L. Cody and J. M. Diamond (eds.), pp. 292–314. Harvard University Press, Cambridge, Massachusetts. (19)

Pianka, E. R. and R. B. Huey. 1971. Bird species density in the Kalahari and Australian deserts. Koedoe 14: 123–130. (19)

Pickard, D. W. and C. E. Stevens. 1972. Digesta flow through the rabbit large intestine. Amer. J. Physiology 222: 1161–1166. (11)

Pickett, S. T. A. 1976. Succession: an evolutionary interpretation. Amer. Natur. 110: 107–119. (13)

Pielou, E. C. and A. N. Arnason. 1966. Correction to one of MacArthur's species-abundance formulas. Science 151: 592. (18)

Pienaar, U. de V. 1963. Predator-prey relationships amongst the larger mammals of the Kruger National Park. Koedoe 12: 68–76. (16)

Pimentel, D. 1968. Population regulation and genetic feedback. Science 159: 1432–1437. (9)

Pimentel, D. 1982. Genetic diversity and stability in parasite-host systems. (Unpublished manuscript) (9)

517

Pimentel, D. and F. A. Stone. 1968. Evolution and population ecology of parasite-host systems. Can. Entomol. 100: 655–662. (1)

Pimentel, D., E. H. Feinberg, D. W. Wood and J. T. Hayes. 1965. Selection, spatial distribution and the coexistence of competing fly species. Amer. Natur. 99: 97–108. (1)

Pleasants, J. M. 1980. Competition for bumblebees in Rocky Mountain plant communities. Ecology 61: 1446–1459. (13)

Pohley, W. G. 1976. Relationships among three species of *Littorina* and their larval Digenea. Mar. Biol. 37: 179–186. (14)

Poignant, A. F. 1977. The Mesozoic red algae: a general survey. In *Fossil Algae*, E. Flugel (ed.), pp. 177–189. Springer-Verlag, New York. (15)

Ponder, W. F. and R. U. Gooding. 1978. Four new eulimid gastropods associated with diadematid echinoids in the shallow-water western Pacific. Pac. Sci. 32: 157–181. (14)

Poole, R. W. and B. J. Rathcke. 1979. Regularity, randomness, and aggregation in flowering phenologies. Science 203: 470–471. (18)

Popper, K. R. 1963. *Conjectures and Refutations: The Growth of Scientific Knowledge*. Harper & Row, New York. (18)

Porsild, A. E., C. R. Harington and G. A. Mulligan. 1967. *Lupinus arcticus* Wats. grown from seeds of Pleistocene age. Science 158: 113–114. (11)

Poulton, E. B. 1898. Natural selection: the cause of mimetic resemblance and common warning colours. Linn. Soc. J. Zool. 26: 558–612. (12)

Powell, J. A. 1980. Evolution of larval food preferences in microlepidoptera. Ann. Rev. Entomol. 26: 133–159. (4)

Powell, J. A. and R. A. Mackie. 1966. Biological interrelationships of moths and *Yucca whipplei*. U. Calif. Publ. Entomol. 42: 1–46. (13)

Prance, G. T. and S. A. Mori. 1978. Observations on the fruits and seeds of neotropical Lecithydaceae. Brittonia 30: 21–33. (11)

Preer, J. R., Jr. 1975. The hereditary symbionts of *Paramecium aurelia*. Symp. Soc. Exp. Biol. 29: 125–144. (6)

Preer, J. R., Jr. 1977. The killer system in Paramecium-Kappa and its viruses. In *Microbiology*, D. Schlessinger (ed.), pp. 576–578. American Society of Microbiologists, Washington, D. C. (6)

Preer, J. R., Jr., L. B. Preer and A. Jurand. 1974. Kappa and other endosymbionts in *Paramecium aurelia*. Bact. Rev. 38: 113–163. (6)

Pregill, G. K. and S. L. Olson. 1981. Zoogeography of West Indian vertebrates in relation to Pleistocene climatic cycles. Ann. Rev. Ecol. and Syst. 12: 75–98. (3)

Price, M. V. 1978. Seed dispersion preferences of coexisting desert rodent species. J. Mammal. 58: 107–110. (11)

Price, M. V. and N. M. Waser. 1979. Pollen dispersal and optimal outcrossing in *Delphinium nelsonii*. Nature 277: 294–297. (13)

Price, P. W. 1973. Parasitoid strategies and community organization. Environmental Entomology 2: 623–626. (19)

Price, P. W. (ed.) 1975. *Evolutionary Strategies of Parasitic Insects and Mites*. Plenum Press, New York. (8)

Price, P. W. 1977. General concepts on the evolutionary biology of parasites. Evolution 31: 405–420. (8)

Price, P. W. 1980. *Evolutionary Biology of Parasites*. Princeton University Press, Princeton, New Jersey. (1, 2, 3, 4, 8, 9)

Price, P. W., C. E. Bouton, P. Gross, B. A. McPherson, J. N. Thompson and A. E. Weis. 1980. Interactions among three trophic levels: influence of plants on interactions between insect herbivores and natural enemies. Ann. Rev. Ecol. Syst. 11: 41–65. (10)

Primack, R. B. and J. A. Silander, Jr. 1975. Measuring the relative importance of different pollinators to plants. Nature 255: 143–144. (13)

Proctor, V. W. 1968. Long distance dispersal of seeds by retention in digestive tracts of birds. Science 166: 321–322. (11)

Provenzano, A. J., Jr. 1971. Rediscovery of *Munidopagurus macrocheles* (A. Milne-Edwards, 1830) (Crustacea, Decapoda, Paguridae), with a description of the first zoeal stage. Bull. Mar. Sci. 21: 256–266. (14)

518

Ptashne, M. 1971. Repressor and its action. In *The Bacteriophage Lambda*, A. D. Hershey (ed.), pp. 221–237. Cold Spring Harbor Laboratory, Cold Spring Harbor, New York. (5)

Puhalla, J. E. 1979. Classification of isolates of *Verticillium dahliae* based on heterokaryon incompatibility. Phytopathology 69: 1186–1189. (7)

Pulliam, H. R. 1974. Coexistence of sparrows: a test of community structure. Science 189: 474–476. (17)

Pulliam, H. R. and T. H. Parker. 1979. Population regulation of sparrows. Fortschritte der Zoologie 25: 137–147. (19)

Pyke, G. H. 1978. Optimal foraging in bumblebees and coevolution with their plants. Oecologia 36: 281–293. (13)

Pyke, G. H., H. R. Pulliam and E. L. Charnov. 1977. Optimal foraging: a selective review of theory and tests. Quart. Rev. Biol. 52: 137–154. (2, 10)

Quackenbush, R. L. 1977. Phylogenetic relationships of bacterial endosymbionts of *Paramecium aurelia* polynucleotide sequence relationships of 51 kappa and its mutants. J. Bacteriol. 129: 895–900. (6)

Quast, J. C. 1971. Fish fauna of the rocky inshore zone. Nova Hedwigia 32: 481–507. (19)

Quayle, D. B. 1964. Distribution of introduced marine mollusca in British Columbia waters. Jour. Fish. Res. Bd. Canada 21: 1155–1181. (19)

Quinlan, R. J. and J. M. Cherret. 1978. Aspects of the symbiosis of the leaf cutting ant *Acromyrmex octospinosus* (Reich) and its fungus food. Ecol. Entomol. 3: 221–230. (3)

Rabinowitz, D., J. K. Rapp, V. L. Sork, B. J. Rathcke, G. A. Reese and J. C. Weaver. 1981. Phenological properties of wind- and insect-pollinated prairie plants. Ecology 62: 49–56. (13)

Racine, C. H. and J. F. Downhower. 1974. Vegetative and reproductive strategies of *Opuntia* (Cactaceae) in the Galapagos Islands. Biotropica 6: 175–186. (11)

Ractliffe, L. H., H. M. Taylor, J. H. Whitlock and W. R. Lynn. 1969. Systems analysis of a host-parasite interaction. Parasitology 59: 649–661. (8)

Radinsky, L. 1978. Evolution of brain size in carnivores and ungulates. Amer. Natur. 112: 815–831. (15)

Radovsky, F. J. 1967. The Macronyssidae and Laelapidae (Acarina: Mesostigmata) parasitic on bats. Univ. Calif. Publ. Ent. 46. (4)

Rahm, U. 1972. Note sur la répartition, l'écologie et le régime alimentaire des sciuridés au Kivu (Zaïre). Rev. Zool. Bot. Afr. 85: 321–339. (11)

Raitt, R. J. and S. L. Pimm. 1976. Dynamics of bird communities in the Chihuahuan Desert, New Mexico. Condor 78: 427–442. (19)

Ramirez, B. W. 1974. Specificity of Agaonidae: the coevolution of *Ficus* and its pollinators. Ph.D. Dissertation, University of Kansas, Lawrence. (4)

Randolph, S. E. 1979. Population regulation in ticks: the role of acquired resistance in natural and unnatural hosts. Parasitology 79: 141–156. (8)

Raper, J. R. 1968. On the evolution of fungi. In *The Fungi: An Advanced Treatise*, Vol. III, *The Fungal Population*, G. C. Ainsworth and A. S. Sussman (eds.), pp. 677–693. Academic Press, New York. (7)

Rasmussen, E. 1973. Systematics and ecology of the Isefjord marine fauna (Denmark). Ophelia 11: 1–495. (14)

Rathcke, B. J. 1976. Competition and co-existence within a guild of herbivorous insects. Ecology 57: 76–87. (10)

Rathore, D. S. 1976. Effect of season on the growth and chemical composition of guava (*Psidium guajava L.*) fruits. J. Hort. Sci. 51: 41–47. (11)

519

Raunkiaer, C. 1934. *The Life Forms of Plants and Statistical Plant Geography.* Clarendon Press, Oxford. (19)

Rausher, M. D. 1980. Host abundance, juvenile survival, and oviposition preference in *Battus philenor.* Evolution 34: 342–355. (10)

Rausher, M. D. 1982. Population differentiation in *Euphydryas editha* butterflies: larval adaptation to different hosts. Evolution 36: 581–590. (10)

Razin, S. 1973. Physiology of mycoplasmas. Adv. Microb. Physiol. 10: 1–80. (6)

Reaka, M. L. 1978. The effect of an ectoparasitic gastropod, *Caledoniella montrouzieri,* upon molting and reproduction of a stomatopod crustacean, *Gonodactylus viridis.* Veliger 21: 251–254. (14)

Real, L. A. 1980. Fitness, uncertainty, and the role of diversification in evolution and behavior. Amer. Natur. 115: 623–638. (13)

Reanney, D. 1976. Extrachromosomal elements as possible agents of adaptation and development. Bacteriol. Rev. 40: 552–590. (5)

Recher, H. F. 1969. Bird species diversity and habitat diversity in Australia and North America. Amer. Natur. 103: 75–80. (19)

Reece, F. 1932. Which side are you on? Folkways Record Album No. FH 5285, 1955. (5)

Rees, A. R. 1963. Some factors affecting the germination of oil palm seeds under natural conditions. J. West African Institute for Oil Palm Research 4: 201–207. (11)

Rees, C. J. C. 1969. Chemoreceptor sensitivity associated with choice of feeding site by the beetle *Chrysolina brunsvicensis* on its food plant, *Hypericum hirsutum.* Ent. exp. appl. 12: 565–583. (10)

Reeves, P. 1972. *The Bacteriocins.* Springer-Verlag, New York. (5)

Regal, P. J. 1977. Ecology and evolution of flowering plant dominance. Science 196: 622–629. (11, 13)

Regenfuss, H. 1967. Untersuchungen zur Morphologie, Systematik und Ökologie der Podapolipidae (Acarina, Tarsonemini) (Unter besonderer Berücksichtigung der Parallelevolution der Gattungen *Entarsopolipus* und *Dorispes* mit ihren Wirten [Coleopt., Carabidae]). Z. wissenschaft. Zool. 177: 183–282. (4)

Rehr, S. S., P. P. Feeny and D. H. Janzen. 1973. Chemical defense in Central American non-ant-acacias. J. Anim. Ecol. 42: 405–416. (10)

Reichman, O. J. 1979. Desert granivore foraging and its impact on seed densities and distributions. Ecology 60: 1085–1092. (11)

Reichman, O. J. and D. Oberstein. 1977. Selection of seed distribution types by *Dipodomys merriami* and *Perognathus amplus.* Ecology 58: 636–643. (19)

Reiner, A. M. 1974. *Escherichia coli* females defective in conjugation and in adsorption of single-stranded deoxyribonucleic acid phage. J. Bact. 119: 183. (5)

Rettenmeyer, C. W. 1970. Insect mimicry. Ann. Rev. Entomol. 15: 43–74. (12)

Rex, M. A. 1976. Biological accommodation in the deep-sea benthos: comparative evidence on the importance of predation and productivity. Deep Sea Res. 23: 975–987. (14)

Rhoades, D. F. 1979. Evolution of plant chemical defense against herbivores. In *Herbivores: Their Interaction with Secondary Plant Metabolites,* G. A. Rosenthal and D. H. Janzen (eds.). Academic Press, New York. (19)

Rhoades, D. F. 1983. Responses of alder and willow to attack by tent caterpillars and webworms: evidence for pheromonal sensitivity of willows. In *Mechanisms of Plant Resistance to Insects,* P. Hedin (ed.). Amer. Chem. Soc. Symp. (In press) (19)

Rhoades, D. F. and R. G. Cates. 1976. Toward a general theory of plant antiherbivore chemistry. In *Biochemical Interaction between Plants and Insects,* J. W. Wallace and R. L. Mansell (eds.), pp. 168–213. Plenum Press, New York. (10, 19)

Ricard, M. D. 1974. Hypothesis for the longterm survival of *Taenia pisiformis* cystocerci in rabbits. Z. Parasitenkd. 44: 203–209. (8)

Rick, C. M. and R. I. Bowman. 1961. Galapagos tomatoes and tortoises. Evolution 15: 407–447. (11)

Ricketts, E. F., J. Calvin and J. W. Hedgpeth. 1968. *Between Pacific Tides,* 4th Ed. Stanford University Press, Stanford, California. (19)

Ricklefs, R. E. and G. W. Cox. 1972. Taxon cycles in the West Indian avifauna. Amer. Natur. 106: 195–219. (3)

Ricklefs, R. E. and K. O'Rourke. 1975. Aspect diversity in moths: a temperate-tropical comparison. Evolution 29: 313–324. (3, 17)

Ridley, H. N. 1930. *The Dispersal of Plants Throughout the World*. L. Reeve, Ashford, Kent, England. (11)

Ritchie, L. E. and J. T. Höeg. 1981. The life history of *Lernaeodiscus porcellanae* (Cirripedia: Rhizocephala) and coevolution with its porcellanid host. J. Crustacean Biol. 1: 334–347. (14)

Robbins, R. J., C. Casbon and G. E. Hattis. 1975. Observations of birds exploiting a central Michigan fruit tree. Jack-Pine Warbler 53: 118–125. (11)

Robertson, R. 1970. Review of the predators and parasites of stony corals, with special reference to symbiotic prosobranch gastropods. Pacific Sci. 24: 43–54. (14)

Robertson, R. 1980. *Epitonium millecostatum* and *Coralliophila clathrata*: two prosobranch gastropods symbiotic with Indo-Pacific *Palythoa* (Coelenterata: Zoanthidae). Pacific Sci. 34: 1–17. (14)

Robinson, R. A. 1976. *Plant Pathosystems*. Springer-Verlag, Berlin. (7)

Robinson, T. 1974. Metabolism and function of alkaloids in plants. Science 184: 430–435. (10)

Rocklin, S. and G. F. Oster. 1976. Competition between phenotypes. J. Math. Biol. 3: 225–261. (9)

Rodman, J. E. and F. S. Chew. 1980. Phytochemical correlates of herbivory in a community of native and naturalized Cruciferae. Biochem. Syst. Ecol. 8: 43–50. (10)

Rodrigues, C. J., Jr., A. J. Bettencourt and L. Rijo. 1975. Races of the pathogen and resistance to coffee rust. Ann. Rev. Phytopathol. 13: 49–70. (7)

Roeske, C. N., J. N. Seiber, L. P. Brower and C. M. Moffitt. 1976. Milkweed cardenolides and their comparative processing by monarch butterflies (*Danaus plexippus* L). Rec. Adv. Phytochem. 10: 93–167. (10)

Rohde, K. 1978a. Latitudinal gradients in species diversity and causes. II. Marine parasitological evidence for a time hypothesis. Biol. Zentralbl. 97: 405–418. (14)

Rohde, K. 1978b. Latitudinal differences in host-specificity of marine Monogenea and Digenea. Mar. Biol. 47: 125–134. (8, 14)

Rohde, K. 1979. A critical evaluation of intrinsic and extrinsic factors responsible for niche restriction in parasites. Amer. Natur. 114: 648–671. (8)

Rohde, K. 1980. Host specificity indices of parasites and their application. Experientia 36: 1370–1371. (8)

Rohde, K. 1981. Niche width of parasites in species-rich and species-poor communities. Experientia 37: 359–361. (9)

Romer, A. S. 1966. *Vertebrate Paleontology*. University of Chicago Press, Chicago, Illinois. (4)

Root, R. B. 1967. The niche exploitation pattern of the blue-gray gnatcatcher. Ecol. Monogr. 37: 317–350. (13, 15, 19)

Rosenthal, G. A., D. L. Dahlman and D. H. Janzen. 1978. L-canaline detoxification: a seed predator's biochemical mechanism. Science 202: 528–529. (10)

Rosenthal, G. A. and D. H. Janzen (eds.). 1979. *Herbivores: Their Interaction with Secondary Plant Metabolites*. Academic Press, New York. (10)

Rosenzweig, M. L. 1968. Net primary productivity of terrestrial communities: prediction from climatological data. Amer. Natur. 102: 67–74. (19)

Rosenzweig, M. L. 1969. Why the prey curve has a hump. Amer. Natur. 63: 81–87. (3)

Rosenzweig, M. L. 1971. Paradox of enrichment: destabilization of exploitation ecosystems in ecological time. Science 171: 385–387. (3)

Rosenzweig, M. L. 1973. Evolution of the predator isocline. Evolution 27: 84–94. (3)

Ross, A. and W. A. Newman. 1973. Revision of the coral-inhabiting barnacles (Cirripedia: Balanidae). Trans. San Diego Soc. Nat. Hist. 17: 137–174. (14)

Ross, D. M. 1970. The commensal association of *Calliactis polypus* and the hermit crab *Dardanus gemmatus* in Hawaii. Canad. J. Zool. 48: 351–357. (14)

521

Ross, D. M. 1971. Protection of hermit crabs (*Dardanus* spp.) from *Octopus* by commensal sea anemones (*Calliactis* spp.). Nature 230: 401–402. (14)

Ross, D. M. 1974. Evolutionary aspects of associations between crabs and sea anemones. In *Symbiosis in the Sea*, W. B. Vernberg (ed.), pp. 111–125. University of South Carolina Press, Columbia. (14)

Ross, H. H. 1962. *A Synthesis of Evolutionary Theory*. Prentice-Hall, Englewood Cliffs, New Jersey. (4)

Ross, H. H. 1972. An uncertainty principle in ecological evolution. In *A Symposium on Ecosystematics*, R. T. Allen and F. C. James (eds.). Univ. of Arkansas Occ. Papers No. 4. (4)

Ross, J. 1982. Myxomatosis: the natural evolution of the disease. In *Animal Disease in Relation to Animal Conservation*. Symp. Zool. Soc. Lond., 26–27, Nov. 1981. (9)

Roth, I. 1974. Desarrollo y estructura anatómica del merey (*Anacardium occidentale* L.). Acta Botánica Venezuelica 9: 197–233. (11)

Roth, I. 1977. *Fruits of Angiosperms*. Gebrüder Borntraeger, Berlin. (11)

Roth, I. and H. Lindorf. 1972. Anatomía y desarrollo del fruto y de la semilla de *Achras sapota* L. (nispero). Acta Botánica Venezuelica 7: 121–141. (11)

Roth, V. L. 1981. Constancy in the size ratios of sympatric species. Amer. Natur. 118: 394–404. (18)

Rothschild, M. 1973. Secondary plant substances and warning colouration in insects. In *Insect/plant Relationships*, H. J. van Emden (ed.). Symp. Roy. Ent. Lond. 6: 59–83. (10)

Rothschild, M. 1975. Recent advances in our knowledge of the order Siphonaptera. Ann. Rev. Entomol. 20: 241–259. (8)

Rothschild, M. 1979. Mimicry, butterflies, and plants. Symp. Bot. Upsal. XXII, 82–99. (12)

Rothschild, M. and T. Clay. 1952. *Fleas, Flukes, and Cuckoos*. Philosophical Library, New York. (4)

Rothschild, M. and B. Ford. 1972. Breeding cycle of the flea *Cediopsylla simplex* is controlled by breeding cycle of host. Science 178: 625–626. (8)

Rothschild, M. and B. Ford. 1973. Factors influencing the breeding of the rabbit flea (*S. cuniculi*): a spring-time accelerator and a kairomone in nestling rabbit urine, with notes on *Cediopsylla simplex*, another "hormone bound" species. J. Zool. 170: 87–137. (8)

Rothschild, M. and N. Marsh. 1978. Some peculiar aspects of danaid/plant relationships. Ent. exp. appl. 24: 437–450. (10)

Roughgarden, J. 1971. Density-dependent natural selection. Ecology 52: 453–468. (2, 3)

Roughgarden, J. 1972. Evolution of niche width. Amer. Natur. 106: 683–718. (3, 9)

Roughgarden, J. 1974. Niche width: biogeographic patterns among *Anolis* lizard populations. Amer. Natur. 108: 429–442. (3, 17)

Roughgarden, J. 1975. Evolution of marine symbiosis—a simple cost-benefit model. Ecology 56: 1201–1208. (3, 14, 19)

Roughgarden, J. 1976. Resource partitioning among competing species—a coevolutionary approach. Theor. Pop. Biol. 9: 388–424. (3, 17)

Roughgarden, J. 1977. Coevolution in ecological systems II. Results from "loop analysis" for purely density-dependent coevolution. In *Measuring Selection in Natural Populations*, F. B. Christiansen and T. Fenchel (eds.), pp. 499–517. Springer-Verlag, New York. (3)

Roughgarden, J. 1979. *Theory of Population Genetics and Evolutionary Ecology: An Introduction*. Macmillan, New York. (1, 2, 3, 9, 18)

Roughgarden, J. and E. R. Fuentes. 1977. The environmental determinants of size in solitary populations of West Indian Anolis lizards. Oikos 29: 44–51. (17, 18)

Roughgarden, J., D. Heckel and E. R. Fuentes. 1982. How coevolutionary theory explains the biogeography and community structure of the *Anolis* lizard communities in the Lesser Antilles. In *Lizard Ecology: Studies of a Model Organism*, R. Huey, E. Pianka and T. Schoener (eds.). Harvard University Press, Cambridge, Massachusetts. (In press) (3, 17, 18)

522

Roughgarden, J., W. Porter and D. Heckel. 1981. Resource partitioning of space and its relationship to body temperature in *Anolis* lizard populations. Oecologia 50: 256-264. (17, 18)

Roughgarden, J., S. Pacala and J. Rummel. 1983. Strong present-day competition between the *Anolis* lizard populations of St. Maarten—A review of evidence. In *Evolutionary Ecology, A Symposium of the British Ecological Society*, B. Shorrocks (ed.), Blackwell Scientific, Oxford. (In press) (17)

Rozin, P., L. Gruss and G. Berk. 1979. Reversal of innate aversions: attempts to induce a preference for chili peppers in rats. J. Comp. Physiol. Psychol. 93: 1001-1014. (11)

Rudman, W. B. 1981. Further studies on the anatomy and ecology of opisthobranch molluscs feeding on the scleractinian coral *Porites*. Zool. J. Linn. Soc. 71: 373-412. (14)

Rühm, W. 1956. Die Nematoden der Ipiden. Parasit. Schriftenreihe 6: 1-437. (4)

Russell, D. E. 1964. Les Mammifères Paléocènes D'Europe. Mém. Mus. Nat. D'Hist. Natur. C. 13: 1-324. (16)

Ryahchikov, A. M. 1968. Hydrothermal conditions and the productivity of plant mass in the principal landscape zones. Vestnik MGU, Geogr., Moscow 5: 41-48. (19)

Ryan, C. A. 1979. Proteinase inhibitors. In *Herbivores: Their Interaction with Secondary Plant Metabolites*, G. A. Rosenthal and D. H. Janzen (eds.), pp. 599-618. Academic Press, New York. (10)

Ryther, J. H. 1955. Potential productivity of the sea. Science 130: 602-608. (19)

Safeeulla, K. M. 1977. Genetic vulnerability: the basis of recent epidemics in India. Ann. N.Y. Acad. Sci. 287: 72-85. (7)

Saimbhi, M. S., K. Gurdeep and K. S. Nanpuri. 1977. Chemical constituents in mature green and red fruits of some varieties of chili (*Capsicum annum* L.). Qualitas Plantarum 27: 171-175. (11)

Sakaguchi, B., K. Oishi and S. Kobayashi. 1965. Interference between "sex-ratio" agents of *Drosophila willistoni* and *Drosophila nebulosa*. Science 147: 160-162. (6)

Sakaguchi, B. and D. F. Poulson. 1960. Transfer of the "sex-ratio" from *Drosophila willistoni* to *D. melanogaster*. Anat. Rec. 138: 381. (6)

Sakaguchi, B. and D. F. Poulson. 1961. Distribution of "sex-ratio" agent in tissues of *Drosophila willistoni*. Genetics 46: 1665-1676. (6)

Salisbury, F. B. 1942. *The Reproductive Capacities of Plants*. George Bell & Sons, London (19)

Salomonson, M. G. 1978. Adaptations for animal dispersal of one-seed juniper seeds. Oecologia 32: 333-339. (11)

Saoud, M. F. A. 1965. Susceptibilities of various snail intermediate hosts of *Schistosoma mansoni* to different strains of the parasite. J. Helminth. 39: 363-376. (8)

Saunders, I. W. 1980. A model for myxomatosis. Math. Biosci. 48: 1-16. (9)

Savile, D. B. O. 1968. Possible interrelationships between fungal groups. In *The Fungi: An Advanced Treatise, Vol. III, The Fungal Population*, G. C. Ainsworth and A. S. Sussman (eds.), pp. 649-675. Academic Press, New York. (7)

Sbordoni, V., L. Bullini, G. Scarpelli, S. Forestiero and M. Rampini. 1979. Mimicry in the burnet moth *Zygaena ephialtes*: population studies and evidence of a Batesian-Müllerian situation. Ecol. Entomol. 4: 83-93. (12)

Schad, G. A. 1963. Niche diversification in a parasitic species flock. Nature 198: 404-406. (9)

Schad, G. A. 1966. Immunity, competition and natural regulation of helminth populations. Amer. Natur. 100: 359-364. (8, 9)

Schaffer, W. M. and E. G. Leigh. 1976. The prospective role of mathematical theory in plant ecology. Syst. Bot. 1: 209-232. (19)

Schaffer, W. M. and M. L. Rosenzweig. 1978. Homage to the Red Queen. I. Coevolution of predators and their victims. Theor. Pop. Biol. 14: 135-157. (3, Ep)

Schaller, G. B. 1972. *The Serengeti Lion.* University of Chicago Press, Chicago. (16)

Schankler, D. 1980. Faunal zonation of the Willwood Formation in the Central Bighorn Basin, Wyoming. Univ. Mich. Pap. Paleont. 24: 99-114. (16)

Schankler, D. 1982. Biostratigraphic zonation of the Willwood Formation. Ph.D. Thesis, Yale University. (16)

Scheltema, R. S. 1971. Larval dispersal as a means of genetic exchange between geographically separated populations of shallow-water, benthic marine gastropods. Biol. Bull. 140: 284-322. (19)

Schemske, D. W. 1981. Floral convergence and pollinator sharing in two bee-pollinated tropical herbs. Ecology 62: 946-954. (13)

Schiel, D. R. and J. H. Choat. 1980. Effects of density on monospecific stands of marine algae. Nature 285: 324-326. (14)

Schimper, A. F. W. 1903. *Plant Geography upon a Physiological Basis.* (Trans. by W. R. Fisher) Clarendon Press, Oxford. (19)

Schmitt, J. 1980. Pollinator foraging behavior and gene dispersal in *Senecio* (Compositae). Evolution 34: 934-943. (13)

Schneider, J. C. 1980. The role of parthenogenesis and female aptery in microgeographic, ecological adaptation in the fall cankerworm, *Alsophila pometaria* Harris (Lepidoptera: Geometridae). Ecology 61: 1082-1090. (10)

Schoenberg, D. A. and R. K. Trench. 1980a. Genetic variation in *Symbiodinium* (= *Gymnodinium*) *microadriaticum* Freudenthal, and specificity in its symbiosis with marine invertebrates. I. Isoenzyme and soluble protein patterns of axenic cultures of *Symbiodinium microadriaticum.* Proc. Roy. Soc. London (B) 207: 405-427. (14)

Schoenberg, D. A. and R. K. Trench. 1980b. Genetic variation in *Symbiodinium* (= *Gymnodinium*) *microadriaticum* Freudenthal, and specificity in its symbiosis with marine invertebrates. II. Morphological variation in *Symbiodinium microadriaticum.* Proc. Roy. Soc. London (B) 207: 429-444. (14)

Schoenberg, D. A. and R. K. Trench. 1980c. Genetic variation in *Symbiodinium* (= *Gymnodinium*) *microadriaticum* Freudenthal, and specificity in its symbiosis with marine invertebrates. III. Specificity and infectivity of *Symbiodinium microadriaticum.* Proc. Roy. Soc. London (B) 207: 445-460. (14)

Schoener, T. W. 1965. The evolution of bill size differences among sympatric congeneric species of birds. Evolution 19: 189-213. (18)

Schoener, T. W. 1967. The ecological significance of sexual dimorphism in size in the lizard *Anolis conspersus.* Science 155: 474-477. (18)

Schoener, T. W. 1969a. Models of optimal size for solitary predators. Amer. Natur. 103: 277-313. (18)

Schoener, T. W. 1969b. Size patterns in West Indian *Anolis* lizards. I. Size and species diversity. Syst. Zool. 18: 386-401. (18)

Schoener, T. W. 1970. Size patterns in West Indian *Anolis* lizards. II. Correlation with the sizes of particular sympatric species—displacement and convergence. Amer. Natur. 104: 155-174. (18)

Schoener, T. W. 1971a. Large-billed insectivorous birds: a precipitous diversity gradient. Condor 73: 154-161. (19)

Schoener, T. W. 1971b. Theory of feeding strategies. Ann. Rev. Ecol. Syst. 2: 369-404. (2)

Schoener, T. W. 1974a. Resource partitioning in ecological communities. Science 185: 27-39. (3)

Schoener, T. W. 1974b. Some methods for calculating competition coefficients from resource-utilization spectra. Amer. Natur. 109: 332-340. (17)

Schoener, T. W. 1982. The controversy over interspecific competition. Amer. Sci. 70: 586-595. (17)

Schoener, T. W. 1983. Field experiments on interspecific competition. (In preparation) (17)

Schoener, T. W. and G. C. Gorman. 1968. Some niche differences in three lesser Antillean lizards of the genus *Anolis.* Ecology 49: 819-830. (18)

Schoener, T. W. and A. Schoener. 1971. Structural habitats of West Indian *Anolis* lizards. II. Puerto Rican uplands. Breviora 375: 1-39. (18)

Schom, C., M. Novak and W. S. Evans. 1981. Evolutionary implications of *Tribolium confusum-Hymenolepis citelli* interactions. Parasitology 83: 77-90. (8)

524

Schoonhoven, L. M. 1968. Chemosensory bases of host plant selection. Ann. Rev. Ent. 13: 115–136. (10)

Schoonhoven, L. M. 1969. Gustation and foodplant selection in some lepidopterous larvae. Ent. exp. appl. 12: 555–564. (10)

Schoonhoven, L. M. 1972. Plant recognition by lepidopterous larvae. Symp. Roy. Ent. Soc. Lond. 6: 87–99. (10)

Schoonhoven, L. M. and V. G. Dethier. 1966. Sensory aspects of host-plant discrimination by lepidopterous larvae. Arch. Neerl. Zool. 16: 497–530. (10)

Schoonhoven, L. M. and J. Meerman. 1978. Metabolic cost of changes in diet and neutralization of allelochemics. Ent. exp. appl. 24: 489–493. (10)

Schulz-Key, H. and P. Wenk. 1981. The transmission of *Onchocerca tarsicola* (Filaroidea: Onchocercidae) by *Odagmia ornata* and *Prosimulium nigripes* (Diptera: Simuliidae). J. Helminth. 55: 161–166. (8)

Schultz, J. C. 1978. Competition, predation and the structure of phytophilous insect communities: a study of convergent evolution. Ph.D. Thesis, University of Washington, Seattle. (19)

Schwartz, J. H., I. Tattersall and N. Eldredge. 1978. Phylogeny and classification of the primates revisited. Yearbook Phys. Anthro. 21: 95–133. (4)

Scora, R. W. and C. Adams. 1973. Effects of oleocellosis, desiccation, and fungal infection upon the terpenes of individual oil glands in *Citrus latipes*. Phytochemistry 12: 2347–2350. (11)

Scott, G. D. 1973. Evolutionary aspects of symbiosis. In *The Lichens*, V. Ahmadjian and M. E. Hale (eds.), pp. 581–598. Academic Press, New York. (7)

Scott, W. B. 1892. A revision of the North American Creodonta with notes on some genera which have been referred to that group. Proc. Natur. Sci. Acad. Phil. 291–323. (16)

Scott, W. B. and G. L. Jepsen. 1941. The mammalian fauna of the White River Oligocene. Amer. Philos. Soc. 28: 747–980. (15)

Scriber, J. M. 1978. Cyanogenic glycosides in *Lotus corniculatus*: their effect upon growth, energy budget, and nitrogen utilization of the southern armyworm *Spodoptera eridania*. Oecologia 34: 143–155. (10)

Scriber, J. M. 1979. The effects of sequentially switching foodplants upon biomass and nitrogen utilization by polyphagous and stenophagous *Papilio* larvae. Ent. exp. appl. 25: 203–215. (10)

Scriber, J. M. 1983. The evolution of feeding specialization, physiological efficiency and host races in selected Papilionidae and Saturniidae. In *Impact of Variable Host Quality on Herbivorous Insects*, R. F. Denno and M. S. McClure (eds.). Academic Press, New York. (In press) (10)

Scriber, J. M. and P. Feeny. 1979. Growth of herbivorous caterpillars in relation to feeding specialization and to the growth form of their food plants. Ecology 60: 829–850. (10)

Scriber, J. M. and F. Slansky, Jr. 1981. The nutritional ecology of immature insects. Ann. Rev. Entomol. 26: 183–212. (10)

Seaton, A. P. C. and J. Antonovics. 1967. Population interrelationships. I. Evolution in mixtures of *Drosophila* mutants. Heredity 22: 19–33. (1)

Searcy, D. G., D. B. Stein and K. B. Searcy. 1981. A mycoplasma-like archaebacterium possibly related to the nucleus and cytoplasm of eukaryotic cells. Ann. N.Y. Acad. Sci. 361: 312–323. (5)

Sebens, K. P. and R. T. Paine. 1978. Biogeography of anthozoans along the west coast of South America: habitat, disturbances and prey availability. In Proc. Int. Symp. Mar. Biog. Ecol. S. Hemi. Vol. I: 219–237. (19)

Seevers, C. H. 1957. A monograph on the termitophilous Staphylinidae (Coleoptera). Fieldiana, Zool. 40: 1–334. (4)

Segal, A., J. Manisterski, G. Fischbeck and I. Wahl. 1980. How plant populations defend themselves in natural ecosystems. In *Plant Disease*, Vol. 5, J. G. Horsfall and E. B. Cowling (eds.), pp. 75–102. Academic Press, New York. (7)

Seigler, D. and P. W. Price. 1976. Secondary compounds in plants: primary functions. Amer. Natur. 110: 101–105. (10)

Seilacher, A. 1977. Evolution of trace fossil communities. In *Patterns of Evolution as Illustrated by the Fossil Record*, A. Hallam (ed.), pp. 357–376. Elsevier, Amsterdam. (14)

Selander, R. K. and B. R. Levin. 1980. Genetic diversity and structure in *Escherichia coli* populations. Science 210: 545–547. (5)

Self, L. S., E. F. Guthrie and E. Hodgson. 1964. Adaptation of tobacco hornworm to the ingestion of nicotine. J. Ins. Physiol. 10: 907–914. (10)

Selvin, H. C. and A. Stuart. 1966. Data-dredging procedures in survey analysis. Amer. Statistician 20: 20–23. (18)

Shapiro, A. M. 1981. The pierid red-egg syndrome. Amer. Natur. 117: 276–294. (12)

Sharatchandra, H. C. and M. Gadgil. 1975. A year of Bandipur. J. Bomb. Nat. Hist. Soc. 72: 623–647. (11)

Shaw, W. T. 1934. The ability of the giant kangaroo rat as a harvester and storer of seed. J. Mammalogy 15: 275–286. (11)

Shaw, W. T. 1936. Moisture and its relationship to the cone-storing habit of the western pine squirrel. J. Mammal. 17: 337–349. (11)

Sheppard, P. M. 1962. Some aspects of the geography, genetics and taxonomy of a butterfly. In Taxonomy and Geography Syst. Assoc. Publ. No. 4, 135–152. (12)

Sheppard, P. M. 1975. *Natural Selection and Heredity*, 4th Ed. Hutchinson, London. (15)

Sheppard, P. M., J. R. G. Turner, K. S. Brown, W. W. Benson and M. C. Singer. 1983. Genetics and the evolution of Müllerian mimicry in *Heliconius* butterflies. Phil. Trans. Roy. Soc. Lond. B. (In press) (12)

Shiff, C. J. 1974. Seasonal factors influencing the location of *Bulinus (Physopsis) globosus* by miracidia of *Schistosoma haematobium* in nature. J. Parasitol. 60: 578–583. (8)

Short, L. L. 1978. Sympatry in woodpeckers of lowland Malayan forest. Biotropica 10: 122–133. (18)

Shortridge, G. C. 1931. Field notes on two little-known antelopes: the Damarland dik-dik (*Rhynchotragus damarensis*) and the Angolan impala (*Aepyceros petersi*). S. Afr. J. Sci. 28: 412–417. (11)

Siegel, R. K. 1973. An ethologic search for self-administration of hallucinogens. International Journal of the Addictions 8: 373–393. (11)

Signer, E. R. 1969. Plasmid formation: a new mode of lysogeny by phage Lambda. Nature 233: 158–160. (5)

Sikes, S. K. 1971. *The Natural History of the African Elephant*. Weidenfeld and Nicolson, London. (11)

Sill, W. D. 1967. *Proterochampsa barrionuevoi* and the early evolution of the Crocodilia. Bull. Mus. Comp. Zool. 135: 415–446. (4)

Silva, E. N., G. H. Snoeyenbos, O. M. Weinack and C. F. Smyser. 1981. The influence of native gut microflora on the colonization and infection of *Salmonella gallinarum* in chickens. Avian Dis. 25: 68–73. (8)

Silver, B. B., T. A. Dick and H. E. Welch. 1980. Concurrent infection of *Hymenolepis diminuta* and *Trichinella spiralis* in the rat intestine. J. Parasitol. 66: 786–791. (8)

Silvertown, J. W. 1980. The evolutionary ecology of mast seeding in trees. Biol. J. Linn. Soc. 14: 235–250. (11)

Simberloff, D. 1982. Morphological and taxonomic similarity and combinations of coexisting birds in two archipelagoes. In *Ecological Communities: Conceptual Issues and the Evidence*, D. R. Strong, D. Simberloff and L. G. Abele (eds.). Princeton University Press, Princeton, New Jersey. (In press) (18)

Simberloff, D. S. and W. Boecklen. 1981. Santa Rosalia reconsidered: Size ratios and competition. Evolution 35: 1206–1228. (18)

Simberloff, D. and E. F. Connor. 1981. Missing species combinations. Amer. Natur. 118: 215–239. (18)

Simberloff, D., K. L. Heck, E. D. McCoy and E. F. Connor. 1981. There have been no statistical tests of cladistic biogeographic hypotheses. In *Vicariance Biogeography: A Critique*, G. Nelson and D. E. Rosen (eds.), pp. 40–63. Columbia University Press, New York. (4)

526

Simmonds, N. W. 1959. *Bananas*. Longmans, London. (7)

Simpson, G. G. 1944. *Tempo and Mode in Evolution*. Columbia University Press, New York. (16)

Simpson, G. G. 1953. *The Major Features of Evolution*. Simon and Schuster, New York. (2, 16)

Simpson, G. G. 1969. The first three billion years of community evolution. Brookhaven Symp. Biol. 22: 162–177. (19)

Simpson, G. G. 1970. The Argyrolagidae, extinct South American marsupials. Bull. Mus. Comp. Zool. 139: 1–86. (19)

Sinclair, A. R. E. 1977. *The African Buffalo*. University of Chicago Press, Chicago. (16)

Sinclair, A. R. E. 1979. The Serengeti environment. In *Serengeti, Dynamics of an Ecosystem*, A. R. E. Sinclair and M. Norton-Griffiths (eds.), University of Chicago Press, Chicago. (15, 16)

Singer, M. C. 1971. Evolution of food-plant preference in the butterfly *Euphydryas editha*. Evolution 25: 383–389. (10)

Singler, C. R. and M. D. Picard. 1981. Paleosols and the Oligocene of northwest Nebraska. Univ. of Wyoming Contrib. Geol. 20: 57–68. (15)

Skutch, A. F. 1944. Life history of the blue-throated toucanet. Wilson Bull. 56: 133–151. (11)

Skutch, A. F. 1971. Life history of the keel-billed toucan. Auk 88: 381–396. (11)

Sláma, K. 1979. Insect hormones and antihormones in plants. In *Herbivores: Their Interaction with Secondary Plant Metabolites*, G. A. Rosenthal and D. H. Janzen (eds.), pp. 683–700. Academic Press, New York. (10)

Slansky, F., Jr., and P. P. Feeny. 1977. Stabilization of the rate of nitrogen accumulation of larvae of the cabbage butterfly on wild and cultivated food plants. Ecol. Monogr. 47: 209–228. (10)

Slatkin, M. 1977. Gene flow and genetic drift in a species subject to frequent local extinctions. Theor. Pop. Biol. 12: 253–262. (2)

Slatkin, M. 1979. The evolutionary response to frequency- and density-dependent interactions. Amer. Natur. 114: 384–398. (2, 3)

Slatkin, M. 1980. Ecological character displacement. Ecology 61: 163–178. (3, 17, 18)

Slatkin, M. 1981a. A diffusion model of species selection. Paleobiology 7: 421–425. (2)

Slatkin, M. 1981b. Populational heritability. Evolution 35: 859–871. (2)

Slatkin, M. and J. Maynard Smith. 1979. Models of coevolution. Quart. Rev. Biol. 54: 233–263. (1, 2, 3, 5)

Slobodkin, L. B. 1974. Prudent predation does not require group selection. Amer. Natur. 108: 665–678. (Ep)

Slobodkin, L. B. and H. L. Sanders. 1969. On the contribution of environmental predictability to species diversity. In *Diversity and Stability in Ecological Systems*, G. M. Woodwell and H. H. Smith (eds.), pp. 82–93. Brookhaven Symposium in Biology No. 22. (1)

Slocum, C. J. 1980. Differential susceptibility to grazers in two phases of an intertidal alga: advantages of heteromorphic generations. J. Exp. Mar. Biol. Ecol. 46: 99–110. (19)

Smiley, J. 1978. Plant chemistry and the evolution of host specificity: new evidence from *Heliconius* and *Passiflora*. Science 201: 745–747. (10)

Smith, A. J. 1975. Invasion and ecesis of bird-disseminated woody plants in a temperate forest sere. Ecology 56: 19–34. (11)

Smith, C. C. 1970. The coevolution of pine squirrels (*Tamiasciurus*) and conifers. Ecol. Monogr. 40: 349–371. (11)

Smith, C. C. 1975. The coevolution of plants and seed predators. In *Coevolution of Animals and Plants*, L. E. Gilbert and P. H. Raven (eds.), pp. 53–77. University of Texas Press, Austin. (11, 19)

Smith, C. C. 1981. The indivisible niche of *Tamiasciurus*: an example of nonpartitioning of resources. Ecol. Monogr. 51: 343–363. (11)

Smith, C. C. and R. P. Balda. 1979. Competition among insects, birds and mammals for conifer seeds. Amer. Zool. 19: 1065–1083. (11)

527

Smith, C. L. 1978. Coral reef fish communities: a compromise view. Env. Biol. Fish. 3: 109–128. (18)

Smith, D. A. S. 1973. Batesian mimicry between *Danaus chrysippus* and *Hypolimnas misippus* (Lepidoptera) in Tanzania. Nature 242: 129–132. (12)

Smith, D. A. S. 1975. Genetics of some polymorphic forms of the African butterfly *Danaus chrysippus*. Ent. Scand. 6: 134–144. (12)

Smith, D. A. S. 1976. Phenotypic diversity, mimicry, and natural selection in the African butterfly *Hypolimnas misippus* L. Biol. J. Linn. Soc. 8: 183–204. (12)

Smith, D. A. S. 1979. The significance of beak marks on the wings of an aposematic, distasteful, and polymorphic butterfly. Nature 281: 215–216. (12)

Smith, D. A. S. 1980. Heterosis, epistasis, and linkage disequilibrium in a wild population of the polymorphic butterfly *Danaus chrysippus*. Zool. J. Linn. Soc. 69: 87–109. (12)

Smith, D. C. 1975. Symbiosis and the biology of lichenised fungi. In *Symbiosis*, Symposia of the Society for Experimental Biology XXIX, pp. 373–405. Cambridge University Press, Cambridge. (7)

Smith, D. C. 1976. A comparison between the lichen symbiosis and other symbioses. In *Lichenology: Progress and Problems*, D. H. Brown, D. L. Hawksworth and R. H. Bailey (eds.), pp. 497–513. Academic Press, London. (7)

Smith, D. C., L. Muscarine and D. Lewis. 1969. Carbohydrate movement from autotrophs to heterotrophs in parasitic and mutualistic symbiosis. Biol. Rev. 44: 17–90. (7)

Smith, H. V. and J. R. Kusel. 1979. The acquisition of antigens in the intercellular substance of mouse skin by schistosomula of *Schistosoma mansoni*. Clin. Exp. Immunol. 36: 430–435. (8)

Smith, M. A. and J. A. Clegg. 1979. Different levels of immunity to *Schistosoma mansoni* in the mouse: the role of variant cercariae. Parasitology 78: 311–321. (8)

Smith, N. G. 1968. The advantage of being parasitized. Nature 219: 690–694. (3, 8)

Smith, N. G. 1979. Alternate responses by hosts to parasites which may be helpful or harmful. In *Host-Parasite Interfaces*, B. B. Nickol (ed.), pp. 7–15. Academic Press, New York. (8)

Smith, N. J. 1981. *Man, Fishes and the Amazon*. Columbia University Press, New York. (11)

Smythe, N. 1970. Relationships between fruiting seasons and seed dispersal methods in a neotropical forest. Amer. Natur. 104: 25–36. (11)

Snow, B. K. 1970. A field study of the bearded bellbird in Trinidad. Ibis 112: 299–329. (11)

Snow, B. K. 1972. A field study of the calfbird *Perissocephalus tricolor*. Ibis 114: 139–142. (11)

Snow, B. K. and D. W. Snow. 1971. The feeding ecology of tanagers and honeycreepers in Trinidad. Auk 88: 291–322. (11, 13)

Snow, B. K. and D. W. Snow. 1972. Feeding niches of hummingbirds in a Trinidad valley. J. Anim. Ecol. 41: 471–485. (13)

Snow, D. W. 1961. The natural history of the oilbird, *Steatornis caripensis*, in Trinidad, W.I. I. General behavior and feeding habits. Zoologica 46: 27–48. (13)

Snow, D. W. 1962. The natural history of the oilbird (*Steatornis caripensis*) in Trinidad, W.I. II. Population, breeding ecology and food. Zoologica 47: 199–221. (11, 13)

Snow, D. W. 1966. A possible selective factor in evolution of fruiting seasons in tropical forests. Oikos 15: 274–281. (11, 13)

Snow, D. W. 1971. Evolutionary aspects of fruit-eating by birds. Ibis 113: 194–202. (11)

Snow, D. W. 1976. *The Web of Adaptation. Bird Studies in the American Tropics*. Quadrangle, New York Times Book Co., New York. (11)

Snow, D. W. 1980. Regional differences between tropical floras and the evolution of frugivory. Proc. XVII Int. Congr. Ornithol., pp. 1192–1198. (11)

Snow, D. W. 1981. Tropical frugivorous birds and their food plants: a world survey. Biotropica 13: 1–14. (11)

Snow, D. W. and A. Lill. 1974. Longevity records for some neotropical birds. Condor 76: 262–267. (11)

Snow, D. W. and B. K. Snow. 1980. Relationships between hummingbirds and flowers in the Andes of Colombia. Bull. Brit. Mus. (Nat. Hist.) 38: 105–139. (13)

528

Snyder, W. C. and T. A. Tousson. 1965. Current status of taxonomy in *Fusarium* species and their perfect stages. Phytopathology 55: 833–837. (7)

Soegeng. 1962. The species of *Durio* with edible fruits. Economic Botany 16: 270–282. (11)

Sogandares-Bernal, F. 1959. Digenetic trematodes of marine fishes from the Gulf of Panama and Bimini, British West Indies. Tulane Stud. Zool. 7: 69–117. (8)

Sohl, N. F. 1969. The fossil record of shell boring by snails. Am. Zool. 9: 725–734. (15)

Sokal, R. R. and J. Rohlf. 1969. *Biometry*. W. H. Freeman, San Francisco. (18)

Sonneborn, T. M. 1938. Mating types in *P. aurelia*: diverse conditions for mating in different stocks, occurrence, number and interrelations of the types. Proc. Am. Phil. Soc. 79: 411–434. (6)

Sorensen, A. E. 1981. Interactions between birds and fruit in a temperate woodland. Oecologia 50: 242–249. (11)

Sork, V. L. and D. H. Boucher. 1977. Dispersal of sweet pignut hickory in a year of low fruit production and the influence of predation by a curculionid beetle. Oecologia 28: 289–299. (11)

Sosa, O., Jr. 1981. Biotypes J and L of the Hessian fly discovered in an Indiana wheat field. J. Econ. Entomol. 74: 180–182. (10)

Soulé, M. and B. Stewart. 1970. The "niche variation" hypothesis: a test and alternatives. Amer. Natur. 104: 85–97. (17)

Soulsby, E. J. L. 1965. *Textbook of Veterinary Clinical Parasitology*. F. A. Davis, Philadelphia. (8)

Southwood, T. R. E. 1961. The number of species of insects associated with various trees. J. Anim. Ecol. 30: 1–8. (1, 10)

Southwood, T. R. E. 1973. The insect/plant relationship—an evolutionary perspective. Symp. Roy. Entomol. Soc. London 6: 3–20. (10, 15)

Southwood, T. R. E., V. K. Brown and P. M. Reader. 1979. The relationships of plant and insect diversities in succession. Biol. J. Linn. Soc. 12: 327–348. (19)

Sprague, G. F. and R. G. Dahms. 1972. Development of crop resistance to insects. J. Env. Qual. 1: 28–34. (10)

Sprent, J. F. A. 1962. Parasitism, immunity and evolution. In *The Evolution of Living Organisms*, G. S. Leeper (ed.), pp. 149–165. Melbourne University Press, Melbourne. (8)

Sprent, J. F. A. 1969. Evolutionary aspects of immunity in zooparasitic infections. In *Immunity to Parasitic Animals*, Vol. I., G. J. Jackson, R. Herman and I. Singer (eds.), pp. 3–62. Appleton-Century-Crofts, New York. (8)

Springett, B. P. 1968. Aspects of the relationship between burying beetles, *Necrophorus* spp., and the mite, *Poecilochirus necrophori* Vitz. J. Anim. Ecol. 37: 417–424. (3)

Stakman, E. C. and F. J. Piemeisel. 1917. Biologic forms of *Puccinia graminis* on cereals and grasses. J. Agr. Res. 10: 429–495. (7)

Stakman, E. C., and J. J. Parker and F. J. Piemeisel. 1918a. Can biologic forms of stem rust on wheat change rapidly enough to interfere with breeding for rust resistance? J. Agr. Res. 14: 111–123. (7)

Stakman, E. C., F. J. Piemeisel and M. N. Levine. 1918b. Plasticity of biologic forms of *Puccinia graminis*. J. Agr. Res. 15: 221–250. (7)

Stammer, H. J. 1957. Gedanken zu den parasitophyletischen Regeln und zur Evolution der Parasiten. Zool. Anzeig. 159: 255–267. (4)

Stamp, N. E. 1981. Behavior of parasitized aposematic caterpillars: advantageous to the parasitoid or the host? Amer. Natur. 118: 715–725. (8)

Stamps, J. and S. K. Tanaka. 1981. The influence of food and water on growth rates in a tropical lizard (*Anolis aeneus*). Ecology 62: 33–40. (17)

Stanley, S. M. 1968. Post-Paleozoic adaptive radiation of infaunal bivalve molluscs—a consequence of mantle fusion and siphon formation. J. Paleontol. 42: 214–229. (15)

Stanley, S. M. 1974. What has happened to the articulate brachiopods? Geol. Soc. Amer. Abstr. (with Programs) 6: 966–967. (15)

529

Stanley, S. M. 1975. A theory of evolution above the species level. Proc. Natl. Acad. Sci. USA 72: 646–650. (2, 15)

Stanley, S. M. 1977. Trends, rates, and patterns of evolution in the Bivalvia. In *Patterns of Evolution, as Illustrated by the Fossil Record*, A. Hallam (ed.), pp. 209–250. Elsevier, Amsterdam. (15)

Stanley, S. M. 1979. *Macroevolution: Pattern and Process.* W. H. Freeman, San Francisco. (2, 14, 15, 16)

Stanley, S. M. 1981. Patterns of extinction within the Bivalvia. Geol. Soc. Amer. Abstr. (with Programs) 13: 599. (15)

Stanley, S. M. 1982. Macroevolution and the fossil record. Evolution 36: 460–473. (15)

Stanley, S. M., W. O. Addicott and K. Chinzei. 1980. Lyellian curves in paleontology: possibilities and limitations. Geology 8: 422–426. (14)

Stanley, S. M. and L. D. Campbell. 1981. Neogene mass extinction of western Atlantic molluscs. Nature 293: 457–459. (14)

Stanley, S. M. and W. A. Newman. 1980. Competitive exclusion in evolutionary time: the case of the acorn barnacles. Paleobiol. 6: 173–183. (15)

Stapanian, M. A. 1980. Evolution of fruiting display and fruiting strategies among fleshy-fruited species of eastern Kansas. Ph.D. Thesis, Kansas State University, Manhattan. (11)

Stapanian, M. A. and C. C. Smith. 1978. A model for seed scatterhoarding: coevolution of fox squirrels and black walnuts. Ecology 59: 884–896. (11)

Stasz, J. L. 1981a. Predation by the Naticidae: the position of the borehole. (Unpublished manuscript) (15)

Stasz, J. L. 1981b. Predation by the Naticidae within a Miocene fauna. (Unpublished manuscript) (15)

Stebbins, G. L. 1970. Adaptive radiation of reproductive characteristics in angiosperms, I. Pollination mechanisms. Ann. Rev. Ecol. Syst. 1: 307–326. (13)

Stebbins, G. L. 1974. *Flowering Plants: Evolution Above the Species Level.* Harvard University Press, Cambridge, Massachusetts. (1, 10)

Stebbins, G. L. 1981. *Darwin to DNA: Molecules to Humanity.* W. H. Freeman, San Francisco. (1)

Steneck, R. S. 1977. A crustose coralline-limpet interaction in the Gulf of Maine. J. Phycol. 13: 65 (15)

Steneck, R. S. 1982a. A limpet-coralline alga association: adaptations and defenses between a selective herbivore and its prey. Ecology 63: 507–522. (15, 19)

Steneck, R. S. 1982b. Adaptive trends in the ecology and evolution of crustose coralline algae (Rhodophyta, Corallinaceae). Ph.D. Thesis, The Johns Hopkins University, Baltimore. (15)

Steneck, R. S. and L. Watling. 1982. Feeding capabilities and limitations of herbivorous molluscs: a functional group approach. Marine Biology 68: 299–319. (15)

Stent, G. S. 1963. *Molecular Biology of Bacterial Viruses.* W. H. Freeman, San Francisco. (5)

Stent, G. S. and R. Calendar. 1978. *Molecular Genetics: An Introductory Narrative.* W. H. Freeman, San Francisco. (5)

Stephenson, T. A. and A. Stephenson. 1972. *Life Between Tidemarks on Rocky Shores.* W. H. Freeman, San Francisco. (19)

Stewart, F. M. and B. R. Levin. 1973. Partitioning of resources and the outcome of interspecific competition: a model and some general considerations. Amer. Natur. 107: 171–198. (5)

Stewart, F. M. and B. R. Levin. 1977. The population biology of bacterial plasmids: *a priori* conditions for the existence of conjugationally transmitted factors. Genetics 87: 209–228. (5)

Stiles, E. W. 1980. Patterns of fruit presentation and seed dispersal in bird disseminated woody plants in the eastern deciduous forest. Amer. Natur. 116: 670–688. (11)

Stiles, F. G. 1975. Ecology, flowering phenology, and hummingbird pollination of some Costa Rican *Heliconia* species. Ecology 56: 285–301. (13)

Stiles, F. G. 1977. Coadapted competitors: the flowering seasons of hummingbird-pollinated plants in a tropical forest. Science 198: 1177–1178. (13)

Stiles, F. G. 1978a. Ecological and evolutionary implications of bird pollination. Amer. Zool. 18: 715–727. (13)

530

Stiles, F. G. 1978b. Temporal organization of flowering among the hummingbird food-plants of a tropical wet forest. Biotropica 10: 194–210. (13)

Stiles, F. G. 1980. The annual cycle in a tropical wet forest hummingbird community. Ibis 122: 322–343. (13)

Stiles, F. G. and L. L. Wolf. 1979. Ecology and evolution of lek mating behavior in the long-tailed hermit hummingbird. American Ornithologists' Union Ornithological Monograph #27. (13)

Stiling, P. D. 1980. Competition and coexistence among *Eupteryx* leafhoppers (Hemiptera, Cicadellidae) occurring on stinging nettles (*Urtica dioica*). J. Anim. Ecol. 49: 793–806. (10)

Stocker, B. A. D. 1956. Abortive transduction of motility in *Salmonella*; a non-replicated gene transmitted through many generations to a single descendant. J. Gen. Microbiol. 15: 575–598. (5)

Storer, R. W. 1966. Sexual dimorphism and food habits in three North American accipiters. Auk 83: 423–436. (18)

Straw, R. M. 1955. Hybridization, homogamy and sympatric speciation. Evolution 9: 441–444. (13)

Strickler, K. 1979. Specialization and foraging efficiency of solitary bees. Ecology 60: 998–1009. (13)

Strobeck, C. 1975. Selection in a fine grained environment. Amer. Natur. 109: 419–425. (2)

Strong, D. R., Jr. 1974a. Nonasymptotic species richness models and the insects of British trees. Proc. Natl. Acad. Sci. USA 71: 2766–2769. (10)

Strong, D. R., Jr. 1974b. Rapid asymptotic species accumulation in phytophagous insect communities: the pests of cacao. Science 185: 1064–1066. (10)

Strong, D. R. 1979. Biogeographic dynamics of insect-host plant communities. Ann. Rev. Entomol. 24: 89–119. (10)

Strong, D. R. 1980. Null hypotheses in ecology. Synthese 43: 271–286. (18)

Strong, D. R. and D. Simberloff. 1981. Straining at gnats and swallowing ratios: character displacement. Evolution 35: 810–812. (18, 19)

Strong, D. R., L. A. Szyska and D. S. Simberloff. 1979. Tests of community-wide character displacement against null hypotheses. Evolution 33: 897–913. (13, 17, 18)

Struhsaker, T. T. 1975. *The Red Colobus Monkey*. University of Chicago Press, Chicago. (11)

Stuart, C. T. 1976. Diet of the black-backed jackal, *Canis mesomelas*, in the central Namib desert, South West Africa. Zoologica Africana 11: 193–205. (11)

Stunkard, H. W. 1959. Induced gametogenesis of a monogenetic trematode, *Polystoma stellai* Vigueras, 1955. J. Parasitol. 45: 389–394. (4)

Sturgeon, K. B. 1979. Monoterpene variation in ponderosa pine xylem resin related to western pine beetle predation. Evolution 33: 803–814. (10)

Sved, J. A. and O. Mayo. 1970. The evolution of dominance. In *Mathematical Topics in Population Genetics*, K. Kojima (ed.), pp. 289–316. Springer-Verlag, Berlin. (7)

Swynnerton, C. F. M. 1908. Further notes on the birds of Gazaland. Ibis 2: 391–442. (11)

Szidat, L. 1939. Beiträge zum Aufbau eines natürlichen Systems der Trematoden. I. Die Entwicklung vom *Echinocercaria choanophila* U. Szidat zu *Cathaemaria hians* und die Ableitung der Fasciolidae von den Echinostomidae. Z. Parasitenkd. 11: 238–283. (4)

Szidat, L. 1956. Geschichte, Anwendung und einige Folgerungen aus den parasitogenetischen Regeln. Z. Parasitenk. 17: 237–268. (4)

Szidat, L. 1960a. Der marine Charakter der Parasitenfauna der Süsswasserfische des Stromssystems des Rio de la Plata und ihre Deutung als Reliktfauna des Tertiären Tethys-Meeres. Proc. XIV Int. Cong. Zool. 128–138. (4)

Szidat, L. 1960b. *La Parasitología como Ciencia Auxiliar para Develar Problemas Hidrobiológicas, Zoogeográficas, y Geofísicas del Atlántico Sur*. Libro Homenaje Eduardo Caballero, Mexico City. (4)

Taylor, D. L. 1973. The cellular interactions of algal-invertebrate symbionts. Adv. Mar. Biol. 11: 1–56. (14)

Temple, S. A. 1975. Plant-animal mutualism: coevolution with dodo leads to near extinction of plant. Science 197: 885–886. (11, 13)

Tepedino, V. J. 1981. The pollination efficiency of the squash bee (*Peponapis pruinosa*) and the honey bee (*Apis mellifera*) on summer squash (*Cucurbita pepo*). J. Kansas Entomol. Soc. 54: 359–377. (13)

Terborgh, J. and J. Faaborg. 1980. Saturation of bird communities in the West Indies. Amer. Natur. 116: 178–195. (17)

Terwedow, H. A., Jr. and G. B. Craig, Jr. 1977. *Waltonella flexicauda*: development controlled by a genetic factor in *Aedes aegypti*. Exp. Parasitol. 41: 272–282. (8)

Thom, W. S. 1937. The Malayan or Burmese sambor (*Runa unicolor equinus*). Bomb. Nat. Hist. Soc. 39: 309–319. (11)

Thomas, L. 1979. *The Medusa and the Snail: More Notes of a Biology Watcher*. Viking Press, New York. (6)

Thompson, J. N. 1980. Treefalls and colonization patterns of temperate forest herbs. Amer. Mid. Nat. 104: 176–184. (11)

Thompson, J. N. 1981. Elaiosomes and fleshy fruits: phenology and selection pressures for ant-dispersed seeds. Amer. Natur. 117: 104–108. (11)

Thompson, J. N. and M. F. Willson. 1978. Disturbance and the dispersal of fleshy fruits. Science 200: 1161–1163. (11)

Thompson, J. N. and M. F. Willson 1979. Evolution of temperate fruit/bird interactions: phenological strategies. Evolution 33: 973–982. (11)

Thomson, J. D. 1980. Skewed flowering distributions and pollinator attraction. Ecology 61: 572–579. (13)

Thomson, J. D. 1981. Spatial and temporal components of resource assessment by flower-feeding insects. J. Anim. Ecol. 50: 49–60. (13)

Thomson, J. D. and R. C. Plowright. 1980. Pollen carryover, nectar rewards, and pollinator behavior with special reference to *Diervilla lonicera*. Oecologia 46: 68–74. (13)

Thorp, R. W. 1969. Systematics and ecology of bees of the subgenus *Diandrena* (Hymenoptera: Andrenidae). U. Calif. Publ. Entomol. 52: 1–146. (13)

Thorpe, W. H. 1929. Biological races in *Hyponomeuta padella*. J. Linn. Soc. Zool. 36: 621–634. (10)

Thorpe, W. H. 1930. Biological races in insects and allied groups. Biol. Rev. 5: 177–212. (10)

Thorson, G. 1957. Bottom communities (sublittoral or shallow shelf). Geol. Soc. Amer. Mem. 67: 461–534. (19)

Tokeson, J. P. E. and J. C. Holmes. 1982. The effects of temperature and oxygen on the development of *Polymorphus marilus* (Acanthocephala) in *Gammarus lacustris* (Amphipoda). J. Parasitol. 68: 112–119. (8)

Tomback, D. F. 1978. Foraging strategies of Clark's Nutcracker. Living Bird 16: 123–161. (11)

Tomoff, C. W. 1974. Avian species diversity in desert scrub. Ecology 55: 396–403. (19)

Tramer, E. J. 1974. On latitudinal gradients in avian diversity. Condor 76: 123–130. (19)

Trangle, K. L., M. J. Goluska, M. J. O'Leary and S. D. Douglas. 1979. Distribution of blood groups and secretor status in schistosomiasis. Parasite Immunol. 1: 133–140. (8)

Travis, C. C. and W. M. Post. 1979. Dynamics and comparative statics of mutualistic communities. J. Theor. Biol. 78: 553–571. (3)

Trott, L. B. 1970. Contributions to the biology of carapid fishes (Paracanthopterygii: Gadiformes). Univ. of Calif. Publ. Zool. 89: 1–60. (14)

Turcek, F. J. 1963. Color preferences in fruit and seed-eating birds. Proc. XIII International Ornithological Congress, pp. 285–292. (11)

Turner, J. R. G. 1968. Natural selection for and against a polymorphism which interacts with sex. Evolution 22: 481–495. (12)

Turner, J. R. G. 1971. Studies of Müllerian mimicry and its evolution in burnet moths and heliconid butterflies. In *Ecological Genetics and Evolution*, R. Creed (ed.). Blackwell Scientific, Oxford. (12)

532

Turner, J. R. G. 1976. Muellerian mimicry; classical "beanbag" evolution and the role of ecological islands in adaptive race formation. In *Population Genetics and Ecology*, S. Karlin and E. Nevo (eds.), pp. 185-218. Academic Press, New York. (1)

Turner, J. R. G., 1977. Butterfly mimicry: the genetical evolution of an adaptation. In *Evolutionary Biology*, Vol. 10, M. K. Hecht, W. C. Steere and B. Wallace (eds.), pp. 163-206. Plenum Press, New York. (3, 12)

Turner, J. R. G. 1981. Adaptation and evolution in *Heliconius*: a defense of neo-Darwinism. Ann. Rev. Ecol. Syst. 12: 99-121. (3, 4, 12)

Turner, J. R. G., M. S. Johnson and W. F. Eanes. 1979. Contrasted modes of evolution in the same genome: allozymes and adaptive changes in *Heliconius*. Proc. Natl. Acad. Sci. USA 76: 1924-1928. (12)

Tursch, B., E. Tursch, I. T. Harrison, G. B. C. T. C. B. da Silva, H. J. Monteiro, B. Gilbert, W. B. Mors and C. Djerassi. 1963. Terpenoids. LIII. Demonstration of ring conformational changes in triterpenes of the B-amyrin class isolated from *Stryphnodendron coriaceum*. J. Organ. Chem. 28: 2390-2394. (11)

Uchida, T., D. M. Gill and A. M. Pappenheimer. 1971. Mutation in the structural gene for diptheria toxin carried by the temperate phage. Nature New Biol. 233: 8-11. (5)

Udeinya, I. J., J. A. Schmidt, M. Aikawa, L. H. Miller and I. Green. 1981. Falciparum malaria-infected erythrocytes specifically bind to cultured human endothelial cells. Science 213: 555-557. (8)

Udovic, D. and C. Aker. 1981. Fruit abortion and the regulation of fruit number in *Yucca whipplei*. Oecologia 49: 245-248. (13)

Ulmer, M. J. 1971. Site-finding behavior in helminths in intermediate and definitive hosts. In *Ecology and Physiology of Parasites*, A. M. Fallis (ed.), pp. 123-159. University of Toronto Press, Toronto. (8)

Uyenoyama, M. K. 1979. Evolution of altruism under group selection in large and small populations in fluctuating environments. Theor. Pop. Biol. 15: 58-85. (2)

Valentine, D. H. 1978. The pollination of introduced species, with special reference to the British Isles and the genus *Impatiens*. In *The Pollination of Flowers by Insects*, A. J. Richards (ed.), pp. 117-123. Academic Press, London. (13)

Vance, R. R. 1972. The role of shell adequacy in behavioral interactions involving hermit crabs. Ecology 53: 1074-1083. (14)

Vance, R. R. 1978. A mutualistic interaction between a sessile marine clam and its epibionts. Ecology 59: 679-685. (3, 14, 19)

Van den Hoek, C., A. M. Breeman, R. P. M. Bak and G. Van Buurt. 1978. The distribution of algae, corals and gorgonians in relation to depth, light attenuation, water movement and grazing pressure in the fringing coral reef of Curaçao, Netherlands Antilles. Aquat. Bot. 5: 1-46. (15)

Vandermeer, J. H. 1977. Notes on density dependence in *Welfia georgii* Wendl. ex. Burret (Palmae) a lowland rainforest species in Costa Rica. Brenesia 10/11: 9-15. (11)

Vandermeer, J. H. 1980. Indirect mutualism: variations on a theme by Stephen Levine. Amer. Natur. 116: 441-448. (3)

Vandermeer, J. H. and D. H. Boucher. 1978. Varieties of mutualistic interaction in population models. J. Theor. Biol. 74: 549-558. (3)

van der Pers, J. C. N. 1978. Responses from olfactory receptors in females of three species of small ermine moths (Lepidoptera: Yponomeutidae) to plant odours. Ent. exp. appl. 24: 394-398. (10)

van der Pijl, L. 1957. The dispersal of plants by bats. Acta. Bot. Neerl. 6: 291-315. (11)

van der Pijl, L. 1966. Ecological aspects of fruit evolution. Koninkl. Nederl. Akademie van Weten Schappen, Amsterdam, Proc. Series C, 69: 597-640. (11)

533

van der Pijl, L. 1972. *Principles of Dispersal in Higher Plants.* Springer-Verlag, Berlin. (11)

van der Pijl, L. and C. H. Dodson. 1966. *Orchid Flowers: Their Pollination and Evolution.* University of Miami Press, Coral Gables, Florida. (13)

Van der Plank, J. E. 1963. *Plant Diseases: Epidemics and Control.* Academic Press, New York. (7)

Van der Plank, J. E. 1975. *Principles of Plant Infection.* Academic Press, New York. (9)

van Drongelen, W. 1980. Behavioural responses of two small ermine moth species (Lepidoptera: Yponomeutidae) to plant constituents. Ent. exp. appl. 28: 54–58. (10)

van Drongelen, W. and J. A. van Loon. 1980. Inheritance of gustatory sensitivity in F_1 progeny of crosses between *Yponomeuta cagnagellus* and *Y. malinellus* (Lepidoptera). Ent. exp. appl. 28: 199–203. (10)

van Emden, H. F. (ed.). 1973. *Insect/Plant Relationships.* Blackwell Scientific, London. (10)

Van der Wall, S. B. and R. P. Balda. 1977. Coadaptations of the Clark's nutcracker and the piñon pine for efficient seed harvest and dispersal. Ecol. Monogr. 47: 89–111. (11)

Van der Wall, S. B. and R. P. Balda. 1981. Ecology and evolution of food storage behavior in conifer-seed-caching corvids. Z. Tierpsychol. 56: 217–242. (11)

Van Lawick-Goodall, H. and J. Van Lawick-Goodall. 1971. *Innocent Killers.* Houghton Mifflin, Boston. (16)

van Noordwijk, A., J. Van Balen and W. Scharloo. 1980. Heritability of ecologically important traits in the Great Tit. Ardea 68: 193–203. (17)

Van Tyne, J. 1929. The life history of the toucan *Rhamphastos brevicarinatus.* Univ. Mich. Mus. Zool., Misc. Pub. 19. (11)

Van Valen, L. 1965. Morphological variation and the width of the ecological niche. Amer. Natur. 99: 377–390. (17)

Van Valen, L. 1973. A new evolutionary law. Evol. Theory. 1: 1–30. (8, Ep)

Van Valen, L. 1978. The beginning of the Age of Mammals. Evol. Theory 4: 45–80. (16)

Van Valen, L. and P. R. Grant. 1970. Variation and niche width reexamined. Amer. Natur. 104: 589–590. (17)

Van Valkenburgh, B. 1982a. Evolutionary dynamics of terrestrial, large predator guilds. In Third North American Paleontological Convention Proceedings, Vol. II, B. Mamet and M. J. Copeland (eds.), pp. 557–562. (15)

Van Valkenburgh, B. 1982b. Locomotor diversity in fossil and recent predator guilds. Part 2: Evolutionary implications. (In preparation) (15)

Van Valkenburgh, B. 1982c. Carnivore diets and dentition: a multivariate analysis. (In preparation) (15)

Vane-Wright, R. I. 1978. Ecological and behavioral origins of diversity in butterflies. In *Diversity of Insect Faunas*, L. A. Mound and N. Waloff (eds.). Symp. Roy. Ent. Soc. Lond. 9: 56–70. (10)

Vane-Wright, R. I. 1980. Mimicry and its unknown ecological consequences. In *The Evolving Biosphere*, P. H. Greenwood (ed.), pp. 157–168. Cambridge University Press, New York. (12)

Vazquez-Yanes, C. 1977. Germination of a pioneer tree (*Trema guineensis* Ficahlo) from equatorial Africa. Turrialba 27: 301–302. (11)

Vazquez-Yanes C., A. Orozco, G. Francois and L. Trejo. 1975. Observations on seed dispersal by bats in a tropical humid region in Veracruz, Mexico. Biotropica 7: 73–76. (11)

Vaurie, C. 1951. Adaptive differences between two sympatric species of nuthatches (*Sitta*). Proc. X Internat. Ornithol. Congr. Uppsala, 1950, pp. 163–166. (17)

Velimirov, B., J. E. Field, C. L. Griffiths and P. Zoutendyk. 1977. The ecology of kelp bed communities in the Benguela upwelling system. Helgolander wiss. Meeresunters 30: 495–518. (19)

Vermeij, G. J. 1977. The Mesozoic marine revolution: evidence from snails, predators and grazers. Paleobiology 3: 245–258. (14, 15)

Vermeij, G. J. 1978. *Biogeography and Adaptation: Patterns of Marine Life.* Harvard University Press, Cambridge, Massachusetts. (14, 19)

Vermeij, G. J. 1980. Drilling predation of bivalves in Guam: some paleoecological implications. Malacologia 19: 329–334. (14)

Vermeij, G. J. 1982a. Environmental change and the evolutionary history of the periwinkle (*Littorina littorea*) in North America. Evolution 36: 561–580. (14)

Vermeij, G. J. 1982b. Unsuccessful predation and evolution. Amer. Natur. 120: 701–720. (14)

Vermeij, G. J. and A. P. Covich. 1978. Coevolution of freshwater gastropods and their predators. Amer. Natur. 112: 833–843. (14)

Vermeij, G. J. and J. D. Currey. 1980. Geographical variation in the strength of thaidid snail shells. Biol. Bull. 158: 383–389. (14)

Vermeij, G. J., D. E. Schindel and E. Zipser. 1981. Predation through geological time: evidence from gastropod shell repair. Science 214: 1024–1026. (14)

Vermeij, G. J., E. Zipser and E. C. Dudley. 1980. Predation in time and space: peeling and drilling in terebrid gastropods. Paleobiology 6: 352–364. (11)

Verschaffelt, E. 1910. The cause determining the selection of food in some herbivorous insects. Proc. Acad. Sci. Amsterdam 13: 536–542. (10)

Verwey, J. 1930. Coral reef studies. I. The symbiosis between damsel fishes and sea anemones in Batavia Bay. Treubia 12: 305–366. (3, 19)

Vine, P. J. 1974. Effects of algal grazing and aggressive behavior of the fishes *Pomacentrus lividus* and *Acanthurus sohal* on coral-reef ecology. Mar. Biol. 24: 131–136. (16)

Vinson, S. B. 1976. Host selection by insect parasitoids. Ann. Rev. Ent. 21: 109–133. (10)

Vinson, S. B. 1977. *Microplitis croceipes*: inhibition of the *Heliothis zea* defense reaction to *Cardiochiles nigriceps*. Exp. Parasitol. 41: 112–117. (8)

Vinson, S. B. and G. F. Iwantsch. 1980. Host regulation by insect parasitoids. Quart. Rev. Biol. 55: 143–165. (8)

Vogel, V. S. 1969. Chiropterophilie in der Neotropischen Flora, Neue Mitteilungen III. Flora, Abt. B. 158: 289–323. (11)

Voigt, E. 1970. On grazing traces produced by the radula of fossil and recent gastropods and chitons. In *Trace Fossils* 2, T. P. Crimes and J. C. Harper (eds.), pp. 335–347. Seel House Press, Liverpool. (15)

Voorhies, M. 1969. Taphonomy and population dynamics of an Early Pliocene Vertebrate Fauna. Contrib. Geol. Univ. Wyo. Special Paper 1. (16)

Voorhies, M. R. and J. R. Thomasson. 1979. Fossil grass anthoecia within Miocene rhinoceros skeletons: diet in an extinct species. Science 206: 331–333. (11)

Vrba, E. S. 1980. Evolution, species and fossils: How does life evolve? South African J. Sci. 76: 61–84. (2)

Waage, J. K. 1979. The evolution of insect/vertebrate associations. Biol. J. Linn. Soc. 12: 187–224. (8)

Waddington, C. H. 1960. Evolutionary adaptation. In *Evolution After Darwin*, S. Tax (ed.), pp. 381–402. University of Chicago Press, Chicago. (2)

Waddington, K. D. and B. Heinrich. 1981. Patterns of movement and floral choice by foraging bees. In *Foraging Behavior: Ecological, Ethological and Psychological Approaches*, A. Kamil and T. Sargent (eds.), pp. 215–230. Garland Press, New York. (13)

Wade, M. J. 1977. An experimental study of group selection. Evolution 31: 134–153. (2)

Wade, M. J. 1978. A critical review of the models of group selection. Quart. Rev. Biol. 53: 101–114. (2)

Wade, M. J. and D. E. McCauley. 1980. Group selection: the phenotypic and genotypic differentiation of small populations. Evolution 34: 779–798. (2)

Wahl, I. 1970. Prevalence and geographic distribution of resistance to crown rust in *Avena sterilis*. Phytopathology 60: 746–749. (7)

535

Wahl, I., N. Eshed, A. Segal and Z. Sobel. 1978. Significance of wild relatives of small grains and other wild grasses in cereal powdery mildews. In *The Powdery Mildews*, D. M. Spencer (ed.), pp. 83–100. Academic Press, London. (7)

Wakelin, D. 1978a. Genetic control of susceptibility and resistance to parasitic infection. Adv. Parasitol. 16: 219–308. (8)

Wakelin, D. 1978b. Immunity to intestinal parasites. Nature 273: 617–620. (8)

Waldbauer, G. P. and G. Fraenkel. 1961. Feeding on normally rejected plants by maxillectomized larvae of the tobacco hornworm, *Protoparce sexta* (Lepidoptera, Sphingidae). Ann. Ent. Soc. Amer. 54: 477–485. (10)

Wallace, A. R. 1889. *Darwinism*. Macmillan, New York. (12)

Wallace, J. and R. Mansell (eds.). 1976. *Biochemical Interactions Between Plants and Insects*. Rec. Adv. Phytochem. 10. Plenum Press, New York. (10)

Walls, G. L. 1942. *The Vertebrate Eye and its Adaptive Radiation*. Harper and Bros., New York. (19)

Walsberg, G. E. 1975. Digestive adaptations of *Phainopepla nitens* associated with eating of mistletoe berries. Condor 77: 169–174. (11)

Walter, E. 1973. *Vegetation of the Earth in Relation to Climate and the Eco-Physiological Conditions*, Springer-Verlag, New York. (19)

Wanders, J. B. W. 1977. The role of benthic algae in the shallow reef of Curaçao (Netherlands Antilles) III: The significance of grazing. Aquatic Botany 3: 357–390. (15)

Warén, A. 1980a. Revision of the genera *Thyca*, *Stilifer*, *Scalenostoma*, *Mucronalia*, and *Echineulima* (Mollusca, Prosobranchia, Eulimidae). Zool. Scripta 9: 187–210. (14)

Warén, A. 1980b. Descriptions of new taxa of Eulimidae (Mollusca, Prosobranchia), with notes on some previously described genera. Zool. Scripta 9: 283–306. (14)

Waser, N. M. 1978a. Competition for hummingbird pollination and sequential flowering in two Colorado wildflowers. Ecology 59: 934–944. (13)

Waser, N. M. 1978b. Interspecific pollen transfer and competition between co-occurring plant species. Oecologia 36: 223–236. (13)

Waser, N. M. 1982. Competition for pollination and floral character differences among sympatric plant species: A review of evidence. In *Handbook of Experimental Pollination Ecology*, C. E. Jones and R. J. Little (eds.). Van Nostrand Reinhold, New York. (In press) (13)

Waser, N. M. and M. V. Price. 1981. Pollinator choice and stabilizing selection for flower color in *Delphinium nelsonii*. Evolution 35: 376–390. (13)

Waser, N. M. and M. V. Price. 1982. Optimal and actual outcrossing in plants, and the nature of plant-pollinator interaction. In *Handbook of Experimental Pollination Ecology*, C. E. Jones and R. J. Little (eds.). Van Nostrand Reinhold, New York. (In press) (13)

Waser, N. M. and L. A. Real. 1979. Effective mutualism between sequentially flowering plant species. Nature 281: 670–672. (13)

Waser, P. 1977. Figs, wasps, and primates. Africana 6(7): 23–25. (11)

Wasserman, S. D. and D. J. Futuyma. 1981. Evolution of host plant utilization in laboratory populations of the southern cowpea weevil, *Callosobruchus maculatus* Fabricius (Coleoptera: Bruchidae). Evolution 35: 605–617. (10)

Wassom, D. L., C. S. David and G. J. Gleich. 1979. Genes within the major histocompatibility complex influence susceptibility to *Trichinella spiralis* in the mouse. Immunogenetics 9: 491–496. (8)

Watanabe, T. 1963. Infective heredity of multiple drug resistance in bacteria. Bacter. Rev. 27: 87–115. (5)

Waterhouse, W. L. 1929. Australian rust studies, I. Proc. Linn. Soc. N.S.W. 54: 615–680. (7)

Way, M. J. 1963. Mutualism between ants and honeydew-producing Homoptera. Ann. Rev. Entomol. 8: 307–344. (3)

Webb, S. D. 1977. A history of savanna vertebrates in the New World. Ann. Rev. Ecol. Syst. 8: 355–380. (16)

Weigelt, J. 1930. Ueber die vermutliche Nahrung von *Protosaurus*. Leopoldina 6: 269–280. (11)

Weigle, J. 1966. Assembly of phage Lambda in vitro. Proc. Natl. Acad. Sci. USA 55: 1462–1466. (6)

536

Weinack, O. M., G. H. Snoeyenbos, C. F. Smyser and A. S. Soerjadi. 1981. Competitive exclusion of intestinal colonization of *Escherichia coli* in chicks. Avian Dis. 25: 696–705. (8)

Wells, H. W. 1979. Hydroid and sponge commensals of *Cantharus cancellarius* with a "false shell." Nautilus 82: 93–102. (14)

Werger, M. J. A., F. J. Kruger and H. C. Taylor. 1972a. A phytosociological study of the Cape fynbos and other vegetation at Jonkershoek, Stellenbosch. Bothalia 10: 599–614. (19)

Werger, M. J. A., F. J. Kruger and H. C. Taylor. 1972b. Pflanzensoziologische Studie der Fynbos-Vegetation am Kap der guten Hoffnung. Vegetatio 24: 71–89. (19)

Werger, M. J. A. 1973. Phytosociology of the upper Orange River Valley, South Africa: A synecological and syntaxonomical study. Ph.D. Thesis, Catholic University of Nijmegen. (19)

West, N. E. 1968. Rodent-influenced establishment of ponderosa pine and bitterbush seedlings in Central Oregon. Ecology 49: 1009–1011. (11)

Western, D. 1979. Size, life history and ecology in mammals. Afr. J. Ecol. 17: 185–204. (19)

Westman, W. E. 1975. Edaphic climax pattern of the pygmy forest region of California. Ecol. Monogr. 45: 109–135. (19)

Wheeler, H. 1975. *Plant Pathogenesis.* Springer-Verlag, Berlin. (7)

Wheeler, W. M. 1910. *Ants: Their Structure, Development and Behavior.* Columbia University Press, New York. (1)

Wheelwright, N. T. and G. H. Orians. 1982. Seed dispersal by animals: contrasts with pollen dispersal, problems of terminology, and constraints on coevolution. Amer. Natur. 119: 402–413. (11, 13)

White, J. 1979. The plant as a metapopulation. Ann. Rev. Ecol. System. 10: 109–145. (13)

White, M. J. D. 1978. *Modes of Speciation.* W. H. Freeman, San Francisco. (8, 10)

White, S. C. 1974. Ecological aspects of growth and nutrition in tropical fruit-eating birds. Ph.D. Thesis, University of Pennsylvania, Philadelphia. (11)

White, T. C. R. 1974. A hypothesis to explain outbreaks of looper caterpillars, with special reference to populations of *Selidosema suavis* in a plantation of *Pinus radiata* in New Zealand. Oecologia 16: 279–302. (19)

White, T. C. R. 1978. The importance of a relative shortage of food in animal ecology. Oecologia 33: 71–86. (19)

Whitham, T. G. 1978. Habitat selection by *Pemphigus* aphids in response to resource limitation and competition. Ecology 59: 1164–1176. (10)

Whitham, T. G. and C. N. Slobodchikoff. 1981. Evolution by individuals, plant-herbivore interactions, and mosaics of genetic variability: the adaptive significance of somatic mutations in plants. Oecologia 49: 287–292. (13)

Whittaker, R. H. 1977. Evolution of species diversity in land plant communities. Evol. Biol. 10: 1–66. (19)

Whittaker, R. H. and P. P. Feeny. 1971. Allelochemicals: chemical interactions between species. Science 171: 757–770. (10)

Whittaker, R. H. and W. A. Niering. 1965. Vegetation of the Santa Catalina Mountains of Arizona. II. A gradient analysis of the south slope. Ecology 46: 429–452. (19)

Whittaker, R. H. and W. A. Niering. 1968. Vegetation of the Santa Catalina Mountains, Arizona. IV. Limestone and acid soils. J. Ecol. 56: 523–544. (19)

Wickler, W. 1968. *Mimicry in Plants and Animals.* McGraw-Hill, New York. (12)

Wickler, W. and U. Seibt. 1976. Field studies on the African fruit bat *Epomophorus wahlbergi* (Sundevall), with special reference to male calling. Z. Tierpsychol. 40: 345–376. (11)

Wiebes, J. J. 1979. Co-evolution of figs and their insect pollinators. Ann. Rev. Ecol. System. 10: 1–12. (4, 13)

Wiegert, R. G. and D. F. Owen. 1971. Trophic structure, available resources and population density in terrestrial vs. aquatic ecosystems. J. Theoret. Biol. 30: 69–81. (14)

Wiens, D., 1978. Mimicry in plants. Evolutionary Biology 11: 365–403. (12)

Wiens, J. A. 1977. On competition and variable environments. Amer. Scientist 65: 590–597. (17)

Wiens, J. A. and J. T. Rotenberry. 1980. Patterns of morphology and ecology in grassland and shrubsteppe bird populations. Ecol. Monogr. 50: 287–308. (18)

Wiklund, C. 1975. The evolutionary relationship between adult oviposition preferences and larval host plant range in *Papilio machaon* L. Oecologia 18: 185–197. (10)

Wiley, E. O. 1981. *Phylogenetics: The Theory and Practice of Phylogenetic Systematics.* John Wiley & Sons, New York. (4)

Wilkinson, C. R. 1978. Microbial associations in sponges. I. Ecology, physiology and microbial populations of coral reef sponges. Mar. Biol. 49: 161–167. (14)

Wilkinson, C. R. and P. Fay. 1979. Nitrogen fixation in coral reef sponges with symbiotic Cyanobacteria. Nature 279: 527–529. (14)

Wilkinson, C. R. and J. Vacelet. 1979. Transplantation of marine sponges to different conditions of light and current. J. Exp. Mar. Biol. Ecol. 37: 91–104. (14)

Willetts, N. 1972. The genetics of transmissable plasmids. Ann. Rev. Gen. 6: 257–268. (5)

Williams, E. E. 1969. The ecology of colonization as seen in the zoogeography of anoline lizards on small islands. Quart. Rev. Biol. 44: 345–389. (18)

Williams, E. E. 1972. The origin of faunas. Evolution of lizard congeners in a complex island fauna: a trial analysis. Evol. Biol. 6: 47–89. (17, 18)

Williams, G. C. 1966. *Adaptation and Natural Selection.* Princeton University Press, Princeton, New Jersey. (2)

Williams, G. C. 1975. *Sex and Evolution.* Princeton University Press, Princeton, New Jersey. (9)

Williams, H. H. 1960. The intestine in members of the genus *Raja* and host-specificity in the Tetraphyllidea. Nature 188: 514–516. (8)

Williams, H. H. 1961. Observations on *Echeneibothrium maculatum* (Cestoda: Tetraphyllidea). J. Mar. Biol. Assn. U.K. 41: 631–652. (8)

Williams, H. H. 1966. The ecology, functional morphology and taxonomy of *Echeneibothrium* Beneden, 1849, (Cestoda: Tetraphyllidea), a revision of the genus and comments on *Discobothrium* Beneden, 1870, *Pseudanthobothrium* Baer, 1956, and *Phormobothrium* Alexander, 1963. Parasitology 56: 227–286. (8)

Williams, K. S. and L. E. Gilbert. 1981. Insects as selective agents on plant vegetative morphology: egg mimicry reduces egg laying by butterflies. Science 212: 467–469. (10, 12)

Williams, N. H. 1982a. Floral fragrances as cues in animal behavior. In *Handbook of Experimental Pollination Biology,* C. D. Jones and R. J. Little (eds.). Van Nostrand Reinhold, New York. (In press) (13)

Williams, N. H. 1982b. The biology of orchids and euglossine bees. In *Orchid Biology: Reviews and Perspectives* II, J. Arditti (ed.). Cornell University Press, Ithaca, New York. (In press) (13)

Williams, N. H. and C. H. Dodson. 1972. Selective attraction of male euglossine bees to orchid floral fragrances and its importance in long distance pollen flow. Evolution 26: 84–95. (13)

Williams, P. H. 1975. Genetics of resistance in plants. Genetics 79: 404–419. (10)

Williams, T. C. 1968. Nocturnal orientation techniques of a neotropical bat. Ph.D. Thesis, Rockefeller University, New York. (11)

Williams, T. C. and J. M. Williams. 1967. Radio tracking of homing bats. Science 155: 154–155. (11)

Williams Smith, H. 1972. Ampicillin resistance in *Escherichia coli* by phage infection. Nature New Biology 238: 205–206. (5)

Williamson, D. L. 1965. Kinetic studies of "sex-ratio" spirochetes in *Drosophila melanogaster* Meigen females. J. Invert. Path. 7: 493–504. (6)

Williamson, D. L., K. Oishi and D. F. Poulson. 1977. Viruses of Drosophila sex-ratio spiroplasma. In *The Atlas of Plant and Insect Viruses,* K. Maramorosch (ed.), pp. 465–472. Academic Press, New York. (6)

Williamson, D. L. and D. F. Poulson. 1979. Sex ratio organisms (spiroplasmas) of *Drosophila.* In *The Mycoplasmas,* Vol. III, R. Whitcomb and J. Tully (eds.), pp. 175–208. Academic Press, New York. (6)

538

Williamson, D. L. and R. F. Whitcomb. 1975. Plant mycoplasmas: a cultivatable spiroplasm causes corn stunt disease. Science 188: 1018–1020. (6)

Willis, E. O. 1966. Competitive exclusion in birds of fruiting trees in Western Colombia. Auk 83: 479–480. (11)

Willson, M. F. 1972. Seed size preferences in finches. Wilson Bull. 84: 449–455. (17)

Willson, M. F. 1979. Sexual selection in plants. Amer. Natur. 113: 777–790. (13)

Wilson, A. C., S. S. Carlson and T. J. White. 1977. Biochemical evolution. Ann. Rev. Biochem. 46: 573–639. (Ep)

Wilson, D. E. 1978. Reproduction in Neotropical bats. Periodicum Biologorum 75: 215–217. (11)

Wilson, D. E. and D. H. Janzen. 1972. Predation on *Scheelea* palm seeds by bruchid beetles: seed density and distance from the parent palm. Ecology 53: 954–959. (11)

Wilson, D. E. and E. L. Tyson. 1970. Longevity records for *Artibeus jamaicensis* and *Myotis nigricans*. J. Mammal. 51: 203. (11)

Wilson, D. S. 1975a. The adequacy of body size as a niche difference. Amer. Natur. 109: 769–784. (18)

Wilson, D. S. 1975b. A theory of group selection. Proc. Natl. Acad. Sci. USA 72: 143–146. (3)

Wilson, D. S. 1977. Structured demes and the evolution of group advantageous traits. Amer. Natur. 111: 157–185. (3)

Wilson, D. S. 1979. Structured demes and trait-group variation. Amer. Natur. 113: 606–610. (3)

Wilson, D. S. 1980. *The Natural Selection of Populations and Communities*. Benjamin/Cummings, Menlo Park, California. (2, 3, 4, 8, 9)

Wilson, E. O. 1961. The nature of the taxon cycle in the Melanesian ant fauna. Amer. Natur. 95: 169–193. (3, 17)

Wing, B. L. and K. A. Clendenning. 1971. Kelp surfaces and associated invertebrates. Nova Hedwigia 32: 319–341. (19)

Wingate, D. B. 1965. Terrestrial herpetofauna of Bermuda. Herpetologica 21: 202–218. (17)

Wolf, L. L., F. G. Stiles and F. R. Hainsworth. 1976. Ecological organization of a tropical, highland, hummingbird community. J. Anim. Ecol. 45: 349–379. (13)

Wolfe, M. S. and E. Schwarzbach. 1978. The recent history of the evolution of barley powdery mildew in Europe. In *The Powdery Mildews*, D. M. Spender (ed.), pp. 129–157. Academic Press, London. (7)

Wolff, A. 1981. The use of olfaction in food location and discrimination of food by *Carollia perspicillata* (a neotropical fruit bat). MS Thesis, University of Wisconsin, Madison. (11)

Wood, T. K. 1980. Divergence in the *Enchenopa binotata* complex (Homoptera: Membracidae) effected by host plant adaptation. Evolution 34: 147–160. (10)

Woodin, S. A. and J. B. C. Jackson. 1979. Interphyletic competition among marine benthos. Amer. Zool. 19: 1029–1043. (19)

Wrangham, R. W. and P. G. Waterman. 1981. Feeding behavior of vervet monkeys on *Acacia tortilis* and *Acacia xanthophloea*: with special reference to reproductive strategies and tannin production. J. Anim. Ecol. 50: 715–731. (11)

Wray, J. L. 1971. Algae in reefs through time. Proc. North. Amer. Paleont. Conv. 1969, 2: 1358–1373. (14)

Wray, J. L. 1977. Late Paleozoic calcareous red algae. In *Fossil Algae*, E. Flugel (ed.), pp. 167–177. Springer-Verlag, New York. (15)

Wray, J. L. and P. E. Playford. 1970. Some occurrences of Devonian reef-building algae in Alberta. Bull. Can. Petrol. Geol. 18(4): 544–555. (15)

Wright, C. A. 1971. *Flukes and Snails*. George Allen and Unwin, London. (8)

Wright, H. O. 1973. Effect of commensal hydroids on hermit crab competition in the littoral zone of Texas. Nature 241: 139–140. (3)

Wright, S. 1931. Evolution in Mendelian populations. Genetics 16: 97–159. (2)

Wright, S. 1945. Tempo and mode in evolution: a critical review. Ecology 26: 415–419. (2)

Wright, S. 1955. Classification of the factors of evolution. Cold Spring Harbor Symp. Quant. Biol. 20: 16–24. (2)

Wright, S. 1967. *Evolution and the Genetics of Populations*, Vol. 1, *Genetic and Biometric Foundations*. University of Chicago Press, Chicago. (2)

Wynne-Edwards, V. C. 1962. *Animal Dispersion in Relation to Social Behavior*. Oliver and Boyd, Edinburgh. (2)

Yu, P. 1972. Some host-parasite interaction models. Theor. Pop. Biol. 3: 347–357. (3, 7, 9)

Zak, B. 1964. Role of mycorrhizae in root disease. Ann. Rev. Phytopathol. 2: 377–392. (7)

Zamenhof, S. and P. J. Zamenhof. 1971. Steady-state studies of some factors in microbial evolution. In *Recent Advances in Microbiology*. A. Pérez-Miravete and D. Peláez (eds.), pp. 17–24. Proc. Xth Intl. Cong. Microbiol. (5)

Zielke, E. 1973. Untersuchungen zur Vererbung der Empfänglichkeit gegenüber der Hundefilarie *Dirofilaria immitis* bei *Culex pipiens fatigens* und *Aedes aegypti*. Z. Tropenmed. Parasit. 24: 36–44. (8)

Zimmerman, E. C. 1960. Possible evidence of rapid evolution in Hawaiian moths. Evolution 14: 137–138. (2, 10)

Ziswiler, V. and D. S. Farner. 1972. Digestion and the digestive system. In *Avian Biology*, Vol. II, D. S. Farner, J. R. King and K. C. Parkes (eds.), pp. 343–430. Academic Press, New York. (11)

540

INDEX

absolute fitness, 35–36, 38–60
  passim
Acacia, 10, 211, 212, 267–268
Acalymma, 214
Acanthastominae, 74–77
Acanthiza species, 416
Accipiter species, 405
Aconitum, 301
Acraea encedon, 275
Acyrthosiphon pisum, 220
adaptation tolerance, 180
adaptive gap, 361–362, 364, 365, 379
adaptive lag, 362
adaptive radiation, reciprocal, 84–88
adaptive topography, 16–17
adaptive valley, 16, 27–28
additive genetic variance, 15
advancement, evolutionary, 91, 92
Aedes aegypti, 178
Agaonidae, 93, 282, 292, 308
aggressive mimic, 288
alcohol dehydrogenase, 222
alfalfa, 220
algae
  grazer-dependent, 454
  in lichens, 151
  species selection in, 341–346
  as symbionts, 317, 319
  turf assemblages, 451–452
algae, brown, productivity, 449–451
algae, coralline, 312–327 *passim*,
  341–346, 451–452, 454
algae, solenoporacean, 312, 341–346
alkaloid compounds, 212, 213, 215
allelopathic substance, 115–116, 118
Alligator, 69, 75, 76
allometry, 26–27
Alosa pseudoharengus, 175
Amazilia species, 303, 405–406
amino acids, nonprotein, 213
amphicyonid, 359, 364
Amphorophora rubi, 220

Amplexidiscus, 455
anagenesis, 161, 162
Ancalagon saurognathus, 357
Andrena erigeniae, 286
angiosperms
  diversity, 98
  phylogeny, 89
ankle joint, 356–357
Anolis species, 42, 45, 46, 62, 63,
  390–402, 417–422
ant, 10, 62, 212, 267, 427, 444–446
antibiotic resistance, 117–118
anurans, 162
apes, *see specific species*
aphids, 209, 215, 220–221
aposematic coloration, 18, 227, 268,
  275–277, 280
apostatic selection, 52
apparency hypothesis, 216
apple, 209, 214, 220, 221, 226
Aquilegia caerulea, 289, 290, 337
arbuscule, 155
arctocyonids, 352
Arctocyon primaevus, 358
arms race model, 179
armyworms, 224
Artibeus jamaicensis, 256
artiodactyl, 352, 353
Ascaris lumbricoides, 182, 183
Asclepias, 93
aspect diversity, 52
Aspidontus, 277
association by descent, 65–93, 208
associative learning ability, 284,
  285, 303
asteroid, 437
Atrophocaecum, 75
attenuation, 159–160
attractants, helminths and, 165
Avena species, 146
avirulence, 194, 204, *see also* virulence

baboon, 166, 172
bacteria
  antibiotic resistance, 117–118
  coevolution with plasmids,
    113–120
  coevolution with temperate phage,
    107–113
  coevolution with virulent phage,
    101–107
  defense mechanisms, 120–123
  incipient mutualism model and,
    182
  phage infection model, 101–103,
    107–109
  selection in population of, 103–
    107, 109–111, 114
  sexuality, 123–125
  as symbionts, 129–131
bacteriocin, 115–116, 118
bacteriophage, *see* phage
baculovirus, 203–204
balanced school of genetic variation,
  22
Balanus regalis Pilsbry, 347
balsam fir, 222
banana, 149, 210
barley, 138, 139, 142, 149, *see also*
  Hordeum
barnacle, 59, 312, 315, 318, 346–347
Basutoland, 143
bat, 80, 234, 236, 256
Battus philenor, 211, 228
B cell, 169, 170, 172
bear, 330, 331, 333, 334, 353
beaver, 176
bees
  euglossine, 283, 289, 293–298,
    308, 337
  oligolectic, 285–286
beetle, *see also specific species*
  bruchid, 229, 253
  staphylinid, 82–83, 97
behavior, evolution of, 19
Berberis vulgaris, 138
Beta phage, 111
between-phenotype component, 386
bill size, 383–384, 387, 402, 413–
  417, 422–426
biotype, 220
bird, *see also* bill size; chaparral
    birds; phororhacoids; *specific
    species*
  associated parasites, 66, 79
  coloration, 26

phylogeny, 76
seed dispersal by, 234, 238,
  239–240
species diversity, 439
taxon cycle of West Indies, 62
bird community structure, convergence in, 439–440
bird-fruit interaction, 238–239, 240
birth rate, 195, 196
bivalve mollusks, 316, 322–323, 325,
  336, 339–341, *see also specific
  species*
body size, 26, 38–39, 42, *see also*
    constant size ratio; island size
    ratio; minimum size ratio
  critical difference in, 404, 427
  in predator-ungulate system,
    358–359
  significance of, 436–437
  species number and, 396–397
body type, cursorial, phylogenetic
  analysis, 351–354
body weight–prey size, 330–336,
  358–359
Bombylius major, 286
bone-eating, 331–336
boom and bust cycle, 145
borophagine, 353, 359, 364
BPC, *see* between-phenotype component
bracken fern, 446
breeding experiments, 40
bromelin, 235
Brugia pahangi, 179
bryozoans, 315
Buellia stillingiana, 151
Bulbitermes, 83
bumble bee, 282, 298–301, 308
burnet moths, 266
Buteo species, 414
butterfly, *see also specific species*
  adaptive radiation, 230
  associated host plants, 84, 87–93,
    96
  host specificity of, 208–209, 230
  mimicry and, 264–281 *passim*, 337
  phylogeny, 86, 90, 92

cabbage butterfly, 215, 223, 227,
  230
Caedobacter taeniospiralis, 129–131
Caiman, 69, 75, 76
Caimanicola, 75, 76

Calliactus, 314
L-canavanine, 215, 223
cankerworm, 221
Cantius, 378
Capillaria hepatica, 178
caracal, 332, 334
carbon budget, 442–443
carcinoecium, 314–315
Carcinus maenas, 325–326
caribou, 181
carrying capacity, 38–39, 42–44, 398, 399
Caryedes brasiliensis, 215, 223
catnip, 215
Catostomus species, 175, 408–409
cattle, 171
cellulose, 242
Cercopithecus, 67, 74
Cervus elaphus, 164
chaparral birds, 46
character displacement, 4, 299, 300, 384–385, 462, *see also* competition
    in *Anolis* species, 390–402, 417–422
    among coexisting species, 406–430 *passim*
    in *Hydrobia*, 388–390
character release, 405–406
cheetah, 330, 332, 334, 357, 361, 367
cherry, 8, 225, 226
chicken, 27
chitons, 453
chloroplasts, 128–129
Chrysolina brunsvicensis, 226
cichlid fishes, 29, 277
cladogenesis, 161, 162
cladogram, 66
clam, giant, 317, 323–324
Clark's nutcracker, 45
classical school of genetic variation, 22
Claviceps purpurea, 138
Claytonia virginica, 286
clownfishes, 315–316, 323
cnidarians, 314–315, 322
coaccommodation, 162
coevolution, 5–7, *see also* competitors; diffuse coevolution; *specific association types*

biogeography and, 80–84
community stability and, 61–62
contrasted with temporally related adaptations, 96–98
definition, 1–3, 41, 229
literature in, 3–5
of marine organisms, 311–327
mimicry and, 267, 269, 277–279
phylogenetic aspects of, 65–98
predictive theory, 463
rates of, 328–329, 337–339, 348–349, 379–380
sources of evidence for, 8–13
coevolutionarily stable community
    for competitors, 43–46
    for predators and prey, 47–48, 50–51
    theoretical model, 41–42, 61–62
coevolutionary theory, 33–64
coevolution models
    for host-parasite systems, 176–182, 190–198
    for two populations, 41–57, 64
colicins, 118
Colobus species, 67, 72, 74
colonization, 68–73 *passim*, 93–96
coloration
    aposematic, 18, 227
    of birds, 26, 384
    of butterflies, 264–281 *passim*, 337
    defensive, in plants, 211
    of mammals, 28
    in plants, 290
    polymorphism in prey, 52
commensalism, 52–57, 82, 128
common predation pressure, 51–52
community, *see also* coevolutionarily stable community
    coevolution and, 57–64, 61–62
    as superorganism, 20
community convergence, 431–458
community pattern, 281, 460–461
    diffuse coevolution and, 205–206
community productivity, 434
community saturation, 436
community structure, *see also* competition
    invasion and, 397–402
    patterns of, 404–406
compatible phenotypes, 167–168

competition, *see also* character
　　displacement
　　community convergence and, 438,
　　　444–445
　　convergence and, 405
　　in flowering plants, 300, 308
　　interference, 119, 184
　　interspecific, 42–46, 62–64, 228,
　　　383–388, 397–403, 404–405,
　　　426–430
　　selection factor, 346–347
　　species turnover and, 398–401
competition coefficient, 42
competition function, 42–45
competitive release, 384, *see also*
　　resource partitioning
competitors
　　coevolution between, 42–46,
　　　386–387, 402–403
　　fitness functions, 43
conceptacle, 344
consistency, 68
consistency index, 69
constant size ratio, 411–413
Constrictotermes, 83
Conus species, 313, 437
convergence, *see also* community
　　convergence; mimicry
　　competition and, 405
　　in plant defenses, 213–214
　　plant rarity and, 287–289, 294
　　in pursuit predators, 354
　　theoretical difficulties, 433–437
copepods, 324
Coracina species, 416
coralline algae, *see* algae, coralline
Coregonus species, 174, 175
corn, 139, 144, 145
Corotocina, 82, 83
Corvus species, 414
Corynebacterium diptheriae, 111
cospeciation, 162, *see also* parallel
　　cladogenesis
Costus, 288–289
Cottus cognatus, 175
coumarins, 212
cowbird, 181
coyote, 330, 333, 334, 358
crabs, 316, 453
Crocodylus, 75, 76
cross-reactivity among helminths,
　　183, 184
Cryptocercus, 79
CSC, *see* coevolutionarily stable
　　community

Ctena bella, 325
cucumber, 214, 224
cucurbitacin, 212, 214, 226
curculionid weevils, 252
CWD, 131–133
cyanide, 236
cyanogenesis, 217
Cynanthus latirostris, 405–406
Cynips, 93
cytoplasmic male sterility, 145, 146

Dactylogyrus vastator, 178
damsel fish, 54, 455
Danaus (Limnas) chrysippus,
　　276–277
Daphaenodon-Pliocyon, 353, 364
Daphoenus species, 333, 334
deceit, 266
defensive novelties, 85
Delphinium nelsonii, 290
delta, 131
Dendrocopus kizuki, 417
Dendroica species, 414
Descurainia, 228
development time, 29
diaposematism, 268
Dicranotaenia species, 182
Didymictis, 378
diet, in predator guilds, 330–335
differential invasion, 44
diffuse coevolution, 2, 8, 84,
　　205–206, 210, 230, 251–252,
　　257, 338, 460, 464
dingo, 366
Dinictis felina, 333, 334, 335
Dioclea, 223
disease resistance, 138–145
disjunct distribution, 68–69
Dismorphia, 265
dispersal rate, 20
disperser coterie, 244, *see also* seed
　　dispersal
displacement, 44
display ritual, 19
diterpenoid compounds, 212
divergent evolution, 44
diversity, 435
donor cell, 99
Dromocyon vorax, 357
Drosophila, 8, 222
　　fungus-feeding, 221
Drosophila equinoxialis, 134
Drosophila melanogaster, 24, 29,
　　134

Drosophila nebulosa, 134, 135
Drosophila pachea, 223
Drosophila paulistorum, 131–133, 134
Drosophila pseudoobscura, 24, 134
Drosophila tropicalis, 134
Drosophila willistoni, 134–135
dulcitol, 214, 226
Dusicyon species, 408
Dysaphis devecta, 220

earthworm, 21
Echeneibothrium species, 163
Echinococcus granulosus, 171
Echinoecus pentagonus, 315
Echinothrix calamaris, 315
ecological monophagy, 95, 227–229
ecological release, 384, 385
ecology models, 34
ecosystems, selection among, 20–21
ectomycorrhizae, 154–155
egg mimicry, 210, 211, 230, 277–279
Eichler's Rule, 78
Elaeophora schneideri, 181
elk, 181, 358
Enchenopa binotata, 221
endemic infection, 194, 195
endomycorrhizae, 155
endosymbiosis, 128–136
Ensifera ensifera, 282, 304
Enterobius species, 67, 69–74
Enterolobium cyclocarpum, 249–254
Entobdella soleae, 164
environment, role in evolution, 1, 17, 28, 438–439
Ephestia kuehniella, 133
epidemic infection, 193–194
Epilachna, 214
epithallus, 342
ergot, 138
Erica species, 444
Erysiphe graminis, 138, 146, 149, 150, 158
Erysiphe graminis f. sp. *agropyri*, 150
Erysiphe graminis f. sp. *avenae*, 148
Erysiphe graminis f. sp. *hordei*, 139, 142, 148, 149

Erysiphe graminis f. sp. *tritici*, 148, 149, 150
escape behavior, 329–336 *passim*
Escherichia coli, plasmids and phage of, 101, 105–107, 111, 116, 123–124
Esox lucius, 175
ESS, *see* evolutionarily stable strategy
Euglossia species, *see* bee, euglossine
eukaryotic cell, origin, 128–129
Euonymus, 214, 226
Euphydryas editha, 28, 222
Eusmilus dakotensis, 333, 334
Euspilapteryx phasianipennella, 94
Eutoxeres species, 303–304
evolution, *see also* coevolution; convergence
  density-dependent, 37–41
  discontinuous rate of, 355
  dynamics of, 39–41
  factors of, 36
  genetic, 14–21, 23–29
  mechanism, 14–15
  rates of, 29–30, 328–329, 337–339, 348–349, 354–356, 368–374, 379–380, 460, 461
  sequential, 209
  simultaneous, 44
evolutionary equilibrium, 16, 17
evolutionarily stable character, 16
evolutionarily stable ecological equilibrium, 38–41, 58
evolutionarily stable strategy, 31, 36–37
  with density-dependent selection, 38–39
  in host-parasite system, 197
exclusion, 120, 121
exhabitant, 153–154
extinction, 462, 464, *see also* species turnover
  coevolution and, 61–63, 64
  competition and, 45, 62–63, 64
  of distance predator group, 363–365
  group selection and, 55
  of keystone species, 60
  local, 18, 19, 20
  in mutual aggression model, 179
  of predator or prey, 48

species selection and, 338, 339–341

extinction and invasion, 62–63, 398–401

eye, vertebrate, 27

Fahrenholz's Rule, 66, 78, 79

fall webworm, 221

Fasciola hepatica, 171

feeding behavior theory, 30

fertility, 35

Ficus, see fig

fig, 93, 256, 282, 292

fig wasp, 93, 282, 292, 308

finch, Galápagos, 383, 402, 422–426

fishes, *see also* teleost fishes

  mutualisms in, 315–316

  parasites of, 172–176

  size ratios and, 411, 412

fitness, *see also* absolute fitness; relative fitness

  ESS theory and, 36–37

  frequency-dependent, 191–194

  in host-parasite population model, 197

  individual, 15, 16

  in communities, 60

  for competitors, 43

  for predators and prey, 47, 50

  in symbiotic relationships, 53, 55, 57

fitness sets theory, 30, 31, 32

flagellates, wood-digesting, 79

flavonoid compounds, 212

flea, 94, 163

flower, rich, 302

flower-visitor guild, 447, *see also* pollinator

foliage height, 439–440

folivore guild, 446

foot bones, evolution of, 354, 356–358

forma speciales nomenclature system, 148

frog, southern, associated parasite, 81

frugivore, fruit-related adaptations, 236–239, 244–245, 306

fruit

  content analysis, 235

  convergent evolution of, 257

  functions, 232

morphology, 235, 245

optimal, 258

ripening process, 235–236, 242–243

fruit pigeons, 412

functional constraints on variation, 26–27

fungus

  heritable variation associated with host plant, 139–140

  host specialization, 147–150

  induced resistance, 149

  induced susceptibility, 149

fungus-alga associations, 151–154

fungus-plant symbioses, 137–160

fungus-root associations, 154–156

furanocoumarins, 210

Fusarium oxysporum, 148, 149

fusion cell, 342, 344

game theory, 30–32

Gammarus lacustris, 179

gastropods, 312–327 *passim*

Gavialis, 76

gene flow, 28–29

gene-for-gene hypothesis, 56, 140–143, 157–158, 166, 176–177, 190–191, 220, *see also* mutual aggression model

genetic drift, 23

genetic feedback, 192

geometrid moth, 221

Geospiza species, *see* finch

Geotrygon species, 414

germination rate, 247

Glaucis hirsuta, 303

glaucolide-A, 214

Glaucopsyche lygdamus, 218

glucosinolate compounds, 213, 215, 223, 225, 227, 228

glutathione *S*-transferases, 224

glycosides, cardiac, 212, 214, 224, 227

glycosides, cyanogenic, 211–214 *passim*, 224

Gongora species, 296

Gorilla, 67, 70, 71

Grallatotermes, 83

grasshoppers, 228

grazers, 341–346, 438, 451–452, 454

gregarine sporozoans, 78–79

growth rate, density and, 395

guanacaste tree, seed dispersal, 249–254
guild, 329, 434
Gymnatrema, 75

habitat
  effect on host-insect associations, 94
  predator-prey evolution and, 360–361
habitat preference among predators, 329–336 *passim*
habitat segregation, 45–46
Haematoloechus species, 162
Haemonchus contortus, 162
Haplochromis burtoni, 277
harvester ant, 427–429
hawkmoth, 286, 290–291
Hazda tribe, 166
Hedylepta species, 29, 210
heirloom parasites, 182, 183
Heliconia, 224–25, 304
Heliconius species, 91–93, 95, 97, 210, 211, 268, 270–272, 274, 275, 278–279
Heligmosomoides polygyrus, 171
Helminthosporium maydis, 145–146
helminth, parasitic
  adaptations of, 162–165, 169–172
  coevolution of co-occurring parasites, 182–184
  host specificity, 163–165, 172–176
  rate of evolution in, 161–162
  site selection, 182–183
hemoglobin S, 167
heritability, 40
hermit crab, 314, 322, 453
Hessian fly, 8, 209, 220
heterozygote superiority, *see* overdominance
high-K trait, 37
high-r trait, 37
Homo species, 67, 70, 71, 74
Hopkins host selection principle, 286
Hoplophoneus species, 333, 334, 335
Hordeum spontaneum, 146–147
Hordeum spontaneum nigrum, 139
Hordeum vulgare, 148

hormone
  influence on parasites, 163
  insect, mimicry of, 212
horse, ontogenetic changes in, 370
Hospitalitermes, 83
host, animal, adaptations to parasites, 165–172
host genetics, 191–194
host-guest association, reciprocal adaptations in, 314–317
host-parasite systems, 187–188, *see also* helminths, parasitic; plant-fungus symbioses
  coevolution models for, 176–182
  coevolution to mutualism, 267–281
  models, 157–160
  phylogenetic studies, 65–98
host race, 219–220
host range mutation, 105, 121
host-shift model, 85–86, 94–96, 226
host specificity
  in fungi, 147–150
  of marine parasites, 321
  of parasitic helminths, 163–165, 172–176
  of phytophagous insects, 208, 219
  selective factors and, 226–229
host transfer, 76, 77, 79, 81, 87, 94–96
hot desert, 440ff
housefly, 9
hummingbird, tropical, 301–306, 308, 405
Hyaenodon species, 333, 334, 335, 336
hyaenodonts, 352–353, 359, 363–364, 376
hybrid inviability, 132
hybridization, 384
hybrid sterility, 132
Hydrobia species, 325, 388–390, 408
hydroids, 314, 315
hydrothermal productivity potential, 442
hyena, 330, 331, 332, 334, 335, 351, 353, 358, 364, 376
Hyles lineata, see hawkmoth
Hylobates species, 72–73
Hymenolepsis species, 180, 184

hypericin, 225
Hypericum, 225
Hypolimnas misippus, 276–277
hypothallus, 342

immune response, parasitic modulation of, 168–172, 183–184
immunity loss rate, 196
immunosuppression, 171, 184
incipient mutualism model, 177, 181–182
incompatibility, 120, 122–123
induced resistance, 149
induced susceptibility, 149
inhabitant, 153–154
insect
  behavioral responses to plant compounds, 225–226
  diverse fauna on plants, 218–219
  effects of parasitism on, 227
  effects of plant compounds on, 214–216
  folivorous, 446
  host-associated genetic variation, 219–222, 307
  monophagous, 19–20
  physiological responses to plant hosts, 222–225
  phytophagous, 84–93, 208
  polyphagous, 20
insect-baculovirus system, 203–204
insect hormone, mimicry of, 212
insect–plant systems, 207–231
  coevolution in, 229–231
intrinsic rate of increase, 38–39, 43, 50
intrinsic reproductive rate of parasite, 197–198, 200–201, 203–204
invasion, *see also* extinction and invasion
  differential, 44
  taxon cycle and, 62–63, 401
island size ratio, 413–417
isolating mechanisms, 132
Israel, 148
Ituna, 265

Jarra dieback epidemic, 160
Jehlius, 347

Kairomone, 163

kappa, *see Caedobacter taeniospiralis*
kelp, 322, 449–451
keystone species, 59
killer trait, 129–131
K-selection, 39, 41

Labroides, 277
lactones, 212, 213
Lafresnaya lafresnayi, 304
lambda (bacterial symbiont), 130
Lambda (phage), 111
land-race, 139
large continent advantage, 365–366
Laspeyresia pomonella, 209, 229
latitude, 457
leaf shape, 211
leaf size, 443–444
Lemur, 67, 70
leopard, 330, 332, 334, 335, 357, 361, 367
Lepidoptera, *see* butterfly
Leptinotarsa decemlineata, 214, 221
Lernaeodiscus porcellanae, 315
Leucochloridium, 165
lice, avian-biting, 66
lichens, 150–154
life-dinner principle, 318
limb bones, evolution of, 354, 356–358
limnocyonid, 376
limonene, 217–218
limpet, 322, 437, 449, 450, 453
linkage, 25–26
linkage disequilibrium, 222
Linum usitatissimum, 56, 140, 190
lion, 330, 331, 332, 334, 367
Lithophaga curta, 323
Lithops, 211
Littorina species, 325–326
lizard
  community structure, convergence in, 440–441
  phylogeny, 76
Lophocereus schotti, 223
Lota lota, 175
Lotus corniculatus, 217
Lupinus, 218
Lycaenidae, 86
lymphocytes, 169, 170
lynx, 330, 333, 334
lysogen, 100
lytic cycle, 99–100

macroevolution, 187, 338
major gene, 268
Malacosoma species, 225
malaria, 162
malaria–hemoglobin S system, 167
mallards, 176
mammalian phylogeny, 76
Manter's Rules, 78, 81
maple, 221
marine ecosystem, 435–436, 438–
    439
marine organisms
  coevolution, 311–327
  convergence among, 447–454
  geography of intimate associa-
    tions, 320–325
  mutualism in, 454–456
  selection agents for, 317–320
  species selection in, 339–347
  resource utilization patterns,
    448–449
marsupial, 353–354
Mayetiola destructor, 8, 209, 220
Mediterranean scrub, 440
meiotic drive, 18
Melampsora Lini, 56, 140, 190
Melanosuchus, 76
Melithreptus species, 417
Membranipora membranacea, 451
meristem, 342
Mesocestoides corti, 171
mesonychids, 352, 353, 359, 361–
    364, 370–374
metatarsus, 356–358
Metechinorhynchus salmonis,
    174–176
Methona, 265
Mexican whiptail lizard, 402
migration rate, 18
milkweed, 224, 227
mimicry, 263–281, *see also* egg
    mimicry; signal mimicry
  coevolution and, 267, 269, 277–279
  ecological consequences, 280–281
  of insect hormones, 212, 215
  molecular, *see* molecular mimicry
  in plants, 288
  theory, 264–266
mimicry, Batesian, 52, 264, 272–277
mimicry, Müllerian, 52, 264, 267–272
mimicry ring, 288

minimalist morphogenetic model for
    whole-body transmutation, 368–
    374
minimum size ratio, 404–411
mite, 79, 80, 94
mitochondria, 128–129
mixed-function oxidase, 223–224
Mlg gene, 139
mobilization, 115
models, *see also* coevolution models;
    mutual aggression model
  breeding experiments, 40
  ecology, 34
  epidemic, 193
  host genetics, 191
  incipient mutualism, 177, 181–182
  minimalist morphogenetic, 368–
    374
  parasite-host genetics, 190–191
  parasite-host populations, 194–197
  types of, 34
  value of, 34
modifier gene, 24, 268
molecular clock, 461
molecular mimicry, 171, 172, 183
mollusks, 437
monarch butterfly, 224, 227, 266
monkeys, *see specific species*
monophyletic group, 66
Montipora berryi, 322–323
moose, 181, 358
morphological rate of evolution,
    29–30
mortality rate, 195, 196, 199
mosquitos, 164, 178
most effective pollinator principle,
    287, 309
mouse, 168, 169, 171, 172, 184, 205
mu, 129
mule deer, 181
Müllerian mimicry, 97
muskrats, 176
Mus musculus t allele, 18
mussel, 59
mustard oil glycosides, 213
mutation, 22–23
mutation, novel, 49
mutual aggression model, 176–179,
    see also gene-for-gene hypoth-
    esis; Red Queen hypothesis;
    selection, antagonistic

mutualism, 61, 128, 150, *see also*
   incipient mutualism model;
   lichens; mycorrhizae
   coevolution of partners, 52–57
   in marine communities, 315, 319,
      454–456
   origin of, 462
   in plant-pollinators systems, 283–
      284, 292–306, 308
mycorrhizae, 154–156
Myiarchus species, 413
Mytilus species, 325, 449
myxoma virus, 8, 19, 56, 160, 162,
   180, 196, 198–203

nasty competitor, 60–61
Nasutitermes, 83
natural selection, 35–36
nearest-neighbor criterion, 68
nematodes, 162, 164
Nematospiroides dubius, 172, 184
Neoaplectana carpocapsae, 184
neo-Darwinian theory of evolution,
   32
Neodiprion abietis, 222
neofelid, 359
nepetalactone, 215
Neritina virginea, 325
net reproductive value for parasite,
   197
Neuroterus, 93
neutral mutation theory, 22
niche axis complementarity, 45, 394
niche breadth, 386
niche separation, 45, 64
nicotine, 223, 230
Nilaparvata lugens, 220
Ninox species, 416
Nippostrongylus brasiliensis, 171,
   184
N-nitrosodimethylamine, 235
Notobalanus flosculus, 347
Notoplana californica, 315
Notropis hudsonius, 175
Nucella lapillus, 325, 326
nuthatch, 384, 402
nutrient availability in marine environment, 319–320
Nymphalidae, 86, 88

oak, 93, 221, 256
Octopus, 314

Oenothera, 286
Onchocerca tarsicola, 164
Onchorhynchus species, 175, 176
optimal adaptive compromise, 361
optimal foraging theory, 30, 31, 32,
   228, 284
optimization, 30–32
orchid, 155–156, 282, 288, 293–298
Oryctolagus cuniculus, 160, 163,
   199, *see also* rabbit
Osmerus mordax, 175, 176
Osteolaemus, 76
ostrich, 81–82
overdominance, 22, 167, 219
oxyaenids, 353, 359, 367, 374–379

Pachyaena species, 357, 362–363,
   371, 375
palaeoryctids, 352, 353
paleofelid, 353, 359, 367, 376
Paleosuchus, 76
Pan, 67, 70, 71
papain, 235
Papilionidae, 86, 88, 337
Papio, 67, 74
paradox of enrichment, 48
parallel cladogenesis, 9, 208
parallel level bottom community, 435
Paramecium aurelia, 129–131
parasite
   coevolution among, 182–184
   definition, 100
   fitness and pathogenicity, 159–160
   geologic evidence of, 187
   host population size and, 187–188
   intrinsic reproductive rate, 197–
      198, 200–201, 203–204
   net reproductive value, 197
parasite–host system, 4, 19, 128
   coevolution in, 52–57
   conventional view, 189–190
   macroevolution, 187
   outcome, 462–463
Parelaphostrongylus tenuis, 181
parental care, 18
parsimony criterion, 67–68
Passifloraceae, 91–93, 210, 211,
   278–279, 304
pathogenicity, 140–141, 159–160,
   178, *see also* virulence
Patriofelis ulta, 376
pearlfishes, 316, 323
pearl millet, 146

Pedicularis groenlandica, 284
Penstemon species, 290
Peponapis pruinosa, 286
Perca flavescens, 175
perissodactyl, 352, 353
Petrolisthes cabriolloi, 315
Phaethornis superciliosus, 304
phage
  ancestry, 125–126
  over-replication mechanism, 128
phage, temperate, 100
  advantages of temperance, 112–113
  coevolution with host, 107–113
  infection model, 107
  selection in, 109–111
phage, virulent, 99–100
  coevolution with host, 101–107
  infection model, 101–103
  selection in, 103–107
phenacodontids, 352, 353
phenols, 212, 213, 215, 216
phenotypic change, 15, 23, 50–51, 380
  rate of, 356, 363, 368–379
  with speciation, 23
  theoretical approaches, 30–32
phenylpropanes, 212
phloridzin, 214, 226
phlox, 290
phororhacoids, 352, 365–366
photosynthesis, 442
phyletic gradualism, 354
phylogenetic drift, 339
phylogenetics, 65, 66–68
phylogenetic tree, *see* cladogram
physical structure of community, 434
physiologic specialization, 139
physogastry, 82, 97
Phytophthera species, 139, 143, 147, 160
Pieridae, 86, 88, 228, 268, 279
plant
  color morphs, 290
  heritable variation associated with parasitic fungus, 138–139
  insect fauna on, 218–219
  rarity of, and convergence, 287–289, 294, 302
  specialization in, 286–287, 306

plant community structure, 442–444
plant defenses, 27, 84, 86
  macroevolution, 208–211
  mechanisms, 214–216
  microevolution, 216
  phylogenetic analysis, 213–214
  physiological functions, 216–218
  secondary compounds, 212–214
plant-disperser relationship, tight coevolution, 255–256
plant-fungus symbioses, 137–150, *see also* lichens; mycorrhizae
  evolution of, 156–157
  microevolution in, 157–160
plant growth form, 442–443
plant–herbivore relation, 27, 207–228
  coevolution in, 229–231
plant-pollinator system, 447
  selection in, 283–293, 307–310
  specialization in, 284–289
plant species richness, 444, 456
plasmid, 99
  allelopathic substances of, 115–116, 118
  ancestry, 125–126
  associated genes, 100
  coevolution with host, 113–120
  colicinogenic, 118
  over-replication mechanism, 128
plasmid, conjugative, 99
  growth model, 113–114
  selection, 114–115
  transmissibility, 119–120
plasmid, nonconjugative, selection, 115, 119–120
plasmid pCR1, 120
plasmid R1, 116, 119–120
Plasmodium chabaudi, 184
Plasmodium falciparum, 167, 182, 183, *see also* malaria
pleiotropy, 24–25
Pleurastrum, 152
Pocillopora species, 322
point richness, 435
pollinarium, 294
pollination, coevolution and, 282–310
pollinators, 256
  evolutionary responses to plants, 291–292, 306–310
  specialization in, 284–285

551

polymorphism
  of aposematic traits, 275–277, 280
  host-parasite relation and, 56
  of host resistance genes, 166–167
  multiple-niche, 219
  in resistance-virulence systems,
    190–192, 194, 206
Polymorphus species, 176, 179
ponderosa pine, 217–218, 221, 230
Pongo, 67, 70, 71, 74
population genetics theory, 35–36
population size
  coevolution and, 41–57 *passim*
  community interactions and,
    58–60
  of competitors, 42–46
  evolution and, 37–41
  of predators and prey, 46–52
  selection and, 37–41
  theoretical, 35
population variance, 40
positivity-enhancing keystone
  species, 59
potato, 139, 143, 147, 212, 214,
  221–222, *see also Solanum*
  species
predation
  community convergence and, 438
  convergence and, 457–458
  selection factor, 317–320, 339–346
predation coefficient, 47, 48–49
predator
  body weight estimates, 380–382
  competition for kill, 376
  fitness function, 47, 50
predator, ambush
  adaptive changes, 366–368
  phylogeny, 352–354
predator, distance
  distance-running adaptations, 354,
    356–358
  phylogeny, 352–354
  phylogenetic transmutations, 368–
    374
  running speed, 358–360
predator guilds, coevolution in,
  329–336
predator-prey system
  directional selection, 350–354
  in fossil record, 331–336
  in marine organisms, 339–346
  reciprocal adaptations, 312–313,
    318
  theoretical model, 46–52
Presbytis species, 67, 72

prey, *see also* apostatic selection;
    mimicry
  defensive phenotypes, 52
  fitness function, 47, 50
  polymorphism in coloration, 52
  size, 38–39
principle of parsimony, 15, 19
Prochloron, 317
Proctocaecum, 75
Prodoxus, 293
prophage, 100, 108
proteinase inhibitors, 213
prudent parasite model, 177, 179–
  181
prunasin, 213, 226
Pseudomonas species, 101
Pseudomyrmex, 10, 267
Pseudotrebouxia, 152, 154
Puccinia species, 138, 139, 140, 144,
  146, 148, 149, 150, 158
puma, 333, 334
punctuated equilibrium theory, 18,
  20, 29–30, 268, 308–309, 354–
  356, 364
Pungitius pungitius, 175
Pyura praeputialis, 449

r, see intrinsic rate of increase
rabbit, 8, 19, 56, 160, 162, 163, 180,
  198–204
rabies, 204
Raja species, 163
Rana species, 162
raspberry, 220
rat, 171, 181, 184, 189
ratel, 332, 334
receptor size, 106
recipient cell, 99
recovery rate, 193–203 *passim*
rectangular model, 355
Red Queen hypothesis, 47, 463, *see*
  also mutual aggression model
relative fitness, 35, 36
replacement set, 300
reptiles, associated parasites, 69,
  74–77
resistance
  factor in parasite-host system
    models, 190–198 *passim*
  to parasitic helminths, 166–168
resistance-susceptibility alleles,
  141–142
resistance to phage infection, 103–
  107

resistance-virulence gene systems, *see* gene-for-gene hypothesis
resource partitioning, 42, 44, 45–46, 384, 385–386, 460, *see also* competition
resource sequestering, 119
restriction, 120–122
restriction endonucleases, 121, 122
Reticulotermes, 79
reward to vector, helminths and, 165
Rhagoletis species, 209, 221
rhea, 81–82
rhodanese, 8
Rhopalosiphum maidis, 220
rice, 220
richness, 435
robin, 415, 416
rodent, *see also specific species*
  convergence and, 444–446
  heteromyid, 412
R-plasmids, 117
r-selection, 41
r strategy, 178

St. Maarten, 63
salamanders, 162
salinity tolerance, 24
Salmonella typhimurium, 117
Salmo species, 175
Salvelinus namaycush, 175
scale insect, 221
scallop, 341, 454–455
schistosome, 164, 166, 167–168, 171, 174, 184
Sclerospora graminicola, 146
scrub jay, 45
sea anemone, 54, 314–327 *passim*, 453–455
sea cucumber, 316, 323
sea grass, 322
search image, 211, 228–229
seastar, 315, 316, 323, *see also* starfish
sea urchin, 316, 323, 452–453
seed, gut transit time, 237
seed coats, 211–212
seed dispersal
  animals involved, 233–239, 240, 244–245
  caricature of, 241–248
  coevolutionary questions, 254–262

evolutionary changes, 248–249
  literature, 232–233
  seed predation and, 259–260
seed dormancy, 237–238
seed-eating guild of hot deserts, 444–446
seed germination, 247–248
seed morphology and chemistry, functions, 232
seed parasite, 292
seed shadow, animal-generated, 239–240, 242, 246–247
selection, *see also* K-selection; natural selection
  antagonistic, 103–106
  antipredatory, 318–319, 324–325
  artificial, 40
  on bacterial genome, 109, 114
  in bacterial populations, 103
  between character states, 348
  density-dependent, 37–41, 58
  directional, 24, 25, 29
  ecosystem, 20–21
  frequency-dependent, 36–37, 167, 219
  group, 18–19, 23, 55–56, 180
  individual, 15–18, 20–21
  intensity, 15, 23
  interacting species as units of, 60–61
  interdemic, 18, 180
  kin, 18–19
  levels of, 14–21
  in marine organisms, 317–320, 339–347
  in phage populations, 103–107, 109–111
  in plant-pollinator systems, 283–293, 307–310
  plateau phenomenon, 24, 25, 26
  species, 19–20, 23, 338–339, 348
selective death, rate of, 372
Septoria nodorum, 149
"sequential evolution," 85
Serengeti guild, 330–334
serial endosymbiosis theory, 128–129
Sericornis species, 415–416
sesquiterpene lactone, 213, 214
sesquiterpenoid compounds, 212, 213, 214
sex, 192
sexual isolation, 132

shank length, 27
sheep, 162, 171, 181
shell architecture, 312–313, 320
shifting balance theory, 27
Shigella species, 117
shrimp, 312–327 *passim*
sigma, 130
signal mimicry, 278, 281
simultaneity, 1, 2
sinigrin, 215
sister group, 67
site segregation, in parasites,
    182–183
size, *see* body size; constant size
    ratio; island size ratio; minimum
    size ratio
snail, 164, 165, 168, 172, 312–313,
    322, 453
snake, associated parasites, 75–76
solanidine, 212
Solanum species, 139, 221–222, *see
    also* potato
Solea solea, 164
solitary size, 396, 418–419
sorghum, 220
souvenir parasites, 182, 183
speciation
    phenotypic change and, 23, 30
    sympatric, in insects, 219, 222
speciation rate, 19, 20
species
    stability of, 336–337
    as units of selection, 60–61
species abundance, coevolution and,
    58–60
species diversity, 63–64, 461
species interaction, role in evolution,
    16–17
species number, body size and,
    396–397
species richness, 434–436
species turnover, competition and,
    398–401
specificity, 1, 2
Speothos venaticus, 357, 361, 364
Sphaerotheca fuliginea, 149
Spilopsylla cuniculi, 163
spiroplasma, 134–135
sponges, 314–315, 454–455
Sporidesmium folliculatum, 151
spruce, 222
SR agent, 134–135
stability-enhancing keystone
    species, 59–60

starfish, 59, 453, *see also* seastar
Stellar's jay, 45
steroid mimics, 212
Stizostedion vitreum, 175
strawberry, 138
Strigula complanata, 151
strychnine, 235
Stylobates, 314
superinfection immunity, 108, 111
superorganism, 20
survival probability, 35
susceptibility, 166–168, 177–180
    passim, 194
swallowtail butterfly, 210, 225
swans, 182
Symbiodinium microadriaticum,
    317
symbiosis, 128, 137, *see also* endo-
    symbiosis
    in marine organisms, 317, 319,
        320–325
    plant-fungus, 137–160
    theoretical model, 52–57
Szidat's Rule, 78, 79, 80

Taenia taeniaeformis, 171
talon, 27
tannins, 212, 215, 224, 235
tapeworms, 162, 163, 182, *see also
    specific species*
Tasmanian wolf, 359, 366, 376
taxon cycle, 62–63, 401
taxonomic rate of evolution, 29–30
T cell, 169, 170, 172
teleost fishes, 74–77
    associated parasites, 74–77
    phylogeny, 76
temperance, advantages of, 112–
    113
tephritid fly, 222
Terebra affinis, 325
termite, associated parasites, 79,
    82–83
Termitogastrina, 82–83
terpenoid compounds, 212, 215
terrestrial ecosystem, marine eco-
    system and, 438–439
terrestrial guilds, convergence
    among, 439–447
Tetraopes, 93
Texas male-sterile cytoplasm,
    145–146
Therioaphis maculata, 220

554

Thermoplasma acidophilium, 122
thiol compounds, 213
Thyridia, 265
Timoniella, 75
tobacco, 223
tobacco hornworm, 230
tomatine, 214
Tomistoma, 76
tooth, cutting, of oxyaenids, 374–379
tooth size, 26
tooth wear, 362, 368
T phage, 105–107
transmissibility, 56, 182, 190, 193, 198
transmission rate, 193–203 *passim*
transposon, 118
Trebouxia, 151, 152, 154
tree, defense mechanisms, 20
trematodes, 162, 182
    digenean, 74–77
    paramphistonid, 79
Tribolium species, 8, 19, 180
Trichinella spiralis, 171, 172, 184
Trichonympha, 79
Trichuris muris, 184
Trinervitermes, 83
triterpenes, 235
trophic guild, convergence among, 441–447
Trypanosoma lewisi, 181
trypanosomiasis, 189
turf, marine, 452
turtle
    associated parasites, 75–76
    phylogeny, 76
Typhlosaurus species, 408
Tyto species, 416

Umbelliferae, 210–211
ungulates
    body weight estimates, 380–382
    distance-running adaptations, 354, 356–363
    phylogeny, 352–354
    running speed, 358–360
    tooth-wear survivorship curves, 362, 368
Ustilago violacea, 149

variation, genetic, maintenance of, 21–23, 44, *see also* polymorphism
vector, 205
    of helminths, 164–165
Vernonia species, 214
vertebrae, cervical, 67–68
vertebrate phylogeny, 76
Verticillium species, 143, 149
Vireo species, 413, 414
virulence, 56, 190–206 *passim, see also* pathogenicity
virulence–avirulence alleles, 141, 142
virus, host encapsulation and, 184
virus, bacterial, *see* phage
vitamin C, 235

wasp, 9
wheat, 8, 138, 139, 144, 150, 220
wheat stem sawfly, 221
White River Formation, 331
white-tailed deer, 181
wild dog, 330, 332, 334, 351, 353, 357, 364, 365
within-phenotype component, 386
wolf, 333, 334, 350, 351, 352, 357, 358, 361, 364, 365, 376
wolverine, 331, 333, 334, 376
woodpecker, 417
woody Ranales, 87, 88, 94–95
WPC, *see* within-phenotype component
Wuchereria bancrofti, 164, 179

Xanthium strumarium, 210
Xenorhabdus nematophilus, 184

yellow allele, 24
Yellowstone guild, 330–334
Yersinia pestis, 189
Yponomeuta species, 214, 221, 226
yucca, 282, 293
yucca moth, 282, 293

zebra, 358
Zelleriella, 81